Green Biotechnology and Allied Fields

Green Biotechnology and Allied Fields

Edited by **Teresa Brocco**

SYRAWOOD
PUBLISHING HOUSE
New York

Published by Syrawood Publishing House,
750 Third Avenue, 9th Floor,
New York, NY 10017, USA
www.syrawoodpublishinghouse.com

Green Biotechnology and Allied Fields
Edited by Teresa Brocco

International Standard Book Number: 978-1-68286-322-0 (Hardback)

Printed in China.

Contents

Preface

This book has been an outcome of determined endeavour from a group of educationists in the field. The primary objective was to involve a broad spectrum of professionals from diverse cultural background involved in the field for developing new researches. The book not only targets students but also scholars pursuing higher research for further enhancement of the theoretical and practical applications of the subject.

Green biotechnology is focused on devising environment-friendly agricultural practices. It plays a crucial role in dealing with food security issues and reducing the carbon footprint. This book covers the applications of green biotechnology such as plant genetic engineering, producing biofertilizers and biopesticides, hybridization, etc. Students, researchers, biotechnologists and anyone else interested in this field will be assisted by this book.

It was an honour to edit such a profound book and also a challenging task to compile and examine all the relevant data for accuracy and originality. I wish to acknowledge the efforts of the contributors for submitting such brilliant and diverse chapters in the field and for endlessly working for the completion of the book. Last, but not the least; I thank my family for being a constant source of support in all my research endeavours.

Editor

1

Evaluation of four ionic liquids for pretreatment of lignocellulosic biomass

John Gräsvik[1†], Sandra Winestrand[1†], Monica Normark[1], Leif J Jönsson[1*] and Jyri-Pekka Mikkola[1,2]

Abstract

Background: Lignocellulosic biomass is highly recalcitrant and various pretreatment techniques are needed to facilitate its effective enzymatic hydrolysis to produce sugars for further conversion to bio-based chemicals. Ionic liquids (ILs) are of interest in pretreatment because of their potential to dissolve lignocellulosic materials including crystalline cellulose.

Results: Four imidazolium-based ionic liquids (ILs) ([$C{=}C_2C_1$im][$MeCO_2$], [C_4C_1im][$MeCO_2$], [C_4C_1im][Cl], and [C_4C_1im][HSO_4]) well known for their capability to dissolve lignocellulosic species were synthesized and then used for pretreatment of substrates prior to enzymatic hydrolysis. In order to achieve a broad evaluation, seven cellulosic, hemicellulosic and lignocellulosic substrates, crystalline as well as amorphous, were selected. The lignocellulosic substrates included hybrid aspen and Norway spruce. The monosaccharides in the enzymatic hydrolysate were determined using high-performance anion-exchange chromatography. The best results, as judged by the saccharification efficiency, were achieved with [C_4C_1im][Cl] for cellulosic substrates and with the acetate-based ILs for hybrid aspen and Norway spruce. After pretreatment with acetate-based ILs, the conversion to glucose of glucan in recalcitrant softwood lignocellulose reached similar levels as obtained with pure crystalline and amorphous cellulosic substrates. IL pretreatment of lignocellulose resulted in sugar yields comparable with that obtained with acidic pretreatment. Heterogeneous dissolution with [C_4C_1im][HSO_4] gave promising results with aspen, the less recalcitrant of the two types of lignocellulose included in the investigation.

Conclusions: The ability of ILs to dissolve lignocellulosic biomass under gentle conditions and with little or no by-product formation contributes to making them highly interesting alternatives for pretreatment in processes where high product yields are of critical importance.

Keywords: Ionic liquid, Pretreatment, Lignocellulose, Enzymatic saccharification

Background

As petroleum and other fossil resources are not sustainable, renewable alternatives are needed for production of fuels, chemicals, and materials. Lignocellulosic biomass is an abundant non-food renewable resource that can contribute to a shift towards a sustainable bio-based society [1]. The three major constituents of lignocellulosic biomass are cellulose, hemicelluloses, and lignin. Cellulose and hemicelluloses are polysaccharides that can be hydrolyzed to monosaccharide sugars. The monosaccharides obtained can then be utilized for production of ethanol or other commodities [2].

* Correspondence: leif.jonsson@chem.umu.se
†Equal contributors
[1]Department of Chemistry, Umeå University, Umeå SE-901 87, Sweden
Full list of author information is available at the end of the article

Saccharification is typically accomplished through enzymatic hydrolysis, as the high selectivity of enzymatic catalysis gives high sugar yields. A pretreatment step is needed to achieve efficient bioconversion of lignocellulosic feedstocks, as the cellulose fraction is not easily accessible to enzymes [3,4]. However, many pretreatment methods require harsh conditions and give rise to by-products, which decrease the sugar yields and inhibit enzymatic and microbial biocatalysts.

Solvents that dissolve cellulose are of interest for pretreatment of lignocellulosic feedstocks prior to bioconversion. In nature, cellulose occurs mainly as highly crystalline cellulose I [5]. The crystallinity may prevent efficient enzymatic conversion and it is therefore desirable to disrupt the structure of the cellulose. Since cellulose is amphiphilic [6,7], it has been suggested that both hydrophobic and hydrophilic interactions need to be broken to dissolve cellulose. Therefore, a suitable solvent system should contain a co-solvent that weakens the hydrophobic interactions in an aquatic environment [8]. Several solvent systems for cellulose are known [9-14], but many of them suffer from problems like high toxicity and being difficult to recycle. Ionic liquids [ILs] are non-volatile solvents that can dissolve cellulose under gentle conditions.

Several attempts have been made to utilize ILs for pretreatment of cellulose [15-19]. Aspects that have been studied include the inhibitory effect on enzymes [20-22], the influence of water content [23,24] and cellulose load [25], the particle size of the biomass [26-28], the pretreatment temperature [29], and the pretreatment time [18]. However, an extensive study on the pretreatment effects of different ILs on different types of cellulosic substrates is lacking. Comparisons have been made, but they are based on biomass from different sources and treatments performed under different conditions [30]. Thus, the aim of this study was to investigate the pretreatment effect of different ILs on different types of substrates under identical conditions. Four ILs (Table 1) and seven substrates (Table 2) were studied.

The four ILs (Table 1) were chosen since they have been used to dissolve cellulosic substrates and are recognized for their performance in such applications. Also, the set studied would allow for the detection of performance variations between different types of ILs used for this purpose. $[C_4C_1im][Cl]$ is one of the most studied ILs when it comes to dissolving cellulose and serves as a good reference [15,17,30,37]. $[C_4C_1im][HSO_4]$ is highly interesting because it has been reported to dissolve lignin in high water content [10-40% (w/w) water], which is substantially more than $[Cl]^-$- and $[MeCO_2]^-$-based ILs can handle [18]. $[C_4C_1im][MeCO_2]$ has the highest basicity among the selected $[C_4C_1im]$-based ILs, a property that is strongly linked to the capability of the IL to dissolve cellulose [30,37]. Finally, $[C{=}C_2C_1im][MeCO_2]$ was selected because of its high reactivity, low viscosity, and ability to dissolve cellulose [25]. It has been suggested that π-π interactions between the allyl side chain and lignin will improve lignin solubility [38]. The set of ILs chosen also gives the possibility to investigate the effect of the cation, by comparing the results from $[C{=}C_2C_1im][MeCO_2]$ with those from $[C_4C_1im][MeCO_2]$. With regard to dissolution of wood saw dust and thermo-mechanical wood pulp, the performance of 1-allyl-3-methylimidazolium chloride $[C{=}C_2C_1im][Cl]$ even surpassed that of $[C_4C_1im][Cl]$ [38]. The set of ILs studied (Table 1) would also permit a comparison of ILs that are capable of homogeneous dissolution of lignocellulose, i.e. complete dissolution of all the main components of lignocellulose, and ILs known to be restricted to heterogeneous dissolution of lignocellulose [30]. $[C_4C_1im][MeCO_2]$, $[C_4C_1im][Cl]$ and $[C{=}C_2C_1im][MeCO_2]$ should give rise to homogeneous dissolution of lignocellulose (i.e. dissolve cellulose, hemicellulose and lignin). $[C_4C_1im][HSO_4]$ would effect heterogeneous dissolution of lignocellulose by selectively dissolving lignin and hemicellulose, but would not dissolve cellulose.

Table 1 Results from the ionic liquid synthesis

Abbreviation	Structure	Name	H_2O cont.	Cl^- cont.	Yield[a] (%)
$[C_4C_1im][MeCO_2]$	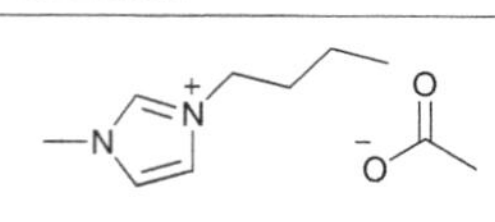	1-butyl-3-methylimidazolium acetate	260 ppm	0.8%	37
$[C_4C_1im][HSO_4]$	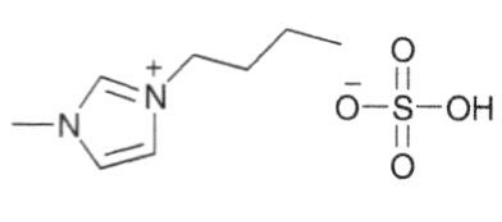	1-butyl-3-methylimidazolium hydrogen sulfate	20%	N.A.[b]	87
$[C_4C_1im][Cl]$	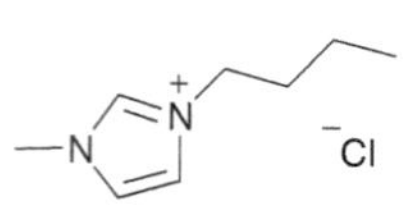	1-butyl-3-methylimidazolium chloride	940 ppm	N.A.	48
$[C{=}C_2C_1im][MeCO_2]$	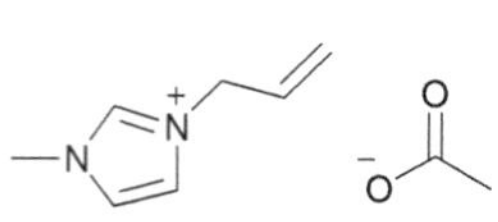	1-allyl-3-methylimidazolium acetate	1500 ppm	0.9%	36

[a]The yields were calculated for the total synthesis (two steps for all of the ILs except $[C_4C_1im][Cl]$, which was synthesized in one step) starting from the substituted imidazoles. [b]Not Applicable.

Table 2 Cellulosic, hemicellulosic and lignocellulosic substrates used in the investigation

Category	Substrate	Description	Source/reference
Crystalline samples	Cotton	Linear cellulosic homopolysaccharide of β-D-glucopyranosyl units (95-99%), crystallinity: 64.9%, typical DP: 8,100 - 15,300.	Selefatrade AB/[31-33]
	Sigmacell Type 20	Linear cellulosic homopolysaccharide of β-D-glucopyranosyl units, crystallinity: 36.1%, DP: 209.	Sigma-Aldrich/[34,35]
Amorphous samples	Regenerated amorphous cellulose	Amorphous cellulosic homopolysaccharide prepared from Sigmacell Type 20 using phosphoric acid.	This work/Prepared using method of Zhang *et al.* [36].
	Beech-wood xylan	Branched hemicellulosic heteropolysaccharide, typical DP of hardwood xylan: 100–200.	Sigma-Aldrich/[32]
	Locust bean gum galactomannan	Branched hemicellulosic heteropolysaccharide, DP: 900–1500.	Sigma-Aldrich/[32]
Wood samples	Hybrid aspen (*Populus tremula x tremuloides*)	Hardwood lignocellulose, see Table 3.	Umeå Plant Science Centre (<www.upsc.se>)/this work
	Norway spruce (*Picea abies*)	Softwood lignocellulose, see Table 3.	SEKAB AB/this work

The study included cellulosic (cotton, Sigmacell Type 20, and regenerated amorphous cellulose), hemicellulosic (xylan and galactomannan), as well as lignocellulosic (hybrid aspen and Norway spruce) substrates (Table 2). The selected collection of substrates differ widely with regard to fundamental properties, such as crystallinity, degree of polymerization (DP), and complexity (Table 2). The most complex substrates, aspen and spruce, represent hardwood and softwood, respectively. Softwood is typically the most recalcitrant type of lignocellulosic biomass [4].

The sets of ILs and substrates used in the investigation would allow a broad evaluation of the ILs with regard to the effects of the anion and the cation as well as the importance of homogeneous and heterogeneous dissolution on the pretreatment efficiency. To permit a comparison of IL pretreatment with a conventional state-of-the-art pretreatment method, one of the lignocellulosic samples was also pretreated using classical sulfuric acid pretreatment. Using in-house prepared ILs and advanced analytical methods such as high-performance anion-exchange chromatography (HPAEC) to quantify different monosaccharides released from the polymeric substrates by the hydrolytic enzymes, this investigation addresses the lack of direct comparisons of the effects of ILs on cellulosic, hemicellulosic, and lignocellulosic samples.

Results and discussion

Ionic liquids

Data on the properties and synthesis of the four ionic liquids synthesized as a part of the investigation are shown in Table 1. Figure 1 shows the results of ^{1}H and ^{13}C NMR analysis of [C_4C_1im][$MeCO_2$], [C_4C_1im][HSO_4], [C_4C_1im][Cl], and [C=C_2C_1im][$MeCO_2$]. The figure shows the peak assignments for the ^{1}H NMR spectra. The clean spectra indicate that the ionic liquids were of high purity. Dialkylated imidazoles, such as [C_4C_1im][$MeCO_2$], [C_4C_1im][HSO_4] and [C_4C_1im][Cl] (Table 1), are among the more common cations used for the dissolution of biomass. In fact, 1-butyl-3-methylimidazolium chloride ([C_4C_1im][Cl]) was one of the first ILs used to dissolve cellulose [15] and lignocellulose [27]. The focus has recently shifted from chloride-based to acetate-based ILs [26], which in this study are represented by [C_4C_1im][$MeCO_2$] and [C=C_2C_1im][$MeCO_2$] (Table 1). However, the stability of acetate-based ILs is not as good as that of chloride-based ILs, such as [C_4C_1im][Cl] (Table 1).

The Kamlet-Taft polarity, or, more precisely, the hydrogen-bond basicity of the IL, is believed to be strongly linked to its capability to dissolve or to swell lignocellulose [23,39]. However, the basicity of an IL is related to its hydrophilicity [40] and poses a problem which has to do with the water content of the IL. In equilibrium with air, ILs with acetate as anion can reach a water content of 27% (w/w) [41]. However, even a relatively low content of water (3-5% (w/w) in a study of *Miscanthus* [28] and 5-10% (w/w) in a study of maple-wood flour [24]) has been shown to have a negative effect on the solubility of lignocellulose. This is not optimal in a process, where the energy used for drying the biomass and the IL needs to be taken in to account.

The water content of the ILs (Table 1) was determent by Karl-Fischer titration after the samples had been exposed to high vacuum (<50 mbar) at 65°C overnight. The water content of all ILs was below 0.16 wt%, except for [C_4C_1im][HSO_4], which on purpose had a high water content. According to previous studies [18], the sugar yield from lignocellulose pretreated with [C_4C_1im][HSO_4] was significantly higher when the water content was 10-40%. The water content of the [C_4C_1im][HSO_4] used in our study, 20% (Table 1), is thus within the desired range.

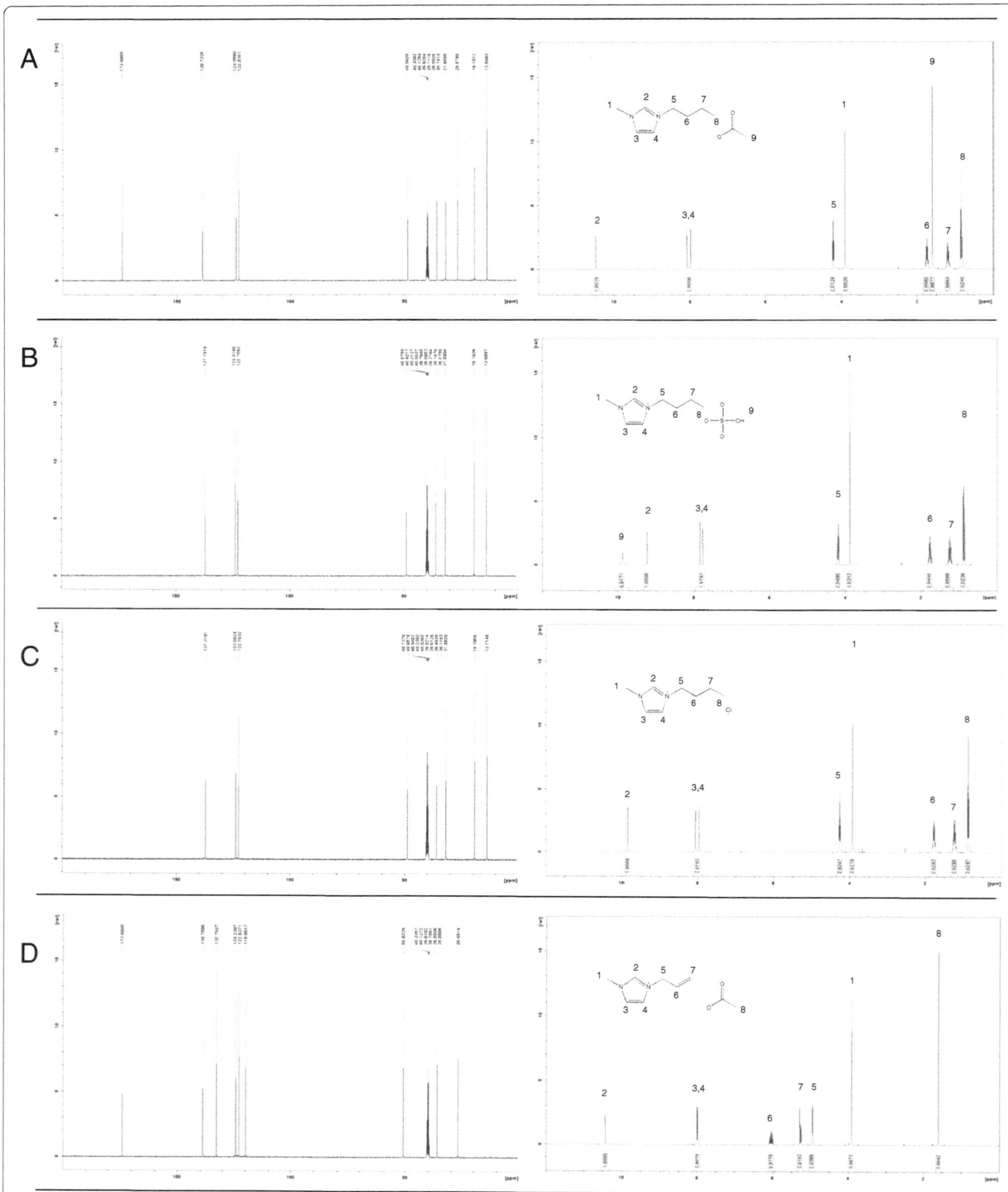

Figure 1 NMR analysis of ionic liqid preparations: (A) 1-butyl-3-methylimidazolium acetate, (B) 1-butyl-3-methylimidazolium hydrogen sulfate, (C) 1-butyl-3-methylimidazolium chloride, and (D) 1-allyl-3-methylimidazolium acetate. Data of ^{13}C NMR (left side) and ^{1}H NMR (right side) analysis are shown.

The yields of the ILs (Table 1) were calculated for the total synthesis starting from the corresponding 1-alkylimidazole (methyl and butyl). This explains why the yields were sometimes relatively low (Table 1). $[C_4C_1im][MeCO_2]$ and $[C{=}C_2C_1im][MeCO_2]$ were made from the corresponding saturated and unsaturated 1-alkyl-3-methylimidazolium chlorides. The ion exchange was tracked using NMR to ensure that the acetate ion was duly present and that the ion exchange was successful. The halide contamination was controlled by using HPAEC and a sodium chloride standard. The chloride contamination was <1% (Table 1).

Biomass samples

The amounts of structural carbohydrates and lignin were determined after extraction of the aspen and the spruce wood (Table 3). The values correspond well with those found in the literature [42]. The glucan content of the aspen and spruce wood was rather similar, around 40% (Table 3). As expected, the aspen wood had a relatively high xylan content (16.6%), while the spruce wood had relatively high mannan content (10.4%). The total lignin content of Norway spruce (28.3%) was higher than that of hybrid aspen (23.4%). The lignin content is highly relevant for enzymatic hydrolysis of cellulose [4], as lignin can obstruct saccharification by forming a physical barrier, by catalytically unproductive binding of hydrolytic enzymes, and by formation of lignin-carbohydrate complexes (LCCs).

Pretreatment and enzymatic saccharification

Cellulosic, hemicellulosic and lignocellulosic samples (5% wt) were pretreated with the four ILs at 100°C for 20 h. The precipitation was carried out using a small amount of methanol and ultra-pure deionized water. The precipitate was collected by centrifugation and washed three times with ultra-pure water and once with citrate buffer prior to the enzymatic saccharification. The enzyme mixture used contains a wide variety of proteins, which are expected to degrade not only cellulose, but also xylan and mannan. The use of a sensitive analytical method for determination of the yields of the monosaccharides (HPAEC) would allow detection of small quantities of sugars such as xylose and mannose. The enzymatic saccharification was carried out analytically in a small scale (reactions with 50 mg sample), which made the processing of multiple parallel samples and the evaluation of the experiments using statistical analysis feasible.

Crystalline samples

Compared to the untreated control, the glucose production rate [GPR, value based on the glucose concentration after 2 h before the reaction levels off (see Section Glucose concentration estimation)] of cotton increased dramatically after treatment with $[C_4C_1im][Cl]$, $[C_4C_1im][MeCO_2]$, and $[C{=}C_2C_1im][MeCO_2]$ (Figure 2). For cellulosic samples, pretreatment with $[C_4C_1im][Cl]$ gave the best results (Figure 2). The GPR increased significantly ($P \leq 5\%$, t-test) also for the $[C_4C_1im][HSO_4]$-pretreated sample compared to that of the control sample. Compared to the untreated control, the glucose yield [value based on the glucose concentration after 72 h, when the reaction has leveled off (see Section Analysis of monosaccharide yields)] increased significantly after pretreatment with all of the ILs except $[C_4C_1im][HSO_4]$ (Figure 3). Pretreatment of cotton with $[C_4C_1im][MeCO_2]$, $[C{=}C_2C_1im][MeCO_2]$ and $[C_4C_1im][Cl]$ increased the yield with 160, 170 and 160%, respectively. The difference can be attributed to that $[C_4C_1im][HSO_4]$ causes heterogeneous dissolution of lignocellulose, while the other three ILs cause homogeneous dissolution. As cotton consists almost of pure cellulose (Table 2), the capacity of $[C_4C_1im][HSO_4]$ to dissolve lignin and hemicellulose would be of little or no help. For $[C_4C_1im][MeCO_2]$, $[C{=}C_2C_1im][MeCO_2]$ and $[C_4C_1im][Cl]$, the increase in GPR (Figure 2) was higher than the increase in yield (Figure 3), which indicates that the effect of the pretreatment is most beneficial for the initial phase of the hydrolysis of the cellulose.

With regard to the microcrystalline Sigmacell Type 20 cellulose, the GPRs of samples pretreated with $[C_4C_1im][Cl]$, $[C_4C_1im][MeCO_2]$ and $[C{=}C_2C_1im][MeCO_2]$ were significantly higher ($P \leq 5\%$, t-test) than that of the control (Figure 2). The glucose yields of pretreated and non-pretreated samples did not exhibit any significant differences (Figure 3). Treatment with $[C_4C_1im][HSO_4]$ did not result in any improvement for either the GPR or the glucose yield, as could be expected since $[C_4C_1im][HSO_4]$ does not dissolve cellulose.

When untreated Sigmacell Type 20 was used as substrate instead of untreated cotton, the GPR increased from 0.41 $g\ l^{-1}\ h^{-1}$ to 2.72 $g\ l^{-1}\ h^{-1}$ and the glucose yield increased from 0.55 $g\ g^{-1}$ to 0.85 $g\ g^{-1}$. Evidently, the

Table 3 Contents of structural carbohydrates and lignin in aspen and spruce[a]

Biomass	Glucan	Arabinan	Galactan	Xylan	Mannan	Total lignin[b]
Hybrid aspen	0.409 ± 0.015	0.008 ± 0.001	0.016 ± 0.001	0.166 ± 0.010	0.030 ± 0.012	0.234 ± 0.002
Norway spruce	0.394 ± 0.003	0.014 ± 0.001	0.020 ± 0.001	0.053 ± 0.001	0.104 ± 0.001	0.283 ± 0.002

[a]Reported as g of carbohydrate or lignin per g of wood (dry weight) with standard deviations.
[b]Klason lignin and acid-soluble lignin.

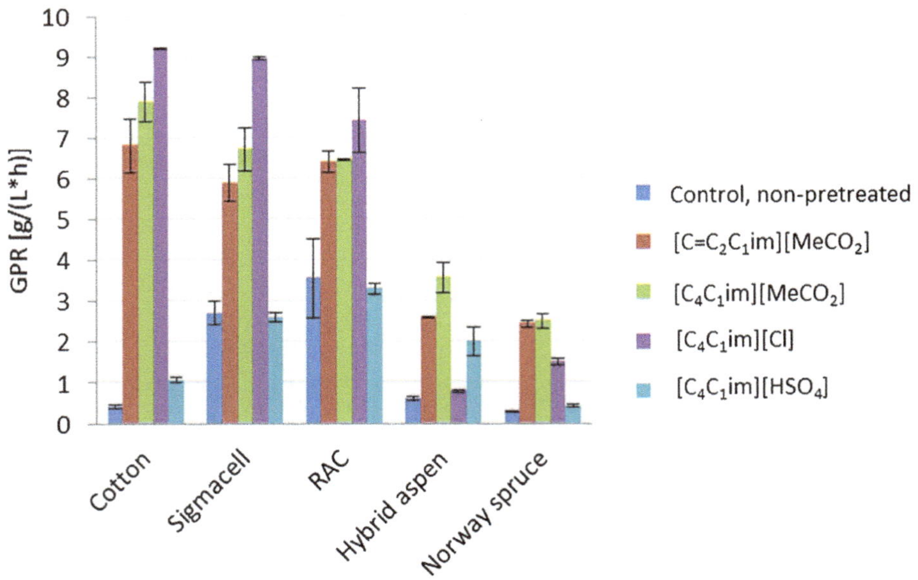

Figure 2 Glucose production rate (GPR) in g glucose l^{-1} h^{-1} after 2 h of enzymatic hydrolysis (before the reaction rates level off) for the ionic-liquid-pretreated samples and the non-pretreated control.

Sigmacell Type 20 cellulose was more susceptible to enzymatic hydrolysis than cotton cellulose. The difference can probably be attributed to the much lower degree of polymerization (DP) of Sigmacell Type 20 compared to cotton (Table 2), which would give the exo-acting cellobiohydrolases a larger number of cellulose chain ends to attack. It is also possible that a lower degree of crystallinity of Sigmacell Type 20 cellulose (Table 2) contributes to facilitating the enzymatic hydrolysis.

After pretreatment of cotton with $[C_4C_1im][Cl]$, $[C_4C_1im][MeCO_2]$ or $[C{=}C_2C_1im][MeCO_2]$, the GPR values were similar or slightly higher than the corresponding GPR values for Sigmacell Type 20 (Figure 2). This was not due to that the substrate had been exhausted already after 2 h, as less than 35% of the substrate had been converted to glucose. The yield values (Figure 3) from samples taken after 72 h were also lower than the theoretical maximum, which would be approx. 1.1 g glucose

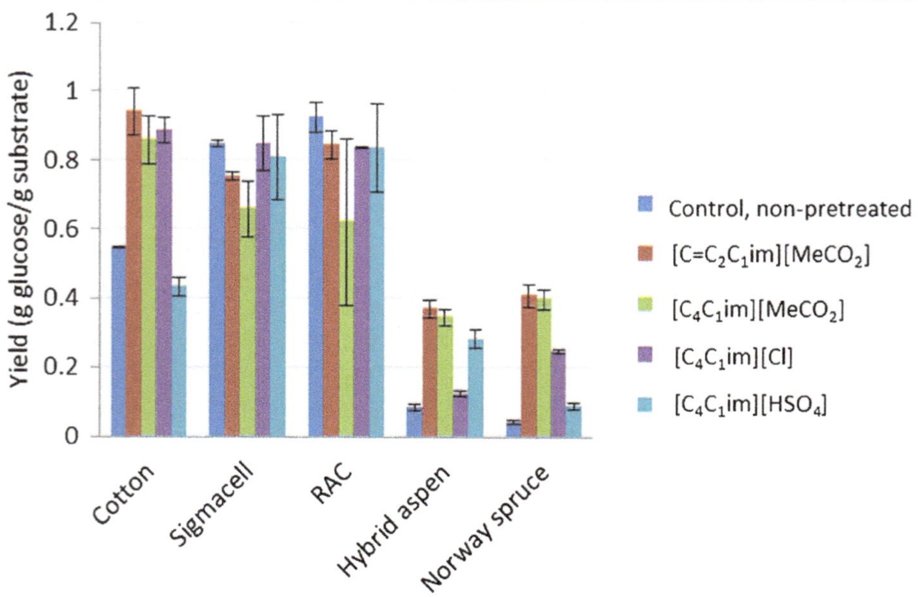

Figure 3 The glucose yield in g glucose g substrate^{-1} after 72 h of enzymatic hydrolysis (at the end of the reactions after that the reaction rates level off) of ionic-liquid-pretreated samples and non-pretreated control.

per g cellulose. The highest yield observed for the two cellulosic crystalline substrates, 0.94 g g^{-1}, was obtained after pretreatment of cotton with [C=C_2C_1im][$MeCO_2$] and corresponds to conversion of around 85% of the substrate to glucose. Thus, although the natural cotton was more recalcitrant than Sigmacell Type 20, pretreatment with all of the ionic liquids except [C_4C_1im][HSO_4] removed that effect.

Amorphous samples

The GPR values (Figure 2) and the glucose yield values (Figure 3) for regenerated amorphous cellulose (RAC) were rather similar to the values of the Sigmacell Type 20 cellulose that the RAC was derived from. This suggests that the crystallinity of the Sigmacell Type 20 cellulose (Table 2) was not an obstacle for the enzymatic hydrolysis. Apart from conversion of crystalline cellulose to amorphous, treatment with acid may potentially also result in hydrolysis which could give decreased cellulose chain length, but evidently this did not lead to improved hydrolysis of RAC compared to Sigmacell Type 20 under the experimental conditions studied.

Enzymatic hydrolysis of untreated or IL-pretreated hemicellulosic samples for 72 h resulted in very low yields (<0.05 g g^{-1}) of glucose and arabinose. Enzymatic hydrolysis of untreated beech-wood xylan gave a xylose yield of 0.55 ± 0.09 g g^{-1}, while untreated locust bean gum galactomannan gave a galactose yield of 0.18 ± 0.01 g g^{-1} and a mannose yield of 0.54 ± 0.03 g g^{-1}. After pretreatment of the beech-wood xylan with ILs, the xylose yields compared to that of the untreated control were 23% for [C=C_2C_1im][$MeCO_2$], 25% for [C_4C_1im][$MeCO_2$], 39% for [C_4C_1im][Cl], and 7% for [C_4C_1im][HSO_4]. For the locust bean gum galactomannan, the galactose yields compared to that of the untreated control were 33% for [C=C_2C_1im][$MeCO_2$], 75% for [C_4C_1im][$MeCO_2$], and 66% for [C_4C_1im][Cl]. No galactose was recovered from galactomannan pretreated with [C_4C_1im][HSO_4]. Finally, the mannose yields compared to that of the untreated galactomannan control were 91% for [C_4C_1im][$MeCO_2$], 73% for [C_4C_1im][Cl], and 3% for [C_4C_1im][HSO_4]. No mannose was recovered after pretreatment of galactomannan with [C=C_2C_1im][$MeCO_2$]. Thus, the recovery of hemicellulose sugars after pretreatment with ILs was always poor.

A possible explanation for poor recovery of hemicellulosic sugars is that previous studies have shown that ILs with a high basicity will hydrolyze hemicellulose to oligomeric sugars [43]. It has also been shown that sulfate-based ILs, especially $[HSO_4]^-$, will hydrolyze hemicellulose to monomeric sugars and even dehydrate pentose sugars to furfural [18]. Poor recovery of hemicellulose sugars has also been indicated in previous studies [30].

Wood samples

As expected considering the recalcitrance of softwood, untreated Norway spruce exhibited lower GPR (0.28 ± 0.02 g l^{-1} h^{-1}) and glucose yield (0.043 ± 0.006 g g^{-1}) than untreated hybrid aspen (GPR, 0.62 ± 0.04 g l^{-1} h^{-1}; glucose yield, 0.085 ± 0.01 g g^{-1}). Pretreatment of the wood samples with the four ILs resulted in significantly higher ($P \leq 5\%$, t-test) GPR values and glucose yields (Figures 2 and 3). The increase of the yields (Figure 3) varied between 150 and 950% compared to that of the untreated wood samples.

The highest yields were obtained upon use of [C_4C_1im][$MeCO_2$] and [C=C_2C_1im][$MeCO_2$] (Figure 3). For aspen, the glucose yields with [C_4C_1im][$MeCO_2$] and [C=C_2C_1im][$MeCO_2$] were 0.35 ± 0.02 g g^{-1} and 0.37 ± 0.02 g g^{-1}, respectively. These yields were slightly higher than that achieved in a control experiment with dilute sulfuric acid pretreatment and enzymatic hydrolysis of aspen, which resulted in a glucose yield of 0.31 ± 0.05 g g^{-1}. For aspen, the glucan conversion after pretreatment with [C_4C_1im][$MeCO_2$] and [C=C_2C_1im][$MeCO_2$] was 76% and 81%, respectively. When spruce was pretreated using [C_4C_1im][$MeCO_2$] or [C=C_2C_1im][$MeCO_2$], more than 90% of the glucan was converted to glucose. That is a higher value than achieved with the pure cellulosic crystalline and amorphous substrates, which indicates the potential of using ILs for pretreatment of lignocellulose.

The glucose yields after pretreatment of aspen with the ILs that did not have acetate as anion were lower and corresponded to 0.28 ± 0.03 g g^{-1} for [C_4C_1im][HSO_4] and 0.13 ± 0.01 g g^{-1} for [C_4C_1im][Cl]. The glucan conversions after pretreatment of aspen with [C_4C_1im][Cl] and [C_4C_1im][HSO_4] were 28% and 62%, respectively. The corresponding values for spruce were 57% and 20%. While [C_4C_1im][HSO_4] did not work very well for pretreatment of the recalcitrant spruce wood, the results obtained with aspen were not far behind the results achieved with acetate-based ILs (Figures 2 and 3).

There were no significant differences between [C_4C_1im][$MeCO_2$] and [C=C_2C_1im][$MeCO_2$] in terms of pretreatment of any of the samples in this study. This indicates that the choice of cation is often not decisive in terms of the efficiency of the pretreatment process. The higher yields obtained upon use of ionic liquids with an acetate anion can most likely be explained by the higher basicity of the acetate compared to the chloride and hydrogen sulfate (reviewed by Brandt *et al.* [30]). The comparatively low yields found for the hydrogen sulfate anion compared to the acetates can be attributed to the fact that it does not dissolve cellulose. However, it does dissolve lignin and hemicelluloses, which would explain the higher yields achieved compared to non-pretreated wood samples.

Conclusions

Four ILs were evaluated for pretreatment of seven cellulosic, hemicellulosic and lignocellulosic substrates prior to enzymatic saccharification. The wide variety of substrates investigated and the use of the same pretreatment conditions for all ILs studied facilitated comparisons of the performance of the different ILs.

Although they were not the most efficient pretreatment agents for simple cellulosic substrates, ILs with acetate as anion gave the best results for complex lignocellulosic substrates from wood. However, among the ILs used in this study the ones with acetate as anion are among the least stable. Moreover, they are cumbersome to synthesize, and they are sensitive to moisture.

The halide-based IL [C_4C_1im][Cl] performed very well for simple cellulosic substrates, but was less efficient for pretreatment of wood. [C_4C_1im][Cl] possesses advantages from the process perspective, since it is easy to synthesize, relatively stable, and less sensitive to moisture than the acetate-based ILs. Disadvantages with [C_4C_1im][Cl] include its high melting point and high viscosity.

The difference between homogeneous and heterogeneous dissolution of lignocellulose was studied by the inclusion of [C_4C_1im][HSO_4] in the set of ILs studied. Homogeneous dissolution was not always better, as [C_4C_1im][HSO_4] gave better results than [C_4C_1im][Cl] with hybrid aspen, the less recalcitrant type of lignocellulose included in the study. This warrants further attention in future studies, for example by optimization of the pretreatment process.

The best choices, as judged by the saccharification efficiency, appeared to be [C_4C_1im][Cl] for cellulosic substrates, acetate-based ILs or [C_4C_1im][HSO_4] for hybrid aspen, and acetate-based ILs for Norway spruce. Pretreatment can, however, involve high water content, something that has always been a problem with homogeneous dissolution of lignocellulose. The possibility to include water and the stability of [C_4C_1im][HSO_4] have potential to decrease the cost for the regeneration process. Therefore other factors than the capacity to dissolve cellulose, such as IL stability, the possibility to include water, and regeneration of the IL, need to be considered when choosing ILs that are most suited for industrial pretreatment.

Conversion of crystalline Sigmacell Type 20 cellulose to RAC did not improve enzymatic hydrolysis. This was shown for both non-pretreated cellulose and cellulose that was pretreated with different ILs. ILs that enhanced hydrolysis of the crystalline form also enhanced hydrolysis of the RAC.

Pretreatment with ILs gave sugar yields comparable with those obtained with acidic pretreatment technique. Pretreatment with ILs can be performed using gentle conditions, while acidic pretreatment is typically performed at high temperature and high pressure, and in a corrosive environment, factors that increase the demands on the quality of the process equipment and contributes to lower yields due to the formation of by-products from the lignocellulose feed. High yields are critical in the production of biofuels and other bulk chemicals. ILs also have low vapor pressure which is advantageous with regard to safety and environmental aspects. Thus, provided that ILs used in industrial processes can be reused, there are distinct advantages that make further research on the use of ILs for conversion of lignocellulose worthwhile. Future areas of research include recycling of ILs, up-scaling, process integration, and techno-economical assessment.

Methods

Ionic liquid synthesis

Purification of starting materials

1-Methylimidazole (Sigma-Aldrich, 99%) and 1-butyli midazole (Aldrich, 99%) were distilled under vacuum from potassium hydroxide (Akzo Nobel, 87.9%). 1-Chlorobutane (Acros Organics, 99%), allyl chloride (Aldrich, 99%) and dimethylsulfate (Fluka, 95%) were distilled from CaO (Sigma-Aldrich). Ethyl acetate (Fisher Chemicals HPLC) and acetonitrile (Fisher Chemicals HPLC) were dried by running them through neutral aluminum oxide 90 (Merck 70–230 mesh ASTM) and toluene (Sigma-Aldrich, 99.7%) through silica (Acros Organics, Ultrapure 40–60 μm, 60A). All starting materials were used within hours of preparation. All reactions and washing procedures were carried out in dry glassware with dried chemicals and under inert atmosphere (argon) if not stated otherwise or in the case where water was used.

[C_4C_1im][Cl]

1-Methylimidazole (99 ml, 1.2 mol) was dissolved in 150 ml ethyl acetate and stirred at room temperature. 1-Chlorobutane (130 ml, 1.2 mol) was added dropwise to the reaction mixture and afterwards the reaction mixture was refluxed at 75°C for 4 days. The reaction vessel was allowed to cool at room temperature and was then thoroughly sealed and was kept at −20°C overnight. The upper organic layer was then removed using Schlenk techniques and the bottom IL layer was washed two times with 75 ml ethyl acetate. After the first wash, the IL precipitated into white crystals that were then dissolved in approximately 50 ml of acetonitrile and were recrystallized overnight at −20°C. The upper organic layer was again removed, and the crystals were washed three times with 50 ml of ethyl acetate and dried overnight using vacuum treatment in order to remove the last of the solvents. The yield was 48%.

$[C{=}C_2C_1im][Cl]$

Allyl chloride (116 ml, 1.43 mol) was added dropwise to 1-methylimidazole (92 ml, 1.14 mol) under stirring and cooled in an ice-bath. Afterwards the reaction mixture was refluxed for 14 h at which point the reflux condenser was replaced with a distillation head and the reaction mixture was vacuum-distilled to give an amber-colored viscous liquid. The yield was 96%.

$[C_4C_1im][MeCO_2]$

$[C_4C_1im][MeCO_2]$ was prepared from $[C_4C_1im][Cl]$ using an ion-exchange column. Amberlite IRA-400 (OH^- form) ion-exchange resin (64 g; enough to support the conversion of 20 g $[C_4C_1im][Cl]$) was dispersed in deionized water and poured into a glass column (3 cm internal diameter). The Amberlite column was then loaded with $[MeCO_2]^-$ by running through 1000 ml of a 1 M aqueous solution of $[Na][MeCO_2]$ using a flow rate of 0.8 ml/min. Once the correct ion was loaded on the column, 15.3 g of $[C_4C_1im][Cl]$ was dissolved in 60 ml of deionized water and poured onto the Amberlite column with a flow-rate of 0.2 ml/min. When all the $[C_4C_1im][Cl]$ aqueous solution had passed into the column, another 10 ml of water was added and were allowed to run through in the same flow-rate. Fractions were collected during this time, and analyses with HPAEC were used to determine the chloride content. All fractions were evaporated and pumped overnight at 65°C under stirring to remove the water eluent. The chloride content was below 0.8% (mol/mol). Afterwards, the Amberlite column was rejuvenated with 1000 ml of a 1 M aqueous solution of NaOH and with a flow-rate of 0.8 ml/min. The yield was 78%

$[C{=}C_2C_1im][MeCO_2]$

$[C{=}C_2C_1im][MeCO_2]$ was prepared from $[C{=}C_2C_1im][Cl]$ by using an ion-exchange column. A dispersion of Amberlite IRA-400 (OH^- form) ion-exchange resin (64 g; enough to support the conversion of 20 g $[C_4C_1im][Cl]$) was prepared in deionized water and poured into a glass column (3 cm internal diameter). The Amberlite column was then loaded with $[MeCO_2]^-$ by running through 1000 ml of a 1 M aqueous solution of $[Na][MeCO_2]$ with a flow-rate of 0.8 ml/min. Once the correct ion was loaded on the column, 15.2 g of $[C{=}C_2C_1im][Cl]$ was dissolved in 100 ml of deionized water and poured on the Amberlite column and the flow-rate was set to 0.2 ml/min. When all the $[C{=}C_2C_1im][Cl]$ aqueous solution had passed into the column, another 10 ml of water was added and were allowed to run through in the same flow-rate. Fractions were collected during this time and analyses with HPAEC were performed to determine the chloride content. All fractions were subjected to a drying process in evaporator and high vacuum overnight at 65°C under stirring to remove the water eluent. The chloride content was found to reside below 0.9% (mol/mol). Afterwards, the Amberlite column was rejuvenated with 1000 ml of a 1 M aqueous solution of sodium hydroxide with a flow-rate of 0.8 ml/min. The yield was 76%.

$[C_4C_1im][MeSO_4]$

1-Butylimidazole (51 ml, 0.39 mol) was diluted in 60 ml toluene and cooled in an ice bath. Dimethyl sulfate (37 ml, 0.39 mol) was then added drop-wise (flow rate 0.5 drops/sec) and under stirring to the reaction mixture. Afterwards the reaction mixture was stirred for another hour at room temperature before the upper toluene layer was removed using Schlenk techniques. The lower IL layer was subsequently washed three times with 50 ml of toluene before it was dried overnight at 65°C under vacuum and stirring. The yield obtained was 89%.

$[C_4C_1im][HSO_4]$

$[C_4C_1im][MeSO_4]$ (29.9 g) was diluted in 5 ml deionized water. The reaction vessel was kept open to the air and put in a heating block at 215°C. The internal temperature of the reaction mixture was continuously monitored and kept between 170 to 180°C for four hours by drop-wise addition of deionized water. Afterwards, the reaction mixture was cooled to 65°C and was kept under vacuum and stirring overnight to remove all residual water. The yield was 98%.

Ionic liquid analysis

NMR

Analyses by using ^1H and ^{13}C NMR were carried out on a Bruker DRX 400 MHz instrument with dimethyl sulfoxide (DMSO) as solvent. In all cases, the chemical shifts were calibrated from the solvent peak [44] (DMSO ^1H 2.54, ^{13}C 40.45).

Karl-Fischer titration

The water content of the ILs was determined using a Metrohm 756 KF coulometer with diaphragm-free titration cell and Hydranal Coulomat E reagent. After exposure to high vacuum at 65°C overnight, approximately 100 μl of the sample was withdrawn into a dry syringe which was then weighed. Subsequently, the sample was injected into the Karl-Fischer reactor and re-weighed to determine the amount of sample injected.

Pretreatment and enzymatic hydrolysis

Cellulosic substrates

Seven cellulosic and hemicellulosic substrates ranging from highly crystalline to amorphous were selected in order to achieve a broad evaluation. Wood samples included Norway spruce (kindly provided by SEKAB, Örnsköldsvik, Sweden) and hybrid aspen [kindly provided by the Umeå Plant Science Centre (UPSC),

Umeå, Sweden]. Crystalline polysaccharides included cotton (Selefatrade AB, Spånga, Sweden) and Sigmacell Cellulose Type 20 (Sigma-Aldrich). Amorphous polysaccharides included beech-wood xylan (Sigma-Aldrich), galactomannan (locust bean gum, Sigma-Aldrich) and regenerated amorphous cellulose (prepared from Sigmacell cellulose Type 20 according to Zhang *et al.* [36]).

The wood samples, spruce and aspen, were milled using an IKA A11 Basic Analytical Mill and sieved (Retsch Analytical AS 200, 500–100 μm sieves). All cellulosic, hemicellulosic and lignocellulosic samples were preparatively dried using a vacuum pump until a dry matter content of ~100% was reached (Mettler Toledo HG63) thus avoiding water upon the IL pretreatment.

Pretreatment

Cellulosic samples (50 mg) were dispersed in 950 mg of ionic liquid and were then incubated for 20 hours at 100°C. The samples were allowed to cool to room temperature followed by the addition of up to 1 g of methanol in order to give a smoother initial precipitation. After that, 10 g of ultra-pure water (Millipore) was added followed by vigorous mixing and centrifugation for 10 min at 14,500 *g* (Eppendorf centrifuge 5810R). The supernatant was decanted and the biomass was subsequently washed using 3 × 10 g ultra-pure water and 1 × 10 g 50 mM citrate buffer pH 5.2 in a similar manner as previously stated. As a negative control, non-pretreated biomass was used. As a positive control, the milled and sieved aspen wood was pretreated at a combined severity of 2.2 using 1% (w/w) sulfuric acid as catalyst.

Enzymatic hydrolysis

The total weight of the reaction mixture was 1,670 mg and it contained 50 mg of the cellulosic/hemicellulosic sample, 50 mg enzyme mixture (a 1:1 ratio of a mixture of Celluclast 1.5 L and Novozyme 188, both of which were obtained from Sigma-Aldrich) and 50 mM citrate buffer (pH 5.2). The enzymatic hydrolysis was performed at 45°C in an orbital shaker set at 170 rpm (Ecotron incubator shaker, Infors). The samples (10 μl) were collected before enzyme addition (0 h), and after 2 h, 48 h and 72 h of enzymatic hydrolysis.

Product analysis

Glucose concentration estimation

During the hydrolysis (samples taken after 0, 2, 48, and 72 h, respectively) the concentration of glucose was estimated using a glucometer (Accu-Chek Aviva, Roche Diagnostics GmBH). The glucose concentrations after 2 h were used for calculation of the GPR (the glucose production rate), which is determined before the hydrolysis reaction levels off.

Analysis of monosaccharide yields

The concentration of separate monosaccharides (arabinose, galactose, glucose, xylose, and mannose) was determined after 72 h of enzymatic hydrolysis using high-performance anion-exchange chromatography (HPAEC). A Dionex ICS-5000 equipped with an electrochemical detector, a CarboPac PA20 (3 × 30 mm) guard column and a CarboPac PA20 (3 × 150 mm) separation column (all from Dionex) were used to perform the analysis. Prior to analysis all samples were diluted using ultra-pure water and filtered through a 0.2 μm nylon membrane (Millipore). The concentrations determined with HPAEC were used for calculation of the monosaccharide yields, as the hydrolysis reactions had leveled off after 72 h.

Determination of structural carbohydrates and lignin in the wood samples

The extractives were determined according to Lestander *et al.* [45] with the exception that the extraction was performed in 15 cycles instead of during 1 h. After the extraction, the contents of carbohydrates (arabinan, galactan, glucan, mannan and xylan) and lignin (acid soluble and acid-insoluble lignin) were determined according to NREL/TP-510-42618 [46] with the exception that the concentration of the monosaccharides was determined using HPAEC (according to the procedure described in Section Analysis of monosaccharide yields).

Abbreviations

HPAEC: High-performance anion-exchange chromatography; ILs: Ionic liquids; NMR: Nuclear magnetic resonance.

Competing interests

The authors declare that they have no competing interests.

Authors' contributions

SW, JG and LJJ conceived and designed the study. The experimental work was carried out by JG, SW and MN under the supervision of LJJ and JPM. JG and SW contributed equally to this work and should therefore be considered as co-first-authors. All authors read and approved the final manuscript.

Authors' information

JG is a doctoral student with focus on utilization of ionic liquids for conversion of lignocellulosic materials. SW is a postdoctoral researcher with interest in enzyme chemistry and technology. MN is a doctoral student with interest in the area of pretreatment and enzymatic saccharification of lignocellulose. LJJ is professor at Umeå University, and is working on biotechnology for biorefining of lignocellulosic feedstocks. He is also leader of the Biochemical Platform of the Bio4Energy research initiative (www.bio4energy.se). JPM is professor at Umeå University and Åbo Akademi University, and is working on ionic-liquid technologies and chemical catalysis for the biorefining of lignocellulose, e.g. within the Bio4Energy program.

Acknowledgements

This research was supported by the Swedish Energy Agency (35367-1), the Swedish Research Council (621-2011-4388), the Kempe Foundations, and the Bio4Energy research initiative (www.bio4energy.se). In Finland, the Academy of Finland is acknowledged. This work is also a part of the activities of the Process Chemistry Centre at Åbo Akademi.

Author details
[1]Department of Chemistry, Umeå University, Umeå SE-901 87, Sweden.
[2]Industrial Chemistry & Reaction Engineering, Åbo Akademi University, Åbo-Turku FI-20500, Finland.

References

1. Perlack RD, Wright LL, Turhollow AF, Graham RL, Stokes BJ, Erbach DC: *Biomass as a feedstock for a bioenergy and bioproducts industry: the technical feasibility of a billion-ton annual supply.* Springfield, VA: U.S. Department of Commerce, National Technical Information Service; 2005.
2. Lichtenthaler FW, Peters S: **Carbohydrates as green raw materials for the chemical industry.** *C R Chim* 2004, **7:**65–90.
3. Mosier N, Wyman C, Dale B, Elander R, Lee YY, Holtzapple M, Ladisch M: **Features of promising technologies for pretreatment of lignocellulosic biomass.** *Bioresour Technol* 2005, **96:**673–686.
4. Chandra RP, Bura R, Mabee WE, Berlin A, Pan X, Saddler JN: **Substrate pretreatment: the key to effective enzymatic hydrolysis of lignocellulosics?** *Adv Biochem Eng Biotechnol* 2007, **108:**67–93.
5. Zhang YHP, Lynd LR: **Toward an aggregated understanding of enzymatic hydrolysis of cellulose: noncomplexed cellulase systems.** *Biotech Bioeng* 2004, **88:**797–824.
6. Biermann O, Hadicke E, Koltzenburg S, Muller-Plathe F: **Hydrophilicity and lipophilicity of cellulose crystal surfaces.** *Angew Chem Int Edit* 2001, **40:**3822–3825.
7. Yamane C, Aoyagi T, Ago M, Sato K, Okajima K, Takahashi T: **Two different surface properties of regenerated cellulose due to structural anisotropy.** *Polym J* 2006, **38:**819–826.
8. Lindman B, Karlström G, Stigsson L: **On the mechanism of dissolution of cellulose.** *J Mol Liq* 2010, **156:**76–81.
9. Liebert TF, Heinze TJ, Edgar KJ: *Cellulose solvents: for analysis, shaping and chemical modification.* Washington: American Chemical Society; 2009.
10. Hermanutz F, Gaehr F, Uerdingen E, Meister F, Kosan B: **New developments in dissolving and processing of cellulose in ionic liquids.** *Macromol Symp* 2008, **262:**23–27.
11. Heinze T, Dicke R, Koschella A, Kull AH, Klohr EA, Koch W: **Effective preparation of cellulose derivatives in a new simple cellulose solvent.** *Macromol Chem Phys* 2000, **201:**627–631.
12. Yan LF, Chen J, Bangal PR: **Dissolving cellulose in a NaOH/thiourea aqueous solution: a topochemical investigation.** *Macromol Biosci* 2007, **7:**1139–1148.
13. Yan LF, Gao ZJ: **Dissolving of cellulose in PEG/NaOH aqueous solution.** *Cellulose* 2008, **15:**789–796.
14. Liu SL, Zhang LN: **Effects of polymer concentration and coagulation temperature on the properties of regenerated cellulose films prepared from LiOH/urea solution.** *Cellulose* 2009, **16:**189–198.
15. Swatloski RP, Spear SK, Holbrey JD, Rogers RD: **Dissolution of cellose with ionic liquids.** *J Am Chem Soc* 2002, **124:**4974–4975.
16. Tan SSY, MacFarlane DR: **Ionic liquids in biomass processing.** *Ionic Liquids* 2009, **290:**311–339.
17. Pinkert A, Marsh KN, Pang SS, Staiger MP: **Ionic liquids and their interaction with cellulose.** *Chem Rev* 2009, **109:**6712–6728.
18. Brandt A, Ray MJ, To TQ, Leak DJ, Murphy RJ, Welton T: **Ionic liquid pretreatment of lignocellulosic biomass with ionic liquid-water mixtures.** *Green Chem* 2011, **13:**2489–2499.
19. Sun N, Rodriguez H, Rahman M, Rogers RD: **Where are ionic liquid strategies most suited in the pursuit of chemicals and energy from lignocellulosic biomass?** *Chem Commun* 2011, **47:**1405–1421.
20. Zhao H, Baker GA, Song ZY, Olubajo O, Crittle T, Peters D: **Designing enzyme-compatible ionic liquids that can dissolve carbohydrates.** *Green Chem* 2008, **10:**696–705.
21. Turner MB, Spear SK, Huddleston JG, Holbrey JD, Rogers RD: **Ionic liquid salt-induced inactivation and unfolding of cellulase from Trichoderma reesei.** *Green Chem* 2003, **5:**443–447.
22. Engel P, Mladenov R, Wulfhorst H, Jager G, Spiess AC: **Point by point analysis: how ionic liquid affects the enzymatic hydrolysis of native and modified cellulose.** *Green Chem* 2010, **12:**1959–1966.
23. Brandt A, Hallett JP, Leak DJ, Murphy RJ, Welton T: **The effect of the ionic liquid anion in the pretreatment of pine wood chips.** *Green Chem* 2010, **12:**672–679.
24. Doherty TV, Mora-Pale M, Foley SE, Linhardt RJ, Dordick JS: **Ionic liquid solvent properties as predictors of lignocellulose pretreatment efficacy.** *Green Chem* 2010, **12:**1967–1975.
25. Gräsvik J, Eliasson B, Mikkola JP: **Halogen-free ionic liquids and their utilization as cellulose solvents.** *J Mol Struct* 2012, **1028:**156–163.
26. Sun N, Rahman M, Qin Y, Maxim ML, Rodriguez H, Rogers RD: **Complete dissolution and partial delignification of wood in the ionic liquid 1-ethyl-3-methylimidazolium acetate.** *Green Chem* 2009, **11:**646–655.
27. Fort DA, Remsing RC, Swatloski RP, Moyna P, Moyna G, Rogers RD: **Can ionic liquids dissolve wood? processing and analysis of lignocellulosic materials with 1-n-butyl-3-methylimidazolium chloride.** *Green Chem* 2007, **9:**63–69.
28. Padmanabhan S, Kim M, Blanch HW, Prausnitz JM: **Solubility and rate of dissolution for Miscanthus in hydrophilic ionic liquids.** *Fluid Phase Equilibr* 2011, **309:**89–96.
29. Fukaya Y, Hayashi K, Wada M, Ohno H: **Cellulose dissolution with polar ionic liquids under mild conditions: required factors for anions.** *Green Chem* 2008, **10:**44–46.
30. Brandt A, Gräsvik J, Hallett JP, Welton T: **Deconstruction of lignocellulosic biomass with ionic liquids.** *Green Chem* 2013, **15:**550–583.
31. Young RA, Rowell RM: *Cellulose: structure, modification and hydrolysis.* New York: John Wiley and Sons; 1986.
32. Fengel D, Wegner G: *Wood: chemistry, ultrastructure, reactions.* New York: Walter de Gruyter; 1989.
33. Krässig HA: *Cellulose: structure, accessibility and reactivity: polymer monographs vol. 11.* Amsterdam: Gordon and Breach Science Publishers; 1996.
34. Van de Vyver S, Thomas J, Geboers J, Keyzer S, Smet M, Dehaen W, Jacobs PA, Sels BF: **Catalytic production of levulinic acid from cellulose and other biomass-derived carbohydrates with sulfonated hyperbranched poly(arylene oxindole)s.** *Energy Environ Sci* 2011, **4:**3601–3610.
35. Zhang Y-HP, Lynd LR: **Determination of the number-average degree of polymerization of cellodextrins and cellulose with application to enzymatic hydrolysis.** *Biomacromolecules* 2005, **6:**1510–1515.
36. Zhang YHP, Cui JB, Lynd LR, Kuang LR: **A transition from cellulose swelling to cellulose dissolution by o-phosphoric acid: evidence from enzymatic hydrolysis and supramolecular structure.** *Biomacromolecules* 2006, **7:**644–648.
37. Wang H, Gurau G, Rogers RD: **Ionic liquid processing of cellulose.** *Chem Soc Rev* 2012, **41:**1519–1537.
38. Kilpeläinen I, Xie H, King A, Granstrom M, Heikkinen S, Argyropoulos DS: **Dissolution of wood in ionic liquids.** *J Agr Food Chem* 2007, **55:**9142–9148.
39. Zavrel M, Bross D, Funke M, Buchs J, Spiess AC: **High-throughput screening for ionic liquids dissolving (ligno-)cellulose.** *Bioresour Technol* 2009, **100:**2580–2587.
40. Brandt A: *Ionic liquid pretreatment of lignocellulosic biomass.* London: Imperial College London; 2012.
41. Troshenkova S, Sashina E, Novoselov N, Arndt K, Jankowsky S: **Structure of ionic liquids on the basis of imidazole and their mixtures with water.** *Russ J Gen Chem* 2010, **80:**106–111.
42. Saddler JN: *Bioconversion of forest and agricultural plant residues.* Wallingford: CAB International; 1993.
43. Arora R, Manisseri C, Li CL, Ong MD, Scheller HV, Vogel K, Simmons BA, Singh S: **Monitoring and analyzing process streams towards understanding ionic liquid pretreatment of switchgrass (Panicum virgatum L.).** *Bioenergy Res* 2010, **3:**134–145.
44. Gottlieb HE, Kotlyar V, Nudelman A: **NMR chemical shifts of common laboratory solvents as trace impurities.** *J Org Chem* 1997, **62:**7512–7515.
45. Lestander TA, Geladi P, Larsson SH, Thyrel M: **Near infrared image analysis for online identification and separation of wood chips with elevated levels of extractives.** *J Near Infrared Spec* 2012, **20:**591–599.
46. Sluiter A, Hames B, Ruiz R, Scarlata C, Sluiter J, Templeton D, Crocker D: *Determination of structural carbohydrates and lignin in biomass. Technical Report NREL/TP-510-42618.* Golden, CO: National Renewable Energy Laboratory; 2011.

2

CBD binding domain fused γ-lactamase from *Sulfolobus solfataricus* is an efficient catalyst for (–) γ-lactam production

Jianjun Wang, Junge Zhu, Cong Min and Sheng Wu*

Abstract

Background: γ-lactamase is used for the resolution of γ-lactam which is utilized in the synthesizing of abacavir and peramivir. In some cases, enzymatic method is the most utilized method because of its high efficiency and productivity. The cellulose binding domain (CBD) of cellulose is often used as the bio-specific affinity matrix for enzyme immobilization. Cellulose is cheap and it has excellent chemical and physical properties. Meanwhile, binding between cellulose and CBD is tight and the desorption rarely happened.

Results: We prepared two fusion constructs of the γ-lactamase gene *gla*, which was from *Sulfolobus solfataricus* P2. These two constructs had Cbd (cellulose binding domain from *Clostridium thermocellum*) fused at amino or carboxyl terminus of the γ-lactamase. These two constructs were heterogeneously expressed in *E. coli* rosetta (DE3) as two fusion proteins. Both of them were immobilized well on Avicel (microcrystalline cellulose matrix). The apparent kinetic parameters revealed that carboxyl terminus fused protein (Gla-linker-Cbd) was a better catalyst. The V_{max} and k_{cat} value of Avicel immobilized Gla-linker-Cbd were 381 U mg^{-1} and 4.7×10^5 s^{-1} respectively. And the values of the free Gla-linker-Cbd were 151 U mg^{-1} and 1.8×10^5 s^{-1} respectively. These data indicated that the catalytic efficiency of the enzyme was upgraded after immobilization. The immobilized Gla-linker-Cbd had a 10-degree temperature optimum dropping from 80°C to 70°C but it was stable when incubated at 60°C for 48 h. It remained stable in catalyzing 20-batch reactions. After optimization, the immobilized enzyme concentration in transformation was set as 200 mg/mL. We found out that there was inhibition that occurred to the immobilized enzyme when substrate concentration exceeded 60 mM. Finally a 10 mL-volume transformation was conducted, in which 0.6 M substrate was hydrolyzed and the resolution was completed within 9 h with a 99.5% ee value.

Conclusions: Cellulose is the most abundant and renewable material on the Earth. The absorption between Cbd domain and cellulose is a bio-green process. The cellulose immobilized fusion Gla exhibited good catalytic characters, therefore we think the cellulose immobilized Gla is a promising catalyst for the industrial preparation of (–) - γ-lactam.

Keywords: Avicel, Cellulose binding domain, γ-lactamase, (–) γ-lactam

Background

γ-lactamase (EC 3.5.2.-) is a small enzyme sub-classification in amidase [1], which could be used for the resolution of (*rac*)-2-azabicyclo [2.2.1] hept-5-en-3-one (γ-lactam) [2]. The well-known utilizations of (–) γ-lactam are in the synthesizing of abacavir [3] and peramivir [4]. These two drugs were applied in the treatment for HIV, hepatitis and influenza pandemic and have brought profits of several billions U.S dollars for the inventors and developers [5]. γ-Lactam could be applied in the syntheses of other leading compounds and drugs too, such as MK-0812 [6], piperidinium compound [7] and melogliptin [8]. In the three preparation methods for the optically pure enantiomers of γ-lactam [1,9-11], the enzymatic method is the most often utilized method because of its high efficiency and productivity. The method was first reported by Taylor et al. [11]. In their work, microbial whole cells were used in bio-enzymatic resolution of racemic γ-lactam. Later, Littlechild and coworkers cloned and purified a (–)-γ-lactamase from *Aurobacterium sp.*

* Correspondence: shengwu@im.ac.cn
State Key Laboratory of Microbial Resources, Institute of Microbiology, Chinese Academy of Sciences, Beijing 100101, PR China

and elucidated its catalytic mechanism by protein crystal structure resolution [12]. Up to now four (+) -γ-lactamases have been reported in the forms of *E. coli* recombinant proteins [1,12-14], and others are used in forms of whole cells [11,15,16].

In our previous research works, we immobilized a (+) γ-lactamase (Gla) from *Sulfolobus solfataricus* fused with a His-tags on nickel-chelating agarose and applied the immobilized enzyme in batch reactions. However, agarose is a fragile material and is much more expensive than other immobilizing matrices such as resins. These disadvantages of the agarose-immobilized enzyme confined its usage in scale application. Recently, the cellulose binding domain (CBD) of cellulose is often used as the bio-specific affinity matrix for enzyme immobilization. Compared with agarose, cellulose is much cheaper, and it has excellent chemical and physical properties. More importantly, binding between cellulose and CBD is tight and the desorption rarely happened [17]. Considering these merits of the cellulose immobilized enzyme, we wonder a cbd-fused-gla, immobilized to cellulose, represent a promising catalyst for the resolution of racemic gamma lactam?

In this paper, we developed a method in which the CBD fusion Gla was immobilized on the microcrystalline cellulose matrix (Avicel). We found that the fusion enzyme which had CBD fused at the C terminus was immobilized well on the cellulose matrix, and the immobilized enzyme was more stable than the free fusion enzyme in terms of thermo-stability. The immobilized enzyme retained most of its activity and had a relatively constant efficiency when used for batch transformations. In addition, γ-lactamase showed absolute enantioselectivity to (+)-γ-lactam. In conclusion, we believe that the fusion enzyme immobilized on cellulose, reported in this paper, is a promising catalyst for the industrial preparation of (−)- γ-lactam.

Methods

Bacterial strains, plasmids, and culture media

Escherichia coli DH5α (Transgen Biotech, Beijing, PRC) cells were used for the cloning studies. *E. coli* rosetta (DE3) (Novagen, Darmstadt, Germany) cells were used for protein expression. pET30a (+) (Invitrogen, California, USA) was used for standard cloning and expression in *E. coli*. The *E. coli* strains were routinely grown in LB medium or on LB agar plates at 37°C. To select bacterial strains carrying the appropriate recombinant plasmids, 35 μg chloramphenicol mL^{-1} or 50 μg kanamycin mL^{-1} was added to the medium. *S. solfataricus* P2 genomic DNA was obtained from the American Type Culture Collection (ATCC, Manasas, VA; catalog number 35092).

Extraction and purification of DNA

Plasmids were isolated using the TIANprep Mini Kit (Tiangen, Beijing, PR China) according to the protocol provided by the manufacturer. DNA was purified using the TIANgel Midi Kit (Tiangen, Beijing, PR China) according to the manufacturer's instructions.

Gene fusion of gla and cip

The *cbd* encoding gene encoding the partial Cbd domain (from 2113 bp to 2581 bp of gene HF912724.1, [18]) was amplified from genomic DNA of *Clostridium thermocellum* ATCC 27405 (ATCC, Manasas, VA) using primer p1 and p2 with *Nco*I and *BamH*I sites respectively (Table 1). A linker sequence GSAGSA was introduced at the C terminus of *cbd*. The PCR protocol consisted of an initial denaturation step of 5 min at 95°C, followed by 30 cycles of 1 min at 95°C, 1 min at 53°C, 1 min at 72°C and ending with a final elongation step for 10 min at 72°C. PCR was performed with Red-*Pfu* DNA Polymerase (Biocolors, Beijing, China). The PCR product was inserted into the according sites of pET30a and the plasmid was named as pET*cbd*.

The gene encoding the γ-lactamase Gla (GenBank accession no AF290611) was amplified from genomic DNA of *Sulfolobus solfataricus* P2 using primer p3 and p4 with *BamH*I and *Xho*I sites respectively. (Table 1). The PCR protocol was same as for *cbd*. The PCR product was inserted into the *BamH*I and *Xho*I sites of pET*cbd* and the final plasmid was named as pET*cbdgla*. This plasmid harbored a *cbd-linker-gla* fusion gene.

In a same way, *gla* was amplified with primer p5 and primer p6 and inserted into the *Nde*I and *BamH*I sites of pET30a (pET*gla*). The linker sequence was introduced at the C terminal of *gla*. Then *cbd* was amplified with primer p7 and p8 and introduced into downstream the linker sequence (*BamH*I and *Xho*I sites). The final plasmid was named as pET*glacbd* and it harbored a *gla-linker-cbd* fusion gene.

Expression of fusion gene in *E. coli* rosetta (DE3) cells

E. coli rosetta (DE3) was chosen as the expression strain because in the strain Gla showed higher specific activity than in other strains such as *E. coli* (DE3) (data not shown). This is because there were many rare codons existed in *gla* gene. The constructs were transformed into *E. coli* rosetta (DE3) cells and has grown overnight at 37°C. Stock culture (5 mL) was transferred into fresh 500-mL culture medium and has grown up to an optical density of 0.8 (OD 580). Isopropylthiogalactoside (IPTG) was then added to a final concentration of 1.0 mM, and after incubation for 2 h, the cells were harvested by centrifugation (4,000 × *g* for 10 min).

Purification of Gla fusion proteins on a nickel (Ni)-chelating column

Purification of the γ-lactamase fusion protein on the Ni-chelating column was carried out using the protocol

Table 1 Primers used in this study

Name	Sequence (5′ → 3′)[a]	Description
p1 forward	CGCCATGGATTTGAAGGTTGAATTCTAC	For construction of pET*cbd*
p2 reverse	CGGGATCC***AGCGGAGCCAGCGGAGCC*** CACCGGGTTCTTTACCC	
p3 forward	GCGGATCCATGGGAATTAAGTTACCCACATTG	For construction of pET*cbdgla*
p4 reverse	CGCTCGAGTTTTTTGATTCTCTCAAATACAT	
p5 forward	CGCATATGGGAATTAAGTTACCCACATTG	For construction of pET*gla*
p6 reverse	GCGGATCC***AGCGGAGCCAGCGGAGCC***TTTTTTGATTCTCTCAAATACAT	
p7 forward	GCGGATCCAATTTGAAGGTTGAATTCTACAAC	For construction of pET*glacbd*
p8 forward	CGCTCGAG CACCGGGTTCTTTACCC	

[a]Restriction sites are underlined, sequence encoding the linker peptide were in italics and in grey shade.

and buffer supplied by Novagen (New Jersey, USA). Cells from the expression culture (500 mL) were suspended in 40 mL binding buffer (50 mM Tris–HCl, 300 mM NaCl, and 10 mM imidazole, pH 8.0) and disrupted by ultrasonication. The crude samples were heated to 60°C for 10 min then centrifuged. The supernatant was applied to a 1-mL Novagen His•Band gravity flow column that had been equilibrated with 20 mL Ni-NTA binding buffer. The column was then washed with 20 mL wash buffer (50 mM Tris–HCl, 300 mM NaCl, and 20 mM imidazole, pH 8.0). The His-tagged proteins were eluted with 10 mL elution buffer (50 mM Tris–HCl, 300 mM NaCl, and 200 mM imidazole, pH 8.0). To remove the imidazole present in the elution buffer, the eluate was applied to a 5-mL HiTrap desalting column (GE Healthcare, New Jersey, USA) connected to an ÄKTA FPLC system (GE Healthcare), and the column was eluted with 50 mM Tris–HCl buffer (pH 7.5, 150 mM NaCl). The protein samples were collected and dialyzed against 5000 mL of 50 mM Tris–HCl buffer (pH 8.0). The buffer was changed three times over a period of 24 h. The samples were then ready for use.

Immobilization of the crude or purified γ-lactamase fusion proteins on Avicel

For crude fusion proteins, cells from the expression culture (500 mL) were suspended in 40 mL ultrasonication buffer (50 mM Tris–HCl, pH 8.0) and disrupted by ultrasonication. The ultrasonication sample was heated to 60°C for 10 min then centrifuged (13,000 × *g*, 10 min). 5 mL of supernatant was added with 1 g of Avicel and kept at 25°C for 120 min with stirring. For purified fusion proteins, 1 mL protein solutions (20 mg protein in 50 mM Tris–HCl, pH 8.0) were added with 100 mg Avicel and kept at 25°C for 120 min with stirring. Finally the immobilized fusion proteins were collected by centrifuge (3,000 × *g*, 10 min). From hereafter the mass of the immobilized enzyme was included in Avicel and the mass of the absorbed protein on Avicel. When protein concentration of immobilized enzyme was involved in calculations, only the mass of the absorbed protein was counted.

Adsorption Isotherm Measurements for crude fusion proteins

Adsorption isotherm measurements were conducted at room temperature (25°C) in 50 mL tubes. Samples containing 2,000 mg of crude fusion proteins (supernatant samples after treated at 60°C, total activity 5500 U) and 1 g of Avicel were mixed in a final aqueous volume of 5 mL. Suspensions were incubated for 120 min to allow the adsorption to saturate. The protein concentration of the supernatant was measured at regular intervals using a BCA protein assay kit (Pierce, USA), and this value was then subtracted from the initial amount of protein and the difference was considered to be amount of proteins adsorbed onto Avicel. Otherwise, the total activity of supernatant was measured and this value was then subtracted from the initial enzyme activity. All values were measured in triplicate.

Standard reaction conditions and γ-lactamase activity assay

γ-Lactamase activity was assayed by adding 1 mg of enzyme (or 25 mg immobilized enzyme) to 200 μL of the 4.6 mM substrate solution (pH 7.0, 50 mM phosphate buffer). If not specified, the reaction solution was incubated at 70°C for 30 min. After transformation the reaction solution was extracted with 100 μL ethyl acetate and analyzed by chiral HPLC. One unit (U) of enzyme activity was defined as the amount of enzyme that catalyzed the conversion of one micromole of substrate per minute. As to immobilized enzyme the absorbed protein was first calculated based on the binding capacity described above, then the activity was determined.

Temperature optimum and temperature stability determination

For temperatue optimum determination, the transformation solutions (free enzyme) or suspensions (immobilized enzyme) as described in standard reaction conditions were incubated and reacted at various temperature from 20°C to 90°C with a 10 degree interval. Then the samples were analyzed as described in standard reaction conditions and activity assay.

For temperature stability determination, 50 μL of enzyme solution (1 mg enzyme in pH 7.0, 50 mM phosphate buffer) or 100 μL of immobilized enzyme suspension (25 mg immobilized enzyme, pH 7.0, 50 mM phosphate buffer) were incubated at various temperature from 20°C to 90°C with a 10 degree interval for 30 min. Then the enzyme solution or suspension was added to the substrate solution as described in standard reaction conditions. The samples were allowed to react and later analyzed as described in standard reaction conditions and activity assay.

Analytical chiral HPLC

The reaction solution (200 μL) was extracted with 100 μL ethyl acetate, and benzamide was added to the extract as the internal standard. The ethyl acetate extract (10 μL) was applied to a Daicel Chiralpak AS-H column (Daicel, Tokyo, Japan) and eluted with a mobile phase consisting of 80% acetonitrile and 20% isopropanol (volume ratio). The UV absorbance of eluted γ-lactam was measured at 230 nm. Benzamide, (+) γ-lactam, and (−) γ-lactam had retention times of 8.7, 11.8, and 13.9 min, respectively, at a flow rate of 0.6 mL/min.

Determination of the apparent kinetic parameters

The apparent kinetic parameters were determined by the Lineweaver-Burk plot. Enzymatic reactions were carried out with substrate solutions of different concentrations (ranging from 4.6 mM to 32.2 mM). To ensure accuracy, substrate conversion was controlled below 10%.

Protein assay and sodium dodecyl sulfate-polyacrylamide gel electrophoresis (SDS-PAGE)

The protein concentration was determined by the BCA Protein Assay Kit from Pierce (Rockford, USA) using bovine serum albumin (BSA) as the standard. SDS-PAGE was performed as described earlier using a 6% stacking gel and 12% separating gel [19].

DNA sequencing

DNA sequencing reactions were performed on a plasmid template using the ABI Prism BigDye terminator cycle sequencing ready reaction kits (Applied Biosystems, California, USA) and an ABI PRISM 377 DNA sequencer (Applied Biosystems).

Software and online service

The BLAST × [20] program was used for protein homology search.

Results and discussion

Expression and purification of γ-lactamase fusion proteins

The γ-lactamase structural gene was a 1,515-bp fragment. The *cbd* encoding gene was a 469-bp fragment [18] and the linker sequence was an 18-bp fragment. The two fusion gene constructs were expressed as two separate proteins respectively. The *cbd-linker-gla* construct expressed a 78-kDa fusion enzyme that had a 4.7-kDa fusion peptide at its amino (N) terminus and a His-tag at its carboxyl (C) terminus (pET30a). The other construct (*gla -linker-cbd*) expressed a 74.2-kDa single fusion enzyme that only had a 0.9-kDa C terminal His-tag (Figure 1A). The recombinant enzymes were expressed in the soluble fraction of the bacterial

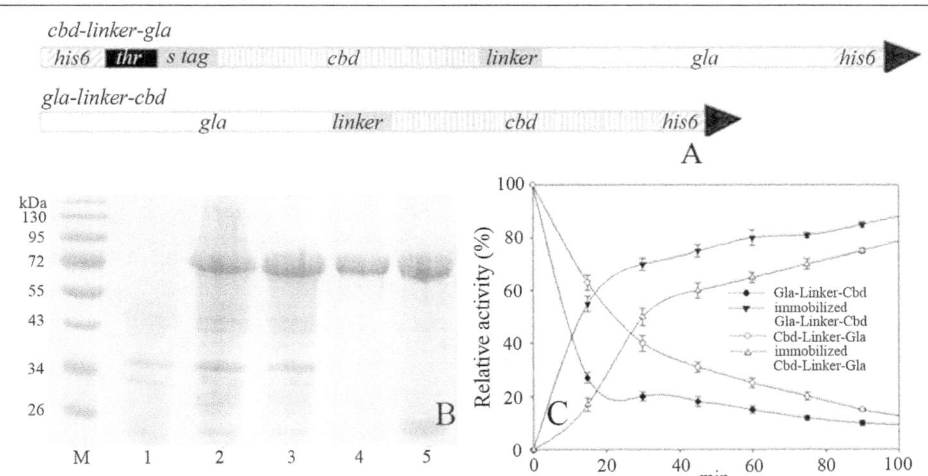

Figure 1 Construction of *cbd-linker-gla* and *gla-linker-cbd* fusion genes (A) and expression of the two constructs (B) and adsorption isotherm measurements for crude fusion proteins (C). A. his6: his-tag encoding gene; thr and s tag: thrombin cleavage site and S peptide tag on pET30; B. 1.*E. coli* rosetta (DE3) harboring pET30; 2.*E. coli* rosetta (DE3) harboring pET*cbdgla*; 3.*E. coli* rosetta (DE3) harboring pET*glacbd*; 4.Purified Cbd-Linker-Gla; 5.Purified Gla-Linker-Cbd C. The total activity of Gla-Linker-Cbd supernatant at 0 h (5500 U) was set as 100% relative activity.

Table 2 Apparent kinetic parameters of the free fusion proteins and their immobilized counterparts

Fusion protein	K_m (mmol l^{-1})	V_{max} (U mg^{-1})	k_{cat} (s^{-1})
Cbd-Linker-Gla	14.3 ± 1.6	71.6 ± 4.2	8.8×10^4
Immobilized Cbd-Linker-Gla	34.0 ± 1.6	344.9 ± 31.2	4.2×10^5
Gla -Linker- Cbd	5.7 ± 0.1	151.5 ± 30.3	1.8×10^5
Immobilized Gla -Linker- Cbd	25.0 ± 1.6	381.6 ± 44.2	4.7×10^5

lysate (Figure 1B lane 2 and 3). After single-step purification on Ni-chelating column, the purified proteins were applied for SDS-PAGE (Figure 1B lane 4 and 5). There was a low-molecular-weight protein contamination in the pure protein of *gla -linker-cbd*, therefore in this paper the enzyme activity of *gla -linker-cbd* was probably lower than the actual value.

Comparison of the apparent kinetic parameters

Analysis of the apparent kinetic parameters revealed obvious differences in the K_m, V_{max} and k_{cat} values of the two forms of free fusion γ-lactamase protein. The changes in these values indicated that the N-terminal Cbd fusion enzyme showed lower substrate affinity than the C-terminal Cbd fusion enzyme, and its V_{max} and k_{cat} values were also much lower than the C-terminal fusion enzyme (Table 2). After immobilization the K_m and V_{max} values of both fusion proteins were improved, and their k_{cat} values were also upgrade significantly. However, there was no obvious difference on enzyme turnover number between the C-terminal Cbd fused enzyme and the N-terminal Cbd fused enzyme.

Adsorption isotherm measurements for crude fusion proteins and purified Gla-Linker-Cbd

To make the method more feasible, first we tried to absorb the crude fusion proteins (supernatants after ultrasonication) on Avicel. The adsorption isotherms of the two crude fusion proteins were measured. However, the protein concentration changes were too small and the experiment errors were far beyond normal values. Therefore, the relative activities of the crude protein supernatants instead of the protein concentration were determined. After absorption the remaining supernatants and the enzyme absorbed Avicels were applied for (*rac*)-γ-lactam transformation at 60°C and then the transformations were analyzed by chiral HPLC respectively. The total activities of the supernatant and immobilized enzyme were calculated. According to Figure 1C, the binding between Gla-Linker-Cbdand Avicel was faster than the binding between Cbd-Linker-Gla and Avicel.

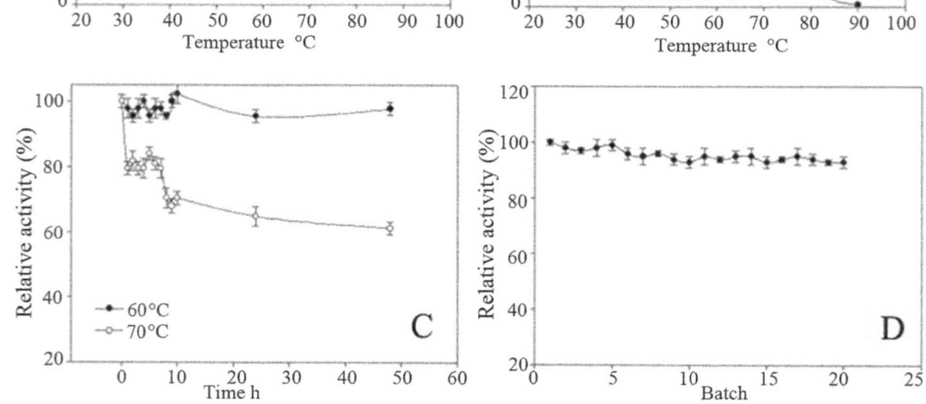

Figure 2 Comparison of the optimal temperatures (A) and thermostabilities (B) between free Gla-Linker-Cbd protein and its immobilized counterpart, and thermostability of immobilized Gla-Linker-Cbd at 60°C and 70°C (C) and batch stability (D). A. Optimal temperature; B. Thermostability. The activity of immobilized Gla-Linker-Cbd at 70°C **(A)** or firstly incubated at 30°C for 30 min (B, 145 ± 12 U mg^{-1}) was set as 100% relative activity. In B, firstly enzymes were incubated at designated temperature for 30 min then applied for catalysis. The activity of immobilized Gla-Linker-Cbd in cubated at 60°C **(C)** or the first batch (D, 116 ± 6 U mg^{-1}) was set as 100% relative activity.

The binding was in a stable state after 120 min absorption. The purified fusion proteins were also used in the absorption experiments. Both the adsorption ratios of the two purified proteins on Avicel were 40 ± 6 mg protein per g Avicel. The binding capacity was calculated to be 0.8 ± 0.1 μM/g, which was in the range for CBD-fusion proteins [21-23]. The binding saturation time of the purified protein was same as the crude protein (data not shown). Based on the above experiment result, Gla-Linker-Cbd immobilized protein was chose for further experiments.

Comparison of the temperature property of free Gla-Linker-Cbd protein and its immobilized counterpart

The temperature optima and thermostability of Gla-Linker-Cbd and its immobilized counterpart were determined. The optimal temperature of Gla protein was at 80°C [2]. However, after fused with Cbd the optimal temperatures of the fusion protein and its immobilized counterpart dropped from 80°C to 70°C (Figure 2A). The immobilized enzyme exhibited higher specific activity than the free enzyme. Meanwhile, according to Figure 2B, after incubated at 80°C for 30 min, the fusion protein lost almost 80% of its original activities while its immobilized counterpart only lost 30% of its original activities.

Protein thermostability was often related to protein rigidity [24]. In *C. thermocellum* most of the Cbd fusion enzymes showed maximum activity at 65°C or 75°C, and had weak thermostabilities at 80°C [25-27]. Therefore, compared to Gla the rigidity of the fusion protein probably was reduced. In this balance process Cbd was the determinant part to affect the protein thermostability.

Stability of immobilized enzymes at 60°C and 70°C

All the other conditions were same as described in standard transformation conditions. The thermostability of the immobilized enzyme was tested at 60°C and 70°C respectively. The immobilized enzyme was stable at 60°C even incubated for 48 h. However, when incubated at 70°C for 1 h the enzyme lost their 20% of the initial activity. If the incubation time extended to 48 h, the immobilized enzyme lost their 40% initial activity (Figure 2C). The batch stability was tested at 60°C for 20 batches reactions. After each transformation, the immobilized enzyme was collected by centrifuge and applied for another transformation. According to Figure 2D, the immobilized enzyme maintained almost 100% of their initial activity after 20 batches reactions.

Optimization of the concentrations of immobilized enzymes and substrate in batch transformation

In term of cost saving, in a transformation the substrate concentration should be as high as possible, and the

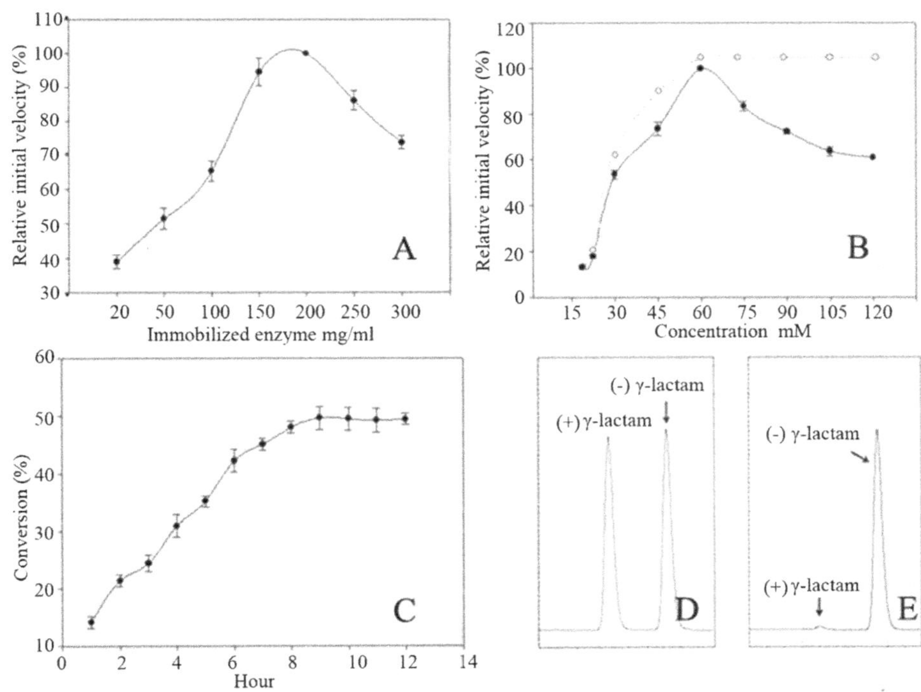

Figure 3 Optimal concentration of immobilized enzyme (A) and substrate (B), and time course of the optimized transformation (C) and chiral HPLC analyses of the (*rac*)-γ-lactam control (D) and sample after the 9 h transformation (E). The activity of 200 mg/ml immobilized Gla-Linker-Cbd at 60°C (A, 120.6 ± 5.2 μmol/min) and immobilized Gla-Linker-Cbd to 60 mM substrate (B, 338.6 ± 15.2 μmol/min) was set as 100% relative activity. The line in dot in B represents no substrate inhibition to enzyme. C: Substrate concentration 0.6 M, immobilized enzyme concentration 200 mg/ml, transformation temperature 60°C.

transformation time should be as short as possible. Taking into account of these two factors, we tested first the highest immobilized enzyme concentration employed in the system, and then the highest substrate concentration. In these optimizations, the initial velocity of transformation was used as the response factor. According to Figure 3A, the optimal enzyme amount in the transformation system was 200 mg/mL. Beyond or below this value, the initial velocity of reaction dropped dramatically. It was easy to understand that lower enzyme amount meant lower catalytic activity. On the other hand, 200 mg immobilized enzyme almost had an autogenous volume of 0.5 mL in transformation buffer, and it was impossible to stir the reaction suspension sufficiently in the situation. Therefore the mass transfer efficiency of the system was probably affected severely [28] and the initial velocity of the enzyme was reduced significantly.

The enzyme initial velocity would be upgraded by increasing the substrate amount till it reached and maintained maximum activity if there was no substrate inhibition [29]. According to Figure 3B, when the substrate concentration exceeded 60 mM, the initial velocity of the enzyme dropped obviously, which indicated that substrate inhibition occurred.

Though there is substrate inhibition, it does not mean that higher concentration can't be applied. We tested the maximum substrate concentration for 200 mg/mL immobilized enzyme in prolonged reaction time. When the substrate concentration was below 0.6 M, the reaction conversion reached 49.8% in various time. If the substrate concentration was beyond this value, the conversion hardly reached 40% even transformed for 96 h, which was very uneconomical (data not shown).

After optimization, we conducted a 10 mL-volume transformation in which 2 g immobilized enzyme was suspended in 0.6 M substrate buffer solution. The reaction was conducted at 60°C and at a certain time interval, 10 μL of sample aliquot was extracted with ethyl-acetate and then applied for chiral HPLC analysis. According to Figure 3C, the transformation was completed within 9 h (49.5% conversion). The *ee* value of the remaining substrate (–) γ-lactam was 99.5% (Figure 3E). The reaction solution was extracted with 30 mL dichloromethane each time for three times. The combined solvent was recovered by using rotary evaporator. At last, 0.30 g (–) γ-lactam in crystal was recovered from the transformation solution, and the recovery rate was 92%.

Conclusions

In our previous work, we employed a single purification and immobilization method to immobilize *Sulfolobus solfataricus* γ-lactamase on Ni-NTA agarose. The method is simple and easy to conduct. And as we reported previously [2] V_{max} and k_{cat} value of the agarose immobilized enzyme were around 450 U mg^{-1} and 4.2×10^5 s^{-1} respectively, which indicated it was also an efficient immobilized catalyst same as the cellulose immobilized enzyme. However because of the fragility of agarose, the activity of immobilized enzyme decreased sharply after multiple usages probably due to the structure crash of agarose. It was clear that the decrease did not come from protein inactivation, because under the same condition the enzyme was not deactivated even for a longer time. On the contrary cellulose is a strong material. Therefore it is a very good candidate as immobilizing material also because it is the most abundant and renewable material on the Earth. And most importantly, unlike glutaraldehyde cross-linked immobilization, the absorption between Cbd domain and cellulose is a bio-green process, in which no extra energy or expensive and harming catalyst is needed. Based on the data provided in this paper we think the cellulose immobilized Gla is a promising catalyst for the industrial preparation of (–)-γ-lactam, which is a very important drug intermediate.

Competing interests

The authors declare that they have no competing interests.

Authors' contributions

JW and CM carried out the cloning, over-expression, purified and immobilization. JZ characterized the immobilized enzyme. SW directed the over-all study and drafted the manuscript. All authors read and approved the final manuscript.

Acknowledgement

This work was supported by National Natural Science foundation of China no. 31070718.

References

1. Toogood HS, Brown RC, Line K, Keene PA, Taylor SJC, McCague R, Littlechild JA: **The use of a thermostable signature amidase in the resolution of the bicyclic synthon (rac)-gamma-lactam.** *Tetrahedron* 2004, **60**(3):711–716.
2. Wang J, Zhang X, Min C, Wu S, Zheng G: **Single-step purification and immobilization of γ-lactamase and on-column transformation of 2-azabicyclo [2.2.1] hept-5-en-3-one.** *Process Biochem* 2011, **46**(1):81–87.
3. Vince R, Hua M: *Synthesis of Carbovir and Abacavir from a Carbocyclic Precursor.* In: Current Protocols in Nucleic Acid Chemistry. John Wiley & Sons, Inc. 2001.
4. Gubareva LV, Webster RG, Hayden FG: **Comparison of the activities of zanamivir, oseltamivir, and RWJ-270201 against clinical isolates of influenza virus and neuraminidase inhibitor-resistant variants.** *Antimicrob Agents Chemother* 2001, **45**(12):3403–3408.
5. Singh R, Vince R: **2-Azabicyclo [2.2.1] hept-5-en-3-one: Chemical Profile of a Versatile Synthetic Building Block and its Impact on the Development of Therapeutics.** *Chem Rev* 2012, **112**(8):4642–4686.
6. Wisniewski T, Bayne E, Flanagan J, Shao Q, Wnek R, Matheravidathu S, Fischer P, Forrest MJ, Peterson L, Song X, Yang L, Demartino JA, Struthers M: **Assessment of chemokine receptor function on monocytes in whole blood: In vitro and ex vivo evaluations of a CCR2 antagonist.** *J Immunol Methods* 2010, **352**(1–2):101–110.
7. Taylor JD: **COPD and the response of the lung to tobacco smoke exposure.** *Pulm Pharmacol Ther* 2010, **23**(5):376–383.
8. Malhotra S, Man SF, Sin DD: **Emerging drugs for the treatment of chronic obstructive pulmonary disease.** *Expert Opin Emerg Drugs* 2006, **11**(2):275–291.

9. King CH, Meckler H, Herr RJ, Trova MP, Glick SD, Maisonneuve IM: **Synthesis of enantiomerically pure (+)- and (–)-18-methoxycoronaridine hydrochloride and their preliminary assessment as anti-addictive agents.** *Bioorg Med Chem Lett* 2000, **10**(5):473–476.
10. Potter GA, Garcia C, McCague R, Adger B, Collet A: **Oscillating Crystallization of (+) and (–) Enantiomers during Resolution by Entrainment of 2-Azabicyclo [2.2.1] hept-5-en-3-one.** *Angew Chem Int Ed Engl* 1996, **35**(15):1666–1668.
11. Taylor SJC, Mccague R, Wisdom R, Lee C, Dickson K, Ruecroft G, Obrien F, Littlechild J, Bevan J, Roberts SM, Evans CT: **Development of the Biocatalytic Resolution of 2-Azabicyclo [2.2.1] Hept-5-En-3-One as an Entry to Single-Enantiomer Carbocyclic Nucleosides.** *Tetrahedron-Asymmetr* 1993, **4**(6):1117–1128.
12. Line K, Isupov MN, Littlechild JA: **The crystal structure of a (–) gamma-lactamase from an Aureobacterium species reveals a tetrahedral intermediate in the active site.** *J Mol Biol* 2004, **338**(3):519–532.
13. Taylor SJC, Brown RC, Keene PA, Taylor IN: **Novel screening methods - The key to cloning commercially successful biocatalysts.** *Bioorgan Med Chem* 1999, **7**(10):2163–2168.
14. Zhu S, Gong C, Song D, Gao S, Zheng G: **Discovery of a Novel (+)-gamma-Lactamase from Bradyrhizobium japonicum USDA 6 by Rational Genome Mining.** *Appl Environ Microbiol* 2012, **78**(20):7492–7495.
15. Li HQ, Su L, Yang L, Wang JJ, Zheng GJ: **Study on the screening of lactamase and its fermentation conditions.** *Wei Sheng Wu Xue Bao* 2006, **46**(4):571–575.
16. Taylor SJC, Sutherland AG, Lee C, Wisdom R, Thomas S, Roberts SM, Evans C: **Chemoenzymatic Synthesis of (–)-Carbovir Utilizing a Whole Cell Catalyzed Resolution of 2-Azabicyclo [2.2.1] Hept-5-En-3-One.** *J Chem Soc Chem Comm* 1990, (16):1120–1121.
17. Bayer EA, Morag E, Lamed R: **The cellulosome–a treasure-trove for biotechnology.** *Trends Biotechnol* 1994, **12**(9):379–386.
18. Yaron S, Morag E, Bayer EA, Lamed R, Shoham Y: **Expression, purification and subunit-binding properties of cohesins 2 and 3 of the Clostridium thermocellum cellulosome.** *FEBS Lett* 1995, **360**(2):121–124.
19. Laemmli UK: **Cleavage of structural proteins during the assembly of the head of bacteriophage T4.** *Nature* 1970, **227**(5259):680–685.
20. Mount DW: **Database searching for similar sequences. Bioinformatics sequence and genome analysis. Chapter 7.** In Edited by Cuddihy J. New York: Cold Spring Harbor Laboratory Press; 2001:300–315.
21. Medve J, Stahlberg J, Tjerneld F: **Isotherms for adsorption of cellobiohydrolase I and II from Trichoderma reesei on microcrystalline cellulose.** *Appl Biochem Biotechnol* 1997, **66**(1):39–56.
22. Santiago-Hernandez JA, Vasquez-Bahena JM, Calixto-Romo MA, Xoconostle-Cazares GB, Ortega-Lopez J, Ruiz-Medrano R, Montes-Horcasitas MC, Hidalgo-Lara ME: **Direct immobilization of a recombinant invertase to Avicel by E-coli overexpression of a fusion protein containing the extracellular invertase from Zymomonas mobilis and the carbohydrate-binding domain CBDCex from Cellulomonas fimi.** *Enzyme Microb Tech* 2006, **40**(1):172–176.
23. Tomme P, Heriban V, Claeyssens M: **Adsorption of two cellobiohydrolases fromTrichoderma reesei to Avicel: Evidence for "exo-exo" synergism and possible "loose complex" formation.** *Biotechnol Lett* 1990, **12**(7):525–530.
24. Radestock S, Gohlke H: **Exploiting the Link between Protein Rigidity and Thermostability for Data-Driven Protein Engineering.** *Engineering in Life Sciences* 2008, **8**(5):507–522.
25. Fernandes AC, Fontes CMGA, Gilbert HJ, Hazlewood GP, Fernandes TH, Ferreira LMA: **Homologous xylanases from Clostridium thermocellum: evidence for bi-functional activity, synergism between xylanase catalytic modules and the presence of xylan-binding domains in enzyme complexes.** *Biochem J* 1999, **342:**105–110.
26. Kataeva I, Li XL, Chen H, Choi SK, Ljungdahl LG: **Cloning and sequence analysis of a new cellulase gene encoding CelK, a major cellulosome component of Clostridium thermocellum: evidence for gene duplication and recombination.** *J Bacteriol* 1999, **181**(17):5288–5295.
27. Kim H, Jung KH, Pack MY: **Molecular characterization of xynX, a gene encoding a multidomain xylanase with a thermostabilizing domain from Clostridium thermocellum.** *Appl Microbiol Biotechnol* 2000, **54**(4):521–527.
28. Fessner WD, Jones JB: **Biocatalysis and biotransformation - From discovery to application - Editorial overview.** *Curr Opin Chem Biol* 2001, **5**(2):103–105.
29. Reed MC, Lieb A, Nijhout HF: **The biological significance of substrate inhibition: a mechanism with diverse functions.** *Bioessays* 2010, **32**(5):422–429.

3

Identification of rapidly induced genes in the response of peanut (*Arachis hypogaea*) to water deficit and abscisic acid

Xiaoyun Li[1†], Jiabao Lu[1†], Shuai Liu[1], Xu Liu[2], Yingying Lin[1] and Ling Li[1*]

Abstract

Background: Peanut (*Arachis hypogaea*) is an important crop, but droughts often affect peanut production. There is a lack of genomic information available for peanut; therefore, little is known about the molecular basis of its drought stress response.

Results: Previously, we found that peanut stomata close rapidly during water deficit and in response to abscisic acid (ABA) treatment, and many genes show changes in their expression levels. To screen for candidate genes involved in the water deficit response, we used the Illumina HiSeq2000/MiSeq sequencing platform to conduct a global transcriptome analysis of peanut seedlings under water deficit with or without an ABA pretreatment. Three peanut tissues (leaves, roots, and stems) collected at each of three developmental stages (four-leaf, flowering, and podding stages) were used to construct sequence libraries. Then, 4.96×10^7 raw sequence reads were generated and the high quality reads were assembled into 47,842 unigenes. We analyzed these sequence libraries to identify differentially expressed genes (DEGs) under water deficit with or without ABA pretreatment. In total, 621 genes were induced rapidly (≥1.5 fold change compared with control) under water deficit, 2,665 genes were induced rapidly under water deficit + ABA pretreatment, and 279 genes overlapped between water deficit and water deficit + ABA pretreatment. Of the 279 overlapping genes, 264 showed the same expression pattern and 15 showed opposite expression patterns. Among the DEGs, 257 were highly induced (>5 fold) by water deficit + ABA pretreatment, while 19 were highly induced (>5 fold) by water deficit alone. The genes induced under water deficit + ABA pretreatment included 100 putative transcription factor (TF) genes, while those induced under water deficit alone included only 22 putative TF genes. To validate the transcriptome results, we conducted quantitative PCR (qPCR) analyses to quantify the transcript levels of nine candidate genes.

Conclusions: The DEGs results show that many genes are rapidly induced in peanut in response to water deficit without or with ABA pretreatment. The results indicate that the main drought response mechanisms in peanut function through an ABA-dependent pathway. Our data provide a comprehensive sequence resource for molecular genetics research on peanut stress responses.

Keywords: *Arachis hypogaea*, Transcriptome, Water deficit, Abscisic acid (ABA)

* Correspondence: liling@scnu.edu.cn
[†]Equal contributors
[1]Guangdong Provincial Key Lab of Biotechnology for Plant Development, College of Life Science, South China Normal University, Guangzhou 510631, P R China
Full list of author information is available at the end of the article

Background

Different plant species have various physiological adaptations to deal with long-term or short-term drought stress. In the short-term drought stress response, the phytohormone abscisic acid (ABA) regulates guard cell movement to close stomata. However, ABA also functions in the long-term stress response by regulating the development of reproductive tissues via massive transcriptional reprogramming events under prolonged drought. This results in growth retardation and decreased crop yields [1,2]. In plants, drought resistance functions through ABA-dependent and ABA-independent regulatory systems, as revealed in studies on drought-induced genes in *Arabidopsis thaliana* and rice (*Oryza sativa*) [3]. In *Arabidopsis*, calcium-dependent and -independent pathways mediate transcriptional reprogramming via an ABA-dependent pathway under water deficit [3].

Peanut is one of the most important oil crops worldwide. It is widely cultivated in Asia, Africa, and the Americas [4]. However, seasonal droughts often affect peanut yields and quality, causing significant economic losses. Little is known about the molecular signaling and regulatory mechanisms of the drought response in peanut because of its complex genetic background, and the lack of systematic research. Therefore, we have focused our research on drought resistance of cultivated peanut for many years. In plants, water deficit triggers the expression of numerous stress-induced genes, which encode products that enhance drought tolerance and allow continued growth [5]. ABA plays a key role in this response, but ABA signaling components and target genes often overlap with other signaling systems [6]. Many drought-responsive transcription factor genes have been identified in various species, such as *AREB1/ABF2* and *DREB2A* in *Arabidopsis*, and *ABP9* in maize (*Zea mays*) [6-10]. In *Arabidopsis*, these drought-responsive transcription factors activate the expression of the downstream target genes *RD29A* and *RD29B*. Consequently, *RD29A* and *RD29B* can serve as marker genes for drought or ABA signaling [11,12]. Large-scale screening of peanut has identified some stress-related genes. For example, fungus-infected peanut seeds showed increased transcript levels of genes encoding late-embryogenesis-abundant (LEA) family proteins, and increased levels of their products [13]. Likewise, a drought-resistant relative of peanut, *Arachis duranensis*, showed increased transcript levels of basic leucine zipper (bZIP) transcription factor genes when subjected to drought (18% water content). A pest-resistant relative of peanut, *Arachis stenosperma,* showed increased transcript levels of MYB family transcription factor genes when infected with fungi [4].

In our previous studies, we cloned and characterized some drought-related genes from peanut, including the ABA synthesis gene *AhNCED1* [14,15], and three dehydration-induced transcription factor genes, *AhAREB1* [16], *AhNAC2* [17], and *AhNAC3* [18]. The expression of these genes rapidly responds to water deficit in peanut. Therefore, similar to *RD29A* and *RD29B* in *Arabidopsis*, the transcript level of *AhNCED1* can serve as a molecular marker for screening drought-resistant varieties of peanut. Furthermore, the promoter region of *AhNCED1* contains many abscisic acid-responsive elements (ABREs). Site-directed mutagenesis of two of these ABREs diminished *AhNCED1* promoter activity [19], suggesting that ABA signaling pathways directly regulate *AhNCED1*. The ABA-responsive element binding protein (AREB) gene *AhAREB1* was cloned from peanut and overexpressed in *Arabidopsis* [16]. A microarray analysis showed that the transcription of *AtNCED3* (the *Arabidopsis* homolog of *AhNCED1*) was significantly higher in the *AhAREB1*-overexpressing plants than in wild type [20]. Recently, we found that AhAREB1 may directly regulate *AhNCED1* (unpublished data).

The response to water deficit or ABA in *Arabidopsis* has been studied using transformation techniques, and the results of such studies have improved our understanding of the *Arabidopsis* drought response. However, similar research on the drought response of peanut has been hampered by the low efficiency of transformation and the complex tetraploid genome of this important crop plant. To address this issue, we used a global transcriptome analysis to study the mechanisms by which peanut copes with drought stress, and to provide a comprehensive sequence resource for further molecular genetic research on this species.

Results and discussion

Sequencing and *de novo* assembly

We generated 44,277,322 (4.41G) paired-end reads with an average length of 905 bp for the peanut cultivar Yueyou7. Of the clean reads, 95.96% had phred-like quality scores at the Q20 level and 88.33% had phred-like quality scores at the Q30 level, with GC contents of 48.09% and 49.45%, respectively. The high-quality reads were assembled into 47,842 unigenes (Table 1). We obtained 7,093,861

Table 1 Summary of short reads and assemblies

Sample	Clean reads	Clean bases	GC (%)	Unigenes	Seq-dup-level (%)
Peanut_R1	44277322	4.41G	48.09	47842	79.72
Peanut_R2	44277322	4.41G	49.45	47842	83.49
ABA_30min	6847397	0.34G	43.83	27969	45.2
PEG_30min	7093861	0.35G	48.43	26917	64.01
Control	7486557	0.37G	49.10	34837	72.6

Peanut_R1 represents left-reads, Peanut_R2 represents right-reads. Seq-dup-level (sequence duplication level) represents the proportion of repeated reads in total reads. ABA_30 min represents 30% PEG6000 treated for 30 min after ABA pretreatment for 30 min. PEG_30 min represents 30% PEG6000 treatment for 30 min. Control, no treatment.

(0.35 G) clean reads from the water deficit without ABA-pretreatment group, 6,847,397 (0.34 G) clean reads from the water deficit + ABA-pretreatment group, and 7,486,557 (0.37 G) clean reads from the control (Table 1).

Genes modulated by water deficit and their predicted functions

To identify genes showing changes in expression in response to water deficit, peanut seedlings were subjected to a 30% PEG6000 treatment. This treatment has been used to simulate water deficit in several other studies [21,22]. The 30-min PEG treatment caused leaf wilting of 2-week-old peanut seedlings. After the PEG treatment, we collected leaves from the same position on the stalk for RNA extraction and sequencing. In total, 621 genes were rapidly induced (≥1.5 fold change compared with their respective expression levels in the control) by water deficit; 173 were up-regulated and 448 were down-regulated (Figure 1, Additional file 1). Also, 19 of these genes showed very large changes in expression levels (≥5-fold change compared with their respective expression levels in the control); 4 were up-regulated and 15 were down-regulated. Among the genes induced by water deficit alone, 22 encoded transcription factor-like proteins.

Next, we grouped the differentially expressed genes (DEGs) into GOseq functional categories (http://www.geneontology.org) based on Wallenius's non-central hyper-geometric distribution [23]. The DEGs were grouped into three main categories (Figure 2); molecular function (42.91%), biological process (50.0%), and cellular components (2.19%). In a similar microarray study of the drought-resistant species *A. duranensis*, most contigs belonged to the molecular function category (49%), with fewer belonging to cellular component (26%) and biological process (39%) categories [4]. These findings suggest that the dynamic changes in gene expression differ between short-term and long-term drought responses.

We further analyzed each functional category to identify which proteins were the most enriched. Proteins related to oxidation reduction processes (17.14% genes) were the most enriched in the biological process category and oxidoreductases (7.73% genes) were the most enriched in the molecular function category. Ribosome (5.67% genes) was the most highly represented in the cellular components category. We used the KEGG (Kyoto Encyclopedia of Genes and Genomes) database to annotate the pathways and found that phenylpropanoid biosynthesis (90 genes) and linoleic acid metabolism (14 genes) were the most enriched pathways in peanut seedlings under water deficit (Table 2).

Genes modulated by water deficit + ABA pretreatment and their predicted functions

In peanut, ABA accumulation and distribution modulate physiological responses during the short-term response to water deficit [24,25]. In our preliminary experiments, exogenous ABA treatment caused stomata to close within 30 min, and ABA treatment for 30 min resulted in changes in the expression levels of numerous stress-related genes. In the present study, 2,665 genes showed altered expression (≥1.5 fold change compared with their respective expression levels in the control) in response to water deficit + ABA pretreatment; 2,425 genes were up-regulated and 240 genes were down-regulated (Figure 1, Additional file 1). Among these genes, those involved in gluconeogenesis (47 genes) and the plant–pathogen interaction (38 genes) were the most enriched (Table 2). Among the genes showing altered expression, 257 were highly induced (≥5 fold induction; 185 up-regulated, 75 down-regulated), and 100 genes encoded putative transcription factors (Additional file 2). The DEGs were classified into three main GO categories; molecular function (35.03%), biological process (2.08%), and cellular components (20.3%) (Figure 2). In the molecular function category, genes encoding proteins with oxido-reductase activity (6.98%), catalytic activity (6.79%) and protein kinase activity (6.87%) were the most enriched. Proteins related to oxidation-reduction processes (12.52%), metabolic process (8.42%), and protein phosphorylation (7.17%) were enriched in the biological process category. In the cellular components category, the proteinaceous extracellular matrix (0.91%) and ubiquitin ligase complex (1.02%) were the most highly enriched categories.

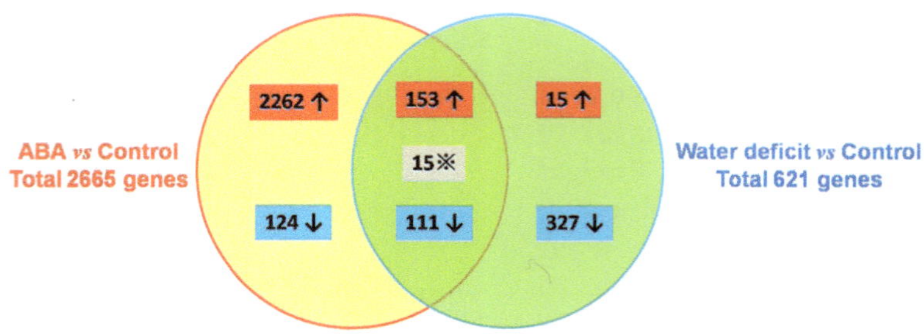

Figure 1 Classification of ABA- and water deficit-inducible genes based on their expression patterns. ABA- or water deficit-inducible up-regulated genes (red) or down-regulated genes (blue); intersection represents genes induced by both ABA and water deficit, "※" indicates genes showing opposite trends in expression between water deficit and ABA treatments.

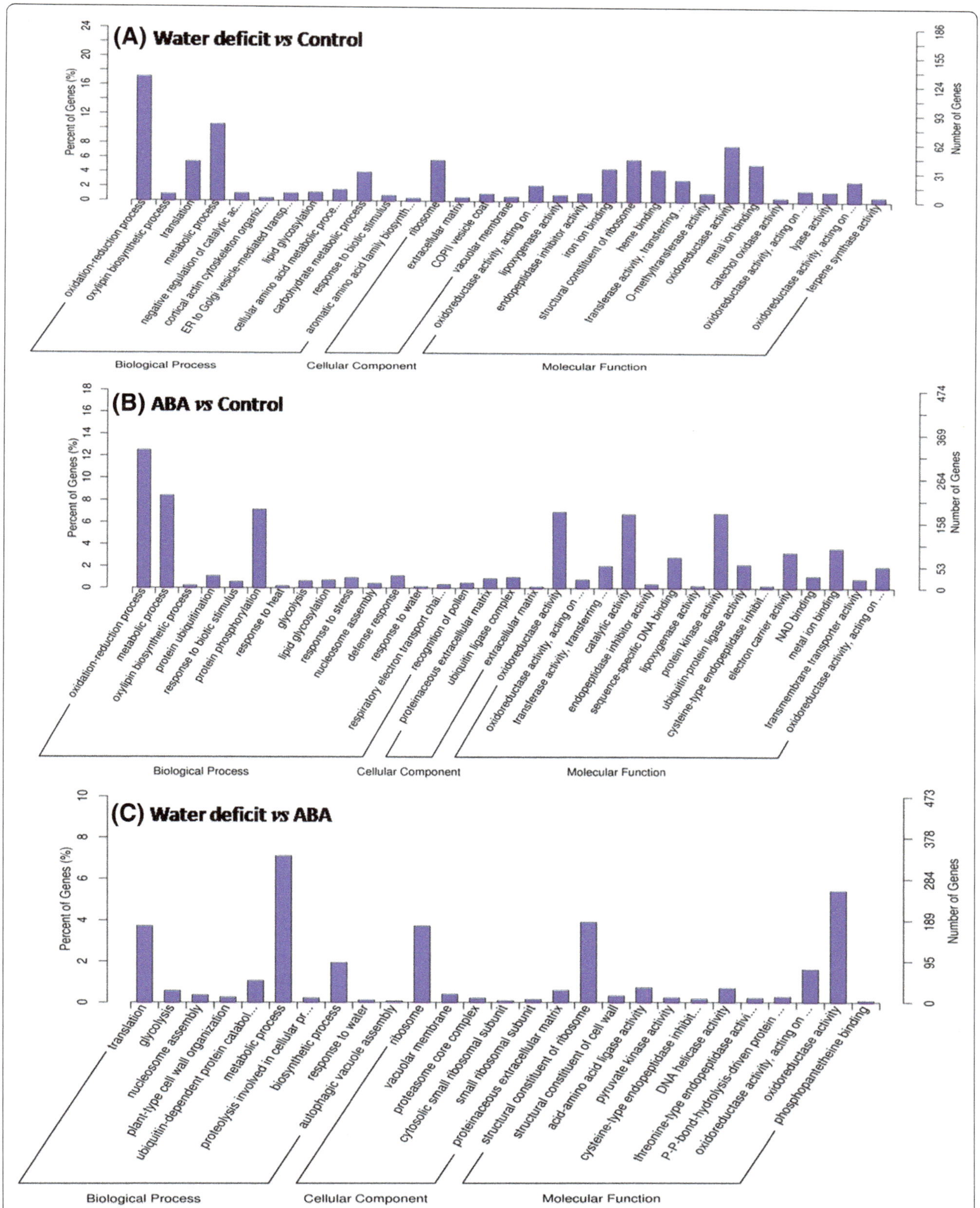

Figure 2 GO annotation of DEGs in the transcriptome. The DEGs were classified into different GO categories. **(A)** GO categories of differentially expressed genes response to water deficit, **(B)** GO categories of differentially expressed genes response to ABA, **(C)** GO categories of differentially expressed genes response to ABA or water deficit independently.

Table 2 KEGG annotations of enriched pathways in different transcriptomes

Term	ID	Sample number	Background number	Q-Value
Water deficit *vs.* Control				
Glycolysis/Gluconeogenesis	Ko00010	47	144	4.20 E-07
Plant-pathogen interaction	Ko04626	38	132	0.000218
ABA *vs.* Control				
Phenylpropanoid biosynthesis	Ko00940	21	90	2.7 E-06
Linoleic acid metabolism	Ko00591	14	45	8.01 E-06
Water deficit *vs.* ABA				
Glycolysis/Gluconeogenesis	Ko00010	62	144	0.003227
protein export	Ko03060	25	44	0.003227

Many ubiquitin-related enzymes are thought to regulate the plant drought stress response, but for many of these candidate regulators, including *MMS21* [26] and *ZmRFP1* [27], their mechanism of action remains unclear. In *Arabidopsis,* AtPUB18 and AtPUB19 negatively regulate the ABA-mediated drought response [28,29]. In peanut, overexpression of the ubiquitin-related gene *AhUBC2* was shown to enhance drought tolerance [30]. Together, these results indicate that protein ubiquitination is linked to ABA-mediated drought responses in peanut. Compared with the water-deficit group, the water deficit + ABA pretreatment group showed dramatic down-regulation of water deficit-induced genes, including those related to translation, metabolic process, ribosome, structural constituents of ribosome, and oxido-reductase activity (Figure 2C).

ABA-dependent and ABA-independent regulation of gene expression during the water deficit response in peanut

To examine the relationship between signaling under water deficit and that under water deficit + ABA pretreatment, we identified genes showing similar or different patterns of regulation between these two treatments. First, we identified 279 genes with significant overlaps in expression between the water deficit only and water deficit + ABA treatment groups. Of these, 153 genes were up-regulated and 111 were down-regulated by both water deficit alone and water deficit + ABA pretreatment, compared with their respective expression levels in the control. Only 81 genes were more strongly induced by water deficit + ABA treatment than by water deficit alone (>1.5 fold difference in water deficit + ABA group compared with their respective levels in water deficit only group). Of these 81 genes, 55 were up-regulated, 16 were down-regulated, and 12 encoded putative transcription factors.

The transcript levels of *comp65973_c0* (predicted to encode a MYC2-like transcription factor) decreased both under water deficit (−5.29 fold) and under water deficit + ABA pretreatment (−5.25 fold). The transcript levels of *comp64977_c0* (predicted to encode a NAC-like transcription factor) increased both under water deficit (2.45 fold) and under water deficit + ABA pretreatment (3.84 fold). The fact that water deficit + ABA pretreatment did not affect their expression more strongly than did water deficit alone suggests that ABA does not regulate the expression of these genes. Thus, these transcripts appear to be regulated by water deficit through an ABA-independent pathway.

Except for *comp64977_c0,* nine NAC or NAC-like genes showed rapid responses to water deficit + ABA pretreatment, but were not induced by water deficit alone. These genes, and the extent of their induction by water deficit + ABA, were as follows: *comp60831_c0/AhNAC3* (5.21 fold), *comp65819_c0/AhNAC2* (3.43 fold), *comp61624_c1* (1.81 fold), *comp65473_c0* (1.85 fold), *comp66232_c1* (1.50 fold), *comp63320_c1* (3.55 fold), *comp65172_c0* (2.45 fold). These findings indicate that these genes are regulated by water deficit through an ABA-dependent pathway. *comp65819_c0/AhNAC2* and *comp60831_c0/AhNAC3* were reported to show significant increases in transcript levels under dehydration, water deficit (simulated by PEG), or ABA treatment [17,18]. We did not identify these loci as differentially expressed under water deficit in this experiment, probably because of the short treatment time (30 min) used in our analyses.

Among the overlapping genes, 15 showed differences in expression between water deficit and water deficit + ABA pretreatment; 10 were down-regulated under water deficit but up-regulated under ABA + water deficit, and 5 were up-regulated under water deficit but down-regulated under ABA + water deficit. This result indicates that these genes are regulated by water deficit through an ABA-dependent pathway, but they could have opposite functions in the water deficit response. For example, *comp66875_c0* and *comp62508_c0* were down-regulated by water deficit but up-regulated by ABA + water deficit. The protein encoded by *comp66875_c0* shows sequence similarity to a low-temperature-induced 65 kDa protein in *Glycine max,* and that encoded by *comp62508_c0* shows sequence similarity to the homeobox-leucine zipper protein ATHB-7. *comp62508_c0* was down-regulated by water deficit (−4.35 fold), but up-regulated by water deficit + ABA pretreatment (3.57 fold). In *Arabidopsis, ATHB-7* is transcriptionally regulated in an ABA-dependent manner and modulates ABA signaling by regulating the activities of PP2C and an ABA receptor gene [31,32]. Consistent with our results in peanut, in *Arabidopsis,* the *ATHB-7* response to ABA was observed in seedlings within 30 min of drought stress, and its transcript levels continued to increase after 12 h or 72 h of drought stress [31].

The genes encoding 9-cis-epoxycarotenoid dioxygenase (*comp69670_c1/AhNCED1,* 3.05 fold) and abscisic acid

8′-hydroxylase-like proteins (*comp54319_c0*, 4.44 fold; *comp65737_c0*, 1.64 fold), which regulate ABA biosynthesis and degradation, respectively, were significantly induced by water deficit + ABA pretreatment. In our previous study, we detected high transcript levels of *AhNCED1* in response to ABA or drought stress [14]. In the present study, all of the genes related to ABA metabolism showed increased transcript levels in response to exogenous ABA treatment, suggesting positive feedback regulation by ABA. However, the transcript levels of AREB-like transcription factors did not change in response ABA or water deficit, although many showed a constitutive high level of transcription under both conditions and in the control. Therefore, NAC transcription factors appear to respond quickly to water deficit + ABA pretreatment (ABA-dependent). These transcription factors, together with AREB-like transcription factors, may regulate ABA biosynthesis enzymes or hydroxylases. However, further experimental evidence is required to confirm this.

In previous studies, oligonucleotide array or cDNA microarray analyses have identified 200 ABA-responsive genes in rice [33] and 245 ABA-responsive genes in *Arabidopsis* [34]. More ABA-responsive genes were identified in peanut than in other plants. Here, 47,842 unigenes were obtained from the drought-resistant peanut cultivar Yueyou7. In another study, 7,723 high contigs were obtained from 2-month-old plants of *A. stenosperma* (V10309), and 12,792 high contigs were obtained from 3-month-old plants of *A. duranensis* (K7988) [4]. In a different study, 59,077 unigenes were obtained from 10 immature seeds of five plants [35]. These findings suggest that the distribution of unigenes differs among plant types to some extent.

We did not detect homologs of *RD29A* and *RD29B* in the peanut seedling transcriptome. We detected *comp66875_c0* and *comp56618_c0*, which encode proteins similar to the low-temperature-induced 65 kDa protein (accession numbers XP_003556105.1 and XP_003536469.1, respectively). Our results suggest that, similar to *RD29B*, *comp66875_c0* and *comp56618* are controlled mainly by ABA. *RD29A* and *RD29B* encode closely related hydrophilic proteins, and are quickly and strongly induced by drought and salt stress [5,36]. They are considered to be drought stress or ABA-signaling activity marker genes in *Arabidopsis*. Their promoters have been used to drive the expression of genes that confer water-stress tolerance, for example, wheat *expansin* [37] and rice *OsbZIP71* [38], without affecting plant growth and development. However, no close homologs of *RD29A* or *RD29B* have been identified in crop species such as rice, sorghum, maize, and soybean [12]. Therefore, *AhNCED1* and *comp66875_c0* represent marker genes for downstream events in the ABA-dependent pathway in peanut.

Quantitative PCR validation of changes in gene expression

To validate the gene expression changes detected in the transcriptome analysis, we used quantitative PCR to quantify the transcript levels of nine candidate genes. We analyzed the transcript levels in plants under water deficit and in those under water deficit + ABA pretreatment at 30 min and 5 h time points (Figure 3). *comp69184_c3*, *comp65209_c0*, and *comp29919_c0*, which encode putative transcription factors, were induced both by water deficit and water deficit + ABA pretreatment in our transcriptome analysis. Similar to the results of the transcriptome analysis, the qPCR results showed that all three of these genes were induced by water deficit or ABA at 30 min and 5 h. The quantitative PCR results also showed that *comp65209_c0* was strongly induced under water deficit at 5 h. *comp65848_c0*, *comp66763_c0*, *comp69670_c1*, *comp54319_c0*, and *comp65737_c0* were induced by ABA at 30 min or 5 h. These results were generally consistent with the results of our transcriptome analysis.

The transcript level of *comp69670_c1* increased under water deficit at 5 h, as reported previously. *comp69670_c1* (also known as *AhNCED1*) encodes a 9-cis-epoxycarotenoid dioxygenase; AhNCED1 is a key enzyme in ABA biosynthesis in peanut and its expression was shown to increase during drought stress [15,24,25]. *comp54319_c0* and *comp65737_c0*, which belong to the P450 family, are homologs of *abscisic acid 8-hydroxylase 3* and *abscisic acid 8-hydroxylase 4* from soybean (*Glycine max*), respectively, suggesting that they encode enzymes involved in ABA catabolism. ABA is rapidly synthesized and accumulates under water deficit to induce stress resistance in plants. However, stress responses also include a protection mechanism to avoid excessive ABA accumulation, and this can retard growth. Thus, ABA catabolism genes are rapidly induced after ABA accumulation. The biosynthesis and catabolism genes play major roles in maintaining a balance of endogenous ABA for stress resistance and growth recovery after drought stress.

In our previous study, we found that *AhAREB1* was induced by ABA or water deficit [16,20]. We did not detect significant expression of *AhAREB1* under water deficit or ABA in our transcriptome data, so we revalidated its expression. We found that *AhAREB1* expression significantly increased after 0.5 h water deficit with ABA pretreatment, but not under water deficit, although it was slightly up-regulated. The different expression patterns detected in our studies may be due to experimental errors in treating plant materials or in transcriptome sequencing. This highlights the importance of validating the results of large-scale transcriptome analyses.

Conclusion

We used RNA-Seq to conduct a global characterization of the peanut transcriptome response to water deficit

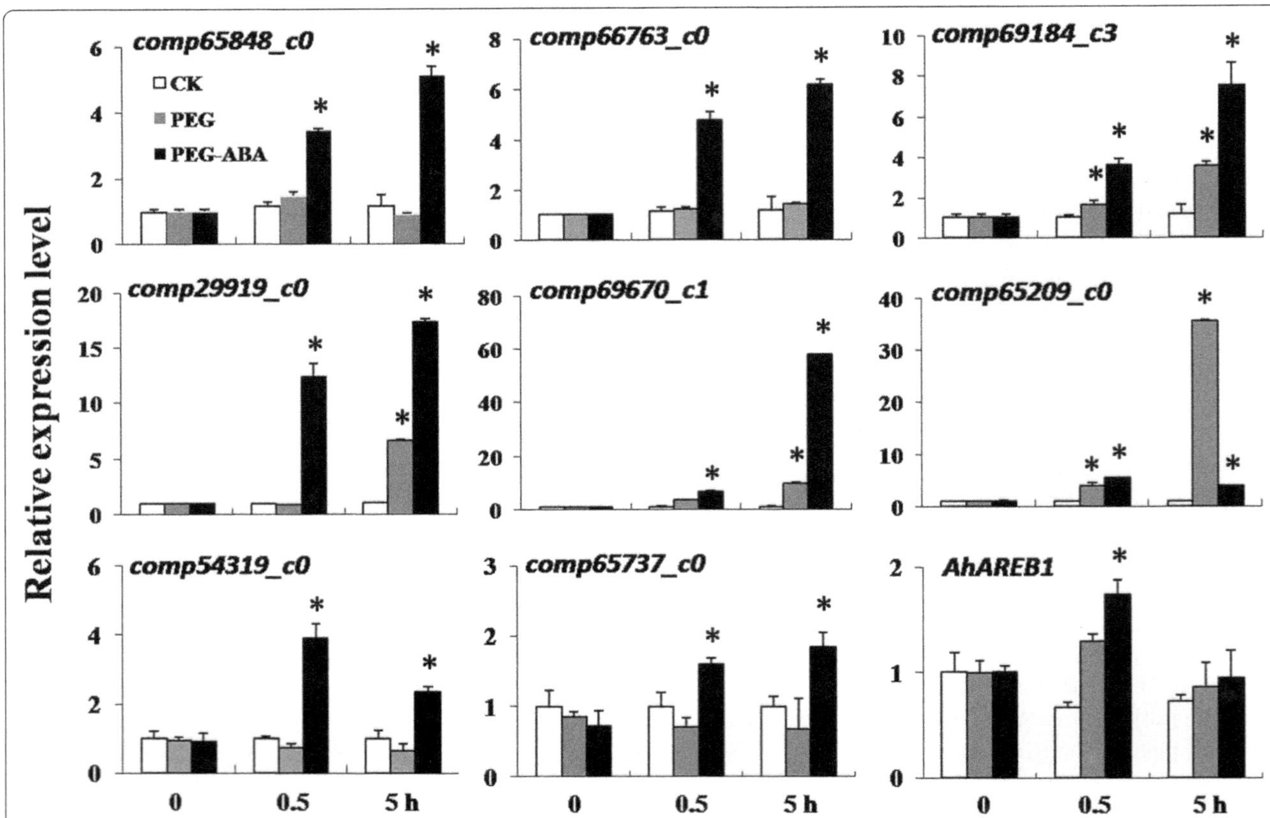

Figure 3 Validation of transcript levels of nine candidate genes under water deficit and ABA by quantitative PCR (qPCR). *comp65848_c0* PREDICTED: histone deacetylase 5 like [*Glycine max*]; *comp66763_c0* PREDICTED: histone deacetylase 6 like [*G. max*]; *comp69184_c3* PREDICTED: WRKY transcription factor WRKY33 like [*G. max*]; *comp65209_c0* PREDICTED: MYB transcription factor MYB92 like [*G. max*]; *comp29919_c0* PREDICTED: putative transcription factor bHLH041 like [*G. max*]; *comp69670_c1* encodes nine-cis-epoxycarotenoid dioxygenase 1 [*Arachis hypogaea*]; *comp54319_c0* PREDICTED: abscisic acid 8-hydroxylase 3 [*G. max*]; *comp65737_c0* PREDICTED: abscisic acid 8-hydroxylase 4 [*G. max*]. * indicates significant difference ($P < 0.05$) between transcript level in PEG treatment group or PEG + ABA treatment group and that in CK (control).

and ABA. We generated 4.96×10^7 raw reads, assembled into 47,842 unigenes, from the peanut cultivar Yueyou7. We identified DEGs between the early response to water deficit or water deficit + ABA pretreatment. These genes were annotated with GO functional categories; under water deficit there were 33 categories, and under water deficit + ABA pretreatment there were 31 categories. Only 19 genes were highly induced by water deficit, but 257 genes were highly induced by water deficit + ABA pretreatment. Now that these genes have been identified, our future research will focus on their functions and relationships. These data will be useful for functional genomic studies on peanut. We have established a biotechnological platform to study the early drought- and ABA-induced transcriptome regulatory network in peanut.

Methods

Plant growth and treatments

We used the South China peanut *(Arachis hypogaea* L.) cultivar Yueyou7 in this study. Yueyou7 is a commercial tetraploid line provided by the Crops Research Institute, Guangdong Academy of Agricultural Sciences (GDAAS), China. This cultivar is resistant to drought, rust, and bacterial infection. At present, there is no reference genome available for peanut, because its entire genome has not yet been sequenced. Therefore, we constructed a sequence library for Yueyou7. Then, we analyzed the sequence library to determine the profiles of DEGs under three different treatments; control, water deficit without ABA, and water deficit + ABA. In our previous study, we found that stomata closed after a 30-min 30% PEG treatment, indicative of a rapid stress response (regulation of stomatal closure and plant adaptation to water deficit) in the leaves [1]. Therefore, in this study, the profiles of DEGs in peanut leaves were determined after subjecting seedlings to water deficit or ABA treatments for 30 min.

To construct the sequencing library, plants were grown under normal conditions as described previously [14,17]. Seeds were planted in soil, and grown in a greenhouse at 25–30°C with ample irrigation. Three tissues (leaves, roots, and stems) were collected at three development stages (four-leaf, flowering, and podding stages).

Then, all of these tissues were mixed to extract total RNA for sequence library construction. For the DEGs study, 2-week-old plants were divided into three groups, each with three independent replicates. The three groups were as follows: (1) water deficit without ABA (seedlings immersed in 30% PEG6000 solution for 30 min); (2) water deficit + ABA (seedlings immersed in 100 μmol/L ABA for 30 min, then in 30% PEG6000 solution for 30 min); (3) control (no treatment). All treatments were conducted in parallel. After treatments, total RNA was extracted from 100 mg leaf tissue, and RNA integrity was checked by gel electrophoresis. The quality and quantity of RNA were evaluated using an Agilent 2100 Bioanalyzer (Agilent technologies, Santa Clara, CA, USA) and a Nanodrop ND-1000 (Thermo Scientific, Waltham, MA, USA).

De novo assembly and analysis of Illumina reads

The raw data (8 G) were pre-processed as previously described [4,38]. First, the adapter sequences, poor-quality sequences containing N (unknown sequences), and low-quality reads were filtered out to obtain clean reads. The clean reads were then combined to form longer fragments. All unigenes were compared with the NR (NCBI non-redundant protein sequences), NT (NCBI non-redundant nucleotide sequences), Pfam (protein family), and KO (KEGG Ortholog) databases using BLASTX v2.2.14. Based on the results of the protein database annotation, blast2GO was used to obtain the functional classifications of the unigenes based on GO terms. All transcription data have been uploaded into the Gene Expression Omnibus (http://www.ncbi.nlm.nih.gov/geo/query/acc.cgi?acc=GSE56439).

Analysis of differences in gene expression

Because there is no high-quality genome sequence available for peanut, we used RSEM (RNA-Seq by Expectation Maximization) software to map and quantify the transcriptome based on our extensive sequence library as described above. We mapped and quantified 87.11%, 91.14%, and 93.09% of reads in the control, ABA, and PEG treatment groups, respectively. Then, we used RPKM (reads per kilobase per million mapped reads) measures to estimate gene expression levels [39]. Roughly, the genes belonged to high- (RPKM > 15), medium- (3.57 < RPKM < 15) or low-level (0.1 < RPKM < 3.57) expression groups. Finally, DEGseq (an R package) was used to identify DEGs using the RNA-seq data from different samples, as described previously [40].

Validation of transcriptome data using quantitative PCR

To validate the results and provide a basis for further transcriptome analysis, nine genes were selected and their transcript levels were analyzed by qPCR. Total RNA was isolated by TRIzol reagent (Invitrogen, Carlsbad, CA, USA). Two-step real-time RT-PCR was performed with an SYBR Premix Ex Taq II kit (Takara, Dalian, China), and qPCR was performed using an Optical 96-well Fast Thermal Cycling Plate with an ABI 7500 real-time PCR system. Each reaction consisted of 10 μL 2 × SYBR Premix Ex Taq, 20 ng cDNA, and 0.1 μmol/L gene-specific primers in a final volume of 20 μL. The thermal cycling conditions were as follows: 95°C for 30 s, then 40 cycles at 95°C for 5 s, and 60°C for 34 s. Gene-specific primers are listed in Additional file 3. The *18S rRNA* gene was used as the internal control as previously described [16,20]. The relative expression levels were calculated using the relative $2^{-\Delta Ct}$ method as described previously [20].

Availability of supporting data

The data sets supporting the results of this article are included within the article and its additional files.

The sequence data sets supporting the results of this article are available in the Gene Expression Omnibus repository [http://www.ncbi.nlm.nih.gov/geo/query/acc.cgi?acc=GSE56439].

Additional files

Additional file 1: Differentially expressed genes. List of genes identified as ABA vs CK (control), water deficit vs CK, water deficit VS ABA, and similar or opposite trend, respectively, with a fold-change of at least 1.5. All differential genes were compared with the NR (NCBI non-redundant protein sequences), NT (NCBI non-redundant nucleotide sequences), Pfam (protein family), and KO (KEGG Ortholog) databases.

Additional file 2: Putative transcription factor genes expressed in peanut in response to ABA and water deficit. List of genes differentially regulated by ABA and water deficit in peanut that were identified as putative transcription factor genes.

Additional file 3: Primers used in this study. List of primers and oligonucleotides used for qRT-PCR.

Competing interests

The authors declare that they have no competing interests.

Authors' contributions

LXY, LX and LL devised the method and designed the study. LXY, LJB, LS and LYY carried on the experimental work. LXY and LJB analyzed the data. LXY and LX wrote the manuscript. All authors contributed to the manuscript. All authors read and approved the final manuscript.

Acknowledgements

This research was supported by the Natural Science Foundation of China (30971715), Guangdong Natural Science Foundation (10251063101000010), and Guangzhou Science and Technology Project Funding (2012 J4100115). The first two authors contributed equally to this study.

Author details

[1]Guangdong Provincial Key Lab of Biotechnology for Plant Development, College of Life Science, South China Normal University, Guangzhou 510631, P R China. [2]Key Laboratory of Plant Resources Conservation and Sustainable Utilization, South China Botanical Garden, Chinese Academy of Science, Guangzhou 510650, P R China.

References

1. Degenkolbe T, Do PT, Zuther E, Repsilber D, Walther D, Hincha DK, Köhl KI: **Expression profiling of rice cultivars differing in their tolerance to long-term drought stress.** *Plant Mol Biol* 2009, **69:**133–153.
2. Sreenivasulu N, Harshavardhan VT, Govind G, Seiler C, Kohli A: **Contrapuntal role of ABA: does it mediate stress tolerance or plant growth retardation under long-term drought stress?** *Gene* 2012, **506:**265–273.
3. Seki M, Umezawa T, Urano K, Shinozaki K: **Regulatory metabolic networks in drought stress responses.** *Curr Opin Plant Biol* 2007, **10:**296–302.
4. Guimarães PM, Brasileiro AC, Morgante CV, Martins AC, Pappas G, Silva OB Jr, Togawa R, Leal-Bertioli SC, Araujo AC, Moretzsohn MC, Bertioli DJ: **Global transcriptome analysis of two wild relatives of peanut under drought and fungi infection.** *BMC Genomics* 2012, **13:**387.
5. Yamaguchi-Shinozaki K, Shinozaki K: **Characterization of the expression of a desiccation-responsive rd29 gene of Arabidopsis thaliana and analysis of its promoter in transgenic plants.** *Mol Gen Genet* 1993, **236:**331–340.
6. Matsui A, Ishida J, Morosawa T, Mochizuki Y, Kaminuma E, Endo TA, Okamoto M, Nambara E, Nakajima M, Kawashima M, Satou M, Kim JM, Kobayashi N, Toyoda T, Shinozaki K, Seki M: **Arabidopsis transcriptome analysis under drought, cold, high-salinity and ABA treatment conditions using a tiling array.** *Plant Cell Physiol* 2008, **49:**1135–1149.
7. Kim S, Kang JY, Cho DI, Park JH, Kim SY: **ABF2, an ABRE-binding bZIP factor, is an essential component of glucose signaling and its overexpression affects multiple stress tolerance.** *Plant J* 2004, **40:**75–87.
8. Sakuma Y, Maruyama K, Qin F, Osakabe Y, Shinozaki K: **Dual function of an Arabidopsis transcription factor DREB2A in water-stress-responsive and heat-stress-responsive gene expression.** *Proc Natl Acad Sci U S A* 2006, **103:**18822–18827.
9. Zheng J, Fu J, Gou M, Huai J, Liu Y, Jian M, Huang Q, Guo X, Dong Z, Wang H, Wang G: **Genome-wide transcriptome analysis of two maize inbred lines under drought stress.** *Plant Mol Biol* 2010, **72:**407–421.
10. Zhang X, Wang L, Meng H, Wen H, Fan Y, Zhao J: **Maize ABP9 enhances tolerance to multiple stresses in transgenic Arabidopsis by modulating ABA signaling and cellular levels of reactive oxygen species.** *Plant Mol Biol* 2011, **75:**365–378.
11. Narusaka Y, Nakashima K, Shinwari ZK, Sakuma Y, Furihata T, Abe H, Narusaka M, Shinozaki K, Yamaguchi-Shinozaki K: **Interaction between two cis-acting elements, ABRE and DRE, in ABA-dependent expression of Arabidopsis rd29A gene in response to dehydration and high-salinity stresses.** *Plant J* 2003, **34:**137–148.
12. Msanne J, Lin J, Stone JM, Awada T: **Characterization of abiotic stress-responsive Arabidopsis thaliana RD29A and RD29B genes and evaluation of transgenes.** *Planta* 2011, **234:**97–107.
13. Wang T, Zhang EH, Chen XP, Li L, Liang XQ: **Identification of seed proteins associated with resistance to pre-harvested aflatoxin contamination in peanut (Arachis hypogaea L).** *BMC Plant Biol* 2010, **10:**267.
14. Wan X, Li L: **Molecular cloning and characterization of a dehydration-inducible cDNA encoding a putative 9-cis-epoxycarotenoid dioxygenase in Arachis hypogaea L.** *DNA Seq* 2005, **16:**217–223.
15. Wan X, Li L: **Regulation of ABA level and water-stress tolerance of Arabidopsis by ectopic expression of a peanut 9-cis-e-poxycarotenoid dioxygenase gene.** *Biochem Biophys Res Commun* 2006, **347:**1030–1038.
16. Hong L, Hu B, Liu X, He CY, Yao Y, Li XL, Li L: **Molecular cloning and expression analysis of a new stress-related AREB gene from Arachis hypogaea.** *Biol Plant* 2013, **57:**56–62.
17. Liu X, Hong L, Li X, Yao Y, Hu B, Li L: **Improved drought and salt tolerance in transgenic Arabidopsis overexpressing a NAC transcriptional factor from Arachis hypogaea.** *Biosci Biotechnol Biochem* 2011, **75:**443–450.
18. Liu X, Liu S, Wu J, Zhang B, Li X, Yan Y, Li L: **Overexpression of Arachis hypogaea NAC3 in tobacco enhances dehydration and drought tolerance by increasing superoxide scavenging.** *Plant Physiol Biochem* 2013, **70:**354–359.
19. Liang J, Yang L, Chen X, Li L, Guo D, Li H, Zhang B: **Cloning and characterization of the promoter of the 9-cis-epoxycarotenoid dioxygenase gene in Arachis hypogaea L.** *Biosci Biotechnol Biochem* 2009, **73:**2103–2106.
20. Li XY, Liu X, Yao Y, Li YH, Liu S, He CY, Li JM, Lin YY, Li L: **Overexpression of arachis hypogaea AREB1 gene enhances drought tolerance by modulating ROS scavenging and maintaining endogenous ABA content.** *Int J Mol Sci* 2013, **14:**12827–12842.
21. Verslues PE, Ober ES, Sharp RE: **Root growth and oxygene relations at low water potentials impact of oxygen availability in polyethylene glycol solutions.** *Plant Physiol* 1998, **116:**1403–1412.
22. Buchanan CD, Lim S, Salzman RA, Kagiampakis I, Morishige DT, Weers BD, Klein RR, Pratt LH, Cordonnier-Pratt MM, Klein PE, Mullet JE: **Sorghum bicolor's transcriptome response to dehydration, high salinity and ABA.** *Plant Mol Biol* 2005, **5:**8699–8720.
23. Young MD, Wakefield MJ, Smyth GK, Oshlack A: **Gene ontology analysis for RNA-seq: accounting for selection bias.** *Genome Biol* 2010, **11:**R14.
24. Hu B, Hong L, Liu X, Xiao SN, Lv Y, Li L: **The higher expression and distribution of 9-cis-epoxycarotenoid dioxygenase1 (AhNCED1) from Arachis hypogaea L. contribute to tolerance to water stress in a drought-tolerant cultivar.** *Acta Physiol Plant* 2013, **35:**1667–1674.
25. Hu B, Hong L, Liu X, Xiao SN, Lv Y, Li L: **Identification of different ABA biosynthesis sites at seedling and fruiting stages in Arachis hypogaea L. following water stress.** *Plant Growth Regul* 2013, **70:**131–140.
26. Zhang S, Qi Y, Liu M, Yang C: **SUMO E3 ligase AtMMS21 regulates drought tolerance in Arabidopsis thaliana.** *J Integr Plant Biol* 2013, **55:**83–95.
27. Liu J, Xia Z, Wang M, Zhang X, Yang T, Wu J: **Overexpression of a maize E3 ubiquitin ligase gene enhances drought tolerance through regulating stomatal aperture and antioxidant system in transgenic tobacco.** *Plant Physiol Biochem* 2013, **73:**114–120.
28. Lyzenga WJ, Stone SL: **Abiotic stress tolerance mediated by protein ubiquitination.** *J Exp Bot* 2012, **63:**599–616.
29. Seo DH, Ryu MY, Jammes F, Hwang JH, Turek M, Kang BG, Kwak JM, Kim WT: **Roles of four Arabidopsis U-box E3 ubiquitin ligases in negative regulation of abscisic acid-mediated drought stress responses.** *Plant Physiol* 2012, **160:**556–568.
30. Wan X, Mo A, Liu S, Yang L, Li L: **Constitutive expression of a peanut ubiquitin-conjugating enzyme gene in Arabidopsis confers improved water-stress tolerance through regulation of stress-responsive gene expression.** *J Biosci Bioeng* 2011, **111:**478–484.
31. Söderman E, Mattsson J, Engström P: **The Arabidopsis homeobox gene ATHB-7 is induced by water deficit and by abscisic acid.** *Plant J* 1996, **10:**375–381.
32. Valdés AE, Overnäs E, Johansson H, Rada-Iglesias A, Engström P: **The homeodomain-leucine zipper (HD-Zip) class I transcription factors ATHB7 and ATHB12 modulate abscisic acid signalling by regulating protein phosphatase 2C and abscisic acid receptor gene activities.** *Plant Mol Biol* 2012, **80:**405–418.
33. Yazaki J, Shimatani Z, Hashimoto A, Nagata Y, Fujii F, Kojima K, Suzuki K, Taya T, Tonouchi M, Nelson C, Nakagawa A, Otomo Y, Murakami K, Matsubara K, Kawai J, Carninci P, Hayashizaki Y, Kikuchi S: **Transcriptional profiling of genes responsive to abscisic acid and gibberellin in rice: phenotyping and comparative analysis between rice and Arabidopsis.** *Physiol Genomics* 2004, **17:**87–100.
34. Seki M, Ishida J, Narusaka M, Fujita M, Nanjo T, Umezawa T, Kamiya A, Nakajima M, Enju A, Sakurai T, Satou M, Akiyama K, Yamaguchi-Shinozaki K, Carninci P, Kawai J, Hayashizaki Y, Shinozaki K: **Monitoring the expression pattern of around 7,000 Arabidopsis genes under ABA treatments using a full-length cDNA microarray.** *Funct Integr Genomics* 2002, **2:**282–291.
35. Zhang J, Liang S, Duan J, Wang J, Chen S, Cheng Z, Zhang Q, Liang X, Li Y: **De novo assembly and characterisation of the transcriptome during seed development, and generation of genic-SSR markers in peanut (Arachis hypogaea L.).** *BMC Genomics* 2012, **13:**90.
36. Bihmidine S, Lin J, Stone JM, Awada T, Specht JE, Clemente TE: **Activity of the Arabidopsis RD29A and RD29B promoter elements in soybean under water stress.** *Planta* 2013, **237:**55–64.
37. Li F, Han Y, Feng Y, Xing S, Zhao M, Chen Y, Wang W: **Expression of wheat expansin driven by the RD29 promoter in tobacco confers water-stress tolerance without impacting growth and development.** *J Biotechnol* 2013, **163:**281–291.
38. Liu C, Mao B, Ou S, Wang W, Liu L, Wu Y, Chu C, Wang X: **OsbZIP71, a bZIP transcription factor, confers salinity and drought tolerance in rice.** *Plant Mol Biol* 2014, **84:**19–36.
39. Mortazavi A, Williams BA, McCue K, Schaeffer L, Wold B: **Mapping and quantifying mammalian transcriptomes by RNA-Seq.** *Nat Methods* 2008, **5:**621–628.
40. Wang L, Feng Z, Wang X, Wang X, Zhang X: **DEGseq: an R package for identifying differentially expressed genes from RNA-seq data.** *Bioinformatics* 2009, **26:**136–138.

Lactococcus lactis carrying the pValac DNA expression vector coding for IL-10 reduces inflammation in a murine model of experimental colitis

Meritxell Zurita-Turk[1†], Silvina del Carmen[2†], Ana CG Santos[3], Vanessa B Pereira[1], Denise C Cara[4], Sophie Y Leclercq[5], Alejandra dM de LeBlanc[2], Vasco Azevedo[1], Jean-Marc Chatel[6], Jean G LeBlanc[2] and Anderson Miyoshi[1*]

Abstract

Background: Inflammatory bowel diseases (IBD) are intestinal disorders characterized by inflammation in the gastrointestinal tract. Interleukin-10 is one of the most important anti-inflammatory cytokines involved in the intestinal immune system and because of its role in downregulating inflammatory cascades, its potential for IBD therapy is under study. We previously presented the development of an invasive strain of *Lactococcus lactis* (*L. lactis*) producing Fibronectin Binding Protein A (FnBPA) which was capable of delivering, directly to host cells, a eukaryotic DNA expression vector coding for IL-10 of *Mus musculus* (pValac:*il-10*) and diminish inflammation in a trinitrobenzene sulfonic acid (TNBS)-induced mouse model of intestinal inflammation. As a new therapeutic strategy against IBD, the aim of this work was to evaluate the therapeutic effect of two *L. lactis* strains (the same invasive strain evaluated previously and the wild-type strain) carrying the therapeutic pValac:*il-10* plasmid in the prevention of inflammation in a dextran sodium sulphate (DSS)-induced mouse model.

Results: Results obtained showed that not only delivery of the pValac:*il-10* plasmid by the invasive strain *L. lactis* MG1363 FnBPA+, but also by the wild-type strain *L. lactis* MG1363, was effective at diminishing intestinal inflammation (lower inflammation scores and higher IL-10 levels in the intestinal tissues, accompanied by decrease of IL-6) in the DSS-induced IBD mouse model.

Conclusions: Administration of both *L. lactis* strains carrying the pValac:*il-10* plasmid was effective at diminishing inflammation in this murine model of experimental colitis, showing their potential for therapeutic intervention of IBD.

Background

Inflammatory bowel diseases (IBD), including ulcerative colitis (UC) and Crohn's disease (CD), are characterized by spontaneous and chronic inflammation of the gastrointestinal tract (GIT). Despite much research in the last years, the exact etiology and pathogenesis of these disorders remain unclear; however, it is nowadays generally accepted that IBD are caused by dysregulation of the mucosal immune system in relation to the native intestinal microbiota in genetically susceptible individuals [1]. Current treatments for IBD are restricted to the use of anti-inflammatory drugs, immunosuppressants and antibiotics, which although showing moderate therapeutic effect, present serious side effects and reveal that better, cheaper and longer lasting drugs are necessary [2].

Interleukin-10 (IL-10) is one of the most important anti-inflammatory cytokines involved in the intestinal immune system [3] and because of its immunosuppressive activity and its central role in downregulating inflammatory cascades [4] it presents itself as a good therapeutic candidate against IBD [5]. Recombinant human IL-10 raised hope when first used in the 90s in CD

* Correspondence: miyoshi@icb.ufmg.br
†Equal contributors
[1]Department of General Biology, Institute of Biological Sciences, Federal University of Minas Gerais, Belo Horizonte, Brazil
Full list of author information is available at the end of the article

patients as the treatment led to remission in patients that were otherwise refractory to treatment [6]; however, two large, multi-centered follow-up studies using subcutaneous dosing were unable to confirm the results [7,8]. Moreover, systemic treatment with IL-10 showed to be quite limiting because of its short half-life (1.1-2.6 h) and requirement of high protein concentration (20 μg/kg), increasing the cost of production, discomfort and secondary effects in the patients [9]. On the other hand, oral treatment with IL-10 has also shown to be limited due to its extreme sensitivity to the environment of the GIT and therefore survival in it [10].

New approaches to yield more specific delivery of IL-10 to the intestinal mucosa and prevent the drawbacks associated to systemic and oral administration led to the development of IL-10-producing *Lactococcus lactis* (*L. lactis*) strains [11]. In 2003, a biological containment system for human IL-10-producing *L. lactis* [12] showed to be safe and improved the disease when tested in CD patients in phase I clinical trials [13]; however, clinical results did not reveal a statistically significant difference in mucosal healing between patients receiving the recombinant strains and placebo.

Since this strain produces IL-10 and releases it in the GIT, its clinical use for IBD is still hindered by IL-10's sensitivity and poor survival in these hostile conditions. In this regard, a new eukaryotic DNA expression vector for delivery using lactococci, called pValac, which allows cloning of an ORF (Open Reading Frame) of interest, expression of the molecule by host cells, replication both in *E. coli* and in *L. lactis* and selection of bacteria, was firstly constructed in 2009 [14]. Its potential to deliver DNA and trigger DNA expression by epithelial cells has already been demonstrated *in vitro* [15] and *in vivo* [16]. This strategy does not only fulfil the aforementioned advantages, but also combines the advantages of mucosal immunity, safety of using non-pathogenic bacteria, technique simplicity and low cost of DNA therapy. Moreover, ingested *L. lactis* strains pose no risk to the individuals as these bacteria are quickly degraded and only around 20-30% reach the sites of inflammation, their transit through the gastrointestinal tract takes between 2 to 3 days and they are incapable of multiplying in the body or become part of the normal gut flora.

Our research group recently evaluated a recombinant invasive *L. lactis* strain expressing the *Staphylococcus aureus* Fibronectin Binding Protein A (FnBPA), harbouring the eukaryotic DNA expression vector pValac coding for the anti-inflammatory cytokine IL-10 of *Mus musculus* (*L. lactis* MG1363 FnBPA + pValac:*il-10*), for *in situ* expression of IL-10 and therefore higher, more efficient and direct production of this cytokine at the sites of inflammation. This strategy showed to be efficient at diminishing inflammation in a TNBS-induced inflammatory mouse model [17].

The aim of the present work was to evaluate and compare the therapeutic capacity of two *L. lactis* strains, the invasive *L. lactis* MG1363 FnBPA + strain and the wt *L. lactis* MG1363, both carrying the pValac:*il-10* plasmid, for the prevention of experimental IBD in a DSS-induced mouse model.

Methods

Bacterial strains, growth conditions and plasmid

The bacterial strains used in this work are listed in Table 1. *E. coli* TG1 was aerobically grown in Luria-Bertani (LB) medium at 37°C with vigorous shaking whereas all *L. lactis* [18] strains were grown in M17 medium (Difco, Sparks, MD, USA) supplemented with 0.5% glucose (GM17) at 30°C without shaking. Recombinant *E. coli* were selected by addition of 10 μg/mL chloramphenicol (Cm) while recombinant *L. lactis* were selected by addition of 10 μg/mL Cm and/or 5 μg/mL of erythromycin (Ery). For animal trials, *L. lactis* cultures grown until an OD_{600} of 1.0-1.2 were previously stocked in glycerol 80% (1:4) and on day of use they were centrifuged in order to eliminate any remaining traces of the antibiotic and medium and resuspended in 100 μL of saline solution (0.15 M NaCl) for animal feedings.

As previously shown by del Carmen and co-workers, the pValac:*il-10* plasmid harbours an eukaryotic region containing the CytoMegaloVirus promoter (pCMV), the IL-10 ORF of *Mus musculus* and the polyadenylation signal of bovine growth hormone (BGH polyA), required for gene expression by eukaryotic host cells, as well as a prokaryotic region containing the RepA/RepC replication origins for both *E. coli* and *L. lactis*, respectively, and a Cm resistance gene for bacterial selection [14].

DNA manipulations

General DNA manipulation techniques were carried out according to standard procedures. Unless otherwise indicated, DNA restriction and modification enzymes were used as recommended by the suppliers. DNA fragments were isolated from agarose gels using the Illustra GFX PCR DNA and Gel Band Purification Kit (GE Healthcare Life Sciences, Chalfont St. Giles, UK). PCR amplifications were performed using AccuPrimePfx DNA Polymerase (Invitrogen, Carlsbad, CA, USA) and/or GoTaq DNA Polymerase (Invitrogen, Carlsbad, CA, USA) in a DNA thermocycler (MJ Research, Inc., Minnesota, USA). Plasmid DNA from *E. coli* and *L. lactis* was isolated as previously described [19] with the following modifications: for plasmid DNA extraction from *L. lactis*, the first step included addition of TES (25% sucrose, 1 mM EDTA, 50 mMTris-HCl, pH 8) containing lysozyme (10 mg/mL) for 30 min at 37°C to prepare protoplasts. Electroporation of *E. coli* and

Table 1 Bacterial strains used in this work

Bacterial Strain	Characteristics	Reference
Escherichia coli (*E.coli*) TG1	[supE, hsd, Δ5, thi, Δlac-proAB), F'(traD36 proAB-lacZΔM15)]	Invitrogen (São Paulo, Brazil)
Escherichia coli (*E.coli*) TG1 (pValac:*il-10*)	(pCMV/Cm^R/RepA/ RepC/*IL-10*)	[17]
Lactococcus lactis (*L. lactis*) MG1363	*L. lactis* subsp. *cremoris*[b]	Laboratory of Cellular and Molecular Genetics (LGCM), Federal University of Minas Gerais (UFMG), Brazil
Lactococcus lactis (*L. lactis*) MG1363 pValac:*il-10*	*L. lactis* MG1363 strain carrying the pValac:*il-10* plasmid	This work
Lactococcus lactis (*L. lactis*) MG1363 FnBPA	*L. lactis* MG1363 strain expressing FnBPA of *S. aureus*	[19]
Lactococcus lactis (*L. lactis*) MG1363 FnBPA pValac:*il-10*	*L. lactis* MG1363 strain expressing FnBPA of *S. aureus* carrying the pValac:*il-10* plasmid	[17]

L. lactis was performed as previously described [20]. *E. coli* transformants were plated on LB agar plates containing the required antibiotic for 24 h at 37°C, whereas *L. lactis* transformants were plated on GM17 agar plates containing the required antibiotic and were counted after 2-days incubation at 30°C.

Induction of intestinal inflammation and feeding procedure in mice

Conventional female C57BL/6 mice [21] weighing approximately 18 g, obtained from the inbred closed colony (CEBIO) maintained at the Federal University of Minas Gerais (UFMG – Belo Horizonte, Brazil), were used to evaluate the therapeutic effect of *L. lactis* MG1363 pValac:*il-10* and *L. lactis* MG1363 FnBPA + pValac:*il-10* in the prevention of intestinal inflammation in a DSS-induced mouse model. Procedures and manipulation of animals followed the rules of the ethics and research committees of the Biological Institute of the UFMG and all animal protocols were approved by the Ethics Committee on Animal Experiments (CETEA). All animals were maintained in collective cages (6 animals/cage) in an environmentally controlled room with a 12-hour light/dark cycle and unlimited free access to water (or DSS solution) and food. After acclimatization for 17 days, colonic inflammation was induced by the addition of 1.5% (w/v) DSS (MW 40,000-50,000; USB Affymetrix, Santa Clara, CA, USA) in drinking water for 7 consecutive days. Liquid consumption was monitored to ensure that all mouse groups consumed similar volumes (4 ± 2 mL) of the DSS solution daily. For experimental procedure, mice were divided in 6 experimental groups: i) control group which received 100 μl NaCl intragastrically while no DSS was added to the drinking water; ii) DSS group which received 100 μl NaCl intragastrically while DSS was added to the drinking water and iii) *L. lactis* MG1363 group (wt group), iv) *L. lactis* MG1363 pValac:*IL-10* group, v) *L. lactis* MG1363 FnBPA + group and vi) *L. lactis* MG1363 FnBPA + pValac:*il-10* group, which all received intragastrically 100 μl of the corresponding bacterial strain as suspension, at a dose of 2×10^9 CFU/100 μl, in NaCl. The intragastric administration of NaCl or the bacterial suspensions was once daily since the day before the beginning of DSS administration until sacrifice. At day 9, mice were sacrificed by cervical dislocation for organ collection.

Macroscopic and microscopic (histological) assessment of intestinal inflammation

On the day of sacrifice (day 9), colons were removed, excised and visually inspected for macroscopic evaluation to assess colonic inflammation [disease activity index (DAI)], using a scoring system in which the following features were graded: body weight loss (0, no loss; 1, 1-5% loss; 2, 5-10% loss; 3, 11-15% loss and 4, >15% loss), diarrhoea (0, absent; 2, moderate; 4, severe) and rectal bleeding (0, absent; 2, moderate; 4, severe). The macroscopic damage score was calculated from the score of all clinical signs with a maximum score of 12. For histological inflammation scoring [histological activity index (HAI)], samples of the colon were fixed in 10% formalin in phosphate-buffered saline (PBS), embedded in paraffin, cut into 3–5-μm sections and stained with haematoxylin–eosin (H&E-staining) for microscopic analysis. These sections were blindly scored based on a semi quantitative scoring system previously described [22] in which the following features were graded: extent of destruction of the mucosa's architecture (0, normal; 1, 2 and 3, light, moderate and extensive damage, respectively), presence and degree of cellular infiltration (0, normal; 1, 2 and 3, light, moderate and transmural infiltration, respectively), extent of muscle thickening (0, normal; 1, 2 and 3, light, moderate and extensive thickening, respectively), presence or absence of crypt abscesses (0, absent and 1, present) and the presence or absence of goblet cell depletion (0, absent and 1, present). The histological damage score was calculated as the addition of the scores corresponding to each feature. High macroscopic and histological damage scores indicate increased damage in the intestines.

Secretory IgA (sIgA) assay

Levels of sIgA in the intestinal fluid were determined by enzyme linked immunosorbent assay (ELISA) as previously described [23]. Briefly, microtitre plates (NUNC, Thermo Scientific, Waltham, MA, USA) were coated with goat anti-mouse UNLB antibody (Southern Biotechnology,

Birmingham, AL, USA) in coating buffer (pH 9.8) overnight at 4°C. Wells were then washed with a saline 0.05% tween solution and blocked with 200 μL of PBS with 0.05% (w/v) casein for 1 h at room temperature. The supernatants obtained from the intestinal fluids after centrifugation at 432 *g* for 20 min at 4°C and the diluted standards in PBS-0.25% casein (1:10) were then added to the plate and incubated for 1 hour at 37°C. After washing, goat anti-mouse IgA HRP (Southern Biotechnology, Birmingham, AL, USA) was added and plates were incubated for 1 h at 37°C. The colour reaction was developed at room temperature with the addition of 100 μL/well of orthophenylenediamine (OPD) (1 mg/mL) (Sigma, St. Louis, MO, USA), 0.04% H_2O_2 substrate in sodium citrate buffer. The reaction was finally stopped by the addition of 20 μL/well of 2 N H_2SO_4. Absorbance was measured at 492 nm using a Bio-Rad Model 450 Microplate Reader. Results were expressed as concentration (μg/mL), according to the standard curve.

Colon tissue preparation and cytokine assays

For cytokine assays, colons were weighed and homogenized in PBS containing 0.05% (v/v) Tween-20, 0.1 mM phenylmethylsulphonyl fluoride, 0.1 mM benzethonium chloride, 10 mM EDTA and 20 KIU AprotininA using a tissue homogenizer (1 mL/0.1 g). Suspensions were centrifuged at 600 *g* for 10 min at 4°C and the supernatants collected for cytokine assay. Concentrations of IL-10, TNF-α, IL-6, IL-17 were measured by ELISA as described previously [23]. Briefly, after coating microtitre plates (NUNC, Thermo Scientific, Waltham, MA, USA) with purified monoclonal antibodies reactive to mouse cytokines IL-10, TNF-α, IL-6, IL-17 (BD, New Jersey, USA), standards and samples were added and incubated overnight at 4°C. Biotinylated monoclonal antibodies anti-mouse IL-10, TNF-α, IL-6, IL-17 were added and incubated for 1 h at room temperature, after which peroxidase-labelled streptavidin (Sigma, St. Louis, MO, USA) was added. A colour reaction was developed at room temperature with 100 μL/well of OPD (1 mg/mL) and 0.04% H_2O_2 substrate in sodium citrate buffer. The reaction was stopped by the addition of 20 μL/well of 2 N H_2SO_4. Absorbance was measured at 492 nm using a Bio-Rad Model 450 Microplate Reader. Results were expressed as concentration of each cytokine (pg/mL), according to the respective standard curve.

Statistical analysis

Statistical analyses were performed using the GraphPad Prism 5.0 software (San Diego, CA, USA) and all results were expressed as mean ± standard deviation (SD). Significance of differences among groups was assessed by Student's t-test or analysis of variance (ANOVA) followed by a Tukey comparison post-hoc test. Means were considered statistically different when $p < 0.05$.

Results

Construction of *L. lactis* MG1363 pValac:*il-10*

The pValac:*il-10* plasmid was previously constructed by del Carmen and co-workers and used to construct the invasive *L. lactis* MG1363 FnBPA + pValac:*il-10* strain. This strain was tested in an experimental IBD mouse model induced by TNBS and showed to diminish inflammation and damage scores [17]. In this work, we constructed the non-invasive *L. lactis* MG1363 pValac:*il-10* strain; the construction was confirmed by PCR (data not shown). This strain was tested together with the invasive strain in the experimental IBD mouse model induced by DSS.

L. lactis strains carrying the pValac:*il-10* plasmid show anti-inflammatory properties in the DSS-induced colitis in mice

DSS is a well-established animal model of mucosal inflammation that has been used for over two decades in the study of IBD pathogenesis and preclinical studies [8] and when administrated in drinking water induces an acute inflammation. C57BL/6 mice were used for the DSS study as these animals are highly susceptible to DSS-induced colitis and do not naturally recover from this disease [21].

Similar to the results previously obtained in a TNBS-induced colitis mouse model, *L. lactis* MG1363 FnBPA + pValac:*il-10* was also capable of diminishing inflammation in the DSS-induced colitis mouse model (Figures 1 and 2). It was also observed that the non-invasive variant, *L. lactis* MG1363 pValac:*il-10*, was also capable of decreasing the inflammation in this model. Overall, both groups that received the *L. lactis* strains carrying the pValac:*il-10* (pValac:*il-10* groups) showed a lower DAI (Figure 1) and therefore significant decrease of inflammation, when compared to the DSS control group and the groups receiving *L. lactis* MG1363 and *L. lactis* MG1363 FnBPA+, which presented more diarrhoea and bleeding, which was almost absent in animals from the pValac:*il-10* groups. The colon length of each animal was also measured as a marker of inflammation-caused injury. As expected, there was a significant shortening of the colon induced by DSS treatment (6.92 cm), and the administration of *L. lactis* MG1363 (6.96 cm) or *L. Lactis* MG1363 FnBPA + (6.93 cm) did not improve it. However, the pValac:*il-10* groups showed a significant improvement in colon length, *L. lactis* MG1363 pValac:*il-10* (7.97 cm) and *L. lactis* MG1363 FnBPA pValac:*il-10* (8.01 cm), when compared to the other groups induced with DSS, resembling more the colon lengths of the control group (8.16 cm). Regarding body weight loss, all groups maintained similar weight during the 8 days of experiment (data not shown).

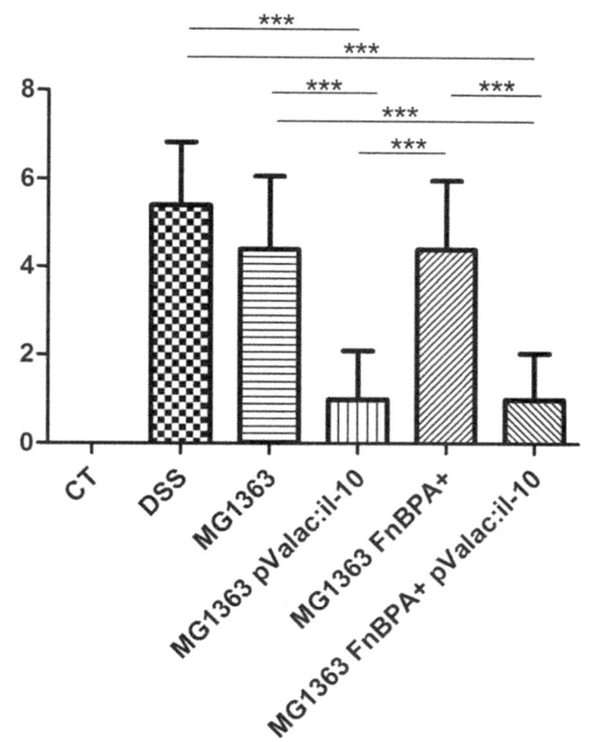

Figure 1 Inflammatory macroscopic score of experimental colitis. Macroscopic score of mice treated with DSS that received the different *L. lactis* strains compared to the DSS group without any specific administration. Disease activity index (DAI) was scored for body weight loss, diarrhoea and rectal bleeding. Bars represent the mean ± SD of 6 mice per group of two independent experiments. Asterisks represent statistical significance (**p < 0.01 or ***p < 0.0001). All groups are statistically significant to the control group (***p < 0.0001).

To evaluate the colons microscopically, a HAI was performed for all groups, and showed that animals from the pValac:*il-10* groups presented a significant decrease of the damage score when compared to the DSS group and the mice that received *L. lactis* MG1363 or *L. lactis* MG1363 FnBPA + (Figure 2A).

Histologically, all samples from the colon of healthy control mice (group without DSS treatment) presented themselves within normal standards. The architecture integrity of the colon was maintained, with goblet cells, thin submucosa without signs of congestion or oedema and absence of inflammatory infiltrate in the mucosa and submucosa (Figure 2B). On the other hand, the colitis control group (DSS group, Figure 2C) presented intense alteration in the mucosal architecture of their colons, with areas of intense ulceration, absence of glands in large parts of the mucosa and accentuated depletion of goblet cells. It was also possible to observe crypt abscesses and the submucosa was marked by an intense oedema and inflammatory infiltrate constituted principally by macrophages. Moreover, thickening of the muscle layer was also observed. No significant differences were observed when animals received *L. lactis* MG1363 (Figure 2D) or *L. lactis* MG1363 FnBPA + (Figure 2F) strains, compared to the DSS treated control. However, groups that received *L. lactis* MG1363 pValac:*il-10* (Figure 2E) or *L. lactis* MG1363 FnBPA + pValac:*il-10* (Figure 2G) presented a milder pathological picture, with tendency to normality, a smaller compromised area and a reduction in the intensity of the lesion. Reduced inflammatory infiltrates without erosion of the epithelium, light oedema and diminished depletion of goblet cells, partially preserving the mucosal architecture, were also observed.

The pValac:*il-10* plasmid increases the production of secretory IgA

Secretory IgA creates a first-line of defence against mucosal compromise that is lost during IBD. Therefore, sIgA levels from intestinal fluids of mice were also analysed. It was observed that only mice that received the *L. lactis* MG1363 FnBPA + pValac:*il-10* strain showed a significant increase in the concentration of sIgA when compared to those from the control group, DSS group and the groups that did not receive the strains carrying the pValac:*il-10* plasmid (Figure 3). Moreover, this strain also induced an increased sIgA production when compared to the non-invasive *L. lactis* MG1363 pValac:*il-10* strain.

The administration of *L. lactis* carrying the pValac:*il-10* plasmid modulates the production of cytokines in the intestinal tissues

Administration of DSS to mice leads to a macrophage-induced inflammation and tissue damage accompanied by a cellular cytotoxic-mediated inflammatory response (macrophage/T_h1/T_h17 chemotactic profile) during the progression of colitis [24]. In this regard, in order to determine if the pValac:*il-10* plasmid could modulate the cytokine level in diseased animal, IL-10, TNF-α, IL-6, IL-17 were measured from colonic tissues. Mice from the *L. lactis* MG1363 pValac:*il-10* group showed significantly higher IL-10 levels compared with the control group, while animals from the *L. lactis* MG1363 FnBPA + pValac:*il-10* group showed higher IL-10 levels than the animals from the control and DSS groups (Figure 4A). Regarding pro-inflammatory cytokines, TNF-α (Figure 4B) and IL-17 (Figure 4D) levels did not show significant differences between the control and test groups, while IL-6 levels showed significant differences between some of the test and control groups (Figure 4C). IL-6 increased significantly (*$p < 0.05$) in the DSS group when compared to the control group, and no significant differences with the groups that received *L. lactis* MG1363 or *L. lactis* MG1363 FnBPA + were observed. The levels of IL-6 were significantly lower (**$p < 0.001$) in the pValac:*il-10* groups

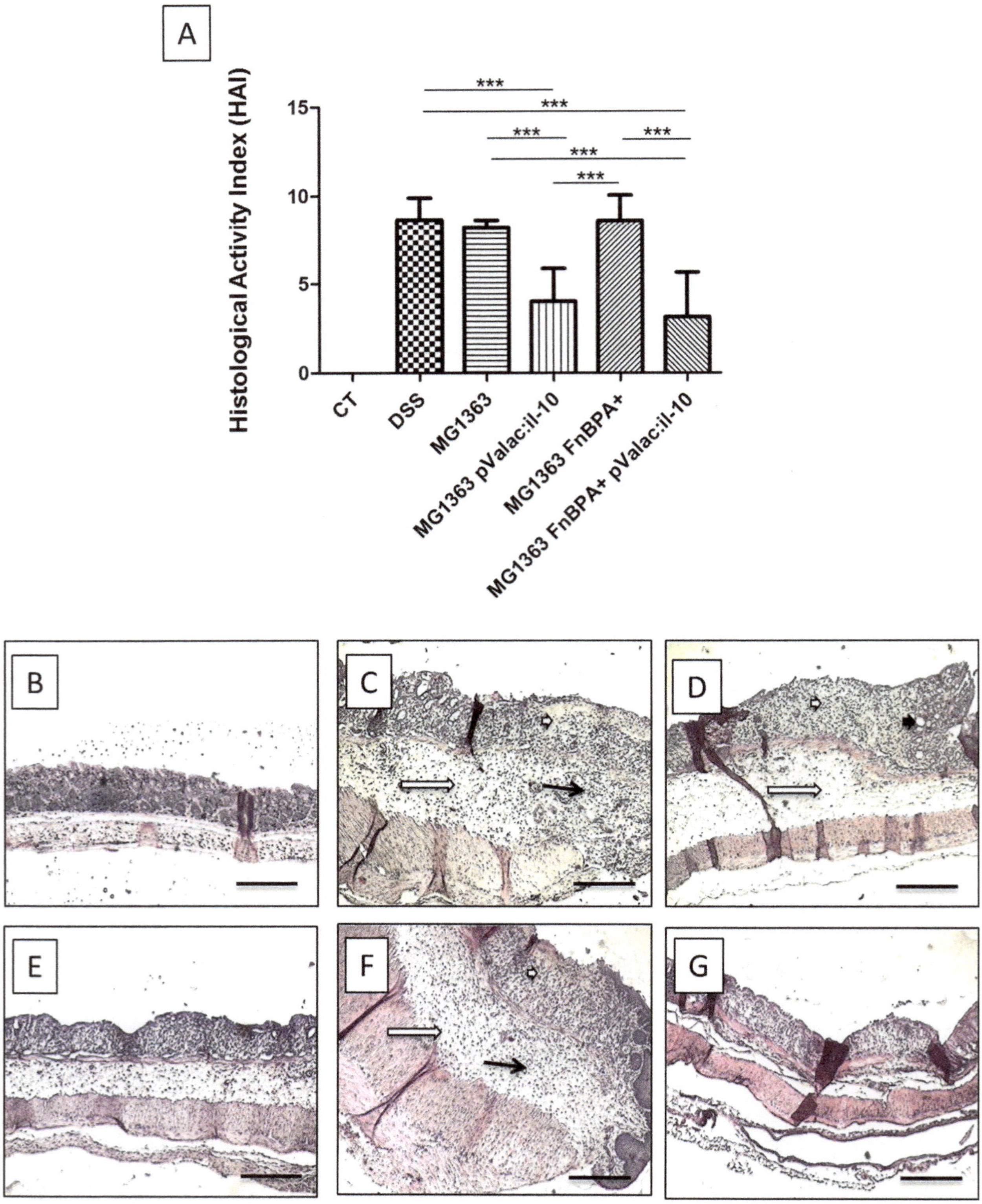

Figure 2 (See legend on next page.)

(See figure on previous page.)
Figure 2 Inflammatory microscopic score of experimental colitis and histopathology of intestinal changes between all groups. Macroscopic score of mice treated with DSS that received the different *L. lactis* strains compared to the DSS group without any specific administration **(A)**. Histological activity index (HAI) was obtained from the colonic tissues stained by H&E. Bars represent the mean ± SD of 6 mice per group of two independent experiments. Asterisks represent statistical significance (**$p < 0.01$ or ***$p < 0.0001$). All groups are statistically significant to the control group (***$p < 0.0001$). Representative photos obtained from the proximal colon of a mouse from control group: Control group **(B)**, DSS group **(C)**, and the groups of mice treated with DSS that received *L. lactis* MG1363 **(D)** *L. lactis* MG1363 pValac:*il-10* **(E)** Proximal colon of diseased mice that received the *L. lactis* MG1363 FnBPA + **(F)** or *L. lactis* MG1363 FnBPA + pValac:*il-10* **(G)**. The bar on each image represents 100 μm. The short white arrows show the depletion of goblet cells; the long white arrow, intense oedema in the submucosa; the short black arrow, crypt abscesses; and the long black arrow, inflammatory infiltrate (especially of macrophages).

compared to the DSS group, resembling the values of the control group.

Discussion

Interleukin 10's interest in the attempt of developing an efficient therapy against IBD has grown since it was observed that IL-10 knockout (IL-10$^{-/-}$) mice develop spontaneous enterocolitis when not maintained in germ-free conditions [25]. Although many advances have been made there still does not exist any treatment for IBD, showing that the major drawback to use IL-10 is its correct administration and directioning to the sites of inflammation.

In order to overcome these problems, new strategies based on local delivery of IL-10 to the intestinal mucosa have been developed; high IL-10 concentration at these sites would outweigh the need for high doses and as such avoid the increase of pro-inflammatory cytokine production and other undesired side effects.

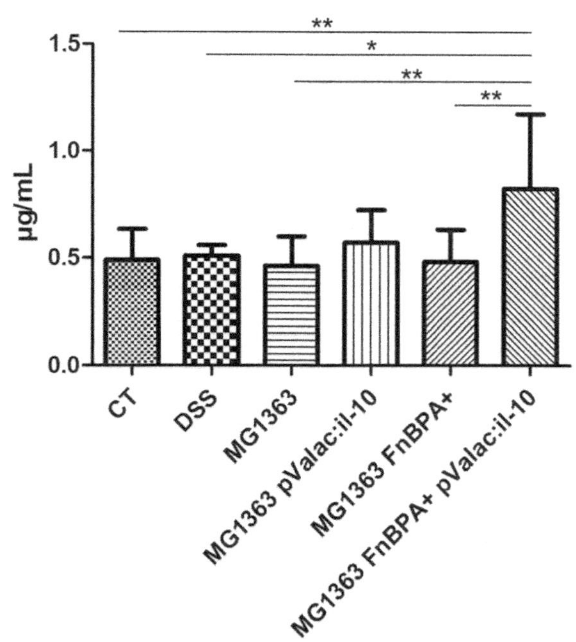

Figure 3 Influence of DSS and different bacterial administrations on gut secretory IgA. Intestinal fluid was collected and total sIgA was measured by ELISA of wild-type mice, DSS wild-type mice and diseased mice that received the *L. lactis* MG1363 strain, *L. lactis* MG1363 pValac:*il-10* strain, *L. lactis* MG1363 FnBPA + strain or *L. lactis* MG1363 FnBPA + pValac:*il-10* strain during treatment. Bars represent the mean N = 6 ± SD of two independent experiments. Asterisks represent statistical significance *$p < 0.05$ or **$p < 0.001$).

One such strategy consists in the use of polymer-based microparticles. These particles encapsulate gelatine nanoparticles containing plasmid DNA expressing murine IL-10 and are capable of releasing these nanoparticles directly at the desired site of action. These IL-10 gene containing particles have shown to successfully reduce the levels of inflammatory cytokines as well as disease activity scores in the TNBS-induced model of colitis [26]. However, despite promising, scale-up to an industrial level appears to be very expensive and complicated.

Another strategy by Yao and collaborators consists in a *Bifidobacterium longum* strain capable of secreting human IL-10 which was capable of alleviating inflammatory damage of colonic tissue in a DSS-induced mouse model by blocking the colitis-activated NF-κB pathways and upregulating CD4 + CD25 + Foxp3+ T_{reg} in blood and mesenteric lymph nodes [27].

As such, in order to improve IL-10 delivery to the sites of inflammation using *L. lactis*, a *L. lactis* strain expressing FnBPA [15,28], that has the capacity to efficiently internalize and trigger recombinant DNA expression by human epithelial cells, was used to deliver a eukaryotic expression vector coding for IL-10 of *Mus musculus*, pValac:*il-10*, directly to host cells in the GIT for recombinant *in situ* IL-10 production. This new strategy recently showed to be capable of diminishing inflammation in a TNBS-induced mouse model [17].

Following these results and in order to evaluate the therapeutic effect of the pValac:*il-10* plasmid in a larger context, in the present work we explored the potential of this therapeutic plasmid in a DSS-induced mouse model delivered by two *L. Lactis* strains as a new therapeutic strategy for the prevention of intestinal inflammation. *L. lactis* MG1363 FnBPA + expresses FnBPA, which confers the strain the capacity to mediate adhesion to host tissue and bacterial uptake by eukaryotic cells. To evaluate if this characteristic results in improved anti-inflammatory effect, it was compared with the non-invasive *L. lactis* MG1363 strain. Our results showed that both strains were capable of delivering the eukaryotic expression vector to host cells directly at the sites of inflammation

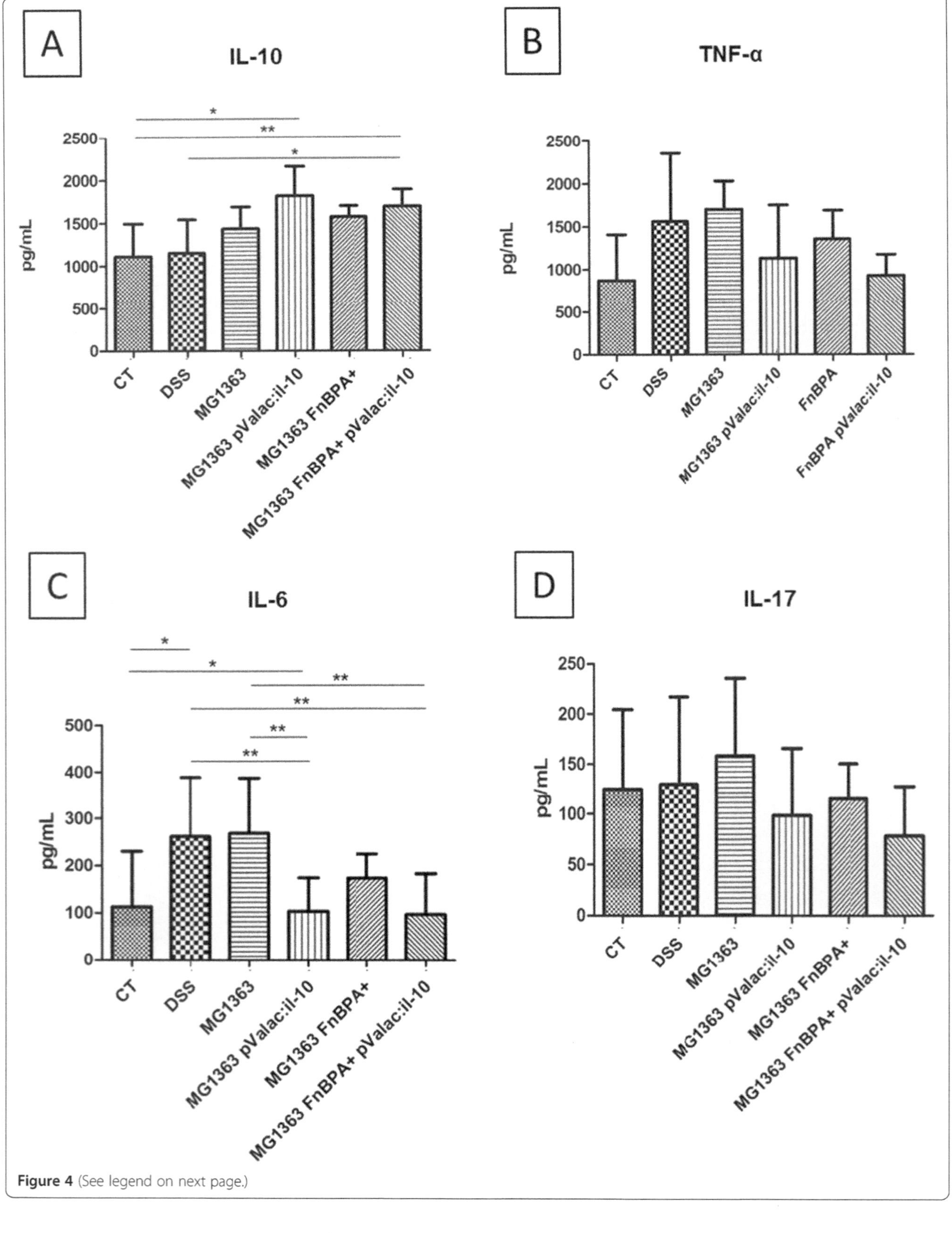

Figure 4 (See legend on next page.)

(See figure on previous page.)
Figure 4 The pValac:*IL-10* plasmid modulates the production of cytokines in intestinal tissues. IL-10 **(A)** TNF-α **(B)**, IL-6 **(C)** and IL-17 **(D)** levels in the intestinal tissues of mice from control group (CT), DSS group and mice that received the *L. lactis* MG1363, *L. lactis* MG1363 pValac:*il-10*, *L. lactis* MG1363 FnBPA + or *L. lactis* MG1363 FnBPA + pValac:*il-10*. Bars represent the mean ± SD of 6 mice per group of two independent experiments. Asterisks represent statistical significance (*p < 0.05 or **p < 0.001).

leading to *in situ* IL-10 production, avoiding strong and undesired side-effects, and diminishing the severity of inflammation by maintaining an anti-inflammatory environment in the gut.

DSS has the intrinsic capacity to disrupt the epithelial cell barrier, causing normal gut substances to activate mucosal macrophages, which in turn produce immunomodulatory cytokines. It is generally accepted that DSS is directly toxic to gut epithelial cells of the basal crypts and affects the integrity of the mucosal barrier [29]. The DSS concentration used in the present model triggered an acute colonic inflammation accompanied by diarrhoea and rectal bleeding and did not lead to mortality. The administration of strains carrying the pValac:*il-10* plasmid statistically lowered the macroscopic score compared to the DSS group, regarding diarrhoea and rectal bleeding. No weight loss was observed within the different groups, but this has already been reported earlier [30]. At histological level, administration of strains carrying the pValac:*il-10* plasmid were capable of decreasing the severity of inflammation, with tendency to normality, demonstrating the anti-inflammatory effects of this strategy.

IgA is the most abundant immunoglobulin produced, in its secretory form, at mucosal sites. In the luminal mucous layer, sIgA protects the intestinal epithelium against colonization and invasion by pathogens or commensals [31] and therefore helps modulating and controlling inflammation. In our model, no significant decrease of sIgA was associated with the inflammation observed in mice treated with DSS that did not receive the pValac:*il-10*. The administration of bacteria carrying the pValac:*il-10* plasmid was accompanied by significantly increased levels of sIgA only in the intestinal fluid of mice that received the *L. Lactis* MG1363 FnBPA + pValac:*il-10* strain, compared to the DSS treated mice. Moreover, this strain also induced an increased sIgA production when compared to the *L. lactis* MG1363 pValac:*il-10* strain. We believe that the difference of IgA production induced by these two strains is due to the different cellular entry mechanism these bacteria use after oral administration. In this regard, *L. lactis* MG1363 FnBPA + pValac:*il-10* strain induced increased IgA production because these bacteria invade enterocytes in the epithelium, while *L. lactis* MG1363 pValac:*il-10* are captured by macrophages (as these bacteria have no invasion capacity), which does not directly lead to the induction of IgA. These results showed that the *L. Lactis* MG1363 FnBPA + pValac:*il-10* strain could not only have an anti-inflammatory effect in this model, but also promote the gut immunological barrier by limiting the penetration of bacteria into host tissues and therefore protect mice by helping to modulate inflammation. The expression of FnBPA by this strain might allow for a higher binding capacity and internalization by eukaryotic cells, thus enhancing the production of IL-10 and its anti-inflammatory properties.

Cytokines produced in the gut mucosa greatly influence the resulting immunological outcome; production of anti-inflammatory cytokines induces mucosal tolerance, while high levels of pro-inflammatory cytokines induce inflammation. Since cytokines are major mediators of inflammation and regulatory activity in the gut mucosa, we analysed the ability of *L. lactis* strains carrying the pValac:*il-10* plasmid to modulate the production of cytokines in the colonic tissues of mice. Mice that received the pValac:*il-10* plasmid, whether delivered by *L. lactis* MG1363 or *L. lactis* MG1363 FnBPA + (pValac:*il-10* groups), showed significantly higher IL-10 levels than the control group (healthy mice), probably due to a constant and higher production of IL-10 by the eukaryotic cells of these animals. It is also important to note that the expression of FnBPA by the bacterial strain carrying the pValac:*il-10* plasmid was associated to significantly higher IL-10 levels in the intestinal tissues compared to the DSS group, where this cytokine was produced as normal immune response against inflammation. However, these levels of IL-10 were not significantly different from those obtained in the *L. lactis* MG1363 pValac:*il-10* group, nor were associated with significant differences in the levels of the other cytokines tested in both pValac:*il-10* groups. The lack of significant difference between both strains carrying the pValac:*il-10* plasmid could be explained by the fact that both cellular entry mechanisms used by these bacteria (enterocytes invasion by the invasive strain and capture by macrophages by the non-invasive strains) lead to efficient and similar IL-10 production; however, by different cell types.

Pro-inflammatory cytokines, including TNF-α, IL-6 and IL-17, can be produced by T and B lymphocytes, macrophages and/or neutrophils, which are massively infiltrated in inflammatory lesions in mice with DSS-induced acute colitis. This DSS colitis switches from a T_h1-T_h17-mediated acute inflammation with increased levels of TNF-α, IL-6 and IL-17, to a predominant T_h2-mediated inflammatory response that shows a decrease in TNF-α, IL-6 and IL-17 while increasing levels of anti-inflammatory cytokines IL-4 and IL-10 [24]. In our experimental acute model, at the time of sacrifice, the results only showed higher IL-6 levels in the DSS group and in the group of mice treated with DSS that received the *L. lactis* MG1363 strain, when compared to the control group (healthy mice). Significant decrease of this cytokine was observed in the intestinal tissues from mice that received the pValac:*il-10* plasmid, compared to the DSS group. These results were associated to the increased IL-10 levels obtained in those animals and confirm the anti-inflammatory capacity of both strains carrying the pValac:*il-10* plasmid at modulating the gut immune response.

Conclusions

The results obtained in the present work confirm and strengthen our previous results that *L. lactis* MG1363 FnBPA + pValac:*il-10* shows to be a good candidate to maintain an anti-inflammatory status in the GIT and diminishing intestinal inflammation. Moreover, we here also showed that not only delivery of the pValac:*il-10* plasmid by the invasive strain *L. lactis* MG1363 FnBPA+, but also by the non-invasive *L. lactis* MG1363 strain, was effective at diminishing intestinal inflammation, showing that this strategy presents potential for therapeutic intervention of IBD. FnBPA expression was not related to more anti-inflammatory capacity; however, it was associated with significantly higher IL-10 levels in the intestinal tissues when compared to animals from the DSS group. Moreover, the highest levels of sIgA in the intestinal fluid where observed in the animals from the *L. lactis* MG1363 FnBPA + pValac:*il-10* group, showing better immunomodulatory effect by the invasive strain.

Competing interests

The authors declare that they have no competing interests.

Authors' contributions

MZT designed the study, carried out the experiments and wrote the paper. SC and ACGS participated with the realization of the experiments and contributed to the scientific discussion. DCCM carried out all histological analyses and contributed with the scientific discussion. SL, AMLB, VA and JGLB helped to draft the manuscript and gave final approval for publication. JMC and AM conceived the study, participated in its design and coordination, helped to draft the manuscript and gave final approval for publication. All authors read and approved the final manuscript.

Authors' information

Jean G LeBlanc and Anderson Miyoshi Share credit in this work for senior authorship.

Acknowledgements

This work was supported by: (i) Conselho Nacional de Desenvolvimento Científico e Tecnológico (CNPq), (ii) Centro Brasileiro-Argentino de Biotecnologia (CBAB), (iii) Coordenação de Aperfeiçoamento de Pessoal de Nível Superior (CAPES), (iv) Consejo Nacional de Investigaciones Científicas y Técnicas (CONICET), (v) Agencia Nacional de Promoción Científica y Tecnológica (ANPCyT) and (vi) Consejo de Investigaciones de la Universidad Nacional de Tucumán (CIUNT).

Author details

[1]Department of General Biology, Institute of Biological Sciences, Federal University of Minas Gerais, Belo Horizonte, Brazil. [2]Reference centre for Lactobacilli (CERELA-CONICET), San Miguel de Tucumán, Argentina. [3]Department of Biochemistry and Immunology, Institute of Biological Sciences, Federal University of Minas Gerais, Belo Horizonte, Brazil. [4]Department of Morphology, Institute of Biological Sciences, Federal University of Minas Gerais, Belo Horizonte, Brazil. [5]Ezequiel Dias Foundation (FUNED), Belo Horizonte, Brazil. [6]INRA, UMR1319 Micalis, Domaine de Vilvert, F-78350 Jouy-en-Josas, France.

References

1. Fiocchi C: **Inflammatory bowel disease: etiology and pathogenesis.** *Gastroenterology* 1998, **115**:182–205.
2. Lakatos L, Lakatos PL: **Changes in the epidemiology of inflammatory bowel diseases.** *Orv Hetil* 2007, **148**(5):223–228.
3. Asadullah K, Sterry W, Volk HD: **Interleukin-10 therapy—review of a new approach.** *Pharmacol Rev* 2003, **55**(2):241–269.
4. Mocellin S, Marincola F, Rossi CR, Nitti D, Lise M: **The multifaceted relationship between IL-10 and adaptive immunity: putting together the pieces of a puzzle.** *Cytokine Growth Factor Rev* 2004, **15**(1):61–76.
5. de Moreno de LeBlanc A, del Carmen S, Zurita-Turk M, Santos Rochat C, van de Guchte M, Azevedo V, Miyoshi A, LeBlanc JG: **Importance of IL-10 modulation by probiotic microorganisms in gastrointestinal inflammatory diseases.** *ISRN Gastroenterol* 2011, **1**(1):1–10.
6. Van Deventer SJH, Elson CO, Fedorak R: **Multiple doses of intravenous interleukin 10 in steroid-refractory crohn's disease.** *Gastroenterology* 1997, **113**:383–389.
7. Colombel FF, Rutgeerts P, Malchow H, Jacyna M, Nielsen OH, Rask-Madsen J, Van Deventer S, Ferguson A, Desreumaux P, Forbes A, Geboes K, Melani L, Cohard M: **Interleukin 10 (Tenovil) in the prevention of postoperative recurrence of Crohn's disease.** *Gut* 2001, **49**:42–46.
8. Schreiber S, Fedorak RN, Nielsen OH: **Safety and efficacy of recombinant human interleukin 10 in chronic active Crohn's disease. Crohn's disease IL-10 cooperative study group.** *Gastroenterology* 2000, **119**:1461–1472.
9. Van Montfrans C, Van de Ende A, Fedorak RN, Gangl A, Elson CO, Rutgeerts P, Schreiber S, Wild G, Hanauer S, Grint P, Van Deventer SJH: **Anti- and proinflammatory effects of interleukin 10 in mild to moderate Crohn's disease.** *Gastroenterology* 1999, **116**:A777.
10. Steidler L, Vandenbroucker K: **Genetically modified Lactococcus lactis: novel tools for drug delivery.** *Int J Dairy Technol* 2006, **59**(2):140–146.
11. Li MC, He SH: **IL-10 and its related cytokins for treatment of inflammatory bowel disease.** *World J Gastroenterol* 2004, **10**(5):620–625.
12. Steidler L, Neirynck S, Huyghebaert N, Snoeck V, Vermeire A, Goddeeris B, Cox E, Remon JP, Remaut E: **Biological containment of genetically modified Lactococcus lactis for intestinal delivery of human interleukin 10.** *Nat Biotechnol* 2003, **21**:785–789.
13. Braat H, Rottiers P, Hommes DW: **A phase I Trial with transgenic bacteria expressing interleukin-10 in Crohn's disease.** *Clin Gastroenterol Hepatol* 2006, **4**(6):754–759.
14. Guimarães V, Innocentin S, Chatel JM, Lefevre F, Langella P, Azevedo V, Miyoshi A: **A new plasmid vector for DNA delivery using lactococci.** *Genet Vaccines Ther* 2009, **10**(7):4.
15. Innocentin S, Guimarães V, Miyoshi A, Azevedo V, Langella P, Chatel JM, Lefevre F: **Lactococcus lactis expressing either Staphylococcus aureus fibronectin-binding protein A or Listeria monocytogenes internalin A can efficiently internalize and deliver DNA in human epithelial cells.** *Appl Environ Microbiol* 2009, **75**:4870–4878.
16. Pontes D, Innocentin S, del Carmen S, Franco Almeida J, LeBlanc JG, de Moreno de LeBlanc A, Blugeon S, Cherbuy C, Lefèvre F, Azevedo V, Miyoshi A, Langella P, Chatel JM: **Production of fibronectin binding protein A at the surface of Lactococcus lactis increases plasmid transfer in vitro and in vivo.** *PLoS One* 2009, **24**:1–9.
17. del Carmen S, Zurita-Turk M, Lima A, Coelho dos Santos JS, Leclercq SY, Chatel JM, Azevedo V, De Moreno de LeBlanc A, Miyoshi A, LeBlanc JG: **A novel interleukin-10 DNA mucosal delivery system attenuates intestinal inflammation in a mouse model.** *EurJ Inflamm* 2013, **11**(3):641–654.
18. Que YA, Francois P, Haefliger JA, Entenza JM, Vaudaux P, Moreillon P: **Reassessing the role of Staphylococcus aureus clumping factor and fibronectin-binding protein by expression in Lactococcus lactis.** *Infect Immunol* 2001, **69**:6296–6302.

19. Sambrook J, Fritsch EF, Maniatis T: *Molecular cloning: a laboratory manual.* Cold Spring Harbor: Cold Spring Harbor Press; 1989.
20. Langella P, Le Loir Y, Ehrlich SD, Gruss A: **Efficient plasmid mobilization by pIP501 in Lactococcus lactis subsp. lactis.** *J Bacteriol* 1993, **175**:5806–5813.
21. Melgar S, Bjursell M, Gerdin AK, Svensson L, Michaëlson E, Bohlooly-Y M: **Mice with experimental colitis show an altered metabolism with decreased metabolic rate.** *Am J Physiol Gastrointest Liver Physiol* 2007, **292**(1):G165–G172.
22. McCafferty DM, Sihota E, Muscara M, Wallace JL, Keith A, Sharkey KA, Kubes P: **Spontaneously developing chronic colitis in IL-10/iNOS double-deficient mice.** *Am J Physiol Gastrointest Liver Physiol* 2000, **279**(1):G90–G99.
23. Maron R, Hancock WW, Slavin A, Hattori M, Kuchroo V, Weiner HL: **Genetic susceptibility or resistance to autoimmune encephalomyelitis in MHC congenic mice is associated with differential production of pro- and anti-inflammatory cytokines.** *Int Immunol* 1999, **11**(9):1573–1580.
24. Alex P, Zachos NC, Nguyen T, Gonzales L, Chen T-E, Conklin LS, Centola M, Li X: **Distinct cytokine patterns identified from multiplex profiles of murine dss and tnbs-induced colitis.** *Inflamm Bowel Dis* 2009, **15**(3):341–352.
25. Kühn R, Lohler J, Rennick D, Rajewsky K, Muller W: **Interleukin-10-deficient mice develop chronic enterocolitis.** *Cell* 1993, **75**:263–274.
26. Bhavsar MD, Amiji MM: **Oral IL-10 gene delivery in a microsphere-based formulation for local transfection and therapeutic efficacy in inflammatory bowel disease.** *Gene Ther* 2008, **15**:1200–1209.
27. Yao J, Wang J-Y, Lai M-G, Li Y-X, Zhu H-M, Shi R-Y, Mo J, Xun A-Y, Jia C-H, Feng J-L, Wang L-S, Zeng W-S, Liu L: **Treatment of mice with dextran sulfate sodium-induced colitis with human interleukin 10 secreted by transformed bifidobacterium longum.** *Mol Pharm* 2011, **8**:488–497.
28. Pontes DS, de Azevedo MS, Chatel JM, Langella P, Azevedo V, Miyoshi A: **Lactococcus lactis as a live vector: heterologous protein production and DNA delivery systems.** *Protein Expr Purif* 2011, **79**(2):165–175.
29. Okayasu I, Hatakeyama S, Yamada M, Ohkusa T, Inagaki Y, Nakaya R: **A novel method in the induction of reliable experimental acute and chronic ulcerative colitis in mice.** *Gastroenterology* 1990, **98**(3):694–702.
30. Hamilton MJ, Sinnamon MJ, Lyng GD, Glickman JN, Wang X, Xing W, Krilis SA, Blumberg RS, Adahi R, Lee DM, Stevens RL: **Essential role for mast cell tryptase in acute experimental colitis.** *Proc Natl Acad Sci* 2011, **108**(1):290–295.
31. Macpherson AJ, Harris NL: **Interactions between commensal intestinal bacteria and the immune system.** *Nat Rev Immunol* 2004, **4**(6):478–485.

Cloning and characterization of a new β-Glucosidase from a metagenomic library of Rumen of cattle feeding with *Miscanthus sinensis*

Yadan Li[1,2], Ning Liu[2,3], Hui Yang[1,2], Fei Zhao[1,2], Ye Yu[1,2,4], Yun Tian[1,2*] and Xiangyang Lu[1,2*]

Abstract

Background: The study on the second generation bio-fuel is a hot area of current research of renewable energy. Among series of key points in this area, the role of β-glucosidase in the degradation of intermediate gluco-oligosaccharides limits the rate of the complete saccharification of lignocellulose.

Results: In this study, a new β-glucosidase gene, *unglu135B12*, which was isolated from a metagenomic library of rumen of cattle feeding with *Miscanthus sinensis* by the function-based screening, encodes a 779 amino acid polypeptide that contains a catalytic domain belonging to glycoside hydrolase family 3 (GH3). It was recombinantly expressed, purified and biochemically characterized. The recombinant β-glucosidase, unglu135B12, displayed optimum enzymatic activity at pH 5.0 at 38°C, and showed the highest specific activity of 2.5×10^3 U/mg under this optimal condition to p-nitrophenyl-β-D-glucopyranoside (pNPG), and its *Km* and *Vmax* values were 0.309 mmol/L and 7.292 μmol/min, respectively. In addition, the presence of Ca^{2+}, K^+, Na^+ slightly improved β-glucosidase activity of unglu135B12 by about 5%, while about 10 ~ 85% loss of β-glucosidase activity was induced by addition of Mn^{2+}, Fe^{3+}, Zn^{2+}, Cu^{2+}. Interestingly, unglu135B12 was activated by glucose at the concentration lower than 40 mM.

Conclusions: Our findings indicate that unglu135B12 is a new β-glucosidase derived from rumen of cattle, and it might be a potent candidate for saccharification of lignocellulose in industrial application.

Keywords: β-glucosidase, Rumen, *Miscanthus sinensis*, Metagenomic library

Background

The study on the second generation bio-fuel, which can be constantly derived from the waste of agriculture and forestry and the widespread distribution of non-food plants, is a hot area of current research of renewable energy [1,2]. Among non-food plants, as the raw material of bio-fuel, *Miscanthus sinensis* has been focused on in recent years as an ideal bio-source plant. This fiber-rich plant is widely distributed around South East Asia, and it's also a common specie in China [3]. However, the complexity of lignocellulose of its cell wall makes the plant very difficult to be degraded [4]. Thus, it is urgent to explore new strong functional enzymes to solve the key problem in degradation of this plant, and to discover an economical way for the production of renewable bio-fuel from *Miscanthus sinensis*.

The conversion of the most abundant part of fiber, cellulose, into bio-fuel will depend on the synergistical hydrolysis of enzymes at least including three main cellulases: endo-β-1,4-glucanase (EC3.2.1.4), cellobiohydrolase (EC3.2.1.91) and β-glucosidase (EC3.2.1.21) [5]. β-glucosidase is widely distributed in nature [6,7], which catalyses the cleavage of the glycosidic bonds existing in disaccharides, oligosaccharides and alkyl- or aryl- β-glucosides [8]. Although β-glucosidase does not act on cellulose directly, it is of great importance in cellulose deconstruction by eliminating cellobiose inhibition on endoglucanases and exoglucanases, allowing the cellulolytic enzymes to function more efficiently [5,9]. Therefore, the activity of β-glucosidase is considered as the rate-limiting factor in cellulose degradation.

* Correspondence: tianyun79616@163.com; xiangyangcn@163.com
[1]College of Bioscience and Biotechnology, Hunan Agricultural University, Changsha 410128, China
[2]Hunan Agricultural Bioengineering Research Institute, Changsha 410128, China
Full list of author information is available at the end of the article

The low efficiency of β-glucosidases discovered in previous studies may be responsible for the high cost of bio-fuel production from lignocellulosic biomass [10], identification of new efficient β-glucosidases is thus important and necessary. Ruminants carry out a foregut fermentation that digests plant polysaccharides materials by a complicated and efficient microbial process [11]. Therefore, the habitat in rumen represents a rich hotspot for exploring diverse functional enzymes which could be used in lignocellulose degradation. Considering the fact that more than 85% of ruminal microbes remains uncultured, metagenomic technology, avoiding from pure culture, has drawn more attention during the study of finding new cellulolytic enzymes from this special environment [12].

Metagenomic strategy had been successfully employed in identification of new biocatalytic enzymes encoding genes from the uncultured component of microbial communities from a variety of environmental samples [13,14]. In general, metagenomic method involves the extraction of metagenomic DNA from uncultured samples, the construction of the metagenomic libraries and the screening for the aimed genes [13]. This culture-independent method is able to avoid the limitation of the pure cultivation method, and many new β-glucosidases encoding genes have been obtained by means of metagenomics [15-17].

It had been demonstrated that the ratio of cellolytic bacteria was increased in large intenstine of *Miscanthus sinensis* feeding cattle [18]. Meanwhile, our unpublished result showed that the ratios of cellolytic bacteria and fungi in rumen were also increased due to the sole forage of *Miscanthus sinensis.* Therefore, exploring new cellolytic enzymes encoding gene from *Miscanthus sinensis* feeding cattle exhibit more potentiality. The aim of the present study was to identify new gene encoding β-glucosidase from total DNA isolated from ruminal microbes of Xiangxi yellow cattle fed on *Miscanthus sinensis* using the metagenomic techniques. Through function-based screening of the metagenomic library, one new β-glucosidase gene was obtained. The recombinant enzyme was overexpressed in Escherichia coli BL21(DE3) and investigated for the enzymatic properties.

Results

Construction and evaluation of the matagenomic library from microbes in Xiangxi yellow cattle rumen

To identify new β-glucosidase from metagenomic library, the strategy was described in Figure 1A. Rumen content samples were collected from Xiangxi yellow cattle in Hunan province. A fosmid library was constructed with metagenomic DNA isolated from microbes of rumen samples, containing 20,160 clones. The average insert

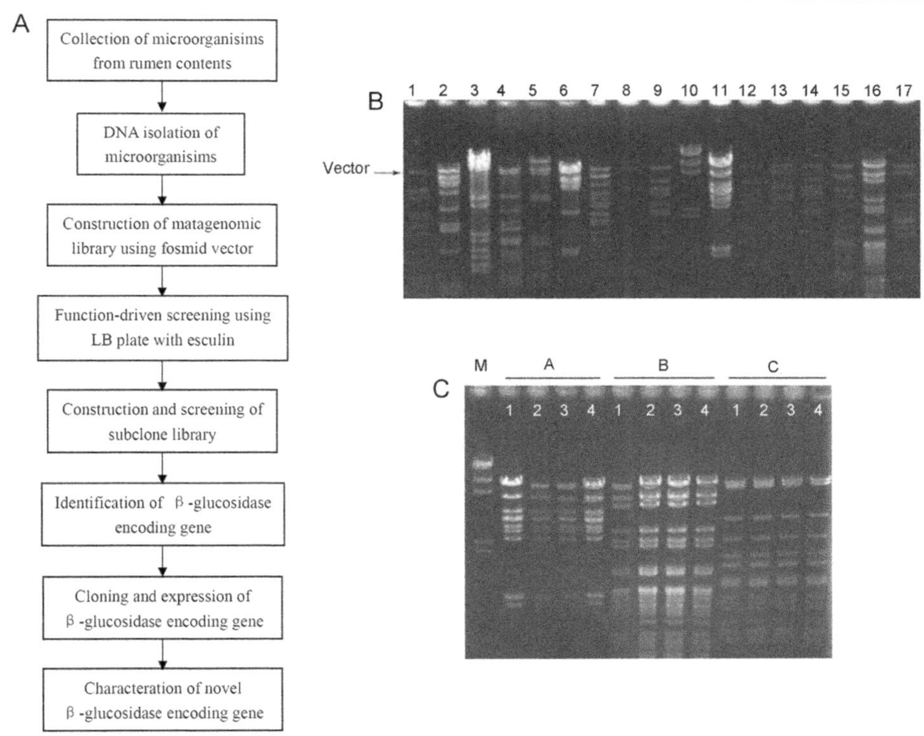

Figure 1 Identification of a positive β-glucosidase functional clone 135B12 from metagenomic library of *Miscanthus sinensis* fed cattle's ruminal inhabitants. A. Strategy of obtaining of new β-glucosidase from metagenomic library. **B**. The volume of metagenomic library. **C**. Stability of exogenous gene in the library after subcultures (**A**, **B**, **C**: digestive pattern of fosmid DNA of three indipendent clones from the library by *EcoR* I and *Hind* III; 1-4: mean four times of subcultures).

size of clones was about 30 kb, and the full size of the library was estimated to be 600 Mb (Figure 1B). The stability of this metagenomic library was reliable according to the result of restriction endonuclease analysis that the insert fragments of three clones which were chosen randomly from the library were in accord with the original one after four times of subcultures (Figure 1C). Furthermore, the reliability of the present metagenomic library was also proved by the results of random sequencing of 92 clones' insert fragments (Additional file 1: Table S1). One singlet of all had ≥97% similarity with *Bacteroides thetaiotaomicron* and two singlets had ≥97% similarity with uncultured *Bacteria*, which demonstrated the present metagenomic library was valid, while, other eighty-nine singlets didn't have high homology with known sequences which perfectly confirmed that the library possessed great novelty (Additional file 1: Table S1).

Annotation of β-glucosidase gene unglu135B12

All of the clones were transferred to the screening plates for β-glucosidase activity. A clone expressing relatively strong β-glucosidase activity was obtained, named 135B12. Due to the average length of insert sequence of 135B12 was about 32 kb, sequence of the new gene encoding β-glucosidase was not fit for direct location. Thus subclone library was further constructed using the pUC19 vector to yield smaller fragments. After the screening by the same activity-based method as mentioned above, the positive subclone was sequenced. The length of predicted ORF was 2340 bp encoding a protein of 779 amino acids (Figure 2A). The new sequence had been submitted to NCBI, and the genebank access number was JX962691. The sequence of 2340 bp was shortly named as *Unglu135B12*, and the encoded protein was Unglu135B12 accordingly.

Analyses of predicted β-glucosidase sequence and domain structures

The theoretical molecular mass of Unglu135B12 is 84.0 kDa, and the value of pI is 6.0 according to the analysis of DNAStar. Sequence analysis with SMART indicated that the putative protein has one glycosyl hydrolase family 3 domain (from 71 to 290 amino acid) and one glycosyl hydrolase family 3 C terminal domain (from 356 to 642 amino acid) (Figure 2B). Additionally, conserved domains search on NCBI also revealed that Unglu135B12 comprised these two conserved domains (figure didn't show in this paper). Thus Unglu135B12 was probably a GH3 β-glucosidase. Information of phylogenetic tree showed it exhibited from 57% to 62% sequence identities to homologous proteins derived from bacteria of Bacteroidetes phylum (e.g. *Prevotella bergensis*) (Figure 2C).

Expression and characterization of Unglu135B12

In order to characterize Unglu135B12, *unglu135B12* was cloned into pET28a (+) and heterologously expressed in BL21 and purified. The western bolt detection showed the positions of expressed protein and purified protein, which were in accordance with the predicted MW value (Figure 3A). Furthermore, the purified Unglu135B12 was tested on the functional screening plate and it was demonstrated that the heterologous protein before/after purification and dialyzed protein were still with the β-glucosidase activity (Figure 3B).

To determine enzymatic characteristics of this purified putative GH3 β-glucosidase, pNPG was used as substrate for testing the activity of β-glucosidase. As shown in the present data, the optimal temperature was 38°C and the optical pH was 5.0 (Figure 4A and C), with the specific activity of 2.5×10^3 U/mg. The relative activity of Unglu135B12 still reserved 60% when placed it under pH5.0-6.0 at 4°C for 24 h. The stability features of Unglu135B12 were not very satisfied, as it was stable only in a narrow range of a battery of temperatures and pHs (Figure 4B and D).

The addition of Ca^{2+}, K^+, Na^+ slightly enhanced β-glucosidase activity of Unglu135B12 by about 5% and about 10-85% loss of β-glucosidase activity was induced by addition of Mn^{2+}, Fe^{3+}, Zn^{2+}, Cu^{2+}. Mg^{2+} had little effect on activity of Unglu135B12 (Figure 4E).

The activity of Unglu135B12 towards pNPG was enhanced by glucose at concentrations lower than 40 mM (Figure 4F). In the presence of 40 mM glucose, the activity of Unglu135B12 increased to a maximum value with 9.6% more than that of the control without glucose. With glucose further increasing, the enzyme activity was gradually inhibited.

The reaction kinetic parameters of the purified enzyme were measured from double reciprocal Lineweaver-Burk plots. This putative β-glucosidase had a K_m value of 0.309 mmol/L, a $Vmax$ value 7.292 μmol/min.

Analysis of 3D structure and substrate docking

Based on structural analysis, Unglu135B12 was composed by three domains: an α/β barrel, an α/β sandwich and a fibronectin-like domain at carboxyl-terminal (Figure 5A). It was supposed that α/β barrel and α/β sandwich were able to contribute each one of the two catalytic residues, while, the function of fibronectin-like domain was still unclearly and this domain was not present in all enzyme of GH3 family [19]. Molecular docking was conducted with the substrate pNPG by the tool of Discovery Studio LibDock based on this lowest-energy model. In the light of the need of clearer description of the relative location between enzyme and substrate, then it was amplified appropriately as Figure 5B. Moreover, Figure 5C indicated that some key residues existed around pNPG in the

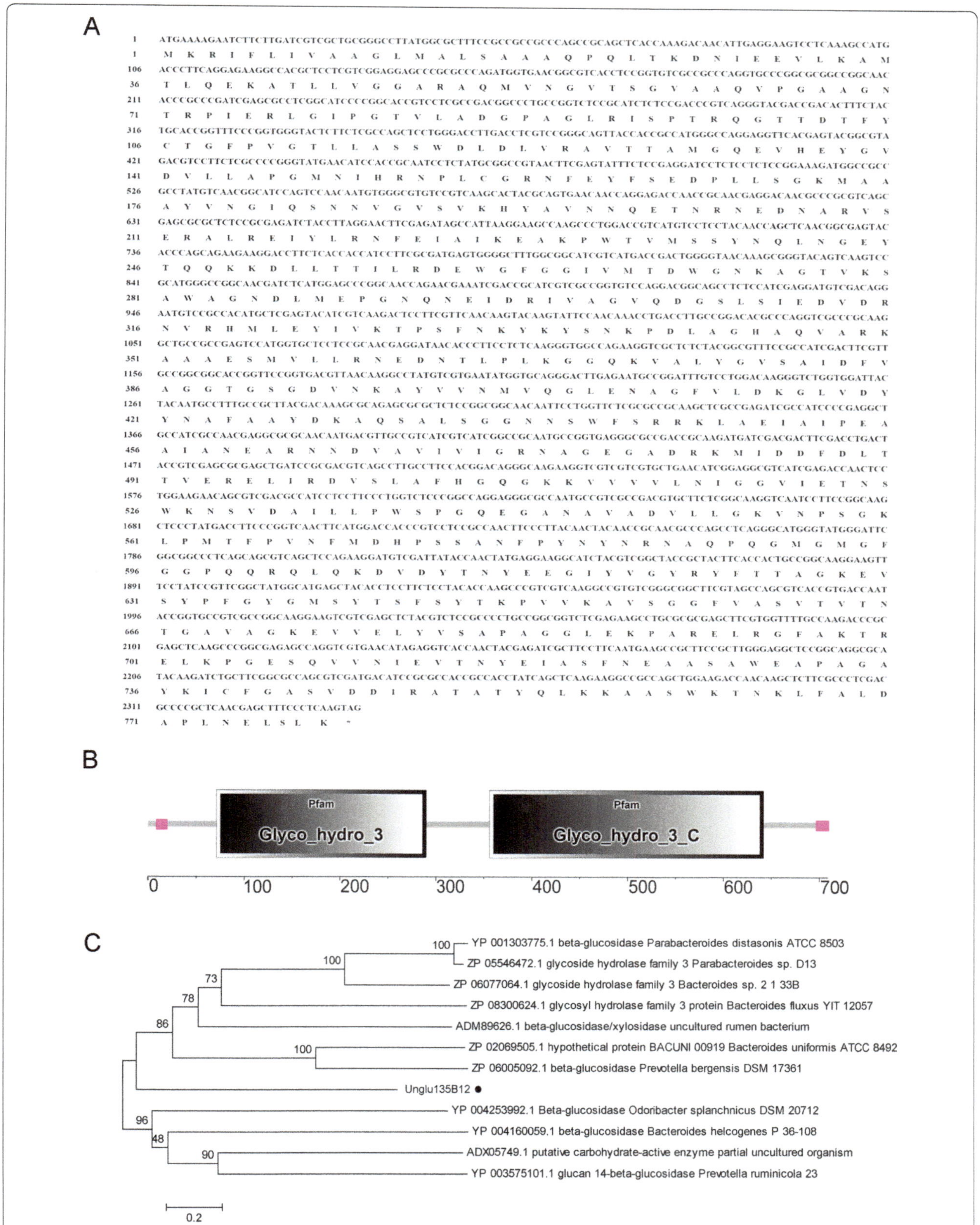

Figure 2 Characterization of amino acid sequence of Unglu135B12. A. amino acid sequence of Unglu135B12. **B.** Predicted modular architecture of Unglu135B12. **C.** Phylogenetic relationship of Unglu135B12 with related proteins. The scale bar corresponds to a genetic distance of 0.2 substitutions.

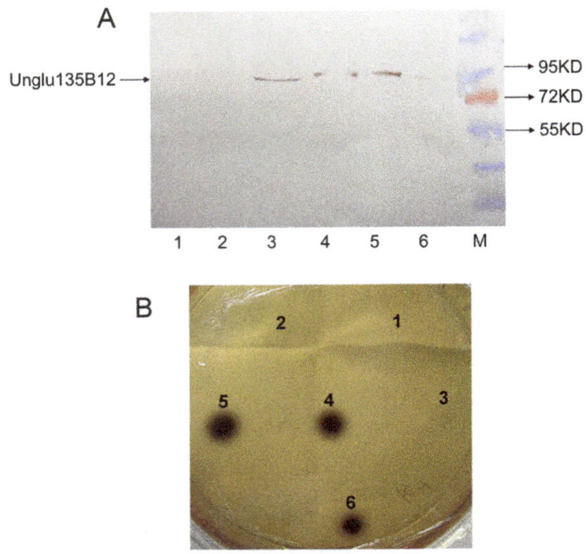

Figure 3 Characterization of Unglu135B12. A. Characterization of Unglu135B12 by western blot. *Lane* 1. Crude extract of BL21 (DE3) carrying pET28a (+) vector; *Lane* 2: IPTG-uninduced crude extract of BL21 (DE3) carrying pET-Unglu135B12; *Lane* 3: IPTG-induced crude extract of BL21 (DE3) carrying pET-Unglu135B12; *Lane* 4: Eluted Unglu135B12 protein after purification by Ni-NTA His · Bind Resins; *Lane* 5: Dialyzed Unglu135B12 protein; *Lane* 6: Flow through; *Lane* M: protein Marker. **B**. Functional verification of purified Unglu135B12. 1. Crude extract of BL21 (DE3); 2: Crude extract of BL21 (DE3) carrying pET28a (+) vector; 3: IPTG-uninduced crude extract of BL21 (DE3) carrying pET-Unglu135B12; 4: IPTG-induced crude extract of BL21 (DE3) carrying pET-Unglu135B12; 5: Eluted Unglu135B12 protein after purification by Ni-NTA His · Bind Resins; 6: Dialyzed Unglu135B12 protein.

model, which could probably form hydrogen bonds and phosphoanhydride bonds and played irreplaceable roles in the catalytic pocket (Figure 5C).

Discussion

In this present work, we used *Miscanthus sinensis*, a promising cellulosic energy crop, as the plant substrate for feeding experimental animal. To explore new β-glucosidase gene from the microbial communities of these *Miscanthus sinensis* fed cattle, a metagenomic library containing about 20,160 clones was constructed, which was screened by function-based method. Consequently, the gene *Unglu135B12* was identified as the potential candidate gene. The purified protein Unglu135B12 showed high specific activity to the substrate pNPG, and its activity could be influenced by different irons. Besides, the inhibition of glucose on Unglu135B12 began to appear only when the concentration of glucose was more than 40 mM. These data provided strong evidence to support the feasibility of searching functional enzyme for mono-/polysaccharides degradation from rumen habitat. Rumen habitats harbor a vast microbial diversity, including various anaerobic bacteria, archaea and fungi that are different from those found in water, soil or other environments. This special microbial community is closely related to the fiber-rich diet, due to the synergistic actions of cellulase, hemicellulase, ligninolytic enzymes and β-glucosidase derived from ruminal microbes, and fullfills the degradation of complex structure of plant in animal's diet [4,20]. Therefore, the metagenomic library in our work distinguished itself as an ideal extensive resource for searching cellulolytic enzymes or other enzymes related to this degradation process.

On the basis of phylogenetic analysis, Unglu135B12 was mostly closed to one beta-glucosidase from *Prevotella bergensis*, belonging to Bacteroidetes (Figure 2C). This indication was in line with the results of our unpublished studies, in which we comparatively analyzed the phylogenetic relationship of ruminal bacteria between *Miscanthus sinensis* fed cattle and mixed forage fed cattle and found that the proportion of Bacteroidetes was increased significantly induced by the sole diet of *Miscanthus sinensis*. So it emphasized again the relevance between feeding of *Miscanthus sinensis* and rise trend of Bacteroidetes in rumen. With the function of glycosyl hydrolase, Bacteroidetes play key roles in the decomposition of carbohydrate polymers and proteins in digestive system of mammal [21,22]. In the light of this point, it's worthy to further detect new glycosyl hydrolase from the present microbial metagenomics library.

In addition, it was said that microbial genomes often contain a substantial number of glycosyl hydrolase (GH) genes, many of which could be induced by different carbon sources. The evidence from conserved domain analysis strongly indicated that Unglu135B12 was a member of glycosyl hydrolase family 3 (GH3). GH3 is one of the largest GH families with more than 4800 sequences in the CAZy database (http://www.cazy.org) and the relative abundance and distribution of GH3 in the metagenome from the bovine rumen microbiome is 18.4% [23].

After the western blot analysis using the specific anti-his tag antibody, the molecular mass of the rumen microbial protein was recognized as about 85 kD, in consistent with the predicted value, which suggested that this protein was probably a monomeric protein. Moreover, the purified β-glucosidase Unglu135B12 from the heterologous expression system were identified that it efficiently hydrolyzed pNPG within a narrow pH range of 5.0-6.0, and had optimal activity at 38°C, which indicated the native Unglu135B12 protein might play an active role under *in vivo* conditions in the cattle rumen. However, these features of the recombinant protein might hinder its industrial application. Given the limitation of Unglu135B12 in industrial production, it's worthy to improve its stability of pH and temperature by

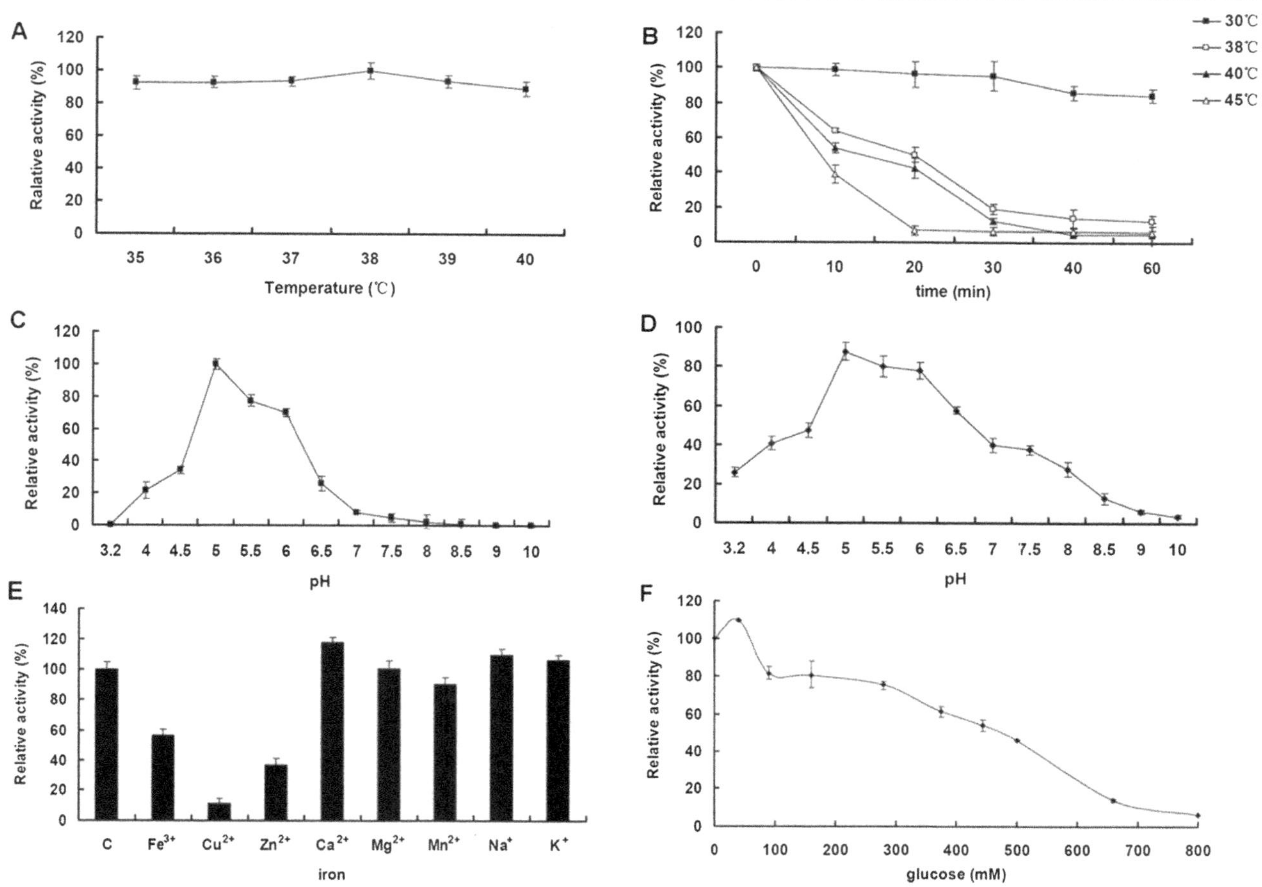

Figure 4 Effect of pH and temperature on the enzyme activity and stability of unglu135B12. A. Effect of temperature on the activity of unglu135B12. **B**. Effect of pH on the activity of unglu135B12. **C**. Temperature stability of unglu135B12. **D**. pH stability of unglu135B12. **E**. Effect of metal ions on activity of unglu135B12. **F**. Influence of glucose on enzyme activity with pNPG as the substrate. Enzymatic reactions without glucose were used as controls. All measurements were analysed in triplicate and error bars were indicated.

biotechnological methods such as site-directed mutagenesis, codon randomisation [24] or immobilization techniques [25]. Therefore, with the help of the prediction by homology modeling and molecular docking, site-directed mutagenesis against the key residues could be considered as a new orientation in further research. While, the specific activity of Unglu135B12 against pNPG was measured as 2.5×10^3 U/mg at pH 5.0 when assayed at 38°C. It was considered as a relatively high value compared with other β-glucosidases found in uncultured microbes or pure-isolated microbes [15,26]. Besides, addition of Ca^{2+}, K^+, Na^+ could enhance its activity to 105%. Based on this information, it's hopefully to achieve more enhancement of Unglu135B12's activity. Furthermore, most of the microbial β-glucosidases reported to date are competitively inhibited by glucose, while several β-glucosidases from a few fungi and yeasts show high glucose tolerance [27,28]. Unglu135B12 is activated by glucose at concentrations below 40 mM and remains more than 75% activity with glucose at concentrations of 280 mM. In further study, the tolerance to glucose of unglu135B12 could be probably improved by protein engineering [29].

Overall, the reason why unglu135B12 exhibited these characteristics described above could be attributed to its source. Consistent with other β-glucosidases obtained by metagenome method from herbivore's digestive tract, their optimal pH and temperature were in a narrow range which similar to their natural environment [15,26]. In addition, it's worth mentioning that the activity of unglu135B12 is significantly higher than that of the β-glucosidases from rabbit cecum [26] and yak rumen [15], moreover, the unglu135B12 still showed tolerance to glucose. These points might indicate more potentiality of further directed mutagenesis on unglu135B12.

Conclusions

The more effective β-glucosidase could definitely convert cellobiose into glucose more conveniently and economically, and the combination of this new β-glucosidase and

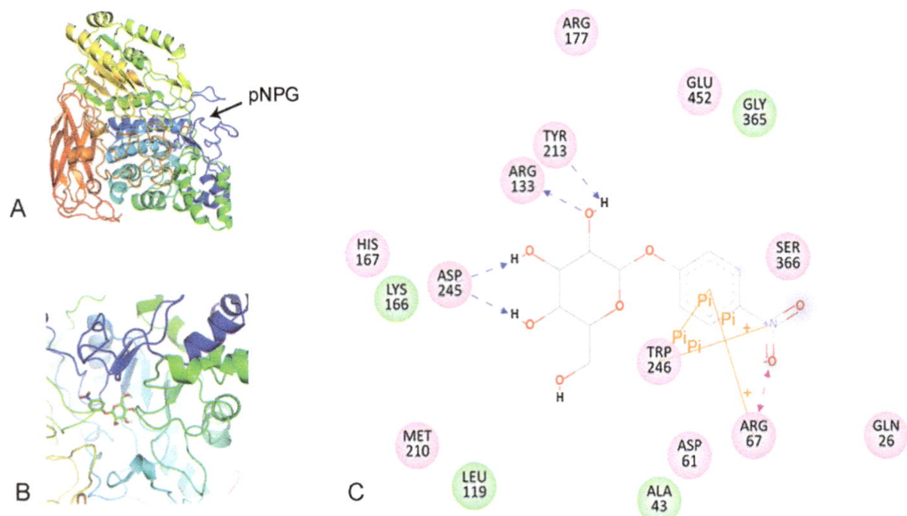

Figure 5 Structural models of β-glucosidases Unglu135B12 characterised in the present work and its substrate docking. A. The pNPG-binding model of Unglu135B12 derived from *Thermotoga neapolitana* β-glucosidase 3B (PDB ID 2X40). The three protein domains: α/β barrel, α/β sandwich and fibronectin-like, were coloured cyan, kelly and red, respectively. **B.** Closeup of the catalytic centre. **C.** The key residues in the catalytic pocket around pNPG.

cellulase or hemicellulase will be beneficial for break the major bottleneck in industrial-scale conversion of cellulosic biomass into biofuels. Furthermore, discovery of other high preformance new monofunctional enzyme or multifuctional enzyme from our library is hopefully believed to provide more possibility for their applications in future industrial production. Besides, heterologously expressed functional enzyme by genetically modifying host organism could also be an effective tool in developing enzyme with desired properties.

Methods

Animal ethics

Care of laboratory animals and animal experimentation were performed in accordance with animal ethics guidelines and approved protocols. All animal experiments were performed according to the Hunan Community Rules of Animal Care with the permission number 38-61 of Hunan Agricultural University Veterinary Services (China).

Metagenomic fosmid library construction and evaluation

Matagenomic DNA was extracted from rumen content of Xiangxi yellow cattle which were fed with sole fresh *Miscanthus sinensis* from 2.5 years old for 18 months. The pooled samples were collected and stored at -80°C until DNA extraction.

Total DNA was extracted from the pooled samples as described before [18]. DNA fragments of about 36 ~ 48 kb were recovered from gel by electroelution for the library construction by using pCC2FOS™ Vector (Epicentre, USA). All of operation was following the manufacturer's instruction. Library clones were duplicated and stored in 96-well plates at -80°C.

Clones of fosmid library were cultured in 96-well plates containing 70 μl lysogeny broth (LB) supplemented with chloramphenicol (22.5 μg/ml) at 37°C overnight. Then LB plates supplemented with esculin hydrate (0.1%), ferric ammonium citrate (0.25%) and chloramphenicol (22.5 μg/ml) were used for screening functional clones with β-glucosidase activity according to the method described by Eberhart et al [30]. Positive clones were detected by the formation of a black halo around the colonies after incubation at 37°C for 20-24 h. The strength of enzyme activity was initially decided by comparing the size of each black halo.

To evaluate the volume and stability of this library, we conducted the analysis of restriction endonuclease map with *EcoR* I and *Hind* III. Moreover, reliability of this library was detected by random sequencing.

Subclone and sequence analysis

Due to the length of insert sequence in each clone of fosmid library was about 40 kb, it was too long to directly sequence the gene of β-glucosidase from positive clone. Therefore, Fosmid DNA of positive clone was extracted and digested by *EcoR* I and *Hind* III. Then DNA fragments around 1 ~ 5 kb length were ligated into the vector pUC19 (Promega, USA), and transformed into *E. coli* DH5α. After screening by the same method mentioned before, a positive subclone was detected and sequenced.

The open reading frames (ORF) of the aimed sequence were analyzed by DNAMAN. The domain architecture and conserved domain was analyzed by the online program SMART (http://smart.embl-heidelberg.de). The phylogenetic tree was constructed with MEGA4.0 software. Reliability of phylogenetic reconstruction was estimated by Boot-strapping values (1,000 replicates).

Gene expression and protein purification

Unglu135B12 was amplified by PCR using the primers 135B12F (5′-CGGAATTC TCCTTTGTCTACCTTTGT CAG-3′) containing a *EcoR* I site (underlined) and 135B12R (5′-ATGCGGCCGCCGAGGTATTTACAAAA CACG-3′) containing a *Not* I site (underlined), then the PCR fragments were digested with restriction endonucleases *EcoR* I and *Not* I and cloned into pET28a (+) digested with the same enzymes, resulting in the recombinant plasmid pET28a/Unglu135B12. For expression of Unglu135B12, pET28a/Unglu135B12 was transformed into *E coli* BL21 (DE3), then was induced with 0.1 mmol/L IPTG at 25°C for 4.5 h. Ni-NTA resin was used to purify the protein, and then dialyzed the protein with 20 mM PBS buffer (pH7.0) overnight at 4°C. The protein before/after purification, flow-through and dialyzed fusion protein were analyzed by SDS-PAGE and western blot.

All protein samples (15 μg) were separated by 8% SDS-PAGE and transferred onto a PVDF membrane (Bio-Rad). Then anti-His Tag (6×) antibody was used to performed the immunoblots. The specifically stained bands were clearly visible by DAB staining.

Enzyme assays and characterization of Unglu135B12

The specific activity measurement was preformed as described by Feng [26], and one unit of β-glucosidase activity was defined as the amount of enzyme releasing 1 μmol p-nitrophenol from p-nitrophenyl-β-glucopyranoside (pNPG). The amount of released p-nitrophenol was measured by the absorbance at 405 nm [31].

The temperature optimum for the freshly purified β-glucosidase activity was measured in a range from 25°C to 65°C by incubating Unglu135B12 at pH 5.0. The effect of pH on enzyme activity was evaluated at pHs ranged from 3.2 to 10 at optimal temperature. The thermostability of Unglu135B12 was evaluated by incubating the recombinant enzyme at 30°C, 40°C, 45°C and optimal temperature at pH 5.5 for 1 h, and then measuring the residual activity. The stability of Unglu135B12 against pH was assayed by measuring the residual activity of Unglu135B12 in the buffer at different pHs after incubation at 4°C for 24 h.

Various cations (3.5 mM) were used to investigate the effects of metal ions on the β-glucosidase. Effect of glucose on the β-glucosidase activity was also evaluated, which was studied by determining the activity of β-glucosidase towards pNPG in presence of dose-dependent glucose (0 mM, 40 mM, 90 mM, 160 mM, 280 mM, 375 mM, 444 mM, 500 mM, 660 mM, 800 mM).

To determine the K_m and *Vmax* values, reactions were acted under optimal condition with pNPG ranging from 0 to 8 mmol l^{-1} according to the double-reciprocal plot method. All measurements were analyzed in triplicate.

Homology modeling and substrate docking

The translation product of aimed sequence *unglu135B12* was further analyzed for protein domains using the Pfam-A database [32]. Model of β-glucosidases unglu135B12 were obtained from the SWISS-MODEL server (http://swissmodel.expasy.org/) using the coordinates of known β-glucosidase structure (PDB ID 2X40: β-glucosidase 3B from *Thermotoga neapolitana*) as the template. The lowest-energy model was chosen for the docking of the substrate of pNPG using Discovery Studio LibDock.

Additional file

Additional file 1: Table S1. The result of random sequencing.

Competing interests

The authors declare that they have no competing interests.

Authors' contributions

LYD, TY and LXY devised the method and designed the study. LYD, LN, YH and ZH carried on the experimental work. LYD and YY analyzed the data. LYD, YY, TY and LXY wrote the manuscript. All authors contributed to the manuscript. All authors read and approved the final manuscript.

Acknowledgments

This research was supported by the International Scientific and Technological Cooperation Project (2010DFA62510), the Program for Changjiang Scholars and Innovative Research Team in University (IRT0963), Hunan Provincial Innovation Foundation for Postgraduate (CX2013B299) and The Construct Program of the Key discipline in Hunan Province.

Author details

[1]College of Bioscience and Biotechnology, Hunan Agricultural University, Changsha 410128, China. [2]Hunan Agricultural Bioengineering Research Institute, Changsha 410128, China. [3]College of Life Science, Hunan Normal University, Changsha 410181, China. [4]Departments of Physiology, Biochemistry, and Molecular Cell Biology, Institute of Medical Science, Shanghai Jiao Tong University School of Medicine, Shanghai 200025, China.

References

1. Sims RE, Mabee W, Saddler JN, Taylor M: **An overview of second generation biofuel technologies.** *Bioresour Technol* 2010, **101:**1570–1580.
2. Yi Z, Ruihong Z: **Overview of biomass pretreatment for cellulosic ethanol production.** *Int J Agric Biol Eng* 2009, **2:**51–69.
3. Shumny VK, Veprev SG, Nechiporenko NN, Goryachkovskaya TN, Slynko NM, Kolchanov NA, Peltek SE: **A new variety of Chinese silver grass (*Miscanthus sinensis anderss.*): a promising source of cellulose containing raw material.** *Russ J Genet Appl Res* 2011, **1:**29–32.
4. Chesson ASC, Dalgarno K, King TP: **Degradtion of ioslated grass mesophyll, epidermis, and fiber cell walls in the rumen and by cellulolytic rumen bacteria in axenic culture.** *J Appl Bacteriol* 1986, **60:**327–336.
5. Lynd LR, Weimer PJ, vanZyl WH, Pretorius IS: **Microbial cellulose utilization: fundamentals and biotechnology.** *Microbiol Mol Biol Rev* 2002, **66:**506–577.
6. Bokkenheuser VD, Shackleton CH, Winter J: **Hydrolysis of dietary flavonoid glycosides by strains of intestinal bacteroides from humans.** *Biochem J* 1987, **248:**953–956.

7. Freer SN: **Kinetic characterization of a beta-glucosidase from a yeast, Candida wickerhamii.** *J Biol Chem* 1993, **268:**9337–9342.
8. Shallom D, Shoham Y: **Microbial hemicellulases.** *Curr Opin Microbiol* 2003, **6:**219–228.
9. Dahai G, Shishir PSC, Tongjun L, Spencer H, Krishne G, Phillip B, Bruce ED, Venkatesh B: **Strategy for identification of novel fungal and bacterial glycosyl hydrolase hybrid mixtures that can efficiently saccharify pretreated lignocellulosic biomass.** *Bioenerg Res* 2010, **3:**67–81.
10. Saha BC, Freer SN, Bothast RJ: **Production, purification, and properties of a thermostable beta-glucosidase from a color variant strain of aureobasidium pullulans.** *Appl Environ Microbiol* 1994, **60:**3774–3780.
11. Brulc JM, Antonopoulos DA, Miller ME, Wilson MK, Yannarell AC, Dinsdale EA, Edwards RE, Frank ED, Emerson JB, Wacklin P, Countinho PM, Henrissat B, Nelson KE, White BA: **Gene-centric metagenomics of the fiber-adherent bovine rumen microbiome reveals forage specific glycoside hydrolases.** *Proc Natl Acad Sci U S A* 2009, **106:**1948–1953.
12. Kristen MD, Martin A, D'haeseleer P, Julian LF, Amitha R, Philip H, Steven WS, Jean SVG, Whendee LS, Blake AS, Terry CH: **Strategies for enhancing the effectiveness of metagenomic-based enzyme discovery in lignocellulolytic microbial communities.** *Bioenerg Res* 2010, **3:**146–158.
13. Banik JJ, Brady SF: **Recent application of metagenomic approaches toward the discovery of antimicrobials and other bioactive small molecules.** *Curr Opin Microbiol* 2010, **13:**603–609.
14. Tuffin M, Anderson D, Heath C, Cowan DA: **Metagenomic gene discovery: how far have we moved into novel sequence space?** *Biotechnol J* 2009, **4:**1671–1683.
15. Bao L, Huang Q, Chang L, Sun Q, Zhou J, Lu H: **Cloning and characterization of two beta-glucosidase/xylosidase enzymes from yak rumen metagenome.** *Appl Biochem Biotechnol* 2012, **166:**72–86.
16. Jiang CJ, Chen G, Huang J, Huang Q, Jin K, Shen PH, Li JF, Wu B: **A novel beta-glucosidase with lipolytic activity from a soil metagenome.** *Folia Microbiol (Praha)* 2011, **56:**563–570.
17. Wang Q, Qian C, Zhang XZ, Liu N, Yan X, Zhou Z: **Characterization of a novel thermostable beta-glucosidase from a metagenomic library of termite gut.** *Enzyme Microb Technol* 2012, **51:**319–324.
18. Li Y, Hu L, Xue G, Liu H, Yang H, Zhu Y, Tian Y, Lu X: **Comparison of bacterial diversity in large intestine of Xiangxi yellow cattle (Bos taurus) associated with different diet: fresh *Miscanthus sinensis* and mixed forage.** *Afr J Microbiol Res* 2012, **6:**5965–5974.
19. Del-Pozo MV, Fernandez-Arrojo L, Gil-Martinez J, Montesinos A, Chernikova TN, Nechitaylo TY, Waliszek A, Tortajada M, Rojas A, Huws SA, Golyshina OV, Newbold CJ, Polaina J, Ferrer M, Golyshin PN: **Microbial beta-glucosidases from cow rumen metagenome enhance the saccharification of lignocellulose in combination with commercial cellulase cocktail.** *Biotechnol Biofuels* 2012, **5:**73.
20. Mrázek JTK, Avguštin G, Kopečný J: **Diet-dependent shifts in ruminal butyrate-producing bacteria.** *Folia Microbiol (Praha)* 2006, **51:**294–298.
21. Mariat D, Firmesse O, Levenez F, Guimaraes V, Sokol H, Dore J, Corthier G, Furet JP: **The firmicutes/bacteroidetes ratio of the human microbiota changes with age.** *BMC Microbiol* 2009, **9:**123.
22. Martens EC, Koropatkin NM, Smith TJ, Gordon JI: **Complex glycan catabolism by the human gut microbiota: the bacteroidetes sus-like paradigm.** *J Biol Chem* 2009, **284:**24673–24677.
23. Ferrer M, Ghazi A, Beloqui A, Vieites JM, Lopez-Cortes N, Marín-Navarro J, Nechitaylo TY, Guazzaroni ME, Polaina J, Waliczek A, Chernikova TN, Reva ON, Golyshina OV, Golyshin PN: **Functional metagenomics unveils a multifunctional glycosyl hydrolase from the family 43 catalysing the breakdown of plant polymers in the calf rumen.** *PLoS One* 2012, **7:**e38134.
24. Eijsink VG, Gaseidnes S, Borchert TV, van den Burg B: **Directed evolution of enzyme stability.** *Biomol Eng* 2005, **22:**21–30.
25. Cesar Mateo JMP, Gloria FL, Jose MG, Roberto FL: **Improvement of enzyme activity, stability and selectivity via immobilization techniques.** *Enzyme Microb Technol* 2007, **40:**1451–1463.
26. Feng Y, Duan CJ, Liu L, Tang JL, Feng JX: **Properties of a metagenome-derived beta-glucosidase from the contents of rabbit cecum.** *Biosci Biotechnol Biochem* 2009, **73:**1470–1473.
27. Belancic A, Gunata Z, Vallier MJ, Agosin E: **Beta-glucosidase from the grape native yeast debaryomyces vanrijiae: purification, characterization, and its effect on monoterpene content of a Muscat grape juice.** *J Agric Food Chem* 2003, **51:**1453–1459.
28. Zanoelo FF, Polizeli ML, Terenzi HF, Jorge JA: **Beta-glucosidase activity from the thermophilic fungus scytalidium thermophilum is stimulated by glucose and xylose.** *FEMS Microbiol Lett* 2004, **240:**137–143.
29. Fang Z, Fang W, Liu J, Hong Y, Peng H, Zhang X, Sun B, Xiao Y: **Cloning and characterization of a beta-glucosidase from marine microbial metagenome with excellent glucose tolerance.** *J Microbiol Biotechnol* 2010, **20:**1351–1358.
30. Eberhart B, Cross DF, Chase LR: **Beta-glucosidase system of neurospora crassa. I. Beta-glucosidase and cellulase activities of mutant and wild-type strains.** *J Bacteriol* 1964, **87:**761–770.
31. Jatinder Kaur BSC, Badhan AK, Ghatora SK, Harvinder SS: **Purification and characterization of β-glucosidase from Melanocarpus sp. MTCC 3922.** *Electron J Biotechnol* 2007, **10:**260–270.
32. Bateman A, Coin L, Durbin R, Finn RD, Hollich V, Sonnhammer EL: **The Pfam protein families database.** *Nucleic Acids Res* 2004, **32:**D138–D141.

6

The characteristics of bacterial nanocellulose gel releasing silk sericin for facial treatment

Pornanong Aramwit* and Nipaporn Bang

Abstract

Background: Recently, naturally derived facial masks with beneficial biological properties have received increasing interest. In this study, silk sericin-releasing bacterial nanocellulose gel was developed to be applied as a bioactive mask for facial treatment.

Results: The silk sericin-releasing bacterial nanocellulose gel produced at a pH of 4.5 had an ultrafine and extremely pure fiber network structure. The mechanical properties and moisture absorption ability of the gel were improved, compared to those of the commercially available paper mask. Silk sericin could be control-released from the gel. A peel test with porcine skin showed that the gel was less adhesive than the commercially available paper mask, which would be removed from the face more easily without pain. The *in vitro* cytotoxicity test showed that the gel was not toxic to L929 mouse fibroblast and HaCaT human keratinocyte cells. Furthermore, when implanted subcutaneously and evaluated according to ISO10993-6 standard, the gel was not irritant to tissue.

Conclusion: The silk sericin-releasing bacterial nanocellulose gel had appropriate physical and biological properties and safety for the facial treatment application.

Keywords: Bacterial nanocellulose, *Gluconacetobacter xylinus*, Coconut water, Silk sericin, Facial mask

Background

Nowadays, various kinds of facial masks are applied for different purposes: for example, revitalizing, healing, and refreshing, and may yield other temporary or long-term benefits to facial skin. Recently, searching for a new naturally derived facial mask with beneficial biological properties has become highly competitive. Cellulose is an organic polysaccharide consisting of a linear chain of several hundred to over ten thousand β $(1 \rightarrow 4)$ linked D-glucose units [1]. Cellulose is frequently derived from plant sources; however, it can also be synthesized by a variety of microorganisms such as bacteria, algae, and fungi [2]. Biosynthetic bacterial cellulose shows appealing characteristics of high purity and unique physicochemical properties such as high tensile strength, crystallinity, and water absorption capacity [3]. *Gluconacetobacter xylinus* (formerly known as *Acetobacter xylinum*) is a species of gram-negative bacteria that can produce high amounts of cellulose. Glucose was used as the common substrate for the bacterial nanocellulose produced by *G. xylinus* [4]. Therefore, searching for cost-effective sources of glucose that can form bacterial nanocellulose with maintained high purity and quality is important.

Coconut is a fruit grown in tropic and subtropic areas that is known for its great versatility, as seen in the many domestic, commercial, and industrial uses of its different parts. Coconut contains a large quantity of coconut water, which is mostly discarded as waste from various agro industries. Coconut water is rich in sucrose (sugar) and nitrogen-containing compounds. Therefore, it could be a cheap substrate for the production of bacterial nanocellulose [5].

In this study, the bacterial nanocellulose was produced at the surface of coconut water and other suitable media by a gram-negative rod-shaped bacterium, *G. xylinus*. The fermentation was carried out in static cultures at 30°C in acidic pH varying from 3 to 5. We investigated the effect of the processing pH on the formation of bacterial nanocellulose gel. As the aim was to be applied as a bioactive facial mask, our bacterial

* Correspondence: aramwit@gmail.com
Bioactive Resources for Innovative Clinical Applications Research Unit and Department of Pharmacy Practice, Faculty of Pharmaceutical Sciences, Chulalongkorn University, PhayaThai Road, Phatumwan, Bangkok 10330, Thailand

nanocellulose gel was adsorbed with silk sericin. Silk sericin is a biocompatible protein derived from the silkworm cocoon. The biological roles of silk sericin such as an antioxidant, bioadhesive, and bioactive agent, as well as a promising implant for tissue-supporting prosthetics for the human body, have been reported [6,7]. Silk sericin could also activate collagen production and significantly increase epithelialization in wounds [8-10]. Herein, the physical properties of the silk sericin-releasing bacterial nanocellulose gel produced at optimal pH were characterized. A test of the release of silk sericin from the gel was carried out. In addition, *in vitro* cytotoxicity and *in vivo* safety tests according to ISO 10993–6 standards were performed, in comparison to the commercially available paper mask (Facial Paper Mask, Hefei Wenqi Industrial and Trading, Co., Ltd., Anhui, Republic of China). It is hypothesized that silk sericin-releasing bacterial nanocellulose gel would be a good candidate as a bioactive facial mask for facial treatment. Silk sericin released from bacterial nanocellulose gel may help to revitalize, heal, or refresh the skin.

Results and discussion

Physical appearance and structure of the bacterial nanocellulose gel

We first studied the influence of pH condition on the formation of bacterial nanocellulose gel because it was supposed that pH may affect the rate of bacterial growth, which was important for the formation of bacterial nanocellulose gel. The bacterial nanocellulose gels prepared from various pH had different physical appearances, as shown in Figure 1. The gels could be set up at a pH of 3.5 to 5 but did not form at a pH of 3. Table 1 summarizes the physical appearance and thickness of the bacterial nanocellulose gels prepared at different pH. At a pH of 3.5 and 4, hard and stable gels with a smooth surface were produced. The gel produced at a pH of 4.5 was also stable with a smooth surface but softer. In contrast, the production at a pH of 5 resulted in a very soft gel. The thickness of bacterial nanocellulose gels seemed to decrease with the increasing pH. The processing pH at 3.5 produced a 0.7 cm gel while the pH at 5 produced a 0.3 cm gel. A pH of 4.5 would be the optimal condition for bacterial growth, as previously reported by Embuscado *et al.* [3]. The

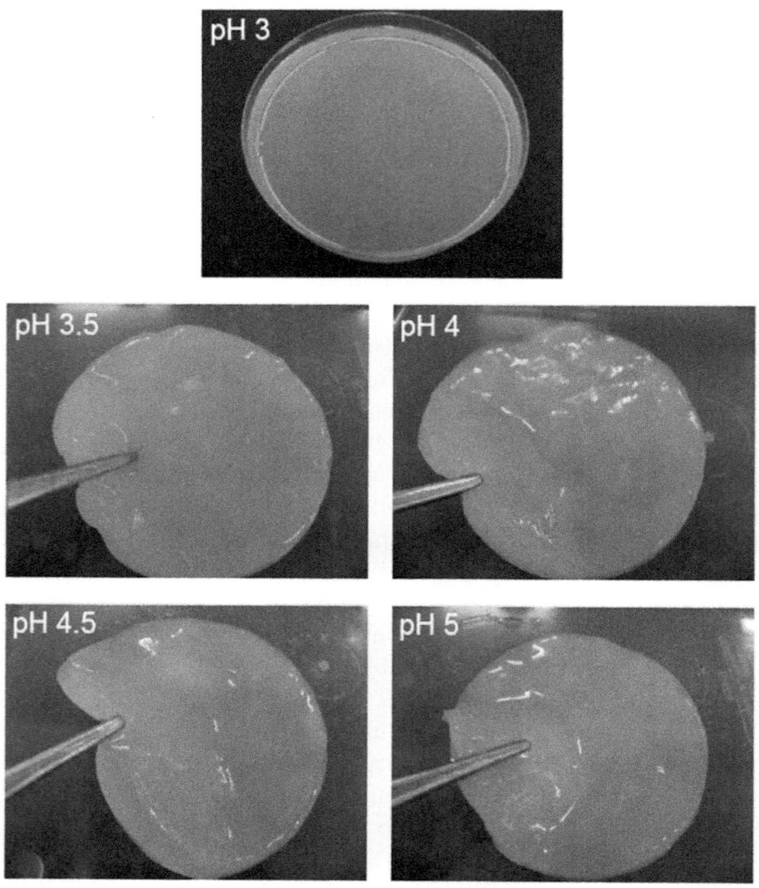

Figure 1 Physical appearance of the bacterial nanocellulose gels prepared at different pH (3, 3.5, 4, 4.5, and 5).

Table 1 Physical appearance and thickness of the bacterial nanocellulose gels prepared at different pH

Processing pH	Physical appearance of the nanocellulose gel	Thickness (cm)
3	Not formed	-
3.5	Hard and stable gel with smooth surface	0.7
4	Hard and stable gel with smooth surface	0.5
4.5	Soft and stable gel with smooth surface	0.6
5	Very soft gel with smooth surface	0.3

0.6 cm soft and stable gel with a smooth surface produced at a pH of 4.5 was selected for further investigation in this study. To improve the biological properties, the bacterial nanocellulose gel prepared at a pH of 4.5 was absorbed with silk sericin solution to obtain the silk sericin-releasing bacterial nanocellulose gel. It is well known that silk sericin has many biological roles such as an antioxidant, bioadhesive, and bioactive activities. It could also activate collagen production and epithelialization, which would be beneficial for facial treatment application [6-10]. The structure of silk sericin-releasing bacterial nanocellulose gel was observed in comparison to the commercially available paper mask, as displayed in Figure 2. We found that the silk sericin-releasing bacterial nanocellulose gel showed a highly fibrous structure with very fine fibers, which corresponded to the report of Vandamme *et al.* [11] that the bacterial cellulose usually has an ultrafine and extremely pure fiber network structure. On the other hand, the cotton fibers of the commercially available paper mask were much larger (~10 μm in diameter).

Mechanical properties of the silk sericin-releasing bacterial nanocellulose gel

To be applied as a facial mask, the silk sericin-releasing bacterial nanocellulose gel should have appropriate mechanical properties. For example, it should be flexible, stretchable, and elongatable in order to adhere completely to the whole of the facial skin. The tensile modulus and percentage of elongation of the dried and wet silk sericin-releasing bacterial nanocellulose gel, compared with those of the commercially available paper mask, are demonstrated in Table 2. Both in the dry and wet conditions, the tensile modulus of the silk sericin-releasing bacterial nanocellulose gel were significantly higher than that of the commercially available paper mask. The percentage of elongation of both samples in the dry state was the same. However, in the wet state, the silk sericin-releasing bacterial nanocellulose gel had a significantly higher elongation percentage than the commercially available paper mask. This could be explained by the fact that our gel is a bacterial nanocellulose, which has more crystalline structure and forms characteristic ribbon-like microfibrils, compared to the plant cellulose of a paper mask [12]. In addition, the silk sericin absorbed on the fibers of our bacterial nanocellulose gel would improve the mechanical properties. Improvement of the mechanical properties by surface medication has been reported elsewhere [13,14].

Moisture absorption capability of the silk sericin-releasing bacterial nanocellulose gel

The facial mask should have good water absorption ability in order to maintain the moisture within its structure during the treatment. Figure 3A shows the moisture absorption capability of the silk sericin-releasing bacterial nanocellulose gel, compared to that of the commercially available paper mask. Initially, the amount of water absorbed on the silk sericin-releasing bacterial nanocellulose gel was increased over time and became stable thereafter. The silk sericin-releasing bacterial nanocellulose gel showed higher moisture absorption capability

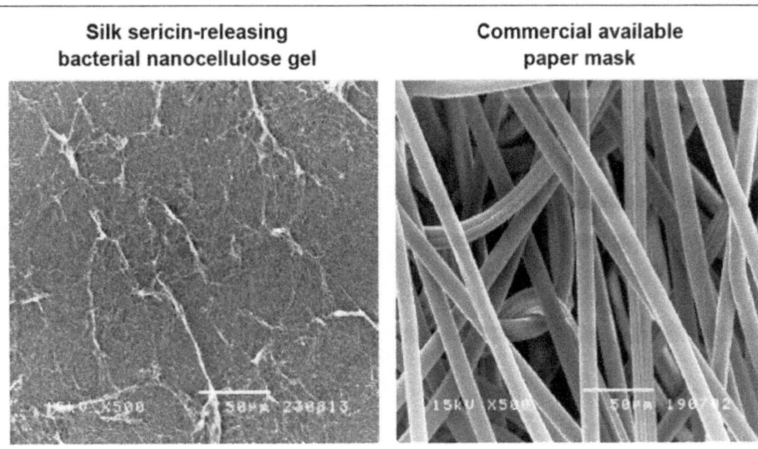

Figure 2 SEM micrographs of the silk sericin-releasing bacterial nanocellulose gel and the commercial available paper mask (scale bar = 50 μm).

Table 2 Mechanical properties of the silk sericin-releasing bacterial nanocellulose gel and the commercial available paper mask at dry and wet conditions

	Tensile modulus (N/mm^2)		Elongation (%)	
	Dry	Wet	Dry	Wet
Silk sericin-releasing bacterial nanocellulose gel	30.84 ± 6.99*	11.23 ± 1.26*	32.89 ± 11.89	67.03 ± 5.72*
Commercial available paper mask	0.86 ± 0.22	0.87 ± 0.05	21.88 ± 8.09	26.9 ± 4.38

*, $p < 0.05$ significant against the value of commercial available paper mask at corresponding condition.

than the commercially available paper mask during the incubation period. This was due to the high hydrophilicity of both bacterial nanocellulose and silk sericin [15,16].

Enzymatic biodegradation rate of the silk sericin-releasing bacterial nanocellulose gel

The biodegradability of a facial mask indicates that it is bioenvironmentally friendly. The percentage of degradation of the silk sericin-releasing bacterial nanocellulose gel, compared to that of the commercially available paper mask, is illustrated in Figure 3B. A total of 90% of the silk sericin-releasing bacterial nanocellulose gel was degraded during the incubation period. On the other hand, the commercially available paper mask was not degraded in the same condition. This would be explained by the fact that our bacterial nanocellulose gel was made of coconut water, bacteria, and silk sericin, which are all quickly biodegradable. In contrast, a commercially available paper mask is mainly cotton made of plant cellulose, waxes, and some fatty substances, which would be degraded more slowly. In addition, the plant bacterial nanocellulose of a paper mask has a mesh-like bulk work structure and tough property while our bacterial nanocellulose is more chemically pure and has a higher water-holding capacity and hydrophilicity. Thus, the bacterial nanocellulose would degrade faster than the plant bacterial nanocellulose [15].

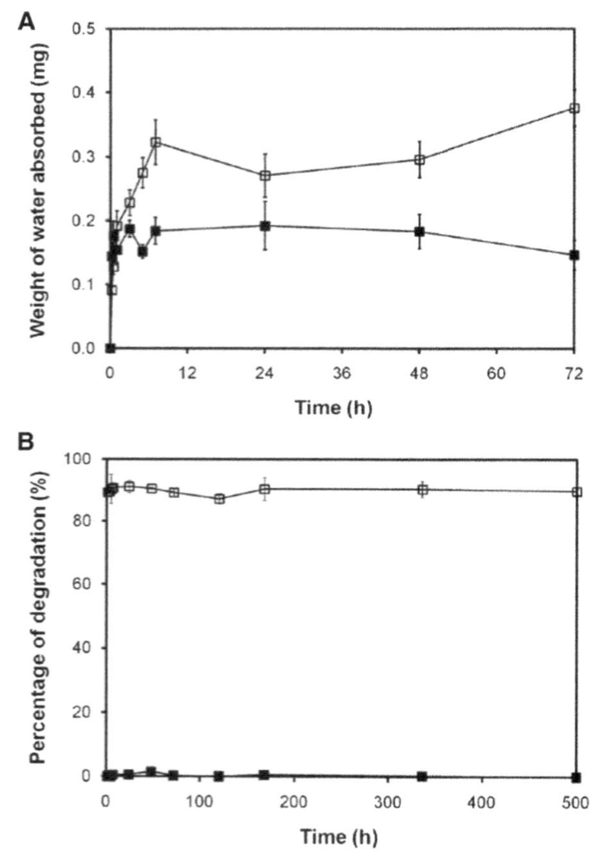

Figure 3 Moisture absorption and enzymatic biodegradation properties of biocelllose gel and paper mask. (A) Weight of water absorbed on the samples after placing in humidified desiccators with a controlled relative humidity at 81.47 ± 1.48%, 25°C for different time periods. **(B)** Percentage of degradation of the samples after incubated in 1.6 μg/ml lysozyme solution (pH 7.4) at 37°C for different time periods. (□) silk sericin-releasing bacterial nanocellulose gel and (■) the commercial available paper mask.

The *in vitro* release profile of silk sericin

We have supposed here that the release of silk sericin from the bacterial nanocellulose gel would support the facial treatment in terms of the activation of collagen production and epithelialization of the skin [8,9]. Figure 4 demonstrates the release profile of silk sericin from the silk sericin-releasing bacterial nanocellulose gel. Initially (0–7 h), silk sericin was burst released from the bacterial nanocellulose gel. Later on, the release of silk sericin from the bacterial nanocellulose gel reached the plateau at 27%. The release of silk sericin was possibly governed by two mechanisms: diffusion and material degradation [17,18]. The initial burst was the diffusional release of silk sericin absorbed on the fiber's surface. Thereafter, the silk sericin interacted with the bacterial nanocellulose fibers that were released during the degradation of the fibers. We supposed that the controlled release behavior of silk sericin from the bacterial nanocellulose gel may be beneficial for facial treatment.

Adhesive property of the silk sericin-releasing bacterial nanocellulose gel

A good facial mask should be less adhesive in order to be removed from face easily without pain. The adhesive property of the silk sericin-releasing bacterial nanocellulose gel and the commercially available paper mask was

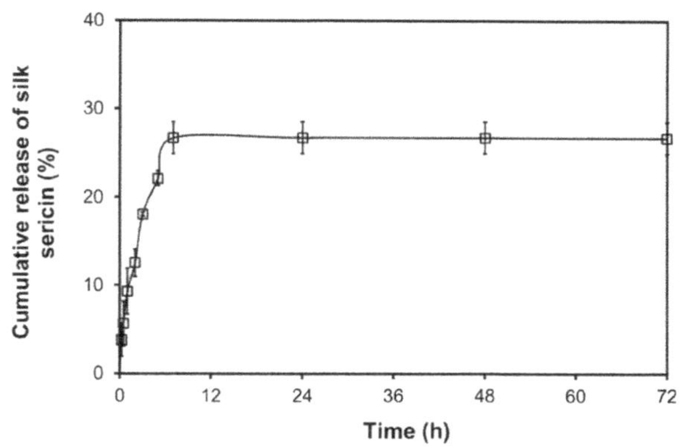

Figure 4 Percentage of cumulative release of silk sericin from the silk sericin-releasing bacterial nanocellulose gel after incubated in PBS solution (pH 7.4) at 37°C for different time periods.

evaluated from a number of cells on the samples after being attached to a wound on porcine skin, namely a peel assay. After being attached to the wound on porcine skin for 4 and 24 h, the number of cells found on the silk sericin-releasing bacterial nanocellulose gel was significantly lower than that found on the commercially available paper mask (Figure 5). Therefore, the silk sericin-releasing bacterial nanocellulose gel was less adhesive than the commercially available paper mask, which may be removed more easily and with reduced pain.

Non-cytotoxicity of the silk sericin-releasing bacterial nanocellulose gel

The silk sericin-releasing bacterial nanocellulose gel must not be cytotoxic when applied as a facial mask in direct contact with facial skin. After culture by a direct contact

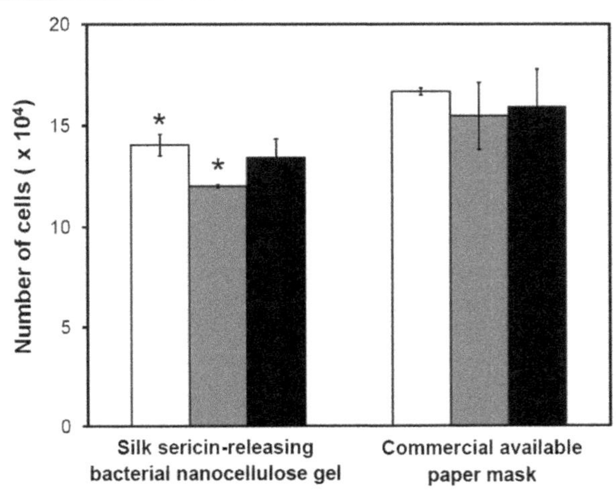

Figure 5 Number of cells on the silk sericin-releasing bacterial nanocellulose gel and the commercial available paper mask after attached to full-thickness wound of porcine skin for 4 (□), 24 (gray square symbol), and 48 h (■).

with the sample, the viable L929 cells were observed on the silk sericin-releasing bacterial nanocellulose gel, similar to those cultured on the HDPE as a negative control (Figures 6A–C). In contrast, the cells were dead when cultured on rubber as a positive control (Figure 6D). Furthermore, HaCat cells cultured in the presence of silk sericin-releasing bacterial nanocellulose gel for 3–12 h showed as high viability (~100%) as those cultured in the presence of a commercially available paper mask (Figure 7A). The apoptotic profile of HaCat cells cultured in the presence of the sample, as analyzed by Annexin V-FITC/PI double staining assay, showed that cells were both Annexin V-FITC- and PI-negative (Figure 7B). This indicated that most of the cells (~100%) were viable, but not in the apoptotic or necrotic state (Figure 7C). The results confirmed that our silk sericin-releasing bacterial nanocellulose gel was non-cytotoxic.

In vivo safety of the silk sericin-releasing bacterial nanocellulose gel

To prove its safety for cosmetic application, the silk sericin-releasing bacterial nanocellulose gel was subcutaneously implanted into rats and the tissue responses were evaluated according to ISO10993-6 standard (Biological evaluation of medical devices – Part 6: Tests for local effects after implantation) in comparison to the commercially available paper mask. The rats that received the implantation of both samples were healthy throughout the implantation period. No signs of inflammation (i.e. redness, swelling, pain, and heat) were observed in any of the rats. Figure 8 shows images of the H&E-stained sections of silk sericin-releasing bacterial nanocellulose gel and the commercially available paper mask implanted. The arrows indicate the interface between the implanted sample and surrounding tissue. No excessive inflammatory reaction was detected around the implantation sites. The infiltration of inflammatory cells into the implanted samples is shown in Figure 9. The number of inflammatory cells that infiltrated into the silk sericin-releasing bacterial nanocellulose gel was comparable to that of the commercially available paper mask during the implantation period. The intensity of inflammatory cells, necrosis, fibrosis, neovascularization, and fatty infiltrate was graded as presented in Table 3. It was observed that polymorphonuclear cells were packed and infiltrated into silk sericin-releasing bacterial nanocellulose gel after 3 and 7 weeks of implantation. However, its intensity was reduced thereafter. Lymphocytes, macrophages, and neovascularization were also found in both samples during the implantation period. The implanted commercially available paper mask showed more intensity of fatty infiltrate and giant cells than the implanted silk sericin-releasing bacterial nanocellulose gel. On the other hand, plasma cells and necrosis were not observed in either sample during the implantation period.

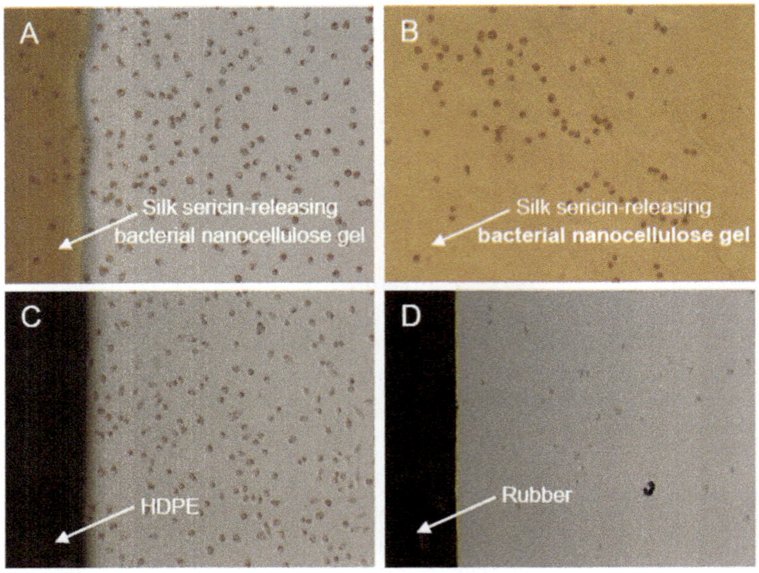

Figure 6 Morphology of L929 mouse fibroblast cells after cultured on materials for 48 h. (A,B) silk sericin-releasing bacterial nanocellulose gel, **(C)** high-density polyethylene (HDPE, negative control), and **(D)** rubber (positive control).

To sum up, the implantation of silk sericin-releasing bacterial nanocellulose gel was evaluated as being nonirritant, relative to the commercially available paper mask. This confirmed that our silk sericin-releasing bacterial nanocellulose gel was safe for further investigation in the application of facial treatment.

Conclusion

Silk sericin-releasing bacterial nanocellulose gel has been developed in this study to be applied as a bioactive mask for facial treatment. We found that a pH of 4.5 was optimal for the production of the bacterial nanocellulose gel, resulting in a 0.6 cm soft and stable gel with a smooth surface. To improve the biological properties, the gel was absorbed with silk sericin. The silk sericin-releasing bacterial nanocellulose gel had an ultrafine and extremely pure fiber network structure. The mechanical properties and moisture absorption ability of the silk sericin-releasing bacterial nanocellulose gel were improved, compared to those of the commercially available paper mask. Silk sericin could be control-released from the gel. Furthermore, the silk sericin-releasing bacterial nanocellulose gel was less adhesive and non-cytotoxic, compared to the commercially available paper mask. The *in vivo* safety test confirmed that our silk sericin-releasing bacterial nanocellulose gel was nonirritant. Further investigation into this novel bioactive facial mask has been carried out in clinic. We supposed that the silk sericin-releasing bacterial nanocellulose gel would be an outstanding candidate in the application of facial treatment.

Methods

Materials

Coconut water was obtained from coconuts purchased locally. Ammonium sulfate ($(NH_4)_2PO_4$), glacial acetic acid (CH_3COOH), and other chemicals used were of analytical grade purchased from Sigma-Aldrich, USA. All chemicals were used without further purification.

Organism

The *G. xylinus* strain (ATCC 23769) was isolated from *nata de coca* and was supplied by Kasetsart University, Bangkok, Thailand.

Culture medium

The coconut water sucrose medium as reported in the work of Verschuren *et al.* was used in this study with a slight modification [19]. The initial pH of sucrose medium was adjusted to 3.0, 3.5, 4.0, 4.5, or 5.0 with CH_3COOH. Then, 10 mL of the *G. xylinus* strain at 10^5 cfu/mL was added into the pH-adjusted coconut water sucrose medium and poured into tissue culture plates for further fermentation process.

Fermentation process

For each pH value of the coconut water sucrose medium, the fermentation was incubated at 30°C under static conditions for 7–10 days until the bacterial nanocellulose gels were formed completely. The formed bacterial nanocellulose gels were washed with 2% aqueous NaOH solution at 70°C for 10 min, and then washed repeatedly until a

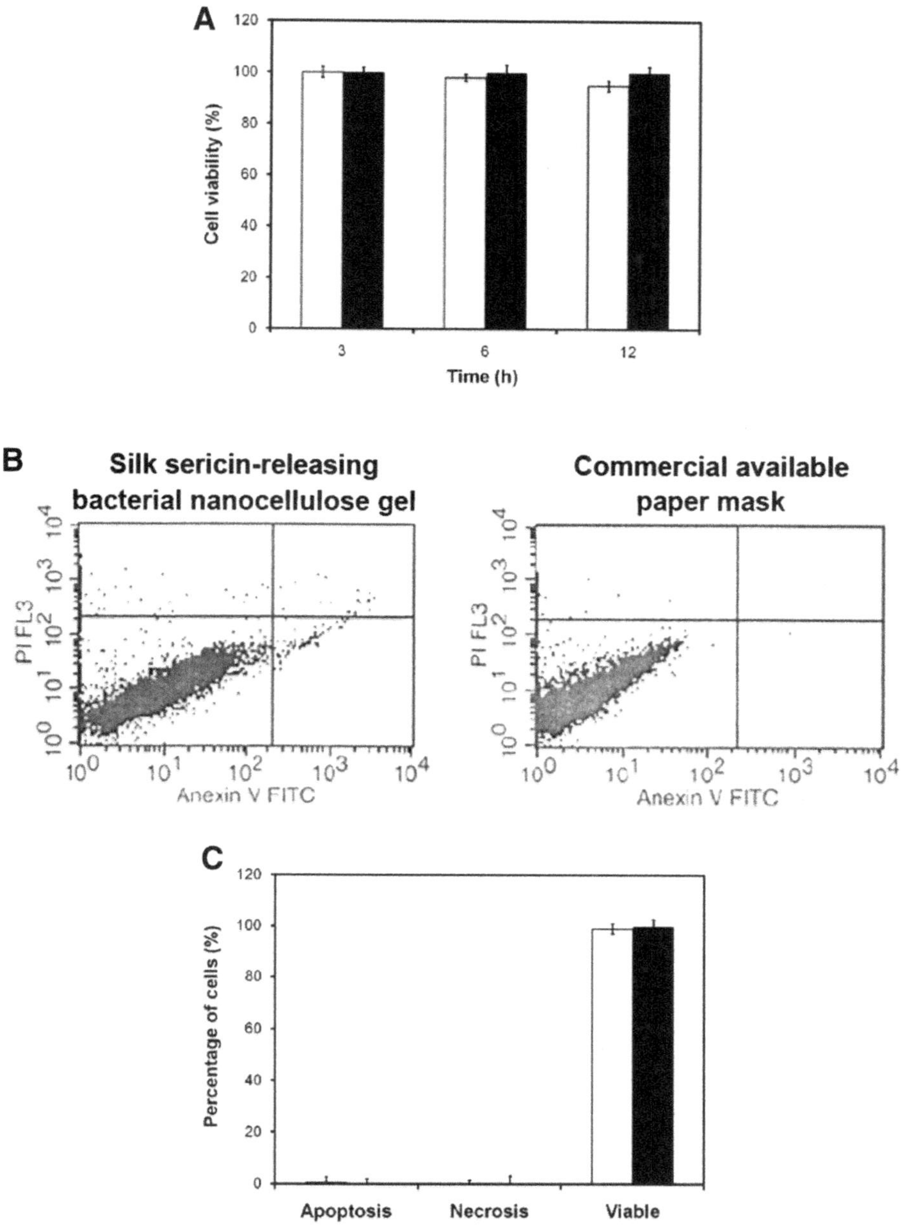

Figure 7 Non-cytotoxicity of silk sericin-releasing bacterial nanocellulose gel and paper mask. (A) Percentage of viability of HaCat cells cultured in the presence of samples for 3, 6, and 12 h. **(B)** Apoptotic profile and **(C)** Viability percentage of HaCat cells cultured in the presence of samples for 3 h, analyzed by Annexin V-FITC/PI double staining assay. (□) silk sericin-releasing bacterial nanocellulose gel (■) and the commercial available paper mask.

neutral pH was obtained. The bacterial nanocellulose gel was sterilized by gamma radiation before the experiment. Silk sericin solution (0.1 g/mL) was prepared in deionized water. The silk sericin solution (0.2 mL) was absorbed into the bacterial nanocellulose gel (diameter 3.5 cm) and incubated at 25°C for 0.5 h to allow the silk sericin impregnated into the surface of bacterial nanocellulose gel. The silk sericin-releasing bacterial nanocellulose gel was obtained.

Morphological observation and thickness measurement

The internal structure of samples was observed on a scanning electron microscope (SEM, JSM 5400, JEOL) at an accelerating voltage of 12–15 kV after sputter-coating with gold. The thickness of the bacterial nanocellulose gels prepared from different pH was measured using ImageJ software (the US National Institutes of Health, USA).

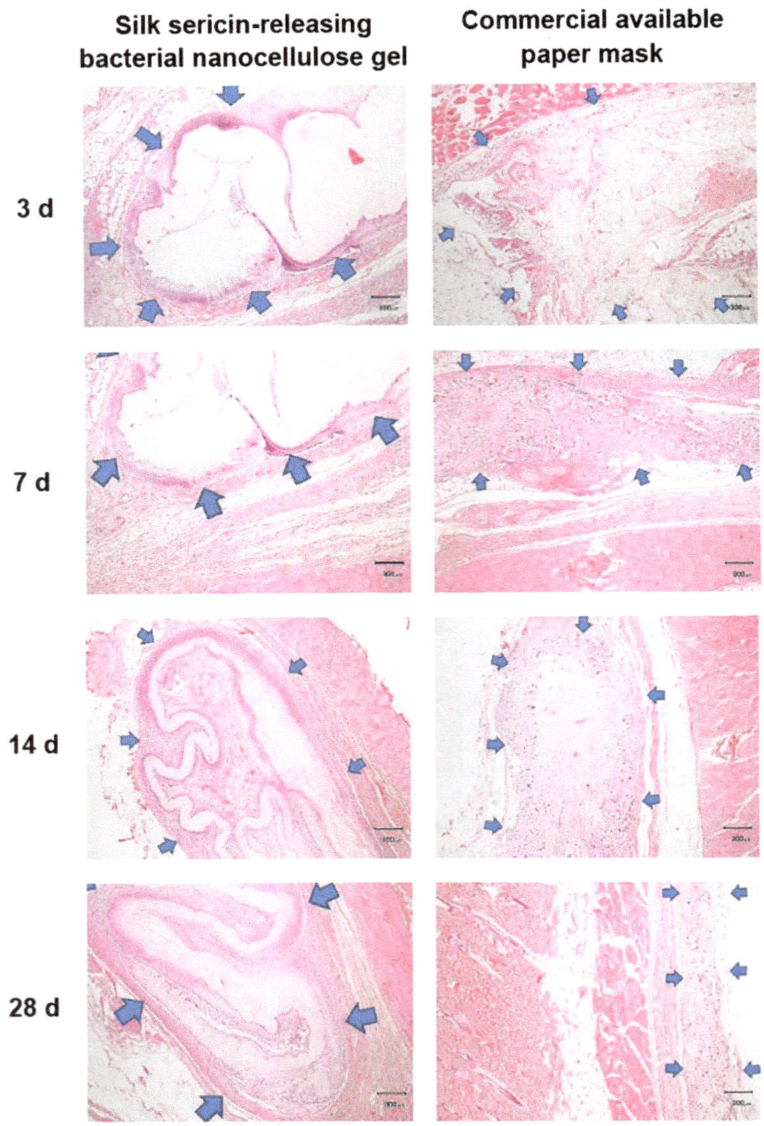

Figure 8 Microscopic images of H&E-stained sections of silk sericin-releasing bacterial nanocellulose gel and the commercial available paper mask after subcutaneous implantation for 3, 7, 14, and 28 days. Scale bar = 300 μm, arrow: interface between sample implanted and surrounding tissue.

Mechanical test

A tensile test was performed on the wet samples (L = 150 mm, W = 25 mm, H = 3 mm) at room temperature using a universal testing machine (Instron, No. 5567) at a constant rate of 300 mm/min. The force curves as a function of deformation (mm) were automatically recorded by the software. The tensile modulus and elongation at break were calculated according to the ASTM D638-01 method (n = 6).

Moisture absorption test

The moisture absorption capability of the samples was evaluated by placing the dried sample in the desiccators in which the relative humidity was controlled by salt solution. Potassium chloride was used to obtain a relative humidity of 81.47 ± 1.48% at 25°C [20]. The sample was removed from the desiccators at the predetermined times and carefully weighed. The percentage of weight increase of the sample was calculated relative to their initial weights (n = 6).

In vitro enzymatic biodegradation test

A known weight sample was incubated in 1.6 μg/mL of lysozyme solution (pH 7.4) at 37°C. The enzyme solution was changed every 2 days to ensure continuous enzyme activity. At each time interval, the remaining

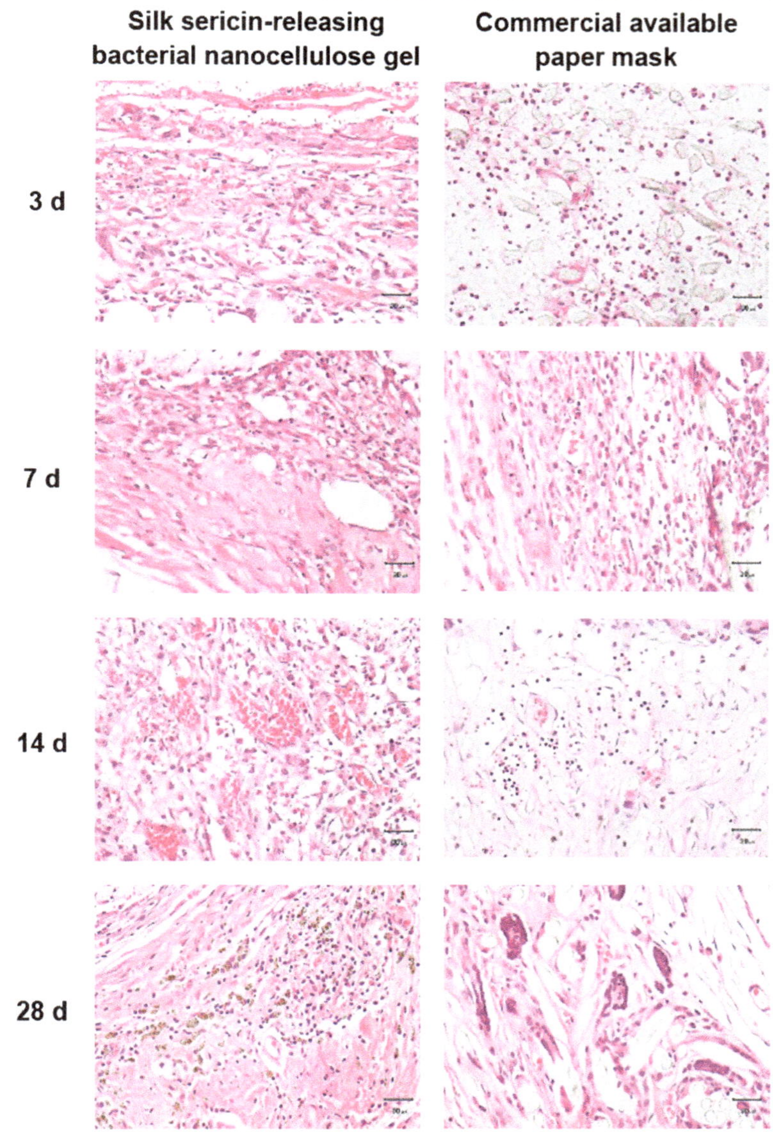

Figure 9 **Microscopic images of H&E-stained sections indicating inflammatory cells in silk sericin-releasing bacterial nanocellulose gel and the commercial available paper mask after subcutaneous implantation for 3, 7, 14, and 28 days.**

sample was taken out of the solution, rinsed repeatedly with deionized water, and freeze dried. The dried sample was weighed and the percentage of weight loss was calculated as follows:

$$\text{Percentage of weight loss } (\%) = \left(\frac{W_0 - W_t}{W_0}\right) \times 100 \quad (1)$$

Where W_0 and W_t represent the initial weight of the sample before degradation and the weight of the sample after degradation at different time periods, respectively (n = 3).

In vitro release test of silk sericin

The samples were placed into phosphate-buffered saline solutions (PBS, pH 7.4) at 37°C with a continuous stirring in a closed container. The PBS solutions (1.5 mL) were collected at the predetermined times and the amount of silk sericin released into the solution was measured using a BCA protein assay kit (Pierce, Rockford, IL, USA). The absorbance of the solution was measured at 562 nm and the amount of silk sericin was determined from a standard

Table 3 Average intensity of inflammatory cells, necrosis, fibrosis, neovascularization, and fatty in silk sericin-releasing bacterial nanocellulose gel and the commercial available paper mask after subcutaneous implantation for 3, 7, 14, and 28 days

	3 days		7 days		14 days		28 days	
	B*	C**	B	C	B	C	B	C
PMN	4 †	2.25	4	1.75	3.25	1.75	3.25	1
Lymphocytes	1	2	2	2	2.75	2	2.5	1.75
Plasma cells	0	0	0	0	0	0	0	0
Macrophages	3	2.5	3	3	3	3	3	3
Giant cells	0	0.5	0	2	0.75	3	1	3
Necrosis	0	0	0	0	0	0	0	0
Fibrosis	4	2.75	4	3.25	4	4	4	3.5
Neovascularization	3	2.5	3	1.25	2.75	1.75	2	3
Fatty infiltrate	0	3	0	1	0	0.75	0	0.5

*B: silk sericin-releasing bacterial nanocellulose gel.
**C: commercial available paper mask.
† Intensity 0–4: 0 = not observed, 1 = rare, 2 = minimal, 3 = heavily infiltrate, and 4 = packed infiltrate.

curve prepared from different concentrations of bovine serum albumin (n = 3).

Peel test with porcine skin

Porcine skin was used within 2 h after sacrifice. Full-thickness wounds were prepared by cutting the skin to a depth of 1 cm. The samples were attached to the wound. After 4, 24, and 48 h, the sample was removed and the number of cells attached to the sample was analyzed by the fluorometric quantification of cellular DNA according to the method reported by Takahashi *et al.* (n = 3) [21].

Cytotoxic test with L929 mouse fibroblast cells

A cytotoxic test of the sample was performed using a direct contact assay. High-density polyethylene (HDPE) and natural rubber containing carbon black were used as negative and positive controls, respectively. L929 mouse fibroblast cells (ECACC no. 85011425) were seeded onto the sterilized samples (1 × 1 cm^2) at a density of 6 × 10^4 cells/sample and cultured in Dulbecco's Modified Eagle Medium (DMEM) supplemented with 10% (v/v) Fetal Bovine Serum (FBS) and 100 U/mL of penicillin/streptomycin at 37°C, 5% CO_2. After 48 h of culture, the cells were stained with 0.01% (w/v) neutral red solution (pH 7.4). The morphology of the cells was observed on a phase contrast microscope.

Viability test with HaCaT human keratinocyte cells

HaCaT human keratinocyte cells were seeded onto a 96-well plate at a density of 5 × 10^3 cells/well and cultured in DMEM supplemented with 10% (v/v) FBS and 100 U/mL of penicillin/streptomycin at 37°C, 5% CO_2. After 24 h of cell attachment, the medium was refreshed and the sterilized sample was placed into the medium. After 3, 6, and 12 h of culture in the presence of the sample, the number of cells was quantified using the conventional 3-(4,5-dimethylthiazol-2-yl)-2,5-diphenyltetra zolium bromide (MTT) assay [22]. The cells cultured on the plate in the presence of a commercially available paper mask in the same conditions were used as a control group (n = 3).

Cell apoptotic test by Annexin V-FITC/PI double staining assay

HaCaT human keratinocyte cells were cultured in the presence of samples for 3 h, as previously described. Apoptosis-mediated cell death was examined using a double staining method with a fluorescein isothiocyanate (FITC)-labeled Annexin V/propidium iodide (PI) Apoptosis Detection kit (BD Pharmingen™, BD Biosciences, USA) according to the manufacturer's instructions. Flow cytometric analysis was performed immediately after staining. Cells in the early stages of apoptosis were Annexin-V-FITC-positive, whereas cells that were both Annexin V-FITC and PI-positive were in the late stages of apoptosis.

In vivo safety test (ISO10993-6 Standard)

An *in vivo* test was approved by the Ethics Committee of the Faculty of Medicine, Chulalongkorn University. The animal experiments were performed according to the Chulalongkorn University Animal Care and Use Committee (CU-ACUC) under standard sterile conditions. The implantation of silk sericin-releasing bacterial nanocellulose gel and the commercially available paper mask (2 × 2 cm^2) into the subcutaneous tissue of female Wistar rats (8 weeks old, 200–300 g) was carried out. Briefly, the rats were anesthetized, and their hair was shaved, and disinfected with 70 vol% ethyl alcohol. A 1 cm skin incision was made to form pockets in the subcutaneous tissue, then the sample was inserted into each pocket. The wound was closed with 6–0 prolene suture and disinfected with Betadine® (povidone-iodine topical antiseptics) solution. After 3, 7, 14, and 28 days of implantation, the rats were sacrificed with an overdose of thiopental sodium. The samples and surrounding tissue were retrieved, fixed with 10 vol% formalin solution, and embedded in paraffin. The paraffin-embedded samples were sectioned and stained with hematoxylin and eosin (H&E). For histological assessment, the H&E slides were semi-quantitatively scored following ISO10993-6. Inflammatory cell types, neovascularization, fibrosis, and fatty infiltrate were scored by one pathologist at two different times. The intensity of inflammatory cells, neovascularization, fibrosis, and fatty infiltrate was recorded using 0–4 scales (0 = not observed, 1 = rare, 2 = minimal, 3 = heavily infiltrated, and 4 = packed infiltrated). The final score was calculated according to Equation 2 and classified as follows:

0.0–2.9 (sample is nonirritant), 3.0–8.9 (sample is slight-irritant), 9.0–15 (sample is moderate-irritant), and >15 (sample is severe-irritant). The level of irritation was compared to that of the commercially available paper mask as a control sample.

$$\text{Final score} = [2I_t + N_t] - [2I_c + N_c] \quad (2)$$

where I_i is the total number of polymorphonuclear cells, lymphocytes, plasma cells, macrophages, giant cells, and necrosis of sample i (i = test sample (t) and control (c)), N_i is the total number of neovascularization, fibrosis, and fatty infiltrate of sample i (i = test sample (t) and control (c)).

Statistical analysis

All quantitative data represent the mean ± standard deviation. The statistical significance was determined by paired and unpaired Student's t-tests and ANOVA. A value of $p < 0.05$ was considered significant.

Competing interests

The authors declare that they have no competing interests.

Authors' contributions

PA conceived of the study, and participated in its design and coordination. NB carried out the experiments. Both authors participated in experimental designs, data analyses, read and approved the final manuscript.

Acknowledgement

This research was supported by the Ratchadapiseksompoch Endowment Fund (2014), Chulalongkorn University (CU-57-091-IC).

References

1. Updegraff DM: **Semimicro determination of cellulose in biological materials.** *Anal Biochem* 1969, **32**:420–424.
2. Ruka DR, Simon GP, Dean KM: **Altering the growth conditions of *Gluconacetobacter xylinus* to maximize the yield of bacterial cellulose.** *Carbohyd Polym* 2012, **89**:613–622.
3. Lin WC, Lien CC, Yeh HJ, Yu CM, Hsu SH: **Bacterial cellulose and bacterial cellulose–chitosan membranes for wound dressing applications.** *Carbohyd Polym* 2013, **94**:603–611.
4. Kurosumi A, Sasaki C, Yamashita Y, Nakamura Y: **Utilization various fruit juices as carbon source for production of bacterial cellulose by *Acetobacter xylinum* NBRC 13693.** *Carbohyd Polym* 2009, **76**:333–335.
5. Kongruang S: **Bacterial cellulose production by *Acetobacter xylinum* strains from agricultural waste products.** *Appl Biochem Biotechnol* 2008, **148**:245–256.
6. Cuttle L, Kempf M, Phillips GE, Mill J, Hayes MT, Fraser JF, Wang XQ, Kimble RM: **A porcine deep dermal partial thickness burn model with hypertrophic scarring.** *Burns* 2006, **32**:806–820.
7. Gibran NS, Boyce S, Greenhalgh DG: **Cutaneous wound healing.** *J Burn Care Res* 2007, **28**:577–579.
8. Aramwit P, Sangcakul A: **The effects of sericin cream on wound healing in rats.** *Biosci Biotechnol Biochem* 2007, **71**:2473–2477.
9. Aramwit P, Kanokpanont S, De-Eknamkul W, Kamei K, Srichana T: **The effect of sericin with variable amino-acid content from different silk strains on the production of collagen and nitric oxide.** *J Biomater Sci Polym Ed* 2009, **20**:1295–1306.
10. Aramwit P, Kanokpanont S, Nakpheng T, Srichana T: **The effect of sericin from various extraction methods on cell viability and collagen production.** *Int J Mol Sci* 2010, **11**:2200–2211.
11. Vandamme EJ, Baets SD, Vanbaelen A, Joris K, Wulf PD: **Improved production of bacterial cellulose and its application potential.** *Polym Degrad Stab* 1998, **59**:93–99.
12. Jonas R, Farah Luiz F: **Production and application of microbial cellulose.** *Polym Degrad Stab* 1998, **59**:101–106.
13. Kanokpanont S, Damrongsakkul S, Ratanavaraporn J, Aramwit P: **Physico-chemical properties and efficacy of silk fibroin fabric coated with different waxes as wound dressing.** *Int J Biol Macromol* 2013, **55**:88–97.
14. Wongputtaraksa T, Ratanavaraporn J, Pichyangkura R, Damrongsakkul S: **Surface modification of Thai silk fibroin scaffolds with gelatin and chitooligosaccharide for enhanced osteogenic differentiation of bone marrow-derived mesenchymal stem cells.** *J Biomed MatEr Res Part B* 2012, **100B**:2307–2315.
15. Klemm D, Schumann D, Udhardt U, Marsch S: **Bacterial synthesized cellulose - artificial blood vessels for microsurgery.** *Prog Polym Sci* 2001, **26**:1561–1603.
16. Zhang YQ: **Applications of natural silk protein sericin in biomaterials.** *Biotechnol Adv* 2002, **20**:91–100.
17. Ratanavaraporn J, Furuya H, Kohara H, Tabata Y: **Synergistic effects of the dual release of stromal cell-derived factor-1 and bone morphogenetic protein-2 from hydrogels on bone regeneration.** *Biomaterials* 2011, **32**:2797–2811.
18. Simmons CA, Alsberg E, Hsiong S, Kim WJ, Mooney DJ: **Dual growth factor delivery and controlled scaffold degradation enhance in vivo bone formation by transplanted bone marrow stromal cells.** *Bone* 2004, **35**:562–569.
19. Verschuren PG, Cardona TD, Nout MJR, Gooijer KDD, Van Den Heuvel JC: **Location and limitation of cellulose production by *Acetobacter xylinum* established from oxygen profiles.** *J Biosci Bioeng* 2000, **89**:414–419.
20. Greenspan L: **Humidity fixed points of binary saturated aqueous solutions.** *Physics* 1977, **81A**:89–96.
21. Takahashi Y, Yamamoto M, Tabata Y: **Osteogenic differentiation of mesenchymal stem cells in biodegradable sponges composed of gelatin and β-tricalcium phosphate.** *Biomaterials* 2005, **26**:3587–3596.
22. Mosmann T: **Rapid colorimetric assay for cellular growth and survival: Application to proliferation and cytotoxicity assays.** *J Immunol Methods* 1983, **65**:55–63.

7

Tobacco BY-2 cell-free lysate: an alternative and highly-productive plant-based in vitro translation system

Matthias Buntru*, Simon Vogel, Holger Spiegel and Stefan Schillberg

Abstract

Background: Cell-free protein synthesis is a rapid and efficient method for the production of recombinant proteins. Usage of prokaryotic cell-free extracts often leads to non-functional proteins. Eukaryotic counterparts such as wheat germ extract (WGE) and rabbit reticulocyte lysate (RLL) may improve solubility and promote the correct folding of eukaryotic multi-domain proteins that are difficult to express in bacteria. However, the preparation of WGEs is complex and time-consuming, whereas RLLs suffer from low yields. Here we report the development of a novel cell-free system based on tobacco Bright Yellow 2 (BY-2) cells harvested in the exponential growth phase.

Results: The highly-productive BY-2 lysate (BYL) can be prepared quickly within 4–5 h, compared to 4–5 d for WGE. The efficiency of the BYL was tested using three model proteins: enhanced yellow fluorescent protein (eYFP) and two versions of luciferase. The added mRNA was optimized by testing different 5′ and 3′ untranslated regions (UTRs). The protein yield in batch and dialysis reactions using BYL was much higher than that of a commercial Promega WGE preparation, achieving a maximum yield of 80 μg/mL of eYFP and 100 μg/mL of luciferase, compared to only 45 μg/mL of eYFP and 35 μg/mL of luciferase in WGEs. In dialysis reactions, the BYL yielded about 400 μg/mL eYFP, representing up to 50% more of the target protein than the Promega WGE, and equivalent to the amount using 5Prime WGE system.

Conclusions: Due to the high yield and the short preparation time the BYL represents a remarkable improvement over current eukaryotic cell-free systems.

Keywords: Cell-free protein synthesis, In vitro translation, *Nicotiana tabacum* cv. BY-2, Wheat germ extract, Protein expression

Background

Cell-free protein synthesis (CFPS) systems based on crude lysates provide several advantages over *in vivo* systems and offer broad applications in protein engineering, biopharmaceutical product development and post-genomic research [1]. Crude lysates contain the necessary components for translation, protein folding, and energy metabolism, so providing them with amino acids, energy substrates, nucleotides and salts allows almost any protein encoded by a RNA template to be synthesized. In coupled transcription/translation systems supplemented additionally with an appropriate RNA polymerase DNA templates can also be used. In contrast to traditional cell-based expression methods, CFPS offers shorter process times, limited protein hydrolysis and the ability to express toxic proteins or proteins containing specific chemical groups or unnatural amino acids at defined positions [2]. Furthermore, the open nature of the system allows the reaction to be controlled and monitored directly. Although chemical synthesis allows the rapid and controlled synthesis of peptides < 40 residues in length, this is not an economically feasible method for the production of larger proteins [3].

The most widely used cell-free systems are based on *Escherichia coli* extract (ECE), wheat germ extract (WGE), rabbit reticulocytes lysate (RLL) and insect cell extract (ICE). These contain diverse cellular components and co-factors that enhance protein expression, folding and modification in different ways. Therefore, the most appropriate

* Correspondence: matthias.buntru@ime.fraunhofer.de
Fraunhofer Institute for Molecular Biology and Applied Ecology IME, Forckenbeckstr. 6, 52074 Aachen, Germany

system will depend on the origin and the biochemical nature of the target protein. The preparation of ECE is simple and inexpensive, and generally achieves the highest protein yields, from hundreds of micrograms to milligrams per milliliter in batch reactions depending on the target protein [4,5]. In contrast, eukaryotic systems are less productive and extract preparation is more laborious, but complex proteins can be produced more efficiently and extended post-translational modifications are supported. WGE normally yields tens of micrograms to milligrams of recombinant protein per milliliter, depending on the protein and reaction format [6-8], but extract preparation takes 4–5 d, and the yield of extract from wheat seeds is low [9]. The yields of RLL systems are typically two orders of magnitude lower than WGE [10] and ICEs prepared from *Spodoptera frugiperda* can achieve yields of up to 50 µg/mL [11]. Recently two further eukaryotic systems based on CHO cells [12] and *Saccharomyces cerevisae* [13] have been described. The CHO extract yield up to 50 µg/mL active firefly luciferase, but the fermentation medium is quit expensive. In contrast the preparation of the yeast extract is inexpensive, but the system produces only low target protein levels of 8 µg/mL active firefly luciferase. The drawbacks of current cell-free systems have therefore created a demand for highly-productive eukaryotic cell-free systems that can be prepared quickly in large amounts.

Here we describe a highly-productive *in vitro* translation system derived from tobacco BY-2 cells. This type of cells was used due to the eukaryotic nature, the simple and cost-effective fermentation and the well-established genetic modification tools. The novel system was tested against commercial WGEs for the production of three model proteins: enhanced yellow fluorescent protein (eYFP), firefly luciferase (FFLuc) and *Renilla reniformis* luciferase (RRLuc). By measuring the activity of these optical reporter proteins, the rate of translation was determined without radioactive substrates. Several different 5' and 3' untranslated regions were also compared for their translation enhancing activity.

Results and discussion

Preparation of the tobacco Bright Yellow-2 lysate (BYL)

At present *in vitro* translation systems suffer from certain shortages like laborious extract preparation, low protein synthesis yields or the inability to support post-translational modifications [14]. The aim of this study was to establish a system, allowing the rapid preparation of lysates for cell-free synthesis of satisfactory recombinant protein yields. For this purpose, we further developed a system based on BY-2 cell suspension cultures designed for the replication of plant RNA viruses [15,16]. In contrast to the original protocol using batch-cultured cells, we used BY-2 cells growing continuously in a stirred-tank fermenter at a constant packed cell volume of 20% and a doubling time of around 32 h, to ensure a reproducible supply of homogeneous cell material. We also replaced the conventional cell wall-digesting enzymes (Cellulase Onozuka RS, Pectolyase Y-23 and Macerozyme R-10) with the liquid enzymes Rohament CL, Rohament PL and Rohapect UF, originally intended for the production of fruit juice and extracts. Rohament CL comprises a cellulase concentrate, Rohament PL is a pectinase concentrate, and Rohapect UF contains an enzyme complex including specialized pectinases and arabanases. This reduced the costs for the protoplastation step more than 100-fold. However, the most important step in the lysate preparation is the removal of the lytic vacuoles (Figure 1). These contain a great part of undesirable enzymes, including proteases and ribonucleases. Lysates prepared without the evacuolation step show nearly no translation activity [15]. The vacuoles were removed by centrifugation in a Percoll gradient using a modified method based on those described by Komoda *et al.* [15] and Ishibashi *et al.* [17]. Due to its low density vacuoles can be separated from the protoplasts yielding high-density evacuolated protoplasts. The continuous Percoll layer described in the reports listed above was replaced with a stepwise density gradient, and the protoplasts were applied directly onto the Percoll-free top layer. After centrifugation, the evacuolated protoplasts concentrated at the interface between the 40% and 70% Percoll layers (Figure 1a), whereas the separated vacuoles floated on the top layer. The purity of the evacuolated protoplasts was verified by microscopy (Figure 1b,c). The yield of the evacuolation step was ~65%, estimated using a Neubauer counting chamber. The evacuolated protoplasts were washed and then disrupted by nitrogen decompression to protect labile cell components from oxidation [18]. After the removal of nuclei and non-disrupted cells, the lysate was incubated with nuclease S7 to destroy endogenous mRNA and to reduce background translation to a minimum [16]. This step leaves the integrity of the 18S and 28S ribosomal RNAs mainly unaffected and therefore increases the translation efficiency of exogenous mRNA substantially [16].

The evacuolation step is comparable to the endosperm removal step during WGE preparation, and is essential because the endosperm contains several translation inhibitors such as tritin and thionin, as well as nucleases and proteases [6,9,19]. However, the evacuolation step takes ~90 min, whereas endosperm removal requires 2–3 d because embryo particles remaining after the grinding and sieving of wheat seeds must be selected manually to eliminate those containing larger amounts of endosperm [9]. Without the preselection of embryo particles by flotation using toxic organic solvents such as carbon tetrachloride, the selection process would take even longer [9].

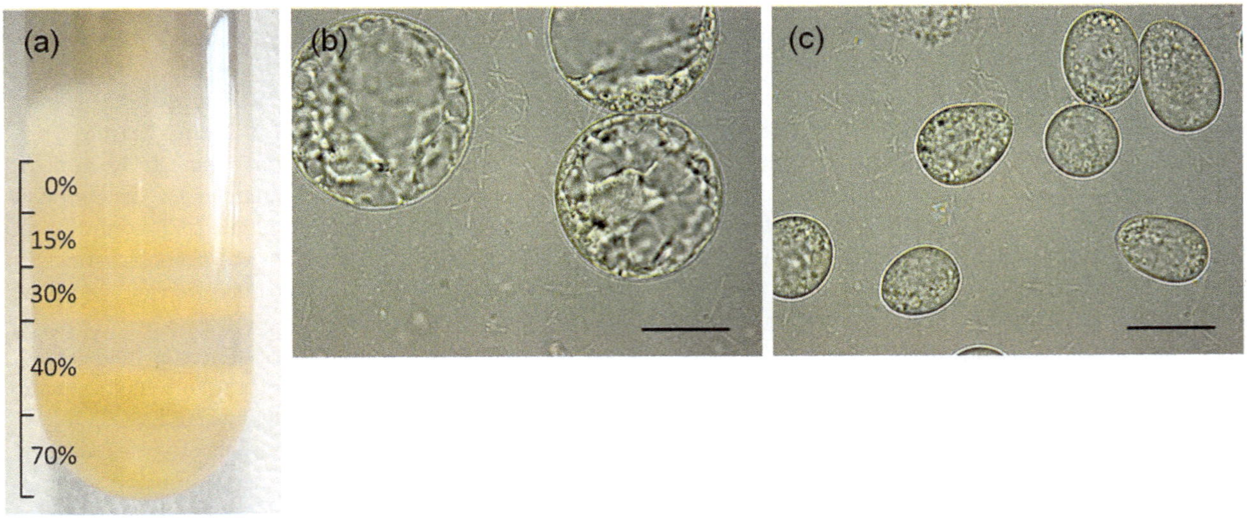

Figure 1 Preparation of evacuolated tobacco BY-2 protoplasts by Percoll gradient centrifugation. (a) Evacuolated protoplasts are concentrated at the interface between 40% and 70% Percoll. **(b)** Representative image of protoplasts before evacuolation (Scale bar = 20 μm). **(c)** Representative image of evacuolated protoplasts (Scale bar = 20 μm).

The lysis of the evacuolated protoplasts takes ~30 min, which again is much faster than the equivalent process during WGE preparation, i.e. the extraction of the wheat embryos which takes 5–7 h [9]. Furthermore, the nitrogen decompression method for protoplast lysis protects labile components from oxidation, in contrast to the crushing of embryo particles in a mill/mixer or the disruption of protoplasts using a Dounce homogenizer as described in the original protocol [15,16]. Figure 2 compares the timelines for the preparation of BYL and WGE, showing that the preparation of BYL takes 4–5 h compared to 4–5 days for the WGE. A 3-L BY-2 cell suspension culture can yield ~30 mL of BYL with a final protein concentration of 14.9 ± 0.3 mg/mL [20].

In vitro translation activity of BYL in batch reactions

The performance of BYL compared to commercial WGEs was investigated by producing the reporter proteins eYFP and luciferase in both systems and measuring fluorescence, which is much faster than standard methods and does not involve radioactive labeling. Kahn *et al.* [21] and Yukawa *et al.* [22] demonstrated that the fluorescence of GFP produced in cell-free systems was comparable with the radioactivity of ^{35}S methionine incorporated during the translation process, showing that translation yield can be quantified by fluorescence measurement. Similarly, a correlation between luciferase activity and the incorporation of ^{14}C leucine has been demonstrated in WGE, RLL and ECE [23,24].

To maximize protein expression, we optimized the mRNA sequence introduced into the BYL translation reaction. Several mRNA structural characteristics affect translation efficiency [25] including the untranslated regions (UTRs) at the 5' and 3' ends of the coding sequence [26,27]. The structure of the 5' UTR influences translational initiation, termination and mRNA stability [28]. One of the rate-limiting steps in translational initiation is the binding of the mRNA to the 43S pre-initiation complex [29]. The translational machinery is recruited by the 5'-cap or translational enhancers in the leader sequence [30-33]. To find the optimal UTRs for the BY-2 system, we constructed eight different eYFP expression constructs with various 5' and 3' UTRs (Figure 3). The 5' UTR in pCITE2a contains an internal ribosomal entry site (IRES) derived from *Encephalomyocarditis virus* (EMCV), which is widely used in eukaryotic host cells and cell-free extracts [34,35]. The pF3A vector was designed for use with wheat germ extract and contains sequences from *Barley yellow dwarf virus* (BYDV) [36]. The pIX4.0 vector used with ICEs contains the 5' UTR from a baculovirus polyhedrin gene and a synthetic 3' UTR including a poly (A) sequence. The polyhedrin promoter achieves the highest yields in the ICE system [37-40], but because it is also compatible with the WGE and RLL systems [38] we assumed it would also be functional in the BYL. The 5' UTR in the pIVEX1.3 vector includes an ARC-1 sequence element, which is complementary to an internal 18S rRNA segment [41] and may promote binding to the 40S ribosomal subunit [42]. This 18S RNA region is highly conserved among eukaryotes [43], so we assumed it would be functional in both the BYL and WGE systems. The *Tobacco mosaic virus* (TMV) 5'-UTR (omega sequence) is a well-known translational enhancer that is included in the vectors pIVEX_Omega_eYFP-His and pIVEX_GAA_Omega_eYFP-His [44-47]. It is functional as an enhancer not only in tobacco plants, but also in the WGE, RLL and

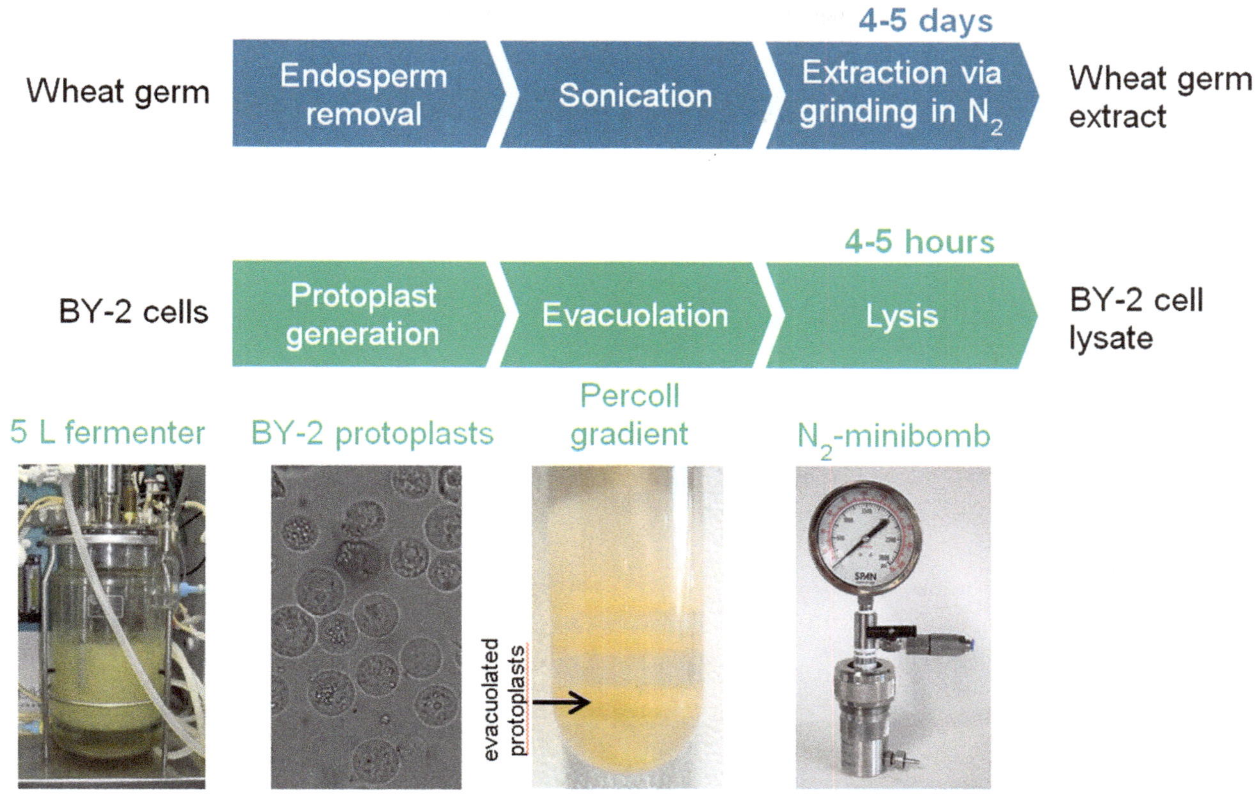

Figure 2 Flowchart showing the preparation of WGE (top panel) and BYL (bottom panel). WGE (top panel). The endosperm is removed from wheat seeds and embryo particles are washed extensively to remove translation inhibitors from the endosperm. To obtain the extract, the washed embryo particles are ground with a mortar under liquid nitrogen or in a Waring blender. BYL (bottom panel). BY-2 cells are cultivated in a 5-L fermenter, harvested in the exponential growth phase (at least 5 days post-inoculation) and converted into protoplasts. The vacuoles are separated from the protoplasts by centrifugation over a stepwise Percoll gradient. The resulting evacuolated protoplasts are lysed using the nitrogen decompression method.

ECE cell-free systems [44]. The TMV omega sequence was improved by adding GAA upstream of the initial GUA triplet [48]. The pIVEX_GAA_E02 construct contains a synthetic sequence which was selected by its ability to promote the formation of polysomes in WGEs [49].

The eYFP expression constructs described above were used as templates to produce capped mRNA *in vitro* in the presence of the cap analog m^{7}G[5']ppp[5']G. Non-incorporated nucleotides and cap analogs were removed by gel filtration, and 2 µg of the purified mRNA was translated in the BYL and WGE (Promega) systems for 18 h. The optimal reaction conditions were defined by monitoring eYFP fluorescence using capped GAA_Omega_eYFP-His mRNA as a template. The highest eYFP fluorescence was detected in the presence of 1.44 mM magnesium acetate and 61 mM potassium acetate (Figure 4a,b). Because the optimal potassium concentration is dependent on the template, each construct was evaluated in the same manner. The optimum potassium concentration ranged from 61 mM (Figure 3; constructs 1, 2, 4a, 4b, 6) to 68 mM (Figure 3; constructs 3, 5, 7). The influence of the template concentration was determined using capped GAA_Omega_eYFP-His mRNA, and saturation was observed at ~40 ng/µL (Figure 4c). Polyadenylation sequences are known to increase translational efficiency and mRNA stability [50]. So we exchanged the TMV 3' UTR sequence in pIVEX1.3_eYFP-His and GAA_Omega_eYFP-His with the equivalent sequence in pIX4.0_His-eYFP, containing a synthetic poly(A) sequence, but this exchange actually reduced the yield of the target protein (Figure 4d). Similarly, Sawasaki *et al.* [48] found that translation in WGE does not depend on a specific 3' UTR or a poly(A) sequence, but rather on the length of the 3' UTR.

In the WGE system, we used 2.1 mM magnesium according to the manufacturer's instructions. The optimal concentrations of potassium and mRNA were determined as described for the BYL system (data not shown). The highest fluorescence intensities were achieved at potassium concentrations of 103 mM (Figure 3, construct 2), 133 mM (constructs 1, 4a, 4b, 5, 6 and 7) and 163 mM (construct 3). As for the BYL system, 40 ng/µL of capped GAA_Omega_eYFP-His mRNA was sufficient. When the optimal concentrations were used with both systems, BYL

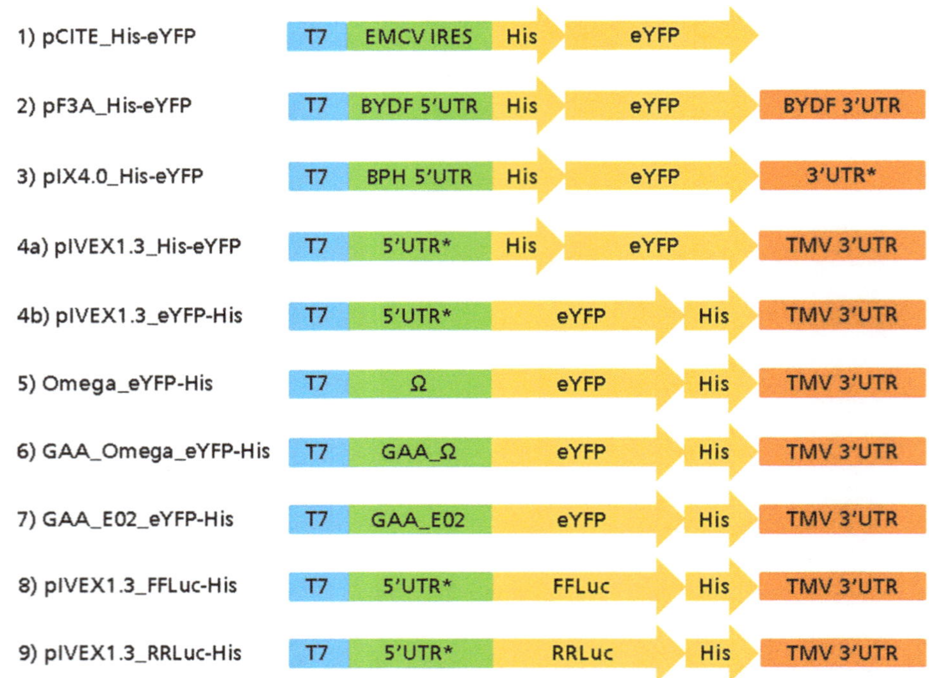

Figure 3 Schematic representation of different eYFP and luciferase expression constructs with various 5' and 3' UTRs. EMCV IRES, *Encephalomyocarditis virus* internal ribosomal entry site; *BYDV* 5' UTR, *Barley yellow dwarf virus* 5' untranslated region; BPH 5' UTR, Baculovirus polyhedrin gene 5' untranslated region; Ω, *Tobacco mosaic virus* 5' leader sequence (omega); GAA_Ω, *Tobacco mosaic virus* 5' leader sequence (omega) with GAA as first nucleotide triplet; GAA_E02, synthetic 5' leader sequence; TMV, *Tobacco mosaic virus*; 5' UTR*, repetition of a sequence complementary to the 18S ribosomal RNA; 3' UTR*, trailer sequence with synthetic poly(A) signal; FFLuc, firefly luciferase; RRLuc, *Renilla reniformis* luciferase.

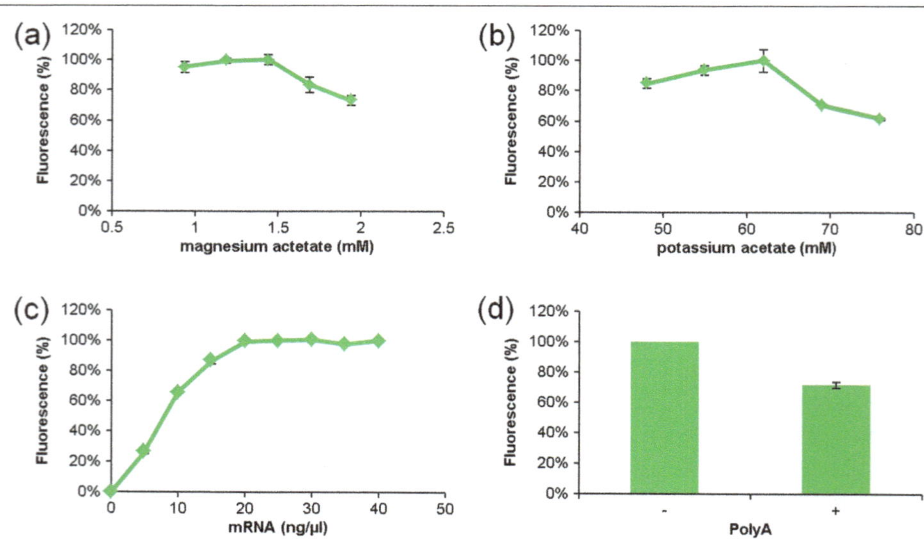

Figure 4 Optimization of reaction conditions in batch mode. Effects on eYFP yield caused by **(a)** magnesium acetate concentration, **(b)** potassium acetate concentration, **(c)** mRNA concentration, **(d)** presence (+) or absence (–) of poly(A) sequence. Translation reactions were carried out using capped GAA_Omega_eYFP-His mRNA as the template at 25°C and 500 rpm for 18 h. The relative fluorescence intensities are shown. Mean values were calculated from three independent translation experiments.

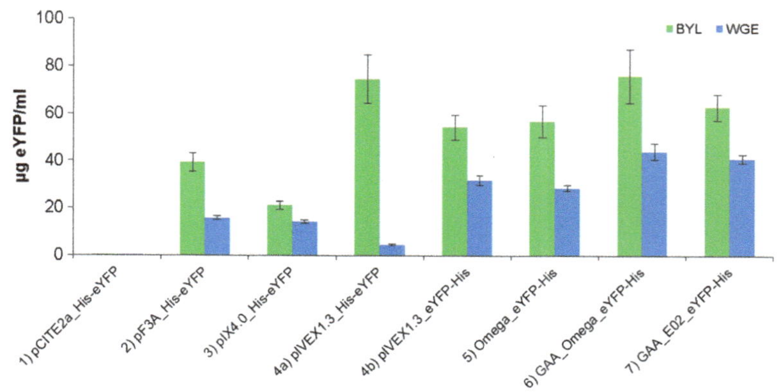

Figure 5 Translation of eYFP mRNAs with various 5′ and 3′ UTRs in BYL and WGE systems in batch mode. Reactions were carried out with 2 µg capped mRNA as the template at 25°C and 500 rpm for 18 h. The amount of fluorescent protein was calculated by comparing with an eYFP standard curve generated by measuring different eYFP concentrations in BYL translation reactions without the mRNA template. Data represent the averages and standard deviations of six independent translation experiments.

yielded significantly more eYFP than WGE ($p < 0.01$) with all eight constructs (Figure 5). GAA_Omega_eYFP-His was the most productive construct in both systems, achieving maximum yields of ~78 ± 9 µg/mL eYFP in the BYL system and 44 ± 4 µg/mL eYFP in WGE batch reactions, after incubation at 25°C and 500 rpm for 18 h. Based on our findings the pCITE 5′ UTR is deemed unsuitable for both the BYL and WGE systems, even though it has previously been used for the replication of plant RNA viruses in a similar BY-2 system [16]. This may reflect the absence of specific translation factors. As expected, construct pIVEX1.3_eYFP-His worked well in both systems, probably due to the strong conservation of the 18S rRNA sequence among eukaryotes. The lysate batch variations of four different preparations tested in quadruplicates using capped GAA_Omega_eYFP-His mRNA as template was ±9% (Additional file 1).

We also compared the expression of luciferase genes from firefly (FFLuc) and *Renilla reniformis* (RRLuc) in the BYL and WGE systems, by exchanging the eYFP sequence in pIVEX1.3_eYFP-His for the corresponding luciferase genes (Figure 3). Again, we used 2 µg of capped mRNA as a template and the reactions were incubated at 25°C and 500 rpm for 18 h. The translation of both the firefly and *Renilla* luciferase RNAs was 350% higher than in the WGE, corresponding to more than ~100 µg/ml of the target protein (Figure 6a,b). These data showed that the target protein yields were similar with three different model proteins.

In vitro translation activity of BYL in dialysis reactions

Although the BYL system is highly productive, sustainable *in vitro* protein synthesis can only be achieved in dialysis mode when substrates and energy components are continuously supplied across a semi-permeable membrane. Dialysis mode also dilutes potentially inhibitory reaction byproducts, which contributes to the higher protein yields. To evaluate the BYL system in dialysis mode, we used RTS Wheat Germ continuous exchange cell-free (CECF) devices (5Prime) and compared the BYL system to the two commercial WGEs from Promega and 5Prime. The reaction mixtures for the BYL and Promega WGE systems were prepared as described above for batch mode production. The feeding solutions were used in 20 fold excess and contained the same components as the reaction mixtures with the exception of lysate, mRNA, RNase inhibitor and creatine phosphokinase. The reaction mixture for the 5Prime WGE system was prepared according to the manufacturer's instructions. We restricted the comparison to those constructs yielding the highest amounts of eYFP in the batch reactions, i.e. constructs 4a, 4b and 6 (Figure 5). We used 60 ng/µl of $m_2^{7,3'-O}G[5']ppp[5']G$ capped mRNA as the template. For all three constructs, the BYL system was substantially more productive than the Promega WGE system (Figure 7a,b), producing 150% more eYFP when tested with the vectors pIVEX1.3_eYFP-His and GAA_Omega_eYFP-His. Surprisingly vector pIVEX1.3_His-eYFP was largely inactive in the Promega WGE system, but worked fine in the BYL. The BYL system was also more productive than the 5Prime WGE system, yielding ~130% more eYFP when tested with pIVEX1.3_His-eYFP and ~150% more eYFP when tested with pIVEX1.3_eYFP-His. The eYFP yield from vector GAA_Omega_eYFP-His was comparable in both systems. For construct GAA_Omega_eYFP-His in each case 10 µl of the BYL and WGE translation reactions were analysed by SDS-PAGE (Figure 7b). Notably, eYFP expression in the BYL system could be enhanced more than five-fold by switching from batch production to dialysis mode.

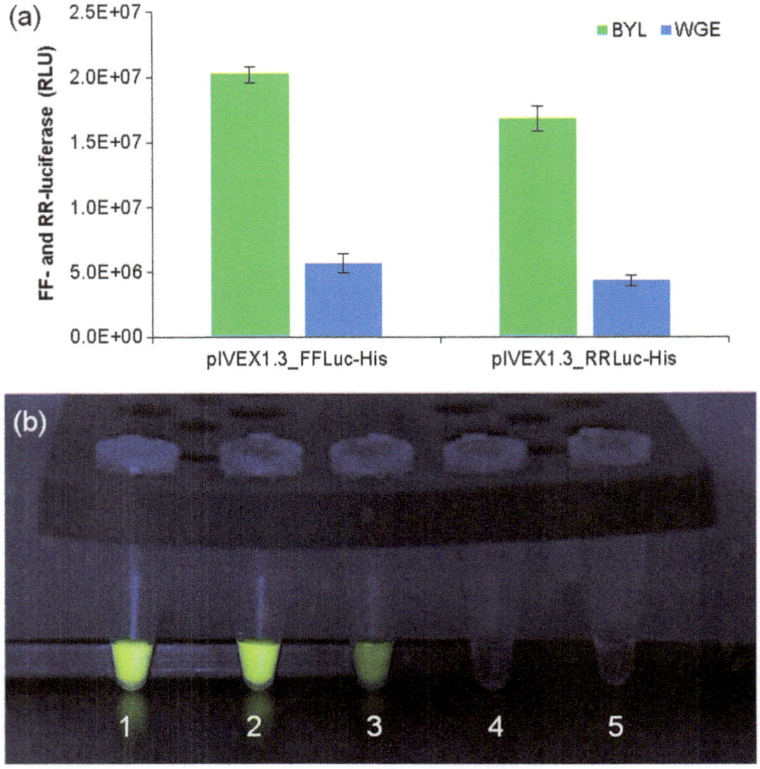

Figure 6 Translation of firefly and *Renilla reniformis* mRNAs in the BYL and WGE systems. Reactions were carried out with 2 µg capped mRNA as the template at 25°C and 500 rpm for 18 h. **(a)** Bar graphs show luciferase activity expressed as relative luciferase units (RLU). Data represent the averages and standard deviations of six independent translation experiments. **(b)** Typical result of FFLuc translation in the BYL and WGE systems, in which 1 µl of reaction mixture was combined with 50 µl luciferase assay buffer: FFLuc standard with 0.1 µg/µl (1), pIVEX1.3_FFLuc-His in BYL (2), pIVEX1.3_FFLuc-His in WGE (3), no template control in BYL (4), no template control in WGE (5).

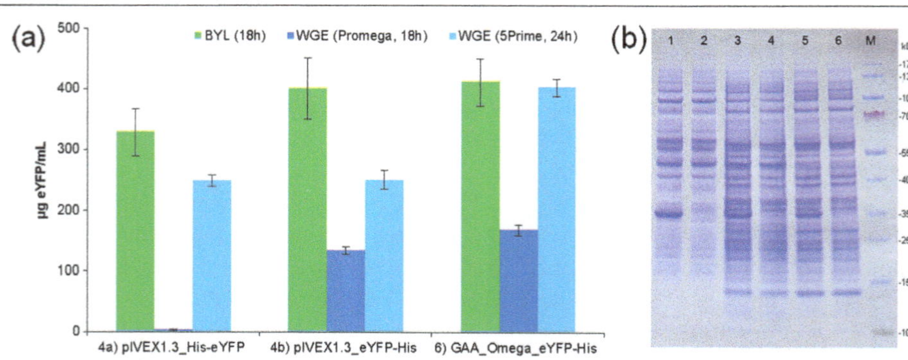

Figure 7 Translation of eYFP mRNAs in BYL and WGE systems in dialysis mode. Reactions were carried out with 3 µg capped mRNA as the template at 25°C and 900 rpm for 18 h and 24 h, respectively. The amount of fluorescent protein **(a)** was determined by comparison with an eYFP standard curve generated by measuring different eYFP concentrations in BYL translation reactions without a mRNA template. Data represent the averages and standard deviations of six independent translation experiments. **(b)** SDS-PAGE of the BYL and WGE translation reactions. In each case 10 µl of the translation reactions were loaded on a 4-12% gradient gel. Lane 1: GAA_Omega_eYFP-His in BYL; lane 2: no template control in BYL; lane 3 GAA_Omega_eYFP-His in 5Prime WGE; lane 4: no template control in 5Prime WGE; lane 5: GAA_Omega_eYFP-His in Promega WGE; lane 6: no template control in Promega WGE; M: molecular weight marker.

Conclusions

In summary, the BYL is a promising system with high translational activity, short preparation time and high potential for scaling-up. It is likely that the efficiency of the BYL system can be improved further, e.g. by optimizing reaction conditions through a factorial design approach [51]. The BYL system is based on eukaryotic components and should therefore be suitable for the expression of eukaryotic proteins, but further studies are required to determine whether the BYL system is capable of carrying out modifications such as glycosylation and the formation of disulfide bonds. A coupled transcription/translation system based on the BYL shall also be addressed in further studies.

Methods

Plant material

Tobacco cells (*Nicotiana tabacum* L. cv. Bright Yellow 2, BY-2) were cultivated continuously in a 5-L fermenter (Type 100e, Applicon Biotechnology, AC Schiedam, Netherlands) with a 20% packed cell volume at 26°C in the dark in Murashige-Skoog liquid medium (Murashige and Skoog Basal Salt Mixture, Duchefa Biochemie, Haarlem, Netherlands) supplemented with 3% (w/v) sucrose, 1 mg/L thiamine-HCl, 0.2 mg/L 2,4 dichlorophenoxyacetic acid and 100 mg/L myo-inositol.

Preparation of the BY-2 cell lysate

The preparation of lysate from evacuolated BY-2 protoplasts was carried out as described by Komoda *et al.* [15] and Gursinsky *et al.* [16] with significant modifications. Protoplasts were prepared from cells in the exponential growth phase of a continuous fermentation at a constant packed cell volume of 20% by treating the cells with 3% (v/v) Rohament CL, 2% (v/v) Rohament PL, and 0.1% (v/v) Rohapect UF (all from AB Enzymes, Darmstadt, Germany) in 0.37 mM mannitol, 5 mM $CaCl_2$, 12.5 mM sodium acetate (pH 5.8) for 1.5 h. The resulting protoplasts were layered onto a discontinuous Percoll gradient containing (from bottom to top) 70% (v/v, 3 ml), 40% (v/v, 5 ml), 30% (v/v, 3 ml), 15% (v/v, 3 ml) and 0% (3 ml) Percoll (GE Healthcare, Munich, Germany) in 0.7 M mannitol, 20 mM $MgCl_2$, and 5 mM PIPES-KOH (pH 7.0). After centrifugation at 12,000 g for 1 h at 25°C in a swinging bucket rotor, evacuolated protoplasts were recovered from the 40-70% (v/v) Percoll solution interface. The evacuolated protoplasts were suspended in three volumes of TR buffer (30 mM HEPES-KOH (pH 7.4), 80 mM potassium acetate, 0.5 mM magnesium acetate, 2 mM DTT) supplemented with one tablet per 50 ml of Complete EDTA-free Protease Inhibitor Mixture (Roche Diagnostics, Mannheim, Germany) and disrupted using the nitrogen decompression method in a cell disruption vessel (Parr Instrument, Frankfurt, Germany) for 30 min at 10 bar. Nuclei and non-disrupted cells were removed by centrifugation at 500 g for 10 min at 4°C. The supernatant was supplemented with 0.5 mM $CaCl_2$ and treated with 75 U/ml nuclease S7 (Roche Diagnostics) for 15 min at 20°C. The lysate was supplemented with 2 mM EGTA as chelating agent for the Ca^{2+} ions to inactivate the nuclease and frozen at −80°C in 1-ml aliquots.

DNA template preparation

Plasmid pIVEX1.3_eYFP-His was kindly provided by Dr. Stefan Kubick (Fraunhofer Institute for Biomedical Engineering, Potsdam-Golm, Germany). Plasmid pIVEX_-GAA_Omega_eYFP-His was prepared by inserting annealed oligonucleotides 1 and 2 (Additional file 2) containing the T7 promoter and the TMV 5' leader sequence (omega) into pIVEX1.3_eYFP-His using the NspI and NcoI sites. Plasmid pIVEX_GAA_E02_eYFP-His was prepared by inserting annealed oligonucleotides 3 and 4 (Additional file 2) containing the T7 promoter and the synthetic 5' UTR E02 into pIVEX1.3_eYFP-His using the NspI and NcoI sites. Plasmids pIX4.0_His-eYFP, pIVEX1.3_His-eYFP, pF3A_-His-eYFP and pCITE2a_His-eYFP were generated by amplifying the His-eYFP sequence from pIX3.0_His-eYFP (kindly provided by Dr. Stefan Kubick) using the following primers (Additional file 2): 5 and 6 for cloning into pIX4.0_eCFP (kindly provided by Dr. Stefan Kubick), 5 and 7 for cloning into pIVEX1.3_eYFP-His and pF3A (Promega, Mannheim, Germany), and 5 and 8 for cloning into pCITELuc (kindly provided by Dr. Sven-Erik Behrens, Institute of Biochemistry and Biotechnology, Halle/Saale, Germany). The PCR products were digested with BspHI and NotI, BspHI and KpnI, and BspHI and XbaI, respectively, and introduced into the NcoI and NotI sites of pIX4.0_eCFP, the NcoI and KpnI sites of pIVEX1.3_eYFP-His and pF3A, and the NcoI and XbaI sites of pCITELuc. Omega_eYFP-His was generated using a two-step PCR procedure (Sawasaki et al., 2007). In the first step, eYFP-His was amplified using pIVEX1.3_eYFP-His as a template and the two gene-specific primers 9 and 10 (Additional file 2). In the second step, the T7 promoter and omega sequence were fused to eYFP-His using primers 11 (5' end of T7 promoter), 12 (3' end of T7 promoter and omega sequence) and 10 as the previous step. For vector pIVEX1.3_FFLuc-His, the *FFLuc* gene was amplified by PCR using pCITELuc as the template and primers 17 and 18 (Additional file 2). The product was digested with NcoI and XhoI and reintroduced into pCITELuc in-frame with the His-tag to generate pCITE2a_F-FLuc-His. Then the FFLuc-His gene was amplified using primers 17 and 19 and introduced into the NcoI and KpnI sites of pIVEX1.3_eYFP-His. The *RRLuc* gene was amplified by PCR using pSP_RRLuc (kindly provided by Dr. Sven-Erik Behrens) as the template and primers 20 and 21 (Additional file 2). The product was digested with NcoI and XhoI and inserted into pCITELuc to generate pCITE2a_RRLuc-His. The RRLuc-His sequence was then amplified using primers 20 and 21 and inserted into the NcoI and KpnI sites of pIVEX1.3_eYFP-His to generate pIVEX1.3_RRLuc-His. Plasmids pIVEX1.3_eYFP-His_Poly (A) and pIVEX_GAA_Omega_eYFP-His_Poly(A) were generated by amplifying the 3'-UTR from pIX4.0_His-eYFP by PCR using primers 22 and 23 and inserting the product into the KpnI and EcoRI sites of pIVEX1.3_eYFP-His and pIVEX_GAA_Omega_eYFP-His.

In vitro transcription

Capped mRNA was transcribed *in vitro* in the presence of the cap analogs $m^7G[5']ppp[5']G$ or $m_2^{7,3'-O}G[5']ppp[5']G$ (New England Biolabs, Ipswich, MA, USA) using the T7 High Yield RNA Synthesis Kit (New England Biolabs) and DNA templates amplified by PCR from the constructs described above using Phusion Polymerase (New England Biolabs) and the following primers (Additional file 2): 10 and 13 for pIVEX1.3_eYFP-His, pIVEX1.3_His-eYFP, pCITE2a_His-eYFP, pIVEX1.3_F-FLuc-His and pIVEX1.3_RRLuc-His; 10 and 14 for pIVEX_GAA_Omega_eYFP-His, pIVEX_GAA_E02_eYFP-His and pIX4.0_His-eYFP; 10 and 11 for Omega_eYFP-His, and 15 and 16 for pF3A_His-eYFP. The RNA was purified using the DyeEx 2.0 Spin Kit (Qiagen, Hilden, Germany).

In vitro translation

Batch reactions contained 50% (v/v) BYL, 30% (v/v) modified TR buffer (30 mM HEPES-KOH, pH 7.6, 80 mM potassium acetate, 2 mM DTT), 0.75 mM ATP, 0.1 mM GTP, 25 mM creatine phosphate, 50 μM of each amino acid, 80 μM spermine and 0.2 mg/mL creatine phosphokinase (Roche Diagnostics). The magnesium concentration was adjusted with magnesium acetate to 1.44 mM and the potassium concentration was adjusted with potassium acetate to 61 mM or 68 mM, according to the expression construct. Any magnesium and potassium already present in the BY-2 cell extracts was ignored. The batch translation reactions were carried out at 25°C and 500 rpm for 18 h in a thermomixer (Eppendorf, Hamburg, Germany). WGE (Promega) translation reactions were carried out according to the manufacturer's instructions.

Dialysis reactions were carried out in RTS 100 Wheat Germ continuous exchange cell-free (CECF) devices (5Prime, Hamburg, Germany) using the same mixtures described for the batch reactions except that we also added 40 units of murine RNase Inhibitor (New England Biolabs). Feeding solutions contained 67.5% (v/v) modified TR buffer. For the BYL system this was 30 mM HEPES-KOH (pH 7.6), 90 mM potassium acetate, 2.1 mM magnesium acetate and 2 mM DTT, supplemented with 0.75 mM ATP, 0.1 mM GTP, 25 mM creatine phosphate, 50 μM of each amino acid and 80 μM spermine. For the Promega WGE system this was 30 mM HEPES-KOH (pH 7.6), 200 mM potassium acetate, 3.1 mM magnesium acetate and 7.4 mM DTT, supplemented with 1.2 mM ATP, 0.1 mM GTP, 10 mM creatine phosphate, 80 μM of each amino acid and 500 μM spermidine. Translation reactions in dialysis mode were carried out at 25°C and 900 rpm for 18 h in a thermomixer (Eppendorf). Translation reactions with the RTS 100 Wheat Germ CECF Kit (5Prime) were carried out according to the manufacturer's instructions.

Product analysis

The fluorescence signal from eYFP was quantified using a Synergy HT Multi-Mode Microplate Reader (Biotek, Bad Friedrichshall, Germany) with 485/20 nm excitation and 528/20 nm emission filters. The quantity of eYFP was determined by generating a standard curve based on different concentrations of eYFP in BYL translation reactions without a mRNA template. The eYFP standard was produced in a home-made *E. coli in vitro* translation system and purified by immobilized metal-affinity chromatography (IMAC) and size exclusion chromatography (SEC). The concentration of protein was determined using a *colorimetric assay* [20].

Firefly and *Renilla* luciferase activities were measured with the Luciferase Assay System (Promega) and a GENios Pro microplate reader (Tecan, Mainz-Kastel, Germany). Firefly luciferase purchased from Roche Diagnostics was used as a standard.

Sodium dodecyl sulfate polyacrylamide gel electrophoresis (SDS-PAGE)

SDS-PAGE was performed using precast NuPAGE 4-12% polyacrylamide Bis-Tris gels (Life Technologies, Carlsbad, CA, USA). The gel was stained with Coomassie brilliant blue R-250. PageRuler Prestained Protein Ladder (Thermo Scientific, Waltham, MA, USA) was used as molecular weight marker.

Availability of supporting data

The data sets supporting the results of this article are included within the article and its additional file.

Additional files

Additional file 1: Lysate batch variations. The table shows the comparison of translation activities of four lots of BYL preparations. Translation reactions were carried out using capped GAA_Omega_eYFP-His mRNA as the template at 25°C and 500 rpm for 16 h, and fluorescence intensity was measured. Average and standard deviation were calculated from four translation experiments.

Additional file 2: Oligonucleotides used in this study. The table shows the oligonucleotides used for the cloning and preparation of DNA templates for *in vitro* transcription.

Abbreviations

BYDF: *Barley yellow dwarf virus*; BYL: BY-2 lysate; CECF: Continuous exchange cell-free; CFPS: Cell-free protein synthesis; EMCV: *Encephalomyocarditis virus*; eYFP: Enhanced yellow fluorescent protein; FFLuc: Firefly luciferase; ICE: Insect cell extract; IMAC: Immobilized metal-affinity chromatography; RLL: Rabbit reticulocyte lysate; RRLuc: *Renilla reniformis* luciferase; SEC: Size exclusion chromatography; TMV: *Tobacco mosaic virus*; UTR: Untranslated region; WGE: Wheat germ extract.

Competing interests

We have filed a patent application relating to the use of continuous fermentation for providing cell material for preparation of the tobacco BY-2 cell-free lysate.

Authors' contributions

MB carried out the *in vitro* translation experiments, the product analysis and drafted the paper. SV cultured the BY2 cells and prepared the BY2 lysate. HS participated in the design of the study. SS participated in the design of the study and helped to draft the manuscript. All authors read and approved the final manuscript.

Acknowledgements

We would like to thank Dr. Stefan Kubick (Fraunhofer Institute for Biomedical Engineering, Potsdam-Golm, Germany) for providing the vectors pIVEX1.3_eYFP-His, pIX4.0_eCFP and pIX3.0_His-eYFP. We also thank Dr. Sven-Erik Behrens (Institute of Biochemistry and Biotechnology, Halle/Saale, Germany) for providing the plasmids pCITELuc and pSP_RRLuc. Furthermore we would like to thank Dr. Tanja Holland (Fraunhofer Institute for Molecular Biology and Applied Ecology IME/IPP, Aachen, Germany) for her help with the BY-2 cell cultures. This work was funded by the Federal Ministry of Education and Research (BMBF, FKZ 0315942) and the Fraunhofer Society.

References

1. Swartz JR: **Transforming biochemical engineering with cell-free biology.** *AIChE Journal* 2012, **58:**5–13.
2. White ER, Reed TM, Ma Z, Hartman MCT: **Replacing amino acids in translation: Expanding chemical diversity with non-natural variants.** *Methods* 2013, **60:**70–74.
3. Nilsson BL, Soellner MB, Raines RT: **Chemical synthesis of proteins.** *Annu Rev Biophys Biomol Struct* 2005, **34:**91–118.
4. Kim HC, Kim TW, Kim DM: **Prolonged production of proteins in a cell-free protein synthesis system using polymeric carbohydrates as an energy source.** *Process Biochem* 2011, **46**(6):1366–1369.
5. Zawada JF, Yin G, Steiner AR, Yang J, Naresh A, Roy SM, Gold DS, Heinsohn HG, Murray CJ: **Microscale to manufacturing scale-up of cell-free cytokine production–a new approach for shortening protein production development timelines.** *Biotechnol Bioeng* 2011, **108**(7):1570–1578.
6. Madin K, Sawasaki T, Ogasawara T, Endo Y: **A highly efficient and robust cell-free protein synthesis system prepared from wheat embryos: plants apparently contain a suicide system directed at ribosomes.** *Proc Natl Acad Sci USA* 2000, **97**(2):559–564.
7. Slater MR, Hurst R, Pferdehirt B, White D, Niles A, Betz N, Schenborn E: **Expression of Soluble Native Human Proteins in Cell-Free Extracts.** *Promega Notes* 2005, **91:**22–25.
8. Zhao KQ, Hurst R, Slater MR, Bulleit RF: **Functional protein expression from a DNA based wheat germ cell-free system.** *J Struct Funct Genomics* 2007, **8**(4):199–208.
9. Takai K, Sawasaki T, Endo Y: **Practical cell-free protein synthesis system using purified wheat embryos.** *Nature protocols* 2010, **5**(2):227–238.
10. Jackson M, Boutell J, Cooley N, He M: **Cell-free protein synthesis for proteomics.** *Briefings in functional genomics and proteomics* 2004, **2:**308–319.
11. Ezure T, Suzuki T, Shikata M, Ito M, Ando E: **A cell-free protein synthesis system from insect cells.** *Methods Mol Biol* 2010, **607:**31–42.
12. Brodel AK, Sonnabend A, Kubick S: **Cell-free protein expression based on extracts from CHO cells.** *Biotechnol Bioeng* 2014, **111**(1):25–36.
13. Hodgman CE, Jewett MC: **Optimized extract preparation methods and reaction conditions for improved yeast cell-free protein synthesis.** *Biotechnol Bioeng* 2013, **110**(10):2643–2654.
14. Carlson ED, Gan R, Hodgman CE, Jewett MC: **Cell-free protein synthesis: applications come of age.** *Biotechnol Adv* 2012, **30**(5):1185–1194.
15. Komoda K, Naito S, Ishikawa M: **Replication of plant RNA virus genomes in a cell-free extract of evacuolated plant protoplasts.** *Proc Natl Acad Sci USA* 2004, **101**(7):1863–1867.
16. Gursinsky T, Schulz B, Behrens SE: **Replication of Tomato bushy stunt virus RNA in a plant in vitro system.** *Virology* 2009, **390**(2):250–260.
17. Ishibashi K, Komoda K, Ishikawa M: *In vitro translation and replication of tobamovirus RNA in a cell-free extract of evacuolated tobacco BY-2 protoplasts.* Berlin: Springer; 2006.
18. Loewus MW, Loewus F: **The Isolation and Characterization of d-Glucose 6-Phosphate Cycloaldolase (NAD-Dependent) from Acer pseudoplatanus L. Cell Cultures: Its Occurrence in Plants.** *Plant Physiol* 1971, **48**(3):255–260.
19. Endo Y, Sawasaki T: **High-throughput, genome-scale protein production method based on the wheat germ cell-free expression system.** *Biotechnol Adv* 2003, **21**(8):695–713.
20. Bradford MM: **A rapid and sensitive method for the quantitation of microgram quantities of protein utilizing the principle of protein-dye binding.** *Anal Biochem* 1976, **72:**248–254.
21. Kahn TW, Beachy RN, Falk MM: **Cell-free expression of a GFP fusion protein allows quantitation in vitro and in vivo.** *Curr Biol* 1997, **7**(4):R207–208.
22. Yukawa M, Kuroda H, Sugiura M: **A new in vitro translation system for non-radioactive assay from tobacco chloroplasts: effect of pre-mRNA processing on translation in vitro.** *Plant J* 2007, **49**(2):367–376.
23. Kolb VA, Makeyev EV, Spirin AS: **Folding of firefly luciferase during translation in a cell-free system.** *The EMBO journal* 1994, **13**(15):3631–3637.
24. Kolb VA, Makeyev EV, Spirin AS: **Co-translational folding of an eukaryotic multidomain protein in a prokaryotic translation system.** *J Biol Chem* 2000, **275**(22):16597–16601.
25. Kozak M: **Regulation of translation via mRNA structure in prokaryotes and eukaryotes.** *Gene* 2005, **361:**13–37.
26. Fan Q, Treder K, Miller WA: **Untranslated regions of diverse plant viral RNAs vary greatly in translation enhancement efficiency.** *BMC Biotechnol* 2012, **12:**22.
27. Sonenberg N: **mRNA translation: influence of the 5' and 3' untranslated regions.** *Curr Opin Genet Dev* 1994, **4**(2):310–315.
28. Sachs AB, Sarnow P, Hentze MW: **Starting at the beginning, middle, and end: translation initiation in eukaryotes.** *Cell* 1997, **89**(6):831–838.
29. Herrshey JWB, Merrick WC: *The pathway and mechanism of initiation of protein synthesis.* New York: Cold Spring Harbor; 2000.
30. Dreher TW, Miller WA: **Translational control in positive strand RNA plant viruses.** *Virology* 2006, **344**(1):185–197.
31. Kneller EL, Rakotondrafara AM, Miller WA: **Cap-independent translation of plant viral RNAs.** *Virus Res* 2006, **119**(1):63–75.
32. Nicholson BL, White KA: **3' Cap-independent translation enhancers of positive-strand RNA plant viruses.** *Curr Opin Virol* 2011, **1**(5):373–380.
33. Walsh D, Mohr I: **Viral subversion of the host protein synthesis machinery.** *Nat Rev Microbiol* 2011, **9**(12):860–875.
34. Elroy-Stein O, Fuerst TR, Moss B: **Cap-independent translation of mRNA conferred by encephalomyocarditis virus 5' sequence improves the performance of the vaccinia virus/bacteriophage T7 hybrid expression system.** *Proc Natl Acad Sci USA* 1989, **86**(16):6126–6130.
35. Parks GD, Duke GM, Palmenberg AC: **Encephalomyocarditis virus 3C protease: efficient cell-free expression from clones which link viral 5' noncoding sequences to the P3 region.** *J Virol* 1986, **60**(2):376–384.
36. Wang S, Miller WA: **A sequence located 4.5 to 5 kilobases from the 5' end of the barley yellow dwarf virus (PAV) genome strongly stimulates translation of uncapped mRNA.** *J Biol Chem* 1995, **270**(22):13446–13452.
37. Ding H, Griesel C, Nimtz M, Conradt HS, Weich HA, Jager V: **Molecular cloning, expression, purification, and characterization of soluble full-length, human interleukin-3 with a baculovirus-insect cell expression system.** *Protein Expr Purif* 2003, **31**(1):34–41.

38. Suzuki T, Ito M, Ezure T, Kobayashi S, Shikata M, Tanimizu K, Nishimura O: **Performance of expression vector, pTD1, in insect cell-free translation system.** *J Biosci Bioeng* 2006, **102**(1):69–71.
39. Wickham TJ, Davis T, Granados RR, Shuler ML, Wood HA: **Screening of insect cell lines for the production of recombinant proteins and infectious virus in the baculovirus expression system.** *Biotechnol Prog* 1992, **8**(5):391–396.
40. Yamaji H, Hirakawa D, Tagai S, Fukuda H: **Production of protein kinase C-delta by the baculovirus-insect cell system in serum-supplemented and serum-free media.** *J Biosci Bioeng* 2003, **95**(2):185–187.
41. Akbergenov RZ, Zhanybekova SS, Kryldakov RV, Zhigailov A, Polimbetova NS, Hohn T, Iskakov BK: **ARC-1, a sequence element complementary to an internal 18S rRNA segment, enhances translation efficiency in plants when present in the leader or intercistronic region of mRNAs.** *Nucleic Acids Res* 2004, **32**(1):239–247.
42. Matveeva OV, Shabalina SA: **Intermolecular mRNA-rRNA hybridization and the distribution of potential interaction regions in murine 18S rRNA.** *Nucleic Acids Res* 1993, **21**(4):1007–1011.
43. Sarge KD, Maxwell ES: **Evidence for a Competitive-Displacement Model for the initiation of protein synthesis involving the intermolecular hybridization of 5 S rRNA, 18 S rRNA and mRNA.** *FEBS Lett* 1991, **294**(3):234–238.
44. Gallie DR, Sleat DE, Watts JW, Turner PC, Wilson TM: **The 5'-leader sequence of tobacco mosaic virus RNA enhances the expression of foreign gene transcripts in vitro and in vivo.** *Nucleic Acids Res* 1987, **15**(8):3257–3273.
45. Zaccomer B, Haenni AL, Macaya G: **The remarkable variety of plant RNA virus genomes.** *J Gen Virol* 1995, **76**(Pt 2):231–247.
46. Gallie DR, Walbot V: **Identification of the motifs within the tobacco mosaic virus 5'-leader responsible for enhancing translation.** *Nucleic Acids Res* 1992, **20**(17):4631–4638.
47. Gallie DR: **Translational control of cellular and viral mRNAs.** *Plant Mol Biol* 1996, **32**(1–2):145–158.
48. Sawasaki T, Ogasawara T, Morishita R, Endo Y: **A cell-free protein synthesis system for high-throughput proteomics.** *Proc Natl Acad Sci USA* 2002, **99**(23):14652–14657.
49. Kamura N, Sawasaki T, Kasahara Y, Takai K, Endo Y: **Selection of 5'-untranslated sequences that enhance initiation of translation in a cell-free protein synthesis system from wheat embryos.** *Bioorg Med Chem Lett* 2005, **15**(24):5402–5406.
50. Kozak M: **How strong is the case for regulation of the initiation step of translation by elements at the 3' end of eukaryotic mRNAs?** *Gene* 2004, **343**(1):41–54.
51. Vasilev N, Gromping U, Lipperts A, Raven N, Fischer R, Schillberg S: **Optimization of BY-2 cell suspension culture medium for the production of a human antibody using a combination of fractional factorial designs and the response surface method.** *Plant Biotechnol J* 2013, **11**(7):867–874.

Saccharification and liquefaction of cassava starch: an alternative source for the production of bioethanol using amylolytic enzymes by double fermentation process

Sidra Pervez, Afsheen Aman, Samina Iqbal, Nadir Naveed Siddiqui and Shah Ali Ul Qader*

Abstract

Background: Cassava starch is considered as a potential source for the commercial production of bioethanol because of its availability and low market price. It can be used as a basic source to support large-scale biological production of bioethanol using microbial amylases. With the progression and advancement in enzymology, starch liquefying and saccharifying enzymes are preferred for the conversion of complex starch polymer into various valuable metabolites. These hydrolytic enzymes can selectively cleave the internal linkages of starch molecule to produce free glucose which can be utilized to produce bioethanol by microbial fermentation.

Results: In the present study, several filamentous fungi were screened for production of amylases and among them *Aspergillus fumigatus* KIBGE-IB33 was selected based on maximum enzyme yield. Maximum α-amylase, amyloglucosidase and glucose formation was achieved after 03 days of fermentation using cassava starch. After salt precipitation, fold purification of α-amylase and amyloglucosidase increased up to 4.1 and 4.2 times with specific activity of 9.2 $kUmg^{-1}$ and 393 $kUmg^{-1}$, respectively. Concentrated amylolytic enzyme mixture was incorporated in cassava starch slurry to give maximum glucose formation (40.0 gL^{-1}), which was further fermented using *Saccharomyces cerevisiae* into bioethanol with 84.0% yield. The distillate originated after recovery of bioethanol gave 53.0% yield.

Conclusion: An improved and effective dual enzymatic starch degradation method is designed for the production of bioethanol using cassava starch. The technique developed is more profitable due to its fast liquefaction and saccharification approach that was employed for the formation of glucose and ultimately resulted in higher yields of alcohol production.

Keywords: Amylases, *Aspergillus fumigatus*, Biofuel, Saccharification, *Saccharomyces cerevisiae*, Starch

Background

Emerging environmental issues raised due to combustion of petroleum-based fossil fuel and emission of toxic gases have diverted the attention of scientists and researchers towards the utilization of various renewable resources for the production of bioethanol. In addition to these global concerns, other important factors that have been kept in preference are the mounting prices of the fuels and the current political scenario among the oil producing nations. Bioprocessing of renewable resources available in a particular region can help in resolving these issues. Various renewable resources in terms of agricultural biomass have been investigated for the production of bioethanol and this development proved beneficent for the biotechnological industries. Amongst various starchy materials available throughout the world; corn, sugarcane, wheat, potato [1], corn stover [2,3], molasses [4] and purified starch [5] have been successfully utilized for the commercial production of bioethanol. As the demand and the cost of these starchy crop materials is

* Correspondence: ali_kibge@yahoo.com
The Karachi Institute of Biotechnology & Genetic Engineering (KIBGE), University of Karachi, Karachi -75270, Pakistan

increasing day by day, it has become indispensible to use substitute raw resources.

Cassava is a tropical root crop which is an economically available fermentable source and is produced by numerous countries [6]. It is incorporated into animal feed (20.0%) and about similar proportion is converted into starch for industrial purposes whereas; some of the portion is also used as food source in several developing countries. About 50.0 to 70.0% starch content is recovered from the cassava root and due to the low ash content and rich organic nature it can be used as an ideal substrate for bioethanol production [7-9]. In addition, it can also be easily hydrolyzed by various techniques. As cassava starch does not have much industrial application in food industries as compared to corn starch, therefore it also lacks competition in terms of price and is available throughout the year due to its flexibility in terms of planting and harvesting [7,10,11].

In recent years, bioprocessing of various value-added products using microbial factories have been potentially explored with reference to extracellular enzymes. Agricultural biomass used as a substrate for the production of bioethanol has several limitations including high fiber content which requires high temperature for hydrolysis and this energy intensive procedure also does not provide desired yields of fermentable sugars. Hydrolysis of lingocellulosic mass by other expensive pre-treatment techniques is also time consuming. Industries also have concerns regarding the availability of the biomass throughout the year and most of the time its storage in bulk quantities is not possible due to space shortage. The development of an ideal pre-treatment method for hydrolyzing poly-phenolic lignin in the feedstock is expensive with several aforementioned limitations thus, enzymatic treatment is more preferable. Conventional method used for the production of bioethanol from cassava starch usually requires the basic gelatinization step followed by liquefaction and saccharification. The sugar formed during these processes is further fermented using either yeast or bacteria. Since, starch derived from any plant source is a complex molecule, it require various hydrolytic enzymes for its conversion into simple fermentable sugars. Among many extracellular hydrolases available, microbial amylases are frequently used for its conversion. For commercial production of amylases *Aspergillus* and *Rhizopus* species are considered most significant sources because the enzymes from these sources are generally thermostable and are available in excessive quantities [12-14].

Despite several advantages of simultaneous saccharification and fermentation using multiple organisms, there are also few shortcomings. In the initial steps, the amylolytic enzymes are produced using fungi and the starch present in the medium is allowed to hydrolyze into simpler sugars and afterwards another microbial factory (yeast or bacteria) is incorporated in the same fermentation flask to produce ethanol. In this case the primary organism (specifically fungal specie) along with amylolytic enzymes also excretes other toxic substances and proteases which in result inhibit the growth and performance of the second ethanol-producing microorganism. Along with this, the establishment of appropriate temperature for starch hydrolysis, enzymatic activity and ethanol production also plays an important role. Current research deals with the production of bioethanol from hydrolysis of an inexpensive renewable resource known as cassava starch, which is commonly available in Pakistan. The methodology used for the production of ethanol was based on double fermentation technique using partially purified fungal amylolytic enzymes for the liquefaction and saccharification of this starchy material. Keeping all disadvantages in view, this study was designed in two separate steps. In first step, amylolytic enzymes (α-amylase and amyloglucosidase) were produced using indigenously isolated filamentous fungi and were partially purified to hydrolyze cassava starch into simple fermentable sugars. In the next step, the sugar cocktail was fermented using *S. cerevisiae* to acquire maximum bioethanol yield.

Results and discussion

In the present study, several different fungal isolates with amylolytic activities were purified from different soil samples and preliminary identification was based on microbiological studies including cultural and microscopic characterization followed by 18S rDNA sequence analysis. Colonial and microscopic characteristics indicate that all isolates belong to genera *Aspergillus*. Microscopic morphology of *A. fumigatus* KIBGE-IB33 showed columnar and uniseriate conidial heads while, conidiophores are short and smooth. On the other hand, *A. niger* KIBGE-IB36 showed large, globose, dark brown conidial heads with hyaline and smooth conidiophores. Likewise, conidial heads of *A. flavus* KIBGE-IB34 are radiate and biseriate whereas, conidiophores are hyaline and coarsely roughened. *A. terreus* KIBGE-IB35 has biseriate and globose conidia with hyaline and smooth conidiophores. *A. versicolor* KIBGE-IB37 showed centrally rising, velvety floccose and slightly blue-green color colony on PDA with conidiophores borne from surface or aerial hyphae.

Screening of amylolytic property of the strains was based on starch hydrolysis method. Initially, 07 fungal strains were selected and among them 05 filamentous fungi including *A. fumigatus* KIBGE-IB33, *A. flavus* KIBGE-IB34, *A. terreus* KIBGE-IB35, *A. niger* KIBGE-IB36 and *A. versicolor* KIBGE-IB37showed production for amylolytic enzymes. When these isolates were cultivated in the starch containing production medium, highest titers of α-amylase (11.0 $kUmg^{-1}$) and amyloglucosidase (142.0 $kUmg^{-1}$) were produced by *A. fumigatus* KIBGE-IB33 (Figure 1). This strain was also capable of producing considerable amount of glucose (81.0 gL^{-1}) which can be used for the production of ethanol. The fermentable sugar produced by this isolate can be easily metabolized by *S. cerevisiae*. However, other filamentous fungi produced lower titers of both α-amylase and amyloglucosidase along with lower concentration of glucose. Although maximum α-amylase was produced by *A. flavus* KIBGE-IB34 but it did not showed higher glucose formation rate as compared to *A. fumigatus* KIBGE-IB33 therefore, this isolate was selected for further studies.

Optimization of various cultivation parameters plays an important role. For effective bioethanol production, fermentation time of the microbial culture and the type of the renewable carbon source used for the production of starch hydrolyzing enzymes and ethanol are among the most important factors. The aforementioned factors will ultimately direct the overall process cost for the development or the scale up of a new methodology in any bioethanol producing industry. Hence, fermentation time for the production of α-amylase and amyloglucosidase and different types of carbon sources were studied. Varieties of carbon source have been tested and can play an important role during microbial fermentation because they are the integral components for the production of cellular material and most of the time they are also associated with microbial growth [15]. Much interest has been diverted towards the utilization of economically available carbon sources in order to fulfill the industrial requirements. In the current study, to improve the production of α-amylase and amyloglucosidase for starch hydrolysis and bioethanol production, seven different carbon sources were utilized (Figure 2). The induction pattern for both amylolytic enzymes was different in various carbon sources suggesting that these hydrolases are inducible. Among all, cassava starch proved to be the

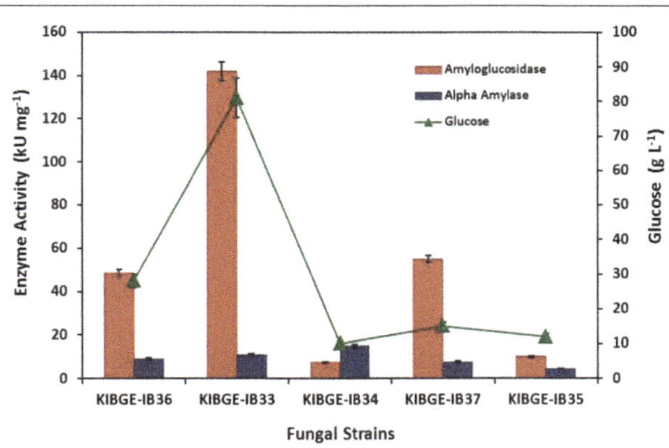

Figure 1 Production of amylolytic enzymes and glucose formation by various filamentous fungi. KIBGE-IB33: *A. fumigatus*; KIBGE-IB34: *A. flavus*; KIBGE-IB35: *A. terreus*; KIBGE-IB36: *A. niger*; KIBGE-IB37: *A. versicolor*.

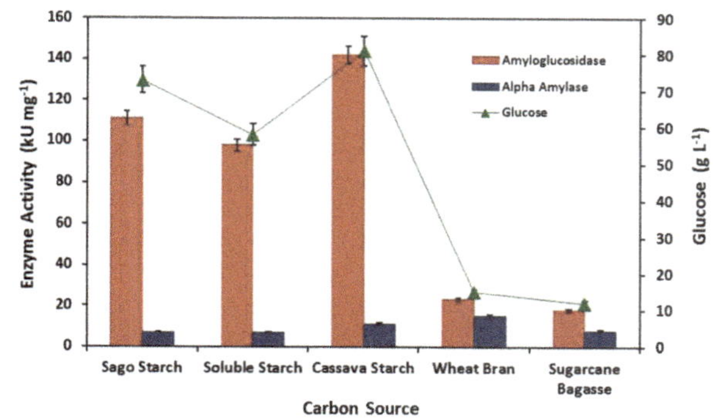

Figure 2 Production of amylolytic enzymes from *Aspergillus fumigatus* KIBGE-IB33 and glucose formation using various carbon sources.

most favorable inducer and contribute highest amount of enzyme units (α-amylase: 11.0 $kUmg^{-1}$; amyloglucosidase: 142.0 $kUmg^{-1}$) and glucose formation (81.0 gL^{-1}) as compared to the other carbon sources tested. These results suggest that pure starch based carbon sources including sago starch, soluble starch (potato) and cassava starch are more suitable for the production of enzyme and glucose formation as compared to the different complex biomass (wheat bran and sugarcane bagasse) whereas, no enzyme production was detected when wheat starch and rice bran were used. This is because the lingocellulosic tough plant matrix was not pretreated. The cell free filtrate (CFF) collected after fermentation showed negligible titers for cellulase, pectinase and xylanase (data not shown) therefore; the starch content was not accessible for fermentation as compared to the purified starch materials. Fatima and Ali [16] tested sixteen fungal species for the production of amyloglucosidase (activity ranged between: 1.906-12.675 U ml^{-1} min^{-1}) using starch in fermentation medium and the best strain they identified was *A. oryzae* IIB-6 (12.673 ± 0.998 Uml^{-1} min^{-1}). They also noticed a 30% increase in the enzyme activity when some of the process parameters were altered (pH and incubation time). Very recently, Puri *et al.* [17] reported the use of rice bran: wheat bran (1:1), rice bran: paddy husk (1:1) for the production of amylase and amyloglucosidase and the maximum amylase (2.72 IU) and amyloglucosidase (4.11 IU) activity was achieved when rice bran was incorporated in the fermentation medium inoculated with *A. oryzae*. However, *A. fumigatus* NTCC 1222 exhibited 341.7 U/mL amylase activity under solid state fermentation when incubated at 35°C (pH-6.0) for 06 days in nutrient salt solution [18]. In another study, detergent mediated production of glucoamylase in the presence of soluble starch is also reported using *A. niger* FME under shake flask system [19]. Several other researchers have also used cassava starch and cassava pulp as alternative carbon source for bioethanol production using α-amylase and amyloglucosidase [20-23]. The compositional analysis of cassava starch used in this study is presented in Table 1. The type and source of starch based materials plays a crucial role for achieving maximum bioethanol yield. The starch content in variety of biomaterials will govern the cost of bioethanol production. Therefore, before considering the production of bioethanol using a specific source, the nature of the starch molecule (linkage, granule size and shape) and the method of extraction employed must be kept in consideration. However, the values of compositional analysis of starch cannot be compared with other starch sources because of the variation of the plant source and the methods used to analyze the structure or content of starch. After selection of a suitable carbon source, fermentation time for the production of amylolytic enzymes was also studied by incubating *A. fumigatus* KIBGE-IB33 for different time interval ranging from 02 to 07 days. It was observed that production of α-amylase and amyloglucosidase started after 02 days of incubation and both the hydrolases were continuously produced up to day 06 and 07, respectively with a maximum titter secreted at day 03 (Figure 3). Afterwards, it was also noticed that as incubation time increases, amylolytic activity as well as glucose formation decreases. This might be due to the fact, with the passage of time the nutrients become depleted and other secondary metabolites are formed which eventually alters the pH of the medium and inhibits both the growth of the fungi as well as enzyme secretion [24]. Most of the time, secondary metabolites have a catabolic repression effect. Amylase from fungal sources is normally produced after 03 to 07 days of incubation but in some cases, enzyme secretion can be extended up to 15 days [25-28]. Prolong incubation time is one of the drawbacks of using filamentous fungi at industrial scale level which eventually increases process cost.

After fermentation, α-amylase and amyloglucosidase were partially purified from the CFF. Most of the time in CFF some other constitutive enzymes may also be present including various types of proteases which could cause interference during liquefaction and saccharification process of cassava starch. Therefore, to avoid any hindrance, starch hydrolyzing enzymes from the CFF were purified using gradient precipitation in the presence of ammonium sulfate ranging from 20.0% to 80.0% and among it 40% saturation level was selected. Table 2 summarizes the purification profile of α-amylase and amyloglucosidase. Fold purification of both α-amylase and amyloglucosidase increased up to 4.1 and 4.2 times with a specific activity of 9.2 $kUmg^{-1}$ and 393 $kUmg^{-1}$, respectively. Previously, Slivinski *et al.* [29] and da Silva and Peralta [30] have also reported precipitation of amyloglucosidase using ammonium sulfate produced by *A. niger* and *A. fumigatus*, respectively. Ammonium sulfate precipitation method was also used for purification of α-amylase from *Pencillium chrysogenum* and *A. niger* JGI 24, respectively [31,32]. The enzyme kinetic analysis showed that the optimum pH, temperature, V_{max} and K_m values for α-amylase from *A. fumigatus* KIBGE-IB33 were 6.0 (citrate buffer, 50.0 mM), 65°C, 25.0 kU ml^{-1} and 0.5 mg ml^{-1}, respectively. Whereas, optimum pH, temperature, V_{max} and K_m values for amyloglucosidase were 5.0 (citrate buffer, 50.0 mM), 60°C, 105.0 kU ml^{-1} and 2.56 mg ml^{-1}, respectively. The optimum pH and temperature for amyloglucosidase activity isolated from *A. niger* FME was 5.0 and 45°C, respectively whereas, the K_m and V_{max} values were determined using soluble starch as substrate as 94 μg ml^{-1} and 39.02 Umg^{-1}, respectively [19]. In another recent study, the optimum pH of amyloglucosidase was 6.0 and the optimum temperature was 60°C along with K_m and V_{max} values of 0.046 mg ml^{-1} and 769 Umg^{-1} [33]. Looking at the kinetic properties of the other recently studied amylolytic enzymes, it is therefore suggested that both the hydrolases from *A. fumigatus* KIBGE-IB33 could be used for industrial

Table 1 Compositional analysis of commercially available cassava starch

Composition	Cassava content (%, *w/w*)
Total sugar*	85.2 ± 4.26
Total protein*	1.4 ± 0.07
Reducing sugar*	6.2 ± 0.31
Glucose*	nil
Moisture content**	0.9 ± 0.04
Amylose***	10.7 ± 0.53
Amylopectin***	89.3 ± 4.46

*2.0% Cassava solution.
**1.0 g Cassava starch.
***0.4 g Cassava starch.
±Specifies standard deviation (SD) among three equivalent replicates. Values in each set differ significantly: $p \leq 0.05$.

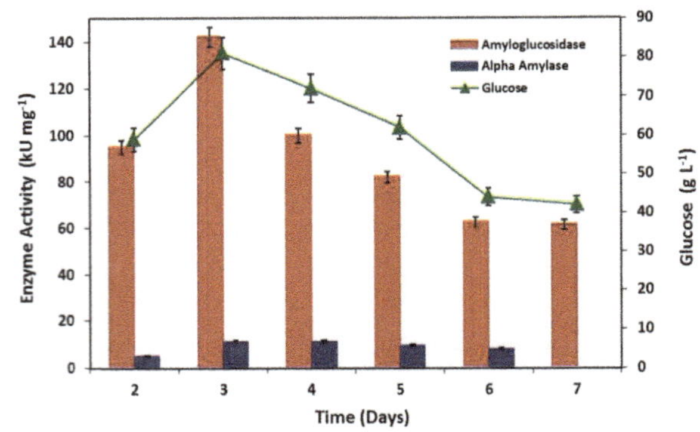

Figure 3 Production of amylolytic enzymes and glucose formation by *Aspergillus fumigatus* KIBGE-IB33 at different incubation times.

Table 2 Purification profile of starch hydrolyzing enzymes produced from *Aspergillus fumigatus* KIBGE-IB33

Steps	Total volume (ml)	Total enzyme units (kU)	Total protein (mg)	Specific activity (kU mg^{-1})	Fold purification
		Alpha amylase			
Crude	500	5500	2500	2.2	1.0
$(NH_4)_2SO_4$ Precipitation	30	1950	210	9.2	4.1
		Amyloglucosidase			
Crude	500	229500	2500	91.8	1.0
$(NH_4)_2SO_4$ Precipitation	30	82500	210	393	4.2

starch saccharification purpose. Both partially purified enzymes were used for bioethanol production.

Conversion of starchy materials into ethanol is an intricate process and several attempts have been made to produce bioethanol in commercially feasible quantities and to easily scale-up the methodology used. Cassava starch is a complex molecule containing amylose and amylopectin and for the production of bioethanol, first the starch molecules must be hydrolyzed into more simple sugars. Some pretreatment techniques including hot water and steam explosion treatment, alkaline and solvent pretreatment, acid hydrolysis and enzymatic degradation for the breakdown of complex starch molecule into simpler sugars have also been studied [34-37]. More recently, a new pre-treatment technique known as popping pretreatment have gained attention for the hydrolysis of starchy feedstock [38]. However, enzymatic degradation using different hydrolases is mostly preferred because during acid hydrolysis the percent conversion of starch into reducing sugars is low as compared to the enzymatic degradation [39-41]. With the progression and advancement in enzymology, amylolytic enzymes are now preferable over conventional methods because enzymatic treatments lead towards high yield of glucose with reduced energy consumption. Therefore, in the current study gelatinized cassava starch was liquefied using α-amylase and was further saccharified by means of amyloglucosidase. However, before breaking starch into simple fermentable sugars, the time required for both the processes to occur effectively was also analyzed by incubating the gelatinized starch slurry with both partially purified amylolytic enzymes for different time intervals. Glucose was the main end-product which is required for production of bioethanol, therefore the concentration of glucose formation as well as percent saccharification was monitored throughout this study. Gelatinized cassava starch was mixed with partially purified α-amylase (9.2 kUmg^{-1}) and amyloglucosidase (393.0 kUmg^{-1}). It was observed that as the reaction time increases, the formation of glucose (40.0 gL^{-1}) as well as percent saccharification (60.0%) also increased up to 90.0 minutes and beyond that both parameters become constant (Figure 4). This glucose containing mixture was further used for the production of ethanol. Efficiency of enzymatic liquefaction and saccharification also depends upon optimum enzyme activity as well as the purity of amylolytic enzymes as crude enzyme takes longer time period to completely hydrolyze starch molecule into glucose as compared to the purified enzyme (Table 3). This percent saccharification (60.0%) could also be further augmented by either improving the purity of enzyme or by incorporating other hydrolyase (xylanases, pectinases or cellulases) along with these amylolytic enzymes [42,43]. As reported earlier further increase in percent saccharification could also be achieved if the starch slurry was autoclaved before addition of amylolytic enzyme [44]. Similarly, Aggarwal *et al.* [45] and Soni *et al.* [46] have also discussed about the role of the purity level of amylolytic enzymes during starch hydrolysis. In the same way, Shanavas *et al.* [20] have also previously analyzed the effect of reaction time on saccharification of cassava starch and have obtained maximum percent saccharification after 30.0 minutes of incubation followed by slight increase when using commercially available starch hydrolyzing enzymes. On the contrary, Aggarwal *et al.* [45] reported maximum percent saccharification using crude amylolytic enzymes after 24 hours of incubation time. Very recently, Gohel *et al.* [47] used simultaneous saccharification and solid state fermentation for the production of ethanol using Indian sorghum feedstock and also incorporated acid fungal protease instead of urea for better ethanol yield.

A large number of microbes including bacteria, yeast and fungi are capable of producing ethanol from fermented

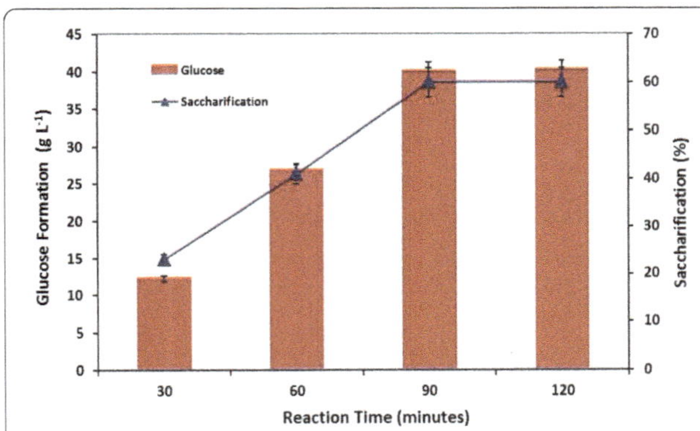

Figure 4 Percent saccharification of cassava starch and glucose formation at different reaction time intervals.

Table 3 Optimized conditions for starch hydrolysis in the presence of crude and partially purified amylolytic enzymes

Enzyme*	Enzyme Units (kU mg^{-1})	Starch** (gL^{-1})	Function	Time (minutes)	Temperature (°C)	pH
Alpha Amylase (Crude)	2.2	20.0	Liquefaction	120	65	5.0
Alpha Amylase (Partially Purified)	9.2	20.0	Liquefaction	15	65	6.0
Amyloglucosidase (Crude)	91.8	20.0	Saccharification	360	60	5.0
Amyloglucosidase (Partially Purified)	393.0	20.0	Saccharification	90	60	5.0

*Both enzymes were produced by *Aspergillus fumigatus* KIBGE-IB33.
**Cassava starch.

sugars [48]. Among them *S. cerevisiae* is widely used for production of bioethanol because it is not only able to produce high amount of ethanol but it can also tolerate and survive higher concentrations of ethanol formed in the medium [49,50]. However, most of the yeast strains are not capable to directly ferment complex starchy materials [48]. Therefore, in the current study *S. cerevisiae* was used for the production of ethanol from glucose which was earlier formed by the action of fungal amylolytic enzymes. Table 4 summarizes the results of bioethanol production after 24 and 48 hours of yeast fermentation and it was noticed that the maximum percent yield of bioethanol (84.0%) was obtained after 48 hours. After 48 hours, the fermented medium was distilled and the percent yield became 53.0%. The concentration and purity of the distilled ethanol was also analyzed using gas chromatography (GC).

Several techniques including direct fermentation, simultaneous saccharification, simultaneous non-thermal saccharification, ultrasound assisted treatment and solid-state fermentation have been studied previously using different starchy materials and microbial sources for the production of bioethanol [20,21,51-56]. Along with this ethanol has also been produced by repeated batch culture through immobilization of *S. cerevisiae* and *S. pastorianus* IFO0751 on calcium alginate and porous cellulose carriers, respectively [57,58]. Nikolic *et al.* [54] used ultrasound-assisted treatment for direct conversion of corn meal into bioethanol but the cost related to this method in amount of energy consumption is very high. Beside this, the pre-treatment of multiple biomass or starch flour will also add extra budget that will eventually affect the feasibility of the bioethanol. The attempt made in the current study by consuming commercially available cheap cassava starch along with saccharification by synergistic effect of fungal amylolytic enzymes had revealed that this two-step based method can be used to achieve higher yields of bioethanol. Further, the process cost can also be reduced by using other inexpensive starchy materials or by establishing pilot programs that will scrutinize the actual feasibility and sustainability of the overall process developed.

Conclusions

In conclusion, an improved and effective enzymatic saccharification of inexpensive cassava starch using amylolytic enzymes from *A. fumigatus* KIBGE-IB33 was developed. The glucose obtained after enzymatic degradation was utilized for the bioethanol production using *S. cerevisiae*. Dual systematic enzyme conversion has advantages in terms of reduced energy consumption as well as increased production of fermentable sugar to achieve maximum bioethanol yield as compared to other processes. In addition, the process developed is more rapid as compared to the previously conducted studies using liquefaction and saccharification of cassava starch.

Methods

Reagents

All reagents were of analytical grade and were obtained from commercial sources. Sago and cassava starch were purchased from local market, Karachi, Pakistan. Peptone (Oxoid, England), yeast extract (Oxoid, England), ammonium sulfate (Serva, Germany) and dipotassium hydrogen phosphate (Serva, Germany) were purchased from a local vendor. Whereas, magnesium sulfate, sodium hydroxide, sodium carbonate, copper sulfate, sodium potassium tartarate, sulfuric acid and potassium dichromate were acquired from Scharlau (Spain). Other chemical used including 3′, 5′- dinitrosalicylic acid (DNS) was purchased

Table 4 Production of bioethanol using *Saccharomyces cerevisiae*

Incubation time (Hour)	Glucose concentration (gL^{-1})	Ethanol (%)	Theoretical yield (g)	Actual yield (g)	Yield** (%)
		Before distillation			
24	40.0	36.0	2.04	1.44	70
48	40.0	43.0	2.04	1.72	84
		After distillation			
48	40.0	27.0	2.04	1.08	53

*Mass of ethanol formed per mass of glucose consumed.

from BDH Chemicals (USA) and anthrone from MP Biomedicals (France) while, soluble starch and folin ciocalteu reagent were purchased from Merck (Germany).

Isolation and identification of filamentous fungi

The natural fungal isolates used in the current study were isolated from different soil samples that were collected aseptically from diverse vegetative fields located in Karachi, Pakistan. All the isolates were obtained after serial platting on potato dextrose agar (PDA) at 30°C for 05 days according to the standard protocols. PDA medium consist of (gL^{-1}): Boiled potato extract, 300.0 ml; dextrose, 20.0 g and agar, 16.0 g. Initially, 07 different fungal species were isolated from different samples. Among them 05 filamentous fungi were selected and characterized based on colonial morphology, 18S rDNA sequence cataloging and microscopic analysis using lactophenol blue staining method [59,60]. After 18S rDNA gene analysis and sequencing, the sequences were analyzed by similarity search using BLAST (http://www.ncbi.nlm.nih.gov/BLAST/) and were submitted to NCBI GenBank database. The confirmed sequences received the following GenBank accession numbers: KF905648, KF905649, KF905650, KF905651, and KF905652 for *A. fumigatus* KIBGE-IB33, *A. flavus* KIBGE-IB34, *A. terreus* KIBGE-IB35, *A. niger* KIBGE-IB36 and *A. versicolor* KIBGE-IB37, respectively. All of these isolates were tested for the amylolytic enzyme production based on starch-iodine plate method. All fungal isolates were plated on starch medium plates containing (gL^{-1}): cassava starch, 10.0; yeast extract, 10.0; peptone, 10.0; K_2HPO_4, 1.0; and $MgSO_4.7H_2O$, 1.0. The cultures were incubated at 30°C for 05 days. After incubation the plates were flooded with potassium-iodide solution for the detection of amylolytic activity. Isolates were selected based on clear halo-zone around the fungal growth. All isolates were preserved on PDA slants at 4°C for further analysis and were sub-cultured routinely. Purified *Sacchromyces cerevisiae* (baker's yeast) was purchased from the local market, Karachi, Pakistan and was grown and maintained in YPD medium (gL^{-1}:yeast extract,10.0; Bacto-peptone, 20.0 and glucose, 20.0).

Inoculum preparation for seed culture

Total viable spores were calculated in order to prepare fungal inoculum. For this purpose spores were transferred using sterile needle from a 05 day old fungal culture grown on PDA slant and re-suspended in 10.0 ml sterile distilled water containing 0.1% Tween-20. Each suspension was serially diluted up to 10^{-5} in order to make homogenous spore suspension of 10^6 to 10^8spores ml^{-1}).

Medium used for the production of amylolytic enzymes

All the selected filamentous fungi were tested for the production of α-amylase and amyloglucosidase in the presence of starch (cassava) under batch conditions using submerged fermentation technique. Production of α-amylase, amyloglucosidase and glucose was monitored at different time intervals (02 to 07 days). Basal medium used for the production of α-amylase and amyloglucosidase consists of (gL^{-1}): Cassava starch, 20.0; yeast extract, 10.0; peptone, 10.0; K_2HPO_4, 1.0; and $MgSO_4.7H_2O$, 1.0. Initial pH of the medium was adjusted at 7.0 before sterilization at 121°C for 15 minutes. Fresh seed culture (10.0 ml) was inoculated in 90.0 ml production medium and incubated at 30°C for 03 days under static and anaerobic conditions. It was then further transferred into 900.0 ml medium and incubated at 30°C up to 07 days. The fungal spores were harvested by centrifuging the fermented broth at 40248 × *g* for 15 minutes at 4°C. The supernatant was filtered using 0.45 μ filter under vacuum. The cell free supernatant containing the amylolytic enzymes was stored at −20°C for further analysis. All the experiments were conducted in triplicates.

Optimization of physicochemical parameters for maximum enzyme yield

For the enhanced production of amylolytic enzymes, different inducing substrates (carbon sources) and fermentation time were optimized. For this purpose seven different carbon sources including sago starch, soluble starch (potato), cassava starch, wheat starch, wheat bran, rice bran and sugarcane bagasse were used in the concentration of 20.0 gL^{-1}. *A. fumigatus* KIBGE-IB33 was incubated for different time intervals ranging from 02 to 07 days at 30°C under static and anaerobic condition for the selection of optimum fermentation time. Enzyme titer in terms of specific activity and glucose formation were monitored in triplicate.

Partial purification of amylolytic enzymes

The cell free supernatant containing α-amylase and amyloglucosidase was precipitated using salt precipitation method. For this purpose, salt gradient precipitation technique was employed ranging from 20.0% to 80.0% saturation using ammonium sulfate. 20% salt was incorporated gradually in CFF with continuous stirring at 4°C and was equilibrated for 18 hours. The precipitates formed were centrifuged at 40248 × *g* for 10 minutes at 4°C and were dissolved in citrate buffer (50.0 mM, pH-5.0). In the next run, again the 20% salt saturation was performed using the same supernatant up to 80% and every time the precipitates were equilibrated for 18 hours at 4°C. During each saturation range, the precipitates were monitored and calculated for both enzymes unit in terms of kU mg^{-1} of protein. The saturation level at which both hydrolyase were precipitated out with maximum unit was selected (40%).

Enzyme assay and total protein estimation

Enzyme activity of α-amylase and amyloglucosidase was estimated using DNS [61] and GOD-PAP method [62,63], respectively. One unit of α-amylase is defined as the "amount of enzyme that liberates 1.0 mM of maltose per minute under standard assay condition". Whereas, one unit of amyloglucosidase is defined as the "amount of enzyme that liberates 1.0 mM of glucose per minute under standard assay condition". The specific units of both amylolytic enzymes are expressed in terms of kilo units per mg of protein (kU mg^{-1}). Total protein was calculated using Lowry's *et al.* [64] method with bovine serum albumin as standard.

Production of bioethanol

Bioethanol was produced using glucose, which was obtained after hydrolysis of starch (cassava) using partially purified amylolytic enzymes. Partially purified α-amylase and amyloglucosidase (30.0 ml) was amalgamated in 2.5 liters of pre-gelatinized cassava starch slurry (20.0 gL^{-1}) which was prepared in citrate buffer (pH-5.0, 50.0 mM). The reaction mixture was kept at 65°C for liquefaction (15.0 minutes) and at 60°C for saccharification (90.0 minutes). The reaction was terminated by boiling the reaction mixture for 10 minutes. The volume was reduced up to 2.5 folds and the liberated glucose was detected before and after concentrating the reaction mixture. However, for optimized liquefaction and saccharification condition, the reactions were monitored from 15 minutes up to 08 hours with an interval of 15 minutes. Percent saccharification of cassava starch was calculated as followed:

$$Saccharification\ (\%) = \frac{Glucose}{Substrate} \times 100$$

In this concentrated hydrolyzate, yeast extract and peptone were incorporated in the concentration of 3.0 gL^{-1} and 10.0 gL^{-1}, respectively and the pH was adjusted up to 7.0. *S. cerevisiae* (2.0%) was cultured in this medium anaerobically at 30°C up to 48 hours. After 48 hours, ethanol was collected through distillation and the distillate was analyzed for the detection of ethanol. Percent yield of bioethanol was calculated as followed:

$$Ethanol\ Yield\ (\%) = \frac{Actual\ Ethanol\ Produced\ (g)}{Theoritcal\ Ethanol\ from\ Sugar\ Consumed\ (g)} \times 100$$

Analytical method for bioethanol analysis

For the determination of bioethanol concentration, the distillate was analyzed using Caputi *et al.* [65] method. The fermented broth was distilled using a Heating Mantle (Barnstead-Electrothermal, Thermo Scientific) at 78°C along with a quick fit distillation apparatus equipped with a Lie-big condenser and the in-let cold water attached to a chiller. All the experiments were run independently in triplicate and the results presented are the mean of three values.

Determination of bioethanol concentration by gas chromatography (GC)

Bioethanol concentration was also verified using gas chromatography system (GC17A, Shimadzu, Japan) equipped with flame ionization detector (FID). Column used was TRB-5 (30 × 0.25 mm × 0.25 μm) with nitrogen as a carrier gas (20 $cmsec^{-1}$). The temperature of the detector and the injector were kept at 200°C and 130°C, respectively. The split ratio was 100:1 and the peak area of the compound was integrated against an external standard of absolute ethanol.

Physico-chemical characteristics of cassava starch

The cassava starch used in this study was purchased from the local market in Karachi, Pakistan. For the determination of total sugar anthrone method was used [66] whereas, reducing sugar was detected using DNS method [61]. Total protein was performed using Lowry's *et al.* [64] method. Glucose content was estimated using GOD-PAP method [62,63]. Moisture content was calculated using standard drying method at 105°C until the weight become constant. Amylose and amylopectin fractions were calculated by iodometric method as suggested previously [67].

Abbreviations

CFF: Cell free filtrate; DNS: 3', 5'-dinitrosalicyclic acid; FID: Flame ionization detector; GC: Gas chromatography; gL^{-1}: Grams per liter; GOD-PAP: Glucose oxidase per oxidase method; kU mg^{-1}: Kilo units per milligrams; mM: Milli moles; PDA: Potato dextrose agar.

Competing interests

Its publication is approved by all authors and they do not have any conflict of interest regarding any financial, personal or other relationships with any other people or organizations.

Authors' contributions

SP carried out the major experimental work. AA supervised in acquisition of laboratory data and interpretation of data along with finalizing the manuscript. SI participated in the analytical analysis of bioethanol. NNS performed enzyme purification. SAQ designed the project and gave the final approval of the manuscript. All authors have read and approved the final manuscript.

Acknowledgements

Authors are obliged to Dr. Asma Ansari for the identification of the natural isolates used in the current study and also gratefully acknowledge the financial support from KIBGE, University of Karachi, Karachi, Pakistan.

References

1. Lindeman LR, Rocchiccioli C: **Ethanol in Brazil: Brief summary of the state of the industry in 1977.** *Biotechnol Bioeng* 1979, **21**:1107–1119.
2. Kadam KL, McMillan JD: **Availability of corn stover as a sustainable feedstock for bioethanol production.** *Bioresour Technol* 2003, **18**:17–25.

3. Wilke CR, Yang RD, Scamanna AF, Freitas RP: **Raw material evaluation and process development studies for conversion of biomass to sugars and ethanol.** *Biotechnol Bioeng* 1981, **23**:163–183.
4. Periyasamy S, Venkatachalam S, Ramasamy S, Srinivasan V: **Production of bioethanol from sugar molasses using *Saccharomyces cerevisiae*.** *modern Appl Sc* 2009, **3**:32–37.
5. Maisch WF, Sobolov M, Petricola AJ: **Distilled beverages.** In *Microbial Technology*. Edited by Peppler HJ, Perlman D. New York: Academic; 1979:79.
6. Sasson A: *Feeding tomorrow's world.* Paris: UNESCO; 1990:500–510.
7. Shetty JK, Chotani G, Duan G, Bates D: **Cassava as an alternative feedstock in the production of transportation fuel.** *Int Sugar J* 2007, **109**:3–11.
8. Pandey A, Soccol CR, Nigam P, Soccol VT, Vanderberghe LPS, Mohan R: **Biotechnological potential of agro-industrial residue. II: Cassava bagasse.** *Bioresour Technol* 2000, **74**:81–87.
9. Emilia-Abraham T, Krishnaswamy S, Ramakrishna SW: **Effect of hydrolysis condition of cassava on oligosaccharide profile and alcohol fermentation.** *Starch-Starke* 1986, **39**:237–240.
10. Nguyen TLT, Gheewala SH, Garivait S: **Energy balance and GHG-abatement cost of cassava utilization for fuel ethanol in Thialand.** *Energy Policy* 2007, **35**:4585–4596.
11. Dai D, Hu Z, Pu G, Li H, Wang CT: **Energy efficiency and potential of cassava fuel ethanol in ghangxi region of China.** *Energy Conver Manage* 2006, **47**:1686–1699.
12. Fogarty WM, Kelly CT: **Microbial amylase.** In *Microbial Enzyme and Biotechnology*. London: Applied Science Publisher; 1983:1–90.
13. Pandey A: **Glucoamylase research: An overview.** *Starch-Starke* 1995, **47**:439–445.
14. Prabakaran M, Thennarasu V, Mangala RA, Bharathidasan R, Chandrakala N, Mohan N: **Comparative studies on enzyme activities of wild and mutant strains isolated from sugarcane field.** *Indian J Sci Technol* 2009, **2**:6846–6849.
15. Dunn GM: **Nutritional requirements of microorganisms.** In *Comprehensive Biotechnology*, Volume 1. Edited by Moo-Young M. New York: Pergamon Press, Oxford; 1985:113–125.
16. Fatima B, Ali S: **Kinetics of improved 1, 4-alpha-D-glucan glucohydrolase biosynthesis from a newly isolated *Aspergillus oryzae* IIB-6 and parameter significance analysis by 2-factorial design.** *Springer Plus* 2012, **1**:32.
17. Puri S, Arora M, Sarao L: **Production and optimization of amylase and glucoamylase using *Aspergillus oryzae* under solid state fermentation.** *Int J Res Pure Appl Microbiol* 2013, **3**:83–88.
18. Singh S, Singh S, Bali V, Sharma L, Mangla J: **Production of fungal amylases using cheap, readily available agriresidues, for potential application in textile industry.** *BioMed Res Int* 2014, http://dx.doi.org/10.1155/2014/215748.
19. Gudi SK, Gurramkonda C, Rather G, Chandra MGS, Mangamuri UK, Podha S, Choi Y: **Glucoamylase from a newly isolated *Aspergillus niger* FME: Detergent-mediated production, purification, and characterization.** *J Korean Soc Appl Biol Chem* 2013, **56**:427–433.
20. Shanavas S, Padmaja G, Moorthy SN, Sajeev MS, Sheriff JT: **Process optimization for bioethanol production from cassava starch using novel eco-friendly enzymes.** *Biomass Bioenergy* 2011, **35**:901–909.
21. Rattanachomsri U, Tanapongpipat S, Eurwilaichitr L, Champreda V: **Simultaneous non thermal saccharification of cassava pulp by multi enzyme activity and ethanol fermentation by *Candida tropicalis*.** *J Biosci Bioeng* 2009, **107**:488–493.
22. Wook CG, Um HJ, Kim M, Kim Y, Kang HW, Chung BW, Kim YH: **Isolation and characterization of ethanol producing *Shizosaccharomyces pombe* CHFY0201.** *J Microbiol Biotechnol* 2010, **20**:828–834.
23. Ziska LH, Runion GB, Tomecek M, Prior SA, Torbet A, Sicher R: **An evaluation of cassava, sweet potato and field corn as potential carbohydrate sources for bioethanol production in Alabama and Maryland.** *Biomass Bioenergy* 2009, **33**:1503–1508.
24. Feroza B, Begum S, Hossain M: **Production of glucoamylase by *Aspergillus niger* in liquid culture and determination of its cultural condition.** *Bangladesh J Sci Ind Res* 1998, **33**:309–311.
25. Ali S, Mahmood S, Alam R, Hossain Z: **Culture condition for production of glucoamylase from rice bran by *Aspergillus terreus*.** *World J Appl Microbiol Biotech* 1989, **5**:525–532.
26. Ghosh A, Chatterjee B, Das A: **Induction and catabolite repression of high affinity glucoamylase in *Aspergillus terreus* strain 4.** *J Gen Microbiol* 1990, **136**:1307–1311.
27. Kumar S, Satyanarayana P: **Statistical optimization of thermostable and neutral glucoamylase produced by a thermophillic mould *Thermomucor indicae-seudaticae* in solid state fermentation.** *Worl J Microbiol Biotechnol* 2004, **20**:895–902.
28. Mohamed L, Zakaria M, Ali A, Senhaji W, Muhamed O, Mohamed E, Hassan BEL, Mohamed J: **Optimization of growth and extracellular glucoamylase produced by *Candida fumata* isolate.** *Afr J Biotechnol* 2007, **6**:2590–2595.
29. Slivinski CT, Machado AVL, Iulek J, Ayub RA, de Almeida MM: **Biochemical characterization of a glucoamylase from *Aspergillus niger* produced by solid-state fermentation.** *Braz Arch Biol Technol* 2011, **54**:559–568.
30. da Silva BW, Peralta WM: **Purification and characterization of thermostable glucoamylase from *Aspergillus fumigatus*.** *Can J Microbiol* 1998, **44**:493–497.
31. Doss A, Anand SP: **Purification and optimization of fungal amylase from litter samples of Western Ghats, Coimbatore, Tamilnadu (India).** *J Sci Res Rev* 2013, **2**:001–004.
32. Varalakshmi KN, Kumudini BS, Nandini BS, Solomon J, Suhas R, Mahesh B, Kavitha AP: **Production and characterization of α-amylase from *Aspergillu sniger* JGI 24 isolated in Banglore.** *Pol J Microbiol* 2009, **58**:29–36.
33. Koç Ö, Metin K: **Purification and characterization of a thermostable glucoamylase produced by *Aspergillus flavus* HBF34.** *Afr J Biotechnol* 2010, **9**:3414–3424.
34. Littlewood J, Wang L, Turnbull C, Murphy RJ: **Techno-economic potential of bioethanol from bamboo in China.** *Biotechnol Biofuels* 2013, **6**:173.
35. Sun Y, Cheng J: **Hydrolysis of lignocellulosic materials for ethanol production.** *Rev Bioresour Tehnol* 2002, **83**:1–11.
36. Kaar WE, Holtzapple MT: **Using lime pretreatment to facilitate the enzymatic hydrolysis of corn stover.** *Biomass Bioenergy* 2000, **18**:189–199.
37. Maiorella BL: **Ethanol.** In *Comprehensive Biotechnology*, Volume 3. New York: Pergamon Oxford; 1985:861–909.
38. Wi SW, Choi IS, Kim KH, Kim HM, Bae H: **Bioethanol production from rice straw by popping pretreatment.** *Biotechnol Biofuels* 2013, **6**:166.
39. Bej B, Basu RK, Ash SN: **Kinetic studies on acid catalyzed hydrolysis of starch.** *J Sci Ind Res* 2008, **67**:295–298.
40. Wang YJ, Truong VD, Wang L: **Structure and rheological properties of corn starch as affected by acid hydrolysis.** *Carbohyd Res* 2003, **52**:327–333.
41. Kunlan L, Lixin X, Jun L, Jun P, Gooying C, Zuwei X: **Salt associated acid hydrolysis of starch to D-glucose under microwave irradiation.** *Carbohyd Res* 2001, **331**:9–12.
42. Thomas KC, Hynes SH, Ingledew WM: **Practical and theoretical considerations in the production of high concentrations of alcohol by fermentations.** *Process Biochem* 1996, **31**:321–324.
43. Jones AM, Ingledew WM: **Fuel alcohol production: Optimization of temperature for efficient very high gravity fermentation.** *Appl Environ Microbial* 1994, **60**:1048–1051.
44. Kimura A, Robyt JF: **Reaction of enzyme with starch granules: Enhanced reaction of glucoamylase with gelatinized starch granules.** *Carbohydr Res* 1996, **288**:233–240.
45. Aggarwal NK, Yadav SK, Dhamija SS, Yadav BS: **Optimization of enzymatic hydrolysis of pearlmillet for glucose production.** *Starch/ Stärke* 2001, **53**:330–335.
46. Soni SK, Rao MW, Das D: **Studies on glucoamylase produced from *Aspergillus awamori* (NRRL-3112) and their effect on saccharification of potato starch.** *Indian J Exp Biol* 1995, **33**:525–532.
47. Gohel V, Duan G, Maisuria VB: **Impact of an acid fungal protease in high gravity fermentation for ethanol production using indian sorghum as a feedstock.** *Biotechnol Prog* 2013, **29**:329–336.
48. Lin Y, Tanaka S: **Ethanol fermentation from biomass resources: Current state and prospects.** *Appl Microbiol Biotechnol* 2006, **69**:627–642.
49. Olsson L, Nielsen J: **The role of metabolic engineering in the improvement of *Saccharomyces cerevisiae*: Utilization of industrial media.** *Enzyme Microb Tech* 2000, **26**:785–792.
50. Olsson L, Hahn-Hägerdal B: **Fermentative performance of bacteria and yeast in lignocellulose hydrolysates.** *Process Biochem* 1993, **28**:249–257.
51. Erdei B, Hancz D, Galbe M, Guido Z: **SSF of steam-pretreated wheat straw with the addition of saccharified or fermented wheat meal in integrated bioethanol production.** *Biotechnol Biofuels* 2013, **6**:169.
52. Nyanga LK, Nout MJR, Smid EJ, Boekhout T, Zwietering MH: **Fermentation characteristics of yeasts isolated from traditionally fermented masau (*Ziziphus mauritiana*) fruits.** *Int J Food Microbiol* 2013, **166**:426–432.
53. Karmakar M, Ray RR: **Saccharification of agro wastes by the endoglucanase of *Rhizopus oryzae*.** *Ann Biol Res* 2011, **2**:201–208.

54. Nikolic S, Mojovic L, Rakin M, Pejin D: **Ultrasound-assisted production of bioethanol by simultaneous saccharification and fermentation of corn meal.** *Food Chem* 2010, **122:**216–222.
55. Wu T, Wang F, Tang Q, Zhu Z: **Arrowroot as a novel substrate for ethanol production by solid state simultaneous saccharification and fermentation.** *Biomass Bioenergy* 2010, **34:**1159–1164.
56. Murado MA, Pastrana L, Vázquez JA, Mirón J, González MP: **Alcoholic chestnut fermentation in mixed culture. Compatibility criteria between *Aspergillus oryzae* and *Sachharomyces cerevisiae* strains.** *Bioresour Technol* 2008, **99:**7255–7263.
57. Göksungur Y, Zorlu N: **Production of ethanol from beet molasses by Ca-alginate immobilized yeast cells in a packed bed reactor.** *Turk J Biol* 2001, **25:**265–275.
58. Sakurai A, Nishida Y, Saito H, Sakakibara M: **Ethanol production by repeated batch culture using yeast cells immobilized within porous cellulose carriers.** *J Biosci Bioeng* 2000, **90:**526–529.
59. Collier L, Balows A, Sussman M: In *Topley & Wilson's Microbiology and Microbial Infections*, Volume 4. 9th edition. Edited by Ajello L, Hay RJ. London: Arnold, Hodder Headline Group; 1998.
60. St-Germain G, Summerbell R: *Identifying Filamentous Fungi - A Clinical Laboratory Handbook.* 1st edition. Belmont, California: Star Publishing Company; 1996.
61. Miller GL: **Use of dinitrosalicylic acid reagent for determination of reducing sugar.** *Annal Chem* 1959, **31:**426–428.
62. Trinder P: **Determination of glucose in blood using glucose oxidase with an alternative oxygen acceptor.** *Ann Clin Biochem* 1969, **6:**24–28.
63. Trinder P: **Determination of glucose in blood using 4-amino phenazone as oxygen acceptor.** *JClin Pathol* 1969, **22:**246.
64. Lowry OH, Rosebrough NJ, Farr AL, Randall RJ: **Protein measurement with the folin phenol reagent.** *J Biol Chem* 1951, **193:**265–375.
65. Caputi A Jr, Veda M, Brown T: **Spectrophotometric determination of ethanol in wine.** *Am J Enol Vitic* 1968, **19:**160–165.
66. Hassid WZ, Abraham S: *Methods in enzymology III.* New York: Academic Press; 1957:34–35.
67. McGrance SJ, Cornell HJ, Rix CJ: **A simple and rapid calorimetric method for the determination of amylase in starch products.** *Starch/Stärke* 1998, **50:**158–163.

Factors affecting polyhydroxybutyrate accumulation in mesophyll cells of sugarcane and switchgrass

Richard B McQualter[1], Maria N Somleva[2], Leigh K Gebbie[1], Xuemei Li[2,3], Lars A Petrasovits[1], Kristi D Snell[2], Lars K Nielsen[1] and Stevens M Brumbley[1,4*]

Abstract

Background: Polyhydroxyalkanoates are linear biodegradable polyesters produced by bacteria as a carbon store and used to produce a range of bioplastics. Widespread polyhydroxyalkanoate production in C_4 crops would decrease petroleum dependency by producing a renewable supply of biodegradable plastics along with residual biomass that could be converted into biofuels or energy. Increasing yields to commercial levels in biomass crops however remains a challenge. Previously, lower accumulation levels of the short side chain polyhydroxyalkanoate, polyhydroxybutyrate (PHB), were observed in the chloroplasts of mesophyll (M) cells compared to bundle sheath (BS) cells in transgenic maize (*Zea mays*), sugarcane (*Saccharum* sp.), and switchgrass (*Panicum virgatum* L.) leading to a significant decrease in the theoretical yield potential. Here we explore various factors which might affect polymer accumulation in mesophyll cells, including targeting of the PHB pathway enzymes to the mesophyll plastid and their access to substrate.

Results: The small subunit of Rubisco from pea effectively targeted the PHB biosynthesis enzymes to both M and BS chloroplasts of sugarcane and switchgrass. PHB enzyme activity was retained following targeting to M plastids and was equivalent to that found in the BS plastids. Leaf total fatty acid content was not affected by PHB production. However, when fatty acid synthesis was chemically inhibited, polymer accumulated in M cells.

Conclusions: In this study, we provide evidence that access to substrate and neither poor targeting nor insufficient activity of the PHB biosynthetic enzymes may be the limiting factor for polymer production in mesophyll chloroplasts of C_4 plants.

Keywords: Polyhydroxybutyrate, Sugarcane, Switchgrass, Biomass, C4 grasses, Acetyl-CoA, Biopolymer

Background

C_4 photosynthesis is more complex than the ancestral C_3 pathway of carbon fixation and involves morphological and biochemical modifications leading to a significant reduction in photorespiration. This is achieved by compartmentalization of the processes of carbon dioxide (CO_2) fixation and reduction between physiologically distinct mesophyll (M) and bundle sheath (BS) cells. The resulting increase in the photosynthesis rate allows C_4 plants to achieve high biomass yields under adverse conditions, such as drought, high temperature, and CO_2 or nitrogen limitation. It also favours their use for the production of biofuels [1,2] and as 'green' chemical biofactories [3-5] to produce replacements for petroleum-derived products.

Polyhydroxyalkanoates (PHAs), linear polyesters produced naturally in bacteria as a means of storing carbon [6,7], are an ideal molecule to produce in C_4 plants. When extracted and processed they have similar properties to many common plastics but are fully biodegradable. Large-scale production of PHAs in high biomass yielding crops would provide a sustainable supply of bioplastics and energy from one plant feedstock that could be processed

* Correspondence: stevens.brumbley@unt.edu
[1]Australian Institute for Bioengineering and Nanotechnology, the University of Queensland, Brisbane QLD 4072, Australia
[4]Current address: Department of Biological Sciences, The University of North Texas, 1155 Union Circle #305220, Denton, TX 76203-5017, USA
Full list of author information is available at the end of the article

economically in a biorefinery [8]. The simplest PHA, polyhydroxybutyrate (PHB), has been successfully produced in a number of plant species [5,8-11] and has potential applications not only as a bioplastic, but also for the production of chemicals and improved animal feeds [11]. In plants, PHB is produced by the heterologous expression of three bacterial genes encoding a β-ketothiolase (PhaA), an acetoacetyl-CoA reductase (PhaB) and a PHB polymerase (PhaC) [7]. The ketothiolase condenses two molecules of acetyl-CoA to form acetoacetyl-CoA which is reduced to 3-hydroxybutyryl-CoA by the acetoacetyl-CoA reductase. PHB polymerase then incorporates 3-hydroxybutyryl-CoA into the growing polymer chain. Thus, appropriate targeting of the PHB biosynthesis enzymes to subcellular compartments where sufficient acetyl-CoA and reducing power are available is necessary.

In C_4 plants, PHB has been successfully produced in the plastids [12-15] and peroxisomes [16,17]. Although PHB accumulation in the chloroplasts of the C_4 crops maize [14] (*Zea mays* L., NADP-ME C_4 subtype), sugarcane [12,13] (*Saccharum sp.*, NADP-ME C_4 subtype) and switchgrass [11,15] (*Panicum virgatum* L., NAD-ME C_4 subtype) has reached more than 5% of leaf dry weight (DW) in individual leaves, these levels are not yet commercially viable [11]. The highest PHB levels to date have been reported in chloroplasts of C_3 plants upon nuclear expression of plastid-targeted enzymes in *Arabidopsis thaliana* [18] (up to ~40% DW) or direct expression of transgenes from the plastome in tobacco [19] (up to 18% DW). Thus the compartmentalization of C_4 photosynthesis between two distinct cell types, i.e. M and BS cells appears to affect polymer production in high yielding C_4 biomass crops. In maize, sugarcane and switchgrass [12-15] less PHB is produced in the chloroplasts of M cells than in BS cells. While the cause of this disparity in product accumulation between cell types is unknown, this presents a significant loss in yield potential for PHB in these C_4 plants. Mesophyll cells occupy on average 79% and 66% of the total chlorenchyma area per vein for the C_4 subtypes NADP-dependent malic enzyme (NADP-ME) and NAD-dependent malic enzyme (NAD-ME), respectively [20]. Significant increases in PHB production are thus theoretically possible if polymer levels in chloroplasts of M cells can be elevated.

The present study explored the distribution of PHB production between M and BS cells in both sugarcane and switchgrass. We also measured some key metabolites to determine if further insight could be gained into the low M production of PHB. We examined the effect of polymer synthesis on the content of fatty acids, starch and soluble sugars (sucrose and glucose). While *de novo* fatty acid biosynthesis occurs in both types of photosynthetic cells in C4 plants [21] starch is produced preferentially in BS where the Benson-Calvin cycle is localized [22]. The cellular localization of sucrose synthesis varies between C_4 grasses [23-26].

Results

Plant material

Sugarcane lines identified in our previous reports are relabelled as follows with PHB values reported earlier [12,13,15] as the mean ± standard error: Wild-type Q117 (WT), TA4 (LP, 0.45 ± 0.02% DW PHB), 4 F1 (MP1, 1.21 ± 0.24% DW PHB), 8C8 (MP2, 1.30 ± 0.11% DW PHB) and 7C3 (HP, 3.11 ± 0.31% DW PHB). These PHB values reflect data measured in whole-leaf blades from the oldest green leaf. The expression of genes encoding PHB pathway enzymes PhaA, PhaB and PhaC in sugarcane is driven by the maize ubiquitin promoter in LP, the rice ubiquitin promoter in MP1, and the maize *Cab-m5* promoter in MP2 and HP. Sugarcane plants were grown from cuttings taken from the original transgenic plant material and grown in a physical containment glasshouse under ambient lighting conditions. Switchgrass lines identified as WT, LP (0.60 ± 0.08% DW PHB, MP1 (1.10 ± 0.12% DW PHB), MP2 (1.60 ± 0.01% DW PHB), HP (2.44 ± 0.43% DW PHB) represent independent transformation events expressing the PHB biosynthesis genes *phaA*, *phaB* and *phaC*, under the control of the maize *cab-m5* promoter [15]. These PHB values reflect data measured in segments approximately 15 cm from the tip of leaf blades adjacent to the node at the base of the stem. Switchgrass plants were grown in a glasshouse at 28°C with supplemental lighting (16 h photoperiod, sodium halide lamps) and maintained by repotting every six months.

Distribution of PHB enzymes between M and BS chloroplasts

Immunolocalization of PhaA and PhaB in transverse leaf sections of sugarcane and switchgrass showed that these two proteins are targeted to the chloroplasts of both M and BS cells (Figure 1). Our highest producing sugarcane plants and all of the switchgrass plants contain a hybrid PhaC [27,28] for which antiserum was not available and so immunolocalization of PhaC was not conducted.

Activity of PHB enzymes in M and BS

Enzyme activities were measured in total protein extracted from whole leaf and from M and BS chloroplast fractions. Both PhaA and PhaB activity could be detected in the whole leaf extracts from sugarcane (Figure 2A) and switchgrass (Figure 2B). In all samples analysed, PhaA (Figure 2C & D) and PhaB (Figure 2E) activities were either equal or higher in M than BS chloroplasts. Detection of PhaB activity in sugarcane BS protein fractions did not yield consistent results. This could be due to the instability of the PhaB enzyme in crude extracts during prolonged and difficult procedures to isolate chloroplasts from BS strands.

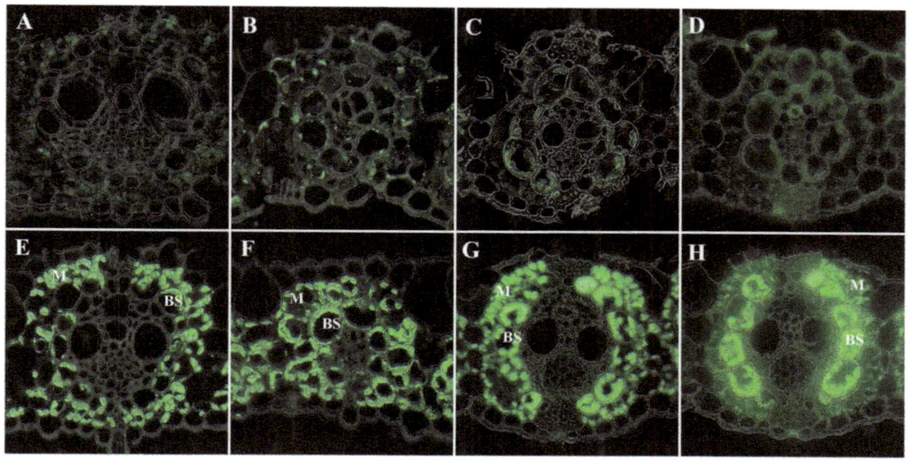

Figure 1 Transverse sections of sugarcane and switchgrass leaves immunolabelled for PHB biosynthesis proteins PhaA and PhaB. Wild type sugarcane **(A, B)**; wild type switchgrass **(C, D)**; high PHB-producing sugarcane **(E, F)** high PHB-producing switchgrass **(G, H)**; PhaA **(A, C, E, G)**; PhaB **(B, D, F, H)**.

Effects on fatty acid synthesis (FAS)

The observed correct targeting of PhaA and PhaB enzymes to M and BS chloroplasts prompted additional studies to examine differences in carbon availability between the two cell types for PHB synthesis. Plastids are the site of FAS, a process that consumes acetyl-CoA as substrate. Acetyl-CoA carboxylase (ACCase) catalyses the first committed step in FAS by converting acetyl-CoA to malonyl-CoA and PhaA would therefore compete with it for the common substrate.

To determine if PHB production affects FAS in our transgenic lines, the total fatty acid content was measured in leaves of low (LP), medium (MP) and high polymer-producing (HP) and wild type switchgrass plants grown in soil for two months. No statistically significant differences in total fatty acid content were observed between the wild type and transgenic lines HP, MP1 and MP2 (Figure 3). The leaves of transgenic line LP, however, contained slightly less fatty acids ($P = 0.006$) than the other samples. This may be due to a position effect of the transgene insertion since all the transgenic lines analysed in the present study are independent transformation events.

We have shown previously that inhibition of fatty acid synthesis results in increased production of PHB in sugarcane [29] but did not determine if this was due to increased production in M chloroplasts. To gain an understanding of how this affects the distribution of the polymer between BS and M, switchgrass plants in tissue culture were treated with quizalofop, a herbicide from the aryloxyphenoxypropionate group which inhibits ACCase activity [30,31]. The plants were regenerated from immature inflorescence-derived callus cultures initiated from two high producers, one of which was also used for other experiments in this study (Figures 1 & 2; HP). A concentration-dependent increase in polymer levels accompanied by a reduction in the content of total fatty acids (≥50%) in plants treated with the highest quizalofop concentrations tested) was measured in the herbicide-treated plants compared to the non-treated plants from both lines (Figure 4A). A pronounced increase in the abundance of PHB granules was observed in both BS and M cells of treated leaves as visualized by Nile Blue A staining (Figure 4D, E).

Possible differences in the effect of inhibiting FAS on polymer production in developing leaf tissue, where demand for fatty acids is high [32], and in mature leaf tissue, where the highest PHB levels are measured [15,33], were also investigated. Whole sugarcane leaves of line LP at increasing stages of maturity, starting from the 1st fully unfurled leaf with visible dewlap down to the fifth and oldest leaf (L1-L5), were analysed for PHB content two weeks post-treatment with fluazifop-p-butyl, a structural analogue of quizalofop. In the non-treated transgenic plants, PHB content was lowest in the youngest leaves and increased with leaf age until L4, while in L5 no significant increase in PHB was observed (Figure 4B). Fluazifop-p-butyl applied at a concentration of 100 μM caused a significant increase in PHB content in L1 and L2 ($P < 0.001$), L3 ($P = 0.016$), and L4 ($P = 0.021$) with the largest increases in PHB accumulation observed in the younger leaves (Figure 4B). PHB content increased in L1 from 0.09 ± 0.01 to 0.38 ± 0.02% DW, a 4.4 fold increase. The polymer content increased 1.8, 1.6, and 1.4 fold in L2, L3, and L4, respectively. The increase in PHB in L5, the oldest leaf, from 0.20 ± 0.01 to 0.23 ± 0.01% DW was not statistically significant. As in switchgrass, the abundance of PHB granules increased in both BS and M cells in treated sugarcane leaves (Figure 4G, H).

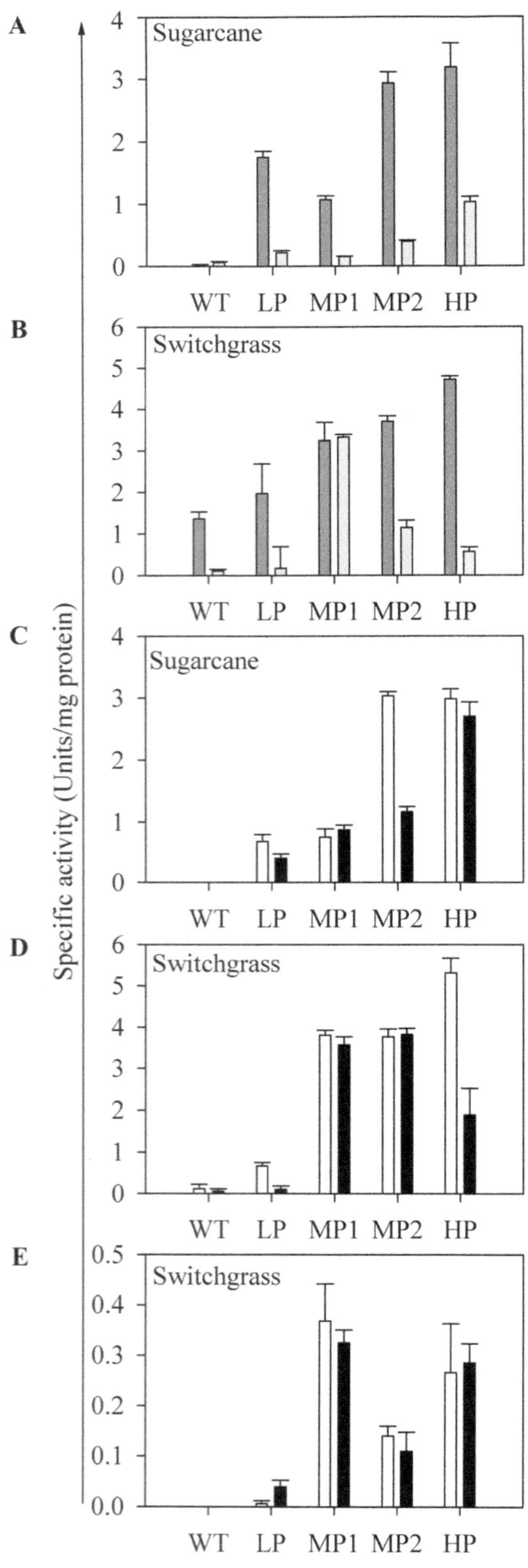

Figure 2 Enzyme activities in total protein extracts from PHB producing transgenic sugarcane and switchgrass. A-B PhaA (dark bars) and PhaB (light bars) activity in whole leaf protein extracts of sugarcane **(A)** and switchgrass **(B)**. **(C, D)** PhaA activity in sugarcane **(C)** and switchgrass **(D)** M (white bars) and BS (black bars) chloroplast protein fractions. **(E)** PhaB activity in switchgrass M and BS chloroplast protein fractions. LP, low PHB producer (≤1% DW); MP, medium producer (>1% ≤ 2% DW); HP, high producer (>2% DW). Error bars represent standard error (n = 3).

Starch and sucrose accumulation

The previously observed lack of starch granules in the BS plastids of polymer producing switchgrass as visualized by TEM [15] suggested that PHB biosynthesis causes a significant reduction in starch accumulation. In this study, starch, sucrose, and glucose contents were measured in four month old PHB producing switchgrass and sugarcane plants (Table 1). In general, both the starch and sucrose levels measured in leaves at approximately midday were inversely correlated to PHB accumulation in transgenic lines (Figure 5). While the content of these two carbohydrates was unaffected in the lowest PHB-producing lines, it was significantly reduced in lines accumulating polymer above 1% DW. In sugarcane leaves, the sucrose content dropped from approximately 5% leaf DW in wild type to about 1% DW in the high PHB-producing lines. Starch levels in these plants decreased by as much as ten-fold. In switchgrass, sucrose levels decreased from 3% DW to 1.5% DW and starch content showed a four-fold decrease in the highest PHB-producing line compared to the wild-type control. In sugarcane, glucose also

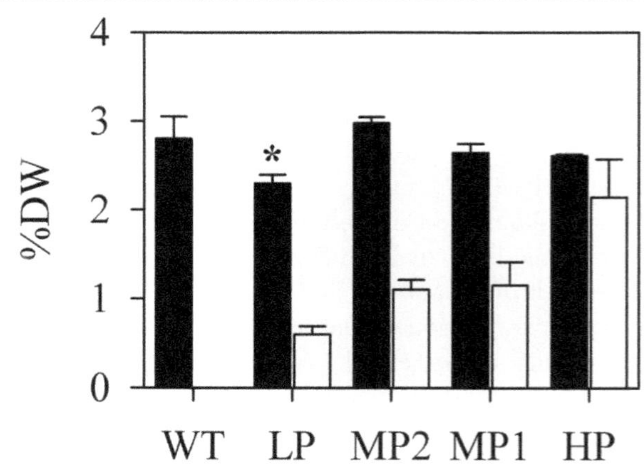

Figure 3 Total fatty acid content and PHB levels in transgenic and wild type switchgrass plants. Bars: black, total fatty acids; white, PHB. Values are presented as a percentage of leaf dry weight (DW). Error is standard deviation of the mean, WT, wild type (n = 3); LP, low producer, ≤1% DW (n = 3); MP1 and MP2, medium producers, >1% ≤ 2% DW (n = 3) and HP, high producer, >2% DW (n = 6). *Indicates total fatty acid content significantly different to the wild-type control ($P < 0.01$).

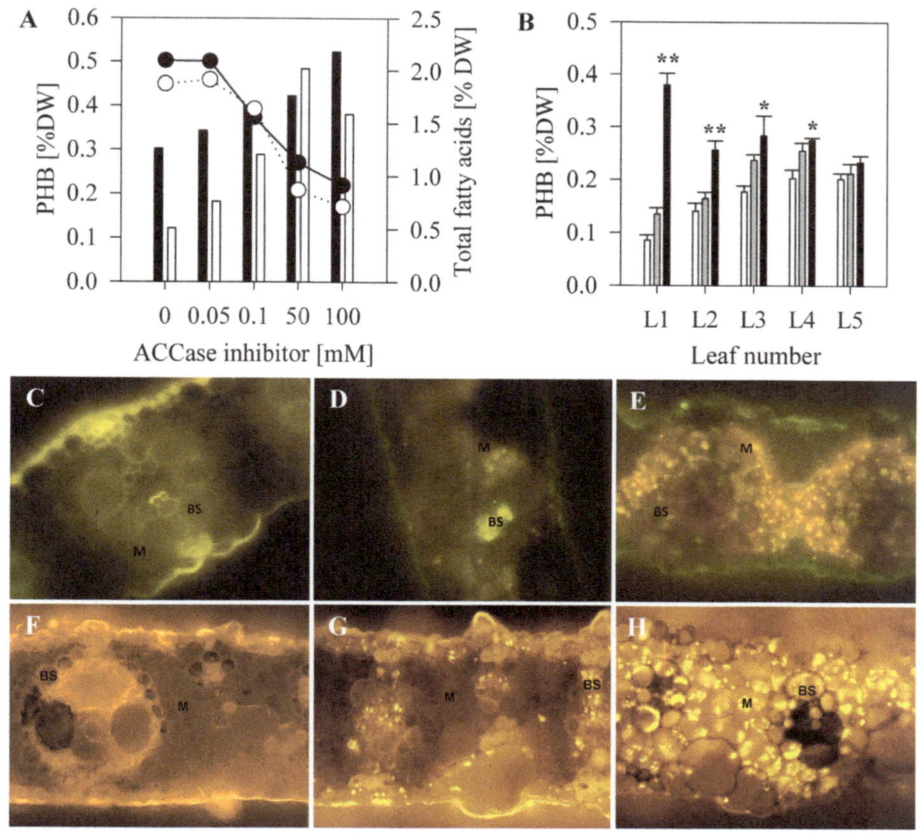

Figure 4 The effect of ACCase inhibition on PHB production in transgenic switchgrass and sugarcane. (A) PHB (bars) and total fatty acid (lines) content in 30 leaves of *in vitro* propagated switchgrass plantlets treated with various concentrations of the ACCase inhibitor quizalofop. Shaded and clear bars or circles represent two independent transgenic events. **(B)** PHB content in leaves of increasing maturity (L1 is the youngest leaf and L5 is the most mature one) from soil grown sugarcane plants treated with various concentrations of the ACCase inhibitor fluazifop-p-butyl (Bars: white = 0 μM, grey = 20 μM, black = 100 μM). PHB levels which are significantly different to the untreated control are indicated: $^{**}P < 0.001$; $^{*}P < 0.05$). **C, D, E** Nile Blue A stained leaf sections from *in vitro* propagated switchgrass plants: **C**, WT; **D**, untreated PHB producing plant; **E**, PHB producing plant treated with 100 μM quizalofop. **F, G, H**, Nile Blue A stained leaf sections from soil-grown sugarcane plants treated with: **F**, WT; **G**, untreated PHB producing plant; **H**, PHB producing plant treated with 100 μM fluazifop. Note the abundance of PHB granules in the M tissue in the Nile Blue A stained sections from treated switchgrass and sugarcane PHB producers.

Table 1 Starch, sucrose and glucose content in the leaves of wild-type and PHB-producing sugarcane and switchgrass plants

Species	ID	Starch (±SD)	Sucrose (±SD)	Glucose (±SD)
Switchgrass	WT	1.19 ± 0.34	3.16 ± 0.50	0.15 ± 0.01
	LP	1.40 ± 0.52	2.83 ± 0.70	0.16 ± 0.04
	MP1	0.83 ± 0.26	1.85 ± 0.24	0.16 ± 0.04
	MP2	0.29 ± 0.06	2.08 ± 0.15	0.13 ± 0.01
	HP	0.20 ± 0.07	1.58 ± 0.39	0.14 ± 0.02
Sugarcane	WT	2.15 ± 0.54	4.68 ± 0.53	0.72 ± 0.09
	LP	2.03 ± 0.65	5.14 ± 0.91	0.77 ± 0.14
	MP1	0.37 ± 0.20	2.27 ± 0.66	0.58 ± 0.13
	MP2	0.15 ± 0.07	1.18 ± 0.07	0.14 ± 0.03
	HP	0.14 ± 0.06	0.97 ± 0.14	0.09 ± 0.01

Values are presented as a percentage of leaf dry weight ± standard deviation (n = 4). WT, wild type; LP, low producer (PHB ≤1% DW); MP, medium producer (PHB >1%, ≤2% DW); HP, high producer (PHB >2% DW).

dropped significantly with increased PHB production whereas glucose content in switchgrass leaves was not significantly affected by PHB production. The sugarcane line MP1 and the switchgrass medium producers MP1 and MP2 with significantly lower sugar and starch contents were similar to wild type plants in terms of leaf chlorophyll and carotenoid contents and biomass yield (results not shown).

Discussion

PHB enzymes are equally abundant and active in M and BS cells

In our first study of plastid-targeted PHB production in sugarcane, where the PHB pathway genes were all driven by the maize ubiquitin promoter, both fluorescent staining patterns and TEMs demonstrated the lack of PHB accumulation in M chloroplasts [12]. This cell specific pattern of polymer accumulation appears to be promoter

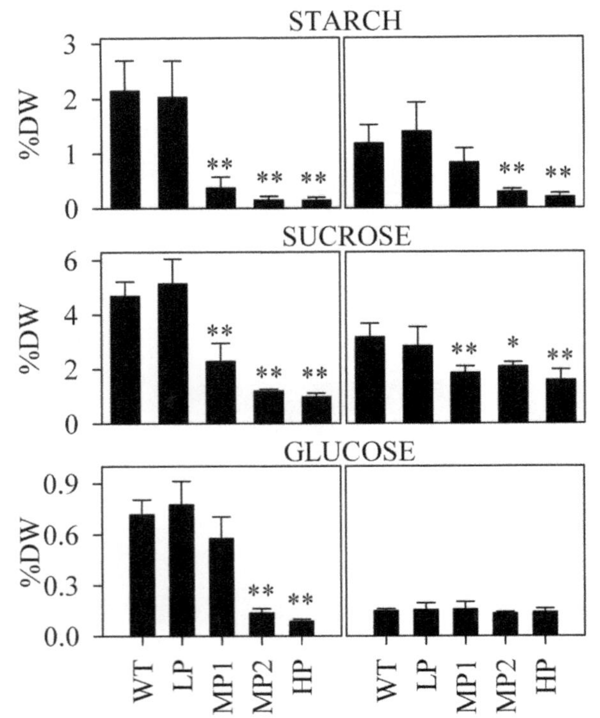

Figure 5 Effect of PHB production on starch, sucrose and glucose content in sugarcane and switchgrass leaves. Values are presented as a percentage of leaf dry weight. Error bars show standard deviation (n = 4). Left panels, sugarcane; Right panels, switchgrass. WT, wild type; LP, low PHB producer (≤1% DW); MP, medium producer (>1%, ≤2% DW); HP, high producer (>2% DW). Significant differences are indicated: $^{**}P < 0.001$; $^{*}P < 0.05$.

independent as it did not occur when PHA production was targeted to the peroxisomes using the same promoter [17]. More recently, we used the maize *cab-m5* promoter to express the PHB genes in switchgrass and sugarcane and although some PHB accumulated in M plastids, it was much less than the polymer detected in BS cells [13-15]. Differential expression of transgenes in M and BS cells is not thought to contribute to this disparity in product accumulation [15] since the *cab-m5* gene is known to be expressed in both types of cells in maize yielding more abundant transcripts in M cells [34]. Greater PHB production in BS chloroplasts has also been observed with the rice ubiquitin 2 promoter [35] in both switchgrass [15] and sugarcane [13]. In the present study, we have confirmed that the signal peptide of the small subunit of Rubisco from pea effectively targets the PHB biosynthesis enzymes to both M and BS chloroplasts of sugarcane and switchgrass (Figure 1) as previously shown in maize [36]. Enzyme assays also showed that PHB enzyme activity is retained following targeting to M plastids and that it is equivalent to that found in the BS plastids (Figure 2). Together these results suggest that anomalous targeting of the PHB biosynthesis enzymes to M plastids is not a contributing factor to low levels of polymer accumulation in this cell type as previously speculated [33].

Diversion of acetyl-CoA to PHB does not affect total fatty acid content

In chloroplasts, PHB biosynthesis competes directly with *de novo* fatty acid biosynthesis for substrate since both pathways use acetyl-CoA. Despite this competition, PHB production in in switchgrass leaves did not reduce the total fatty acid content (Figure 3). This has also been observed in transgenic *Arabidopsis* where substantial production of PHB up to ~40% DW resulted in no net changes in total fatty acids [18]. Nawrath et al. [37] reported that high polymer producing *Arabidopsis* plants (~14% DW) grew normally leading the authors to speculate that a significant increase in the net flux through the plastidic acetyl-CoA pool might occur in response to the introduction of a PHB carbon sink and that mechanisms might exist to enhance synthesis of acetyl-CoA in response to demand.

Inhibition of fatty acid synthesis enhances PHB biosynthesis in M cells

Acetyl-CoA carboxylase (ACCase), which catalyses the first committed step of *de novo* fatty acid biosynthesis and thus is the major flux-controlling enzyme of the pathway [38], competes directly with the thiolase for acetyl-CoA. Chemical inhibition of ACCase has previously been shown to increase PHB production in both C_3 [39] and C_4 plants [29]. When fatty acid biosynthesis was inhibited in developing sugarcane leaves, PHB production increased four-fold (Figure 4B) indicating that substantial flux can be redirected to polymer synthesis. A significant increase in polymer accumulation was most apparent in M chloroplasts as visualized by fluorescence microscopy, but was also detected in BS chloroplasts in both sugarcane and switchgrass (Figure 4E, H). This supports our hypothesis that access to substrate is the main contributing factor to low levels of PHB accumulation in mesophyll chloroplasts.

We found that the rate of polymer accumulation in the leaves of non-treated PHB producing sugarcane is greatest during leaf maturation and plateaus as the leaf completes development (Figure 4B). The state of leaf maturity also affects interaction between fatty acid and PHB biosynthetic pathways. In developing leaves, a large flux of carbon through fatty acid synthesis supports the formation and elongation of new cells [32]. Indeed, ACCase protein abundance was recently shown to peak sharply in the elongating zone of the developing maize leaf, where maximum rates of fatty acid synthesis are required to support plant growth [26]. Inhibition of this pathway during leaf development in PHB-producing transgenic sugarcane allows a large flux of carbon to be diverted to PHB synthesis which normally would not be readily accessible due to substrate competition. In the fully differentiated leaf, where the fatty acid synthesis is lower, its inhibition does not significantly improve PHB production.

PHB production affects gluconeogenesis and starch biosynthesis

The previously reported absence of starch granules in BS chloroplasts of high PHB producing switchgrass plants [15] suggested that polymer synthesis affects carbon partitioning. The results presented here demonstrated significantly decreased levels of starch and sucrose in high PHB producers (Figure 5) suggesting reduced synthesis rate and/or premature degradation. Glucose levels in sugarcane were also significantly decreased with high levels of PHB production, whereas switchgrass retained its low glucose content independent of polymer synthesis.

Taken together, the experimental data suggests that the PHB producing plants have to readjust carbon fluxes to account for the added sink, and that this adjustment comes at the expense of starch, not fatty acids. Reducing the starch pool, that is likely not vital for growth in C_4 plants [22] is a preferable strategy for obtaining the carbon needed to supply the novel carbon sink than limiting fatty acid biosynthesis in leaves, which is critical for plant growth. Since the primary location of starch in differentiated leaves of C_4 plants is BS plastids, this may account, in part, for the difference in PHB synthesis in M and BS cells.

Conclusions

The low production rate of PHB in M is not due to poor targeting or activity of the PHB biosynthetic pathway in M chloroplasts. Evidence suggests that PHB synthesis may not compete effectively with fatty acid synthesis for substrate in M cells as evidenced by increased polymer formation in M cells following chemical inhibition of ACCase. Although sucrose and starch levels are greatly decreased in high PHB producing plants, redirection of carbon from carbohydrates to PHB is probably not substantial enough to account for this decrease. New strategies are required to increase PHB production in C_4 grasses to commercially relevant levels. One strategy could be to enhance the ability of the PHB pathway to capture substrate under competitive conditions. A novel acetoactyl-CoA synthase was recently discovered which could fill the same role as the ketothiolase in the PHB pathway and has properties which suggest it may be a superior catalyst for initiation of PHB production [40]. This might help solve the substrate accessibility issue in M plastids. Another strategy could be to increase flux through C_4 photosynthesis to produce increased levels of polymer while limiting agronomic penalties. We have recently made progress in this area [11] by introducing a cyanobacterial gene encoding a bifunctional enzyme with both fructose-1,6-bisphosphatase (FBPase) and sedoheptulose-1,7-bisphosphatase (SBPase) activities [41,42] into well characterized PHB producing switchgrass lines by re-transformation. Engineering central carbon metabolism is, however, a non-trivial exercise and requires a detailed understanding of the physiology and biochemistry of the plant system at hand. A systems biology approach to PHB biosynthesis in C_4 crops would provide new information about the regulation and function of the C_4 and C_3 photosynthetic pathways in M and BS cells and provide insights on cell-specific strategies to further increase carbon flow to polymer production. This knowledge is critical for optimising plant biofactories for the production of industrial materials, chemicals, and energy.

Methods

Plant material

The production and identification of PHB-producing sugarcane (*Saccharum* sp. cv Q117) and switchgrass (*Panicum virgatum* L.) plants used in this study were described previously [12,13,15]. For simplicity, sugarcane and switchgrass lines have been labelled according to the amount of PHB they produce in mature leaves under glasshouse conditions i.e. wild type (WT), low producers (LP, $\leq$1% DW PHB), medium producers (MP, >1% $\leq$ 2% DW PHB) and high producers (HP, >2% DW PHB).

Separation of chloroplasts from mesophyll and bundle sheath cells

M and BS chloroplasts were isolated according to the method described by Majeran et al. [43] with some modifications. For sugarcane, the first leaf with visible dewlap was harvested from four replicates of each sugarcane line tested. In total two to three grams of leaves were collected. Leaves were cut into small pieces and placed in a small Waring blender with 100 mL of cold grinding buffer (50 mM HEPES, pH 7.9, 350 mM sorbitol, 2 mM EDTA, 5 mM ascorbic acid and 5 mM L-cysteine).

Sugarcane leaves were blended twice (ten second pulse each) on a low setting mode. Released M chloroplasts were filtered through 20 μm nylon mesh (Small Parts Inc., Miramar, FL, USA) and stored on ice while BS chloroplasts were extracted. The residual M attached to BS strands were removed by two or more 30-second pulses on high setting in a small Waring blender with the blade reversed, followed by washing with grinding buffer. Chloroplasts were released from purified BS fibres by blending on high setting, in a small Waring blender with the blades replaced with a razor blade, in 30 second pulses until the grinding buffer turned green. BS chloroplasts were filtered through 20 μm nylon mesh to remove cellular debris. After centrifugation at 1000 g for four minutes, chloroplasts from both fractions were precipitated. Pure chloroplasts were observed after additional washing with grinding buffer and centrifugation.

For isolation of chloroplasts from switchgrass, the second leaf from the base of vegetative tillers at an early elongation stage (three to four tillers per line) from transgenic and wild type plants were harvested. The leaves were cut into small pieces and gently ground by mortar and pestle in the buffer described above. Released M chloroplasts were collected and stored as described for sugarcane. The residual M cells attached to BS strands were removed by vigorous grinding followed by washing. The purity of the BS strands was assessed by microscopy at this step. BS chloroplasts were released by further grinding then passed through a 20 μm nylon mesh. Chloroplasts from both fractions were pelleted and washed as described above for sugarcane.

Enzyme assays

PhaA was assayed in the thiolysis direction using the procedure described by Palmer et al. [44] with some modifications. The assay mixture (1 mL) contained 62.4 mM Tris-Cl pH 8.1, 50 mM $MgCl_2$, 62.5 μM CoA, and 62.5 μM acetoacetyl-CoA. The loss of acetoacetyl-CoA was monitored with time ($\varepsilon 304 = 16.9 \times 10^3$ cm^{-1} M^{-1}). The reaction was initiated by addition of the total protein extract and the absorbance at 304 nm was monitored at 25°C.

PhaB was assayed in the direction of 3-hydroxybutyryl-CoA formation by NADPH reduction using the procedure described by Ploux et al. [45] with modifications. The assay mixture (1 mL) contained 100 mM Tris-Cl pH 8.1, 100 μM NADPH, and 20 μM acetoacetyl-CoA. The loss of NADPH was monitored with time ($\varepsilon 340 = 6.22 \times 10^3$ cm^{-1} M^{-1}). The reaction was initiated by addition of the total protein extract and the absorbance at 340 nm was monitored (25°C). Protein concentrations were determined using Bradford reagent (Sigma-Aldrich). The specific enzyme activity of each sample was calculated as units of enzyme per mg of total protein where one unit is defined as the amount of enzyme that catalyses the transformation of 1 μmol substrate/min.

Immunolocalization

Leaf sections were prepared for immunofluorescence microscopy using the procedure described by Paciorek et al. [46]. Thin transverse leaf sections (6–7 μm) were cut and labelled with primary rabbit antisera specific to *Ralstonia eutropha* β-ketothiolase (1:10000 dilution) or acetoacetyl-CoA reductase (1:5000 dilution) as described above for protein immunoblots. Alexafluor 488 goat anti-rabbit IgG (Invitrogen) was used as a secondary antibody (1:5000 dilution). Images were acquired using a Zeiss LSM 510 META confocal microscope (488 nm excitation; BP 500–550 nm).

PHB analyses

Approximately 10 mg of sugarcane leaf material was analysed for PHB content by HPLC following acid digestion to crotonic acid as previously described in Petrasovits et al. [12]. For switchgrass, 20–80 mg of lyophilized leaf tissue was prepared for analysis by gas chromatography/mass spectrometry (GC/MS) using a simultaneous extraction and butanolysis procedure as described by Kourtz et al. [47].

In vivo ACCase inhibition assay

ACCase inhibitor solution was prepared by diluting the commercially available herbicide Fusilade Forte (128 g L^{-1} of active ingredient fluazifop-P; Syngenta) in distilled water containing Searles Spreadmax wetting agent (700 μl L^{-1}). Approximately 5 mL of 20 μM or 100 μM herbicide solution was applied to the leaves of five week old sugarcane plants grown in a glasshouse. Treated plants were grown for an additional 14 days. All fully expanded leaves were collected separately according to leaf number, dried and analysed for PHB content.

Switchgrass plants were propagated from PHB producing plants following a procedure described previously [11] and grown under *in vitro* conditions. Briefly, immature inflorescence-derived callus cultures were initiated from soil-grown primary transformants producing >2% DW PHB). Plants regenerated from these cultures on MS basal medium supplemented with 30 g L^{-1} maltose and 1.4 μM gibberellic acid [48] were grown at 27°C, with a 16 h photoperiod (cool white fluorescent bulbs, 80 μmol m^{-2} s^{-1}) for about six weeks. Two weeks prior to treatment, plantlets with four to five leaves were transferred to larger tissue culture containers (5 plants per container) and grown under the conditions described above. Switchgrass plantlets were sprayed with approximately 900 μl of quizalofop solution. The herbicide solution was prepared with 90% ethanol (v/v) and diluted to final concentrations of 0.05, 0.10, 50, and 100 μM. Wild-type plantlets treated with 50 nM quizalofop or 90% ethanol (v/v) as well as PHB producing plantlets treated with 90% ethanol (v/v) served as controls. Ten days after the herbicide treatment, leaf segments were stained with Nile Blue A as previously described [28] for visualization of the polymer granules. For measurements of the total fatty acids and polymer contents, whole plantlets without roots were harvested (five plants per treatment) and freeze-dried for three days prior to analyses. At these early stages of their growth under *in vitro* conditions, plants have developing leaves and no differentiated stem tissues.

Starch, sucrose and glucose measurements

Transgenic and wild type plants were grown in a glasshouse under similar light intensity. Sugarcane plants were grown from setts for approximately four months and the first fully unfurled leaf was harvested (four plants per line). Switchgrass plants were grown for about four months after re-potting and the leaf adjacent to the second node from the base of vegetative tillers at an early elongation stage (three tillers per plant) was sampled. Both sugarcane and switchgrass samples were collected between midday and 1 pm on a sunny day and snap frozen in liquid nitrogen. The leaves were freeze-dried, ground to a fine powder and used for analysis of starch, soluble sugars and PHB. PHB content was determined using the methods described above. For measurements of starch content, 50 mg of tissue was extracted with 80% ethanol to remove soluble sugars. The percentage of starch per dry weight was measured as glucose, following digestion with α-amylase and amyloglucosidase using the starch (GO/P) assay kit (Sigma, STA20) according to the manufacturer's instructions. For analysis of soluble sugars, 20–50 mg of sample was extracted with 1 mL of hot water. The extract was analysed by HPLC using a Shodex KS-801 column (Tosho) and a refractive index detector. A column temperature of 80°C and a flow rate of 1 mL/min were used.

Fatty acid content

The amount of total fatty acids in leaf tissue from switchgrass plants was determined by converting fatty acids to their methyl esters and measuring the content by gas chromatography. Methyl esters of the following fatty acids were analysed: palmitic (C16:0), stearic (C18:0), oleic (C18:1), linoleic (C18:2), linolenic (C18:3), arachidic (C20:0), eicosanoic (C20:1), eicosadienoic (C20:2), eicosatrienoic (C20:3), behenic (C22:0), and erucic acid (C22:1). Published procedures were followed [49] with modifications. Briefly, freeze-dried leaf tissue (30 to 50 mg) was placed in screw-cap test tubes and mixed with 1.5 ml of 2.5% (v/v) sulfuric acid in methanol (w/ 0.01% w/v BHT), 400 µl toluene, and 500 µg of a triheptadecanoin (Nu-Chek Prep, Elysian, MN) solution (10 mg/ml in toluene) as internal standard. The tubes were purged with nitrogen, capped, and heated at 90°C for 1 h. Upon cooling, 1 ml of 1 M sodium chloride and 1 ml of heptane were added to each tube. Following mixing and centrifugation, the heptane layer containing fatty acid methyl esters was analysed with an Agilent 7890A gas chromatograph. Fatty acid methyl esters were resolved with a 30 m × 0.25 mm (inner diameter) INNOWax column (Agilent) and detected by flame ionization. The oven temperature was programmed from 185°C (1 min hold) to 235°C (1 min hold) at a rate of 10°C/min (11 min total run time), and the front inlet pressure was 35.8 psi of He. The total fatty acid content was determined by comparison of the detector response from leaf-derived fatty acid methyl esters relative to methyl heptadecanoate from the triheptadecanoin internal standard.

Statistical analysis

Statistical analyses were performed using SigmaPlot for Windows Version 11.0 (Systat Software, Inc.). Multiple comparisons were done by one way ANOVA using the default suggested test at $P < 0.001$.

Abbreviations

ACCase: Acetyl-CoA carboxylase; BS: Bundle sheath; CO2: Carbon dioxide; DW: Dry weight; FAS: Fatty acid synthesis; HP: High producer; PHB: Polyhydroxybutyrate; LP: Low producer; M: Mesophyll; MP: Medium producer; NAD-ME: NAD dependent malic enzyme; NADP-ME: NADP-dependent malic enzyme; PHAs: Polyhydroxyalkanoates; PHB: Polyhydroxybutyrate; Rubisco: Ribulose-1,5-bisphosphate carboxylase/oxygenase; WT: Wild-type.

Competing interests

MNS and KDS are employees of Metabolix, Inc. XL is a former employee of Metabolix, Inc.

Authors' contributions

RBM, LAP, LKG, MNS, KDS, XL, LKN and SMB contributed to research design. RBM, LKG, XL, and MNS performed experiments; RBM performed statistical analyses. RBM, LAP, LKG, MNS, and KDS wrote the paper. All authors discussed the results and commented on the manuscript. All authors read and approved the final manuscript.

Acknowledgment

The authors thank the Queensland node of Metabolomics Australia and Sean Daughtry and Mirel Sharxhi (Metabolix, Inc.) for expert assistance with HPLC, and GC/MS. We thank Sugar Research Australia for supplying office, laboratory, and greenhouse space to the sugarcane team for this project. This study was funded through Australian Research Council linkage grant LP100100659, Metabolix Inc., and the Cooperative Research Centre for Sugar Industry Innovation through Biotechnology and Sugar Research Australia.

Author details

[1]Australian Institute for Bioengineering and Nanotechnology, the University of Queensland, Brisbane QLD 4072, Australia. [2]Metabolix, Inc, 21 Erie St, Cambridge, MA 02139, USA. [3]Current address: Agrivida, 200 Boston Ave, Medford, MA 02155, USA. [4]Current address: Department of Biological Sciences, The University of North Texas, 1155 Union Circle #305220, Denton, TX 76203-5017, USA.

References

1. Byrt CS, Grof CP, Furbank RT: **C4 Plants as Biofuel Feedstocks: Optimising Biomass Production and Feedstock Quality from a Lignocellulosic Perspective.** *J Integr Plant Biol* 2011, **53**(2):120–135.
2. Somerville C, Youngs H, Taylor C, Davis SC, Long SP: **Feedstocks for lignocellulosic biofuels.** *Science* 2010, **329**(5993):790–792.
3. Chong BF, Bonnett GD, Glassop D, O'Shea MG, Brumbley SM: **Growth and metabolism in sugarcane are altered by the creation of a new hexose-phosphate sink.** *Plant Biotechnol J* 2007, **5**(2):240–253.
4. McQualter RB, Chong BF, Meyer K, Van Dyk DE, O'Shea MG, Walton NJ, Viitanen PV, Brumbley SM: **Initial evaluation of sugarcane as a production platform for p-hydroxybenzoic acid.** *Plant Biotechnol J* 2005, **3**(1):29–41.

5. Van Beilen JB, Poirier Y: **Plants as factories for bioplastics and other biomaterials.** In *Plant biotechnology and agriculture: prospects for the 21st century.* Edited by Altman A, Hasegawa PM. London: Academic press; 2012:481–494.
6. Anderson AJ, Dawes EA: **Occurrence, metabolism, metabolic role, and industrial uses of bacterial polyhydroxyalkanoates.** *Microbiol Rev* 1990, **54**(4):450–472.
7. Steinbüchel A, Schlegel HG: **Physiology and molecular genetics of poly (beta-hydroxy-alkanoic acid) synthesis in *Alcaligenes eutrophus*.** *Mol Microbiol* 1991, **5**(3):535–542.
8. Snell KD, Peoples OP: **PHA bioplastic: A value-added coproduct for biomass biorefineries.** *Biofuel Bioprod Bior* 2009, **3**(4):456–467.
9. Mooney BP: **The second green revolution? Production of plant-based biodegradable plastics.** *Biochem J* 2009, **418**(2):219–232.
10. Suriyamongkol P, Weselake R, Narine S, Moloney M, Shah S: **Biotechnological approaches for the production of polyhydroxyalkanoates in microorganisms and plants - A review.** *Biotechnol Adv* 2007, **25**(2):148–175.
11. Somleva MN, Peoples OP, Snell KD: **PHA bioplastics, biochemicals, and energy from crops.** *Plant Biotechnol J* 2013, **11**(2):233–252.
12. Petrasovits LA, Purnell MP, Nielsen LK, Brumbley SM: **Production of polyhydroxybutyrate in sugarcane.** *Plant Biotechnol J* 2007, **5**(1):162–172.
13. Petrasovits LA, Zhao L, McQualter RB, Snell KD, Somleva MN, Patterson NA, Nielsen LK, Brumbley SM: **Enhanced polyhydroxybutyrate production in transgenic sugarcane.** *Plant Biotechnol J* 2012, **10**(5):569–578.
14. Poirier Y, Gruys KJ: **Production of polyhydroxyalkanoates in transgenic plants.** In *Biopolymer*, Volume 3A. Edited by Doi Y, Steinbuchel A. Berlin: Wiley-VHC; 2001:401–435.
15. Somleva MN, Snell KD, Beaulieu JJ, Peoples OP, Garrison BR, Patterson NA: **Production of polyhydroxybutyrate in switchgrass, a value-added co-product in an important lignocellulosic biomass crop.** *Plant Biotechnol J* 2008, **6**(7):663–678.
16. Hahn JJ, Eschenlauer AC, Sleytr UB, Somers DA, Srienc F: **Peroxisomes as sites for synthesis of polyhydroxyalkanoates in transgenic plants.** *Biotechnol Prog* 1999, **15**(6):1053–1057.
17. Tilbrook K, Gebbie L, Schenk PM, Poirier Y, Brumbley SM: **Peroxisomal polyhydroxyalkanoate biosynthesis is a promising strategy for bioplastic production in high biomass crops.** *Plant Biotechnol J* 2011, **9**(9):958–969.
18. Bohmert K, Balbo I, Kopka J, Mittendorf V, Nawrath C, Poirier Y, Tischendorf G, Trethewey RN, Willmitzer L: **Transgenic Arabidopsis plants can accumulate polyhydroxybutyrate to up to 4% of their fresh weight.** *Planta* 2000, **211**(6):841–845.
19. Bohmert-Tatarev K, McAvoy S, Daughtry S, Peoples OP, Snell KD: **High levels of bioplastic are produced in fertile transplastomic tobacco plants engineered with a synthetic operon for the production of polyhydroxybutyrate.** *Plant Physiol* 2011, **155**(4):1690–1708.
20. Hattersley PW: **Characterization of C-4 type leaf anatomy in grasses (Poaceae), mesophyll - bundle sheath area ratios.** *Ann Bot-London* 1984, **53**(2):163–179.
21. Friso G, Majeran W, Huang M, Sun Q, van Wijk KJ: **Reconstruction of metabolic pathways, protein expression, and homeostasis machineries across maize bundle sheath and mesophyll chloroplasts: large-scale quantitative proteomics using the first maize genome assembly.** *Plant Physiol* 2010, **152**(3):1219–1250.
22. Weise SE, van Wijk KJ, Sharkey TD: **The role of transitory starch in C(3), CAM, and C(4) metabolism and opportunities for engineering leaf starch accumulation.** *J Exp Bot* 2011, **62**(9):3109–3118.
23. Ohsugi R, Huber SC: **Light modulation and localization of sucrose phosphate synthase activity between mesophyll cells and bundle sheath cells in C(4) species.** *Plant Physiol* 1987, **84**(4):1096–1101.
24. Edwards GE, Franceschi VR, Ku MS, Voznesenskaya EV, Pyankov VI, Andreo CS: **Compartmentation of photosynthesis in cells and tissues of C(4) plants.** *J Exp Bot* 2001, **52**(356):577–590.
25. Lunn JE, Furbank RT: **Localisation of sucrose-phosphate synthase and starch in leaves of C4 plants.** *Planta* 1997, **202**(1):106–111.
26. Majeran W, Friso G, Ponnala L, Connolly B, Huang M, Reidel E, Zhang C, Asakura Y, Bhuiyan NH, Sun Q, Turgeon R, van Wijk KJ: **Structural and metabolic transitions of C4 leaf development and differentiation defined by microscopy and quantitative proteomics in maize.** *Plant Cell* 2010, **22**(11):3509–3542.
27. Huisman GW, Skraly F, Martin DP, Peoples GE: **Biological systems for manufacture of polyhydroxyalkanoate polymers containing 4-hydroxyacids.** In, Volume US Patent 6,316,262. ; 2001.
28. Kourtz L, Dillon K, Daughtry S, Madison LL, Peoples O, Snell KD: **A novel thiolase-reductase gene fusion promotes the production of polyhydroxybutyrate in Arabidopsis.** *Plant Biotechnol J* 2005, **3**(4):435–447.
29. Petrasovits LA, McQualter RB, Gebbie LK, Blackman DM, Nielsen LK, Brumbley SM: **Chemical inhibition of acetyl coenzyme A carboxylase as a strategy to increase polyhydroxybutyrate yields in transgenic sugarcane.** *Plant Biotechnol J* 2013, **11**(9):1146–1151.
30. Herbert D, Price LJ, Alban C, Dehaye L, Job D, Cole DJ, Pallett KE, Harwood JL: **Kinetic studies on two isoforms of acetyl-CoA carboxylase from maize leaves.** *Biochem J* 1996, **318**(Pt 3):997–1006.
31. Schmidt RR: **HRAC classification of herbicides according to mode of action.** *Proc Brighton Crop* 1997, **3**:1133–1140.
32. Hannapel DJ, Ohlrogge JB: **Regulation of acyl carrier protein messenger RNA levels during seed and leaf development.** *Plant Physiol* 1988, **86**(4):1174–1178.
33. Purnell MP, Petrasovits LA, Nielsen LK, Brumbley SM: **Spatio-temporal characterization of polyhydroxybutyrate accumulation in sugarcane.** *Plant Biotechnol J* 2007, **5**(1):173–184.
34. Sheen JY, Bogorad L: **Differential expression of six light-harvesting chlorophyll a/b binding protein genes in maize leaf cell types.** *Proc Natl Acad Sci U S A* 1986, **83**(20):7811–7815.
35. Wang J, Oard JH: **Rice ubiquitin promoters: deletion analysis and potential usefulness in plant transformation systems.** *Plant Cell Rep* 2003, **22**(2):129–134.
36. Sattarzadeh A, Fuller J, Moguel S, Wostrikoff K, Sato S, Covshoff S, Clemente T, Hanson M, Stern DB: **Transgenic maize lines with cell-type specific expression of fluorescent proteins in plastids.** *Plant Biotechnol J* 2010, **8**(2):112–125.
37. Nawrath C, Poirier Y, Somerville C: **Targeting of the polyhydroxybutyrate biosynthetic pathway to the plastids of *Arabidopsis thaliana* results in high levels of polymer accumulation.** *Proc Natl Acad Sci U S A* 1994, **91**(26):12760–12764.
38. Page RA, Okada S, Harwood JL: **Acetyl-CoA carboxylase exerts strong flux control over lipid synthesis in plants.** *Biochim Biophys Acta* 1994, **1210**(3):369–372.
39. Suzuki Y, Kurano M, Arai Y, Nakashita H, Doi Y, Usami R, Horikoshi K, Yamaguchi I: **Enzyme inhibitors to increase poly-3-hydroxybutyrate production by transgenic tobacco.** *Biosci Biotech Bioch* 2002, **66**(12):2537–2542.
40. Okamura E, Tomita T, Sawa R, Nishiyama M, Kuzuyama T: **Unprecedented acetoacetyl-coenzyme A synthesizing enzyme of the thiolase superfamily involved in the mevalonate pathway.** *Proc Natl Acad Sci U S A* 2010, **107**(25):11265–11270.
41. Tamoi M, Ishikawa T, Takeda T, Shigeoka S: **Molecular characterization and resistance to hydrogen peroxide of two fructose-1,6-bisphosphatases from Synechococcus PCC 7942.** *Arch Biochem Biophys* 1996, **334**(1):27–36.
42. Tamoi M, Nagaoka M, Miyagawa Y, Shigeoka S: **Contribution of fructose-1,6-bisphosphatase and sedoheptulose-1,7-bisphosphatase to the photosynthetic rate and carbon flow in the Calvin cycle in transgenic plants.** *Plant Cell Physiol* 2006, **47**(3):380–390.
43. Majeran W, Cai Y, Sun Q, van Wijk KJ: **Functional differentiation of bundle sheath and mesophyll maize chloroplasts determined by comparative proteomics.** *Plant Cell* 2005, **17**(11):3111–3140.
44. Palmer MA, Differding E, Gamboni R, Williams SF, Peoples OP, Walsh CT, Sinskey AJ, Masamune S: **Biosynthetic thiolase from *Zoogloea ramigera*. Evidence for a mechanism involving Cys-378 as the active site base.** *J Biol Chem* 1991, **266**(13):8369–8375.
45. Ploux O, Masamune S, Walsh CT: **The NADPH-linked acetoacetyl-CoA reductase from *Zoogloea ramigera*. Characterization and mechanistic studies of the cloned enzyme over-produced in Escherichia coli.** *Eur J Biochem* 1988, **174**(1):177–182.
46. Paciorek T, Sauer M, Balla J, Wisniewska J, Friml J: **Immunocytochemical technique for protein localization in sections of plant tissues.** *Nat Protoc* 2006, **1**(1):104–107.
47. Kourtz L, Dillon K, Daughtry S, Peoples OP, Snell KD: **Chemically inducible expression of the PHB biosynthetic pathway in Arabidopsis.** *Transgenic Res* 2007, **16**(6):759–769.

48. Somleva MN, Tomaszewski Z, Conger BV: **Agrobacterium-mediated genetic transformation of switchgrass.** *Crop Sci* 2002, **42**(6):2080–2087.
49. Li YH, Beisson F, Pollard M, Ohlrogge J: **Oil content of Arabidopsis seeds: The influence of seed anatomy, light and plant-to-plant variation.** *Phytochemistry* 2006, **67**(9):904–915.

Highly thermostable GH39 β-xylosidase from a *Geobacillus* sp. strain WSUCF1

Aditya Bhalla[1,3], Kenneth M Bischoff[2] and Rajesh K Sani[1*]

Abstract

Background: Complete enzymatic hydrolysis of xylan to xylose requires the action of endoxylanase and β-xylosidase. β-xylosidases play an important part in hydrolyzing xylo-oligosaccharides to xylose. Thermostable β-xylosidases have been a focus of attention as industrially important enzymes due to their long shelf life and role in the relief of end-product inhibition of xylanases caused by xylo-oligosaccharides. Therefore, a highly thermostable β-xylosidase with high specific activity has significant potential in lignocellulose bioconversion.

Results: A gene encoding a highly thermostable GH39 β-xylosidase was cloned from *Geobacillus* sp. strain WSUCF1 and expressed in *Escherichia coli*. Recombinant β-xylosidase was active over a wide range of temperatures and pH with optimum temperature of 70°C and pH 6.5. It exhibited very high thermostability, retaining 50% activity at 70°C after 9 days. WSUCF1 β-xylosidase is more thermostable than β-xylosidases reported from other thermophiles (growth temperature ≤ 70°C). Specific activity was 133 U/mg when incubated with p-nitrophenyl xylopyranoside, with K_m and V_{max} values of 2.38 mM and 147 U/mg, respectively. SDS-PAGE analysis indicated that the recombinant enzyme had a mass of 58 kDa, but omitting heating prior to electrophoresis increased the apparent mass to 230 kDa, suggesting the enzyme exists as a tetramer. Enzyme exhibited high tolerance to xylose, retained approximately 70% of relative activity at 210 mM xylose concentration. Thin layer chromatography showed that the enzyme had potential to convert xylo-oligomers (xylobiose, triose, tetraose, and pentaose) into fermentable xylose. WSUCF1 β-xylosidase along with WSUCF1 endo-xylanase synergistically converted the xylan into fermentable xylose with more than 90% conversion.

Conclusions: Properties of the WSUCF1 β-xylosidase i.e. high tolerance to elevated temperatures, high specific activity, conversion of xylo-oligomers to xylose, and resistance to inhibition from xylose, make this enzyme potentially suitable for various biotechnological applications.

Keywords: Lignocellulose, Biofuels, β-xylosidase, Thermostable

Background

Xylan represents the second most abundant polysaccharide in the biosphere after cellulose, and therefore is a potential source of sugars for the production of biofuels. Complete degradation of xylan to monomeric sugars requires synergistic action of several xylanolytic enzymes. Endo-1,4-β-xylanases (EC 3.2.1.8) hydrolyze the xylan backbone and β-D-xylosidases (EC 3.2.1.37) sunder the resulting xylooligomers to free xylose [1,2]. Availability of free xylose is immensely important to the overall efficiency of conversion of biomass to biofuels in the fermentation industry.

Xylanases have also been suggested to complement the activity of cellulases for the efficient hydrolysis of lignocellulose [2,3]. Kumar and Wyman [4] reported increased efficiency of cellulases on lignocellulosic biomass after addition of endo-xylanase and β-xylosidase. β-xylosidase degrades xylose oligomers before or during enzymatic hydrolysis of lignocellulosic biomass to reduce inhibition on cellulases by xylose oligomers. But under industrial processes, xylose could accumulate to levels that strongly inhibit performance of the catalyst. To overcome this limitation, a xylose-tolerant xylosidase is desirable which can resist high concentrations of the xylose [5,6]. A number of thermostable β-xylosidases are reported from thermophilic and hyperthermophilic bacteria including species of the genera *Geobacillus*, *Thermoanaerobacter*, *Thermatoga*, and *Thermoanaerobacterium* [5,7-11]. The thermostable

* Correspondence: Rajesh.Sani@sdsmt.edu
[1]Department of Chemical and Biological Engineering, South Dakota School of Mines and Technology, Rapid City, SD 57701, USA
Full list of author information is available at the end of the article

enzymes are of interest in industrial processes due to their long shelf lives, compatibility with heat pretreatment, decreased microbial contamination at high temperatures and activity for longer durations [12].

The bacterial β-D-xylosidases mainly belong to glycosyl hydrolase families GH 3, 30, 39, 43, 52, and 54. Family 39 enzymes belong to the largest glycoside hydrolase clan (GH-A), whose members share a TIM-barrel $(\beta/\alpha)_8$ structure and a conserved catalytic machinery [13]. Reports on characterization of GH39 family β-xylosidases from thermophilic microorganisms are scarce. In this study, we report the cloning and characterization of GH39 thermostable β-xylosidase from *Geobacillus* sp. strain WSUCF1. This strain was isolated from samples collected from a compost facility and is an aerobic spore forming thermophile [14].

Results

Phylogenetic analysis of WSUCF1 β-xylosidase

Genome sequence of WSUCF1 revealed several genes encoding glycoside hydrolases. Out of 865 ORFs for carbohydrate metabolism, 70 ORFs were found to be involved in polysaccharide degradation [15]. It was found that the WSUCF1 genome contained two genes coding for putative β-xylosidases - WP_020755811 (1509 bp) and WP_020755806 (2157 bp). The deduced amino acid sequence analysis of these two β-xylosidases using Pfam showed homology with that of the GH39 and GH52 family, respectively. Literature shows that there is only one report [8] on the cloning and characterization of a GH39 family β-xylosidase from *Geobacillus* spp. Therefore, the WP_020755811 β-xylosidase was selected for further study.

Figure 1 shows the phylogenetic comparison of the putative WSUCF1 β-xylosidase amino acid sequence to other β-xylosidases. It shows that the WSUCF1 enzyme clusters with other GH39 family β-xylosidases. Within the GH39 group, β-xylosidase from WSUCF1 formed sister clades supported by high bootstrap values with sequences belonging to *Geobacillus* spp. (ABI49941, YP_003253769, and ZP_03147822). β-xylosidases from different GH families (e.g., GH3, GH43, and GH52) made separate clusters on the phylogenetic tree (Figure 1).

Alignment of the amino acid sequences of WSUCF1 β-xylosidase (WP_020755811) revealed 98, 95, and 91% identity with GH39 β-xylosidases from *Geobacillus stearothermophilus* strain T-6 (ABI49941), *Geobacillus* sp. Y412MC61 (YP_003253769), and *Geobacillus* sp. G11MC16 (ZP_03147822), respectively. β-xylosidase from *Geobacillus thermoleovorans* (DQ345777) belonging to GH43 family and *Geobacillus stearothermophilus* (ABI49956) belonging to GH52 family showed only 29 and 33% identity respectively.

Expression, purification, and characterization of WSUCF1 β-xylosidase

For functional analysis, the gene encoding WP_020755811 β-xylosidase was amplified from the *Geobacillus* sp. strain WSUCF1 genome. The gene was cloned in the pRham™ N-His SUMO Kan vector, expressed in *E. coli,* and purified using Ni-NTA affinity chromatography. SDS-PAGE

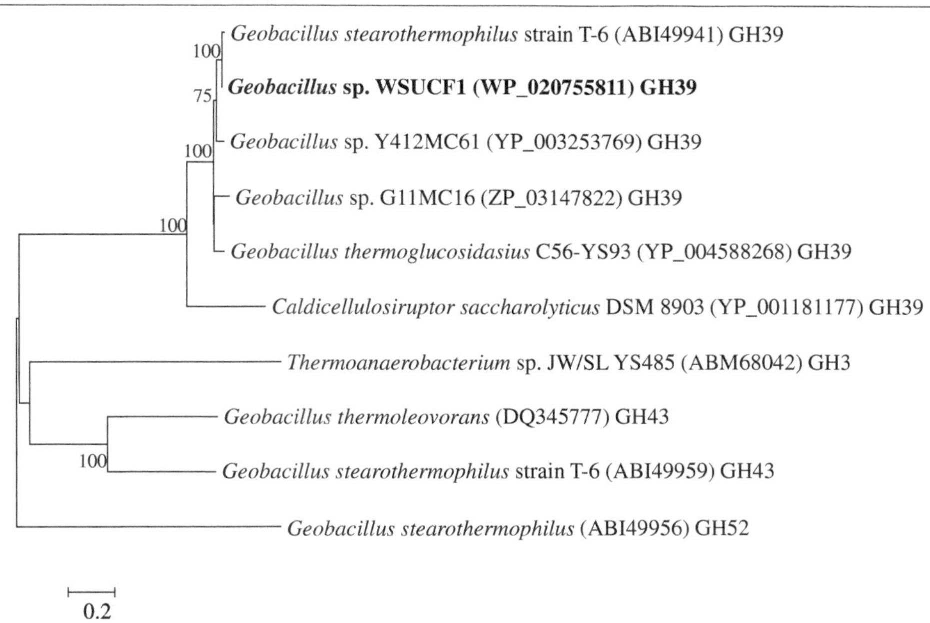

Figure 1 Phylogenetic tree showing relationship between β-xylosidase sequence of *G.* sp. strain WSUCF1 and reference β-xylosidase sequences in GenBank, constructed using neighbor-joining method. Bootstrap values which were ≥75% are indicated at the nodes. The scale bar represents 0.2 substitutions per amino acid position. Short title: Phylogenetic relationship between WSUCF1 β-xylosidase and reference β-xylosidase sequences.

analysis of the purified β-xylosidase showed a prominent band of 58 kDa (Figure 2, Lane 2) which was consistent with the predicted mass of the β-xylosidase enzyme. No enzyme activity was detected by zymogram analysis of the 58 kDa protein. By omitting the heat treatment prior to electrophoresis, the apparent mass of the prominent protein increased to 230 kDa, which displayed β-xylosidase activity in zymogram analysis (Figure 2, Lanes 3 and 5).

The recombinant β-xylosidase from WSUCF1 exhibited activity in a broad range of pH (4.0-9.5) with optima at 6.5 (Figure 3A). β-xylosidase activity increased linearly when pH was increased from 4 to 6.5, but decreased when pH was further increased from 6.5 to 10. It retained more than 80% activity in the pH range of 5.5 - 7. At high pH of 8.6 and 9 (glycine- NaOH buffer), it retained 62 and 46% relative activity respectively. Effect of different temperatures on WSUCF1 xylosidase activity is shown in Figure 3B. It exhibited maximum activity at temperature of 70°C. The enzyme was active between temperatures of 50 to 75°C, with more than 50% relative activity.

WSUCF1 β-xylosidase was highly thermostable in the range of 50-70°C (Figure 4). The enzyme was optimally stable at 50°C for 1 day without any loss of activity while about 93% activity was retained after incubation for 9 days. At 60°C, about 99% residual activity was detectable after 1 day of incubation and it lost only 17% of its original activity after incubation for 9 days. The enzyme retained

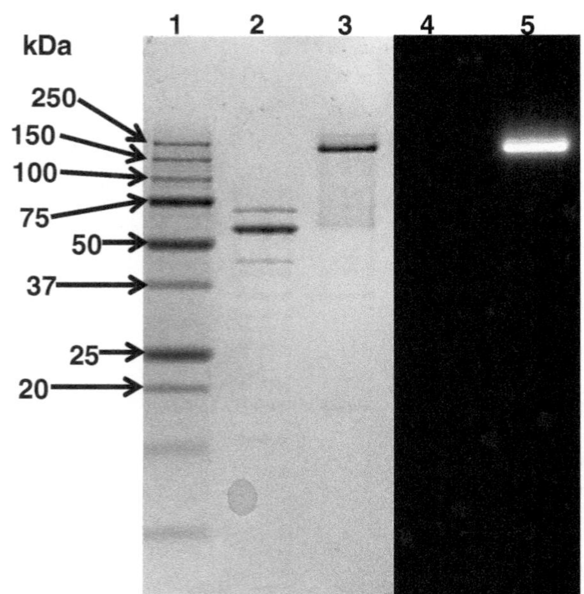

Figure 2 SDS-PAGE (8-16% gradient) and zymogram of WSUCF1 β-xylosidase. Lane 1, Precision Plus Protein Standards (BioRad); Lane 2, β-xylosidase with heat treatment (95°C for 2 min); Lane 3, β-xylosidase without any heat treatment; Lane 4, activity staining of β-xylosidase with heat treatment; and Lane 5, activity staining of β-xylosidase without any heat treatment. Short title: SDS-PAGE and zymogram of WSUCF1 β-xylosidase.

94% of its original activity after incubation of 1 day at 70° C, with a half-life of 9 days. The effect of the presence of metal ions on enzyme activity was studied (Table 1). WSUCF1 β-xylosidase retained its activity in the presence of different metal ions except Cu^{2+} and Hg^{2+}. More than 70% activity was retained in the presence of Zn^{2+}, Co^{2+}, Ni^{2+}, Ca^{2+}, Mn^{2+}, and Mg^{2+} whereas it lost 73 and 100% activity in the presence of Cu^{2+} and Hg^{2+}, respectively.

The kinetic parameters were calculated from Lineweaver-Burke plots of specific activities at various substrate concentrations (data not shown). The K_m and V_{max} values for WSUCF1 β-xylosidase with paranitrophenyl- beta-xylopyranoside (pNP-X) were 2.38 mM and 147.0 U/mg, respectively. The substrate specificity of the purified WSUCF1 β-xylosidase was determined using various substrates. WSUCF1 β-xylosidase was highly active on p-nitrophenyl β-D-xylopyranoside (pNPX). No specificity was observed for other p-NP substrates such as p-nitrophenyl β-D-cellobioside, p-nitrophenyl α-D-xylopyranoside, p-nitrophenyl α-L-arabinofuranoside, and p-nitrophenyl β-D-glucopyranoside (data not shown). In addition, polysaccharides such as oat-spelt xylan, beechwood xylan, and birchwood xylan did not serve as substrates for WSUCF1 β-xylosidase.

Effect of various xylose concentrations on WSUCF1 β-xylosidase activity was investigated. Results showed that concentrations of up to 50 mM xylose did not have much inhibitory effect. The enzyme retained 85 and 66% of its activity with 50 and 210 mM xylose concentrations, respectively (Figure 5). When the xylose concentration was increased to 300 mM, the enzyme still retained 50% activity. These data suggest that the WSUCF1 β-xylosidase is resistant to inhibition from high concentrations of xylose.

Hydrolysis of xylo-oligosaccharides and xylan

Thin layer chromatography (TLC) results showed that WSUCF1 β-xylosidase actively hydrolyzed xylo-oligosaccharides (Figure 6). It was found that enzyme was active on all xylo-oligosaccharides and released xylose and the other xylo-oligosaccharides with a degree of polymerization1 unit smaller than the substrate. For example, after 2 hours of hydrolysis, most of the xylotetraose was converted into smaller units with increase in xylose concentration. In case of xylopentaose hydrolysis, after 6 hours xylotetraose, xylotriose, xylobiose, and xylose were detected. With increase in time of hydrolysis of xylo-oligosaccharides from 2 to 6 hours, xylose yields were increased (Figure 6).

Mixtures of recombinant WSUCF1 endo-xylanase [16] and WSUCF1 β-xylosidase were utilized to hydrolyze birchwood xylan. It was interesting to note that endo-xylanase alone produced different concentrations of oligosaccharides (xylobiose, xylotriose and xylotetraose), but after addition of β-xylosidase, most of these xylo-oligosaccharides were

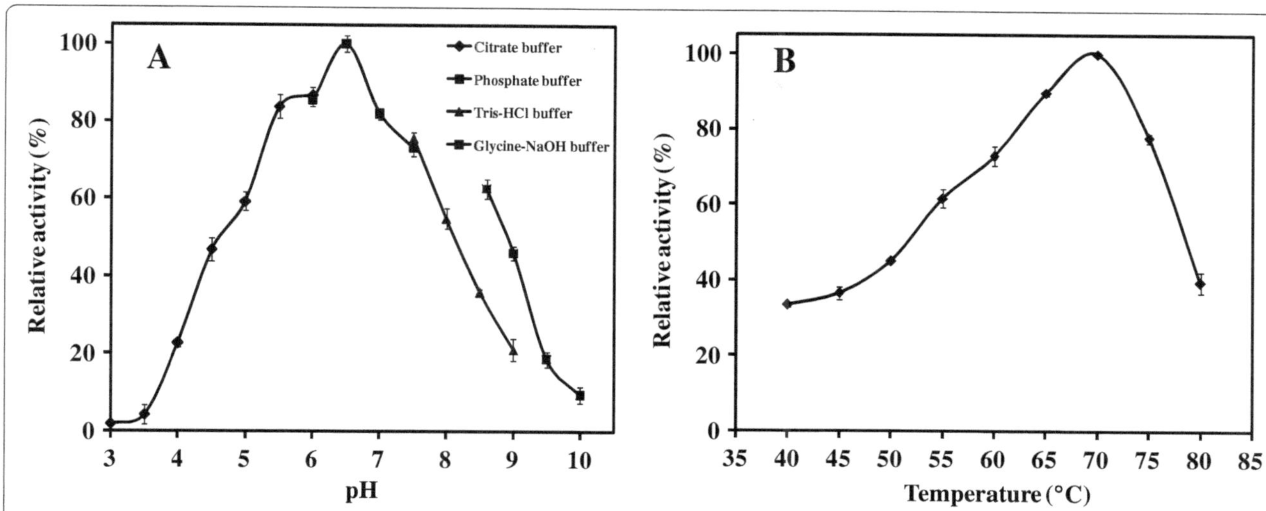

Figure 3 Impact of pH (A) and temperature (B) on the WSUCF1 β-xylosidase activity. The enzyme activities are expressed as percentages of the maximum activity (125.2 U/mg). The points are the averages of triplicates, and error bars indicate ± standard deviations of the means (n = 3). Error bars smaller than the symbols are not shown. Short title: Impact of pH **(A)** and temperature **(B)** on the WSUCF1 β-xylosidase activity.

converted to xylose (Figure 7). Xylan conversion increased with time. Sugar analysis results revealed that mixture of endo-xylanase and β-xylosidase worked efficiently and converted approximately 90% of xylan to xylose.

Discussion

Cellulose and hemicellulose components of lignocellulosic biomass are a renewable source of sugars. The main component of hemicellulose is xylan which can be converted into xylose using an effective enzyme system of endo-xylanase and β-xylosidase. β-xylosidase is known to be the key enzyme for converting xylo-oligosaccharides to xylose. Thermostable β-xylosidases produced from thermophiles are of importance due to their prolonged activity during the hydrolysis which may allow a reduced dose of enzyme and cost-efficient conversion [17]. Out of various GH families for β-xylosidase, reports of GH39 family thermostable β-xylosidases are scarce. To date, most of the reported β-xylosidases from *Geobacillus* spp. are from GH52 and GH43 family. For example, β-xylosidases of *Geobacillus thermodenitrificans* TSAA1 [7], *Geobacillus stearothermophilus* [18], *Geobacillus pallidus* [19], *Bacillus stearothermophilus* 21 [20] belonged to GH52 family and *Geobacillus thermoleovorans* IT-08 [21], *Geobacillus thermoleovorans* IT-08 [22] belonged to GH43 family (Table 2).

A high specific activity of β-xylosidase enzyme for xylo-oligosaccharides is an important attribute for efficient enzymatic hydrolysis of the hemicellulose component of lignocellulose [2]. WSUCF1 β-xylosidase specific activity is 133.0 U/mg, which is high in comparison to other

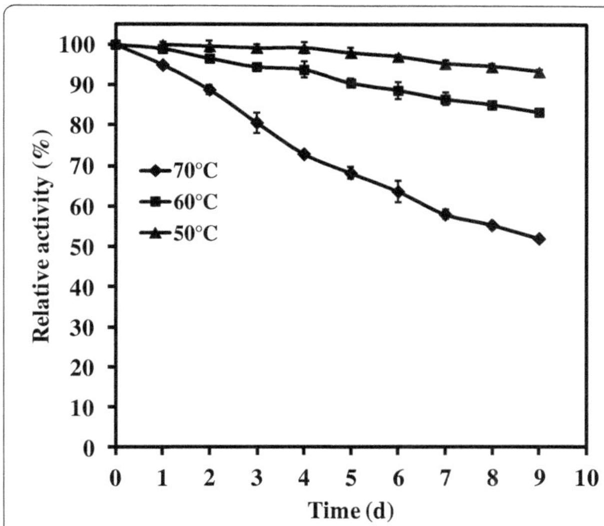

Figure 4 Thermal stability of WSUCF1 β-xylosidase. Enzyme activities are expressed as percentages of the initial activity (100%). The points are the averages of triplicates, and error bars indicate ± standard deviations of the means (n = 3). Error bars smaller than the symbols are not shown. Short title: Thermal stability of WSUCF1 β-xylosidase.

Table 1 Effect of different metal ions on β-xylosidase activity

Metal ions	Relative activity (%)
Control	100.00 ± 2.69
Mn^{2+}	101.18 ± 2.32
Ca^{2+}	96.04 ± 0.95
Mg^{2+}	91.33 ± 2.97
Co^{2+}	84.69 ± 3.20
Ni^{2+}	73.17 ± 2.95
Cu^{2+}	27.79 ± 1.23
Zn^{2+}	74.71 ± 2.37
Hg^{2+}	0

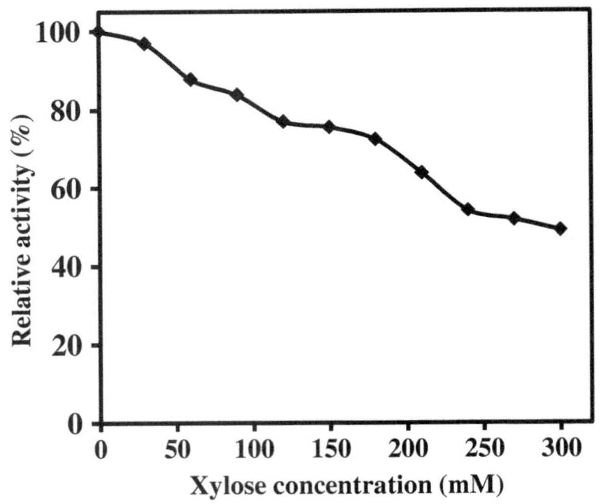

Figure 5 Effect of different concentrations of xylose on WSUCF1 β-xylosidase activity. The enzyme activities are expressed as percentages of the maximum activity (143.4 U/mg). The points are the averages of duplicates, and error bars indicate standard deviations of the means (n = 3). Short title: Effect of xylose on WSUCF1 β-xylosidase activity.

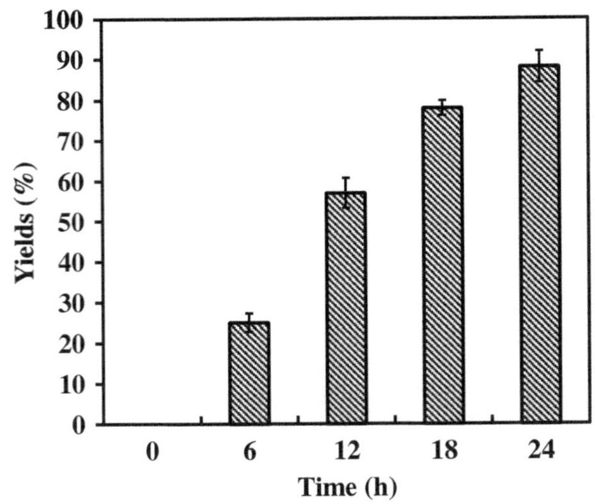

Figure 7 Synergistic action of endoxylanase and β-xylosidase on hydrolysis of xylan. Twenty mg xylan was incubated with WSUCF1 endoxylanase (5 mg/g xylan) and β-xylosidase (2.5 mg/g xylan) at 60°C for 24 h. Yields (100%) is equal to the theoretical maximum concentration of xylose released from xylan. Short title: Synergistic action of endoxylanase and β-xylosidase on xylan hydrolysis.

bacterial β-xylosidases such as *Thermoanaerobacterium saccharolyticum* JW/SL-YS485 [5], *Geobacillus thermoleovorans* IT-08 [21], *G. thermodenitrificans* TSAA1 [7] and *G. pallidus* [19] that showed specific activity of 45.8, 48.4, 68 and 4.75 U/mg respectively. Specific activity was also higher than the β-xylosidases reported from fungal sources *Paecilomyces thermophila* - 43.4 U mg^{-1} [6]; *Aspergillus awamori* X-100 - 4.2 U mg^{-1} [24] and commercial enzymes like CAZyme™ Xylosidase 1 from Lucigen - 31 U mg^{-1}.

Omitting heat denaturation prior to SDS-PAGE resulted in an increase in the apparent mass of the β-xylosidase from 58 kDa to 230 kDa. Interestingly, only the 230 kDa protein displayed xylosidase activity in the zymogram. This suggests that the WSUCF1 β-xylosidase is resistant to full denaturation by the detergent alone, and retains at least some of its quaternary structure which is necessary for successful recovery of catalytic activity following SDS-PAGE. Based on the 4-fold increase

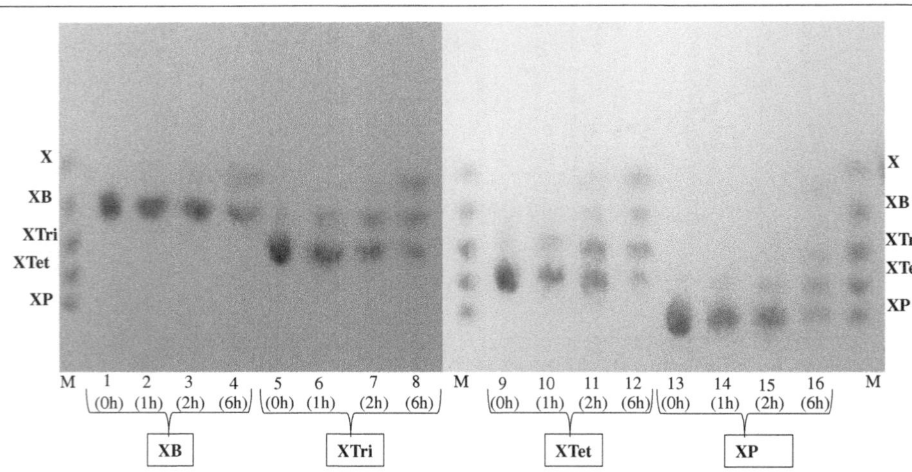

Figure 6 Thin layer chromatography of hydrolysis of xylooligosaccharides. Lane M, mixture of xylooligosaccharides (xylose (X), xylobiose (XB), xylotriose (XTri) and xylotetraose (Xtet), xylopentaose (XP) (1.5% each, wt/vol). Lane 1, Lane 5, Lane 9, Lane 13 - XB, XT, XTet, XP, respectively (Controls without enzyme); Lane 2–4, Lane 6–8, Lane 10–12, Lane 14–16 - XB, XT, XTet, XP (4 µl, 1.5%, wt/vol) each sugar was incubated with WSUCF1 β-xylosidase (0.05 U) for 1 h, 2 h, 6 h respectively. Short title: Thin layer chromatography of hydrolysis of xylooligosaccharides.

Table 2 Properties of thermostable β-xylosidases from thermophiles

Organism	GH family	Optimum pH	Optimum temperature (°C)	Thermal stability	K_m (mM)	V_{max} (U/mg)	References
Geobacillus sp. WSUCF1	39	6.5	70	Half-life of 9 days at 70°C	2.3	147	This study
Thermoanaerobacterium saccharolyticum JW/SL-YS485	Novel	6.0	65	Half-life of 1 h at 67°C	28.0	276	[5]
Geobacillus thermodenitrificans TSAA1	52	7.0	60	Half-life of 3 h at 70°C	2.8	4.16	[7]
Geobacillus stearothermophilus	52	6.5	70	Half-life of 3.4 h at 70°C	0.15	NA	[18]
Bacillus stearothermophilus	NA	6.0	70	Half-life of approx. 1 h at 70°C	1.2	NA	[10]
Bacillus thermantarcticus	NA	6.0	70	Stable for 1 h at 60°C	0.5	NA	[23]
Geobacillus pallidus	52	8.0	70	Half-life of approx. 10 h at 70°C	2.4	4.72	[19]
Geobacillus thermoleovorans IT-08	43	6.0	60	Retained more than 70 % activity after 1 h at 70°C	0.06	NA	[21]
Geobacillus thermoleovorans IT-08	43	5.0	NA	Half-life of 35 min at 57.5°C	0.55	NA	[22]

in apparent mass, it appears that that the WSUCF1 β-xylosidase exists in tetrameric form. GH39 β-xylosidase from *Thermoanaerobacterium* sp. has been reported to exist as a tetramer [1,25]. Property of retaining its activity in the presence of detergent could make it suitable for commercial applications in need for xylosidases resistant to denaturation with detergents.

WSUCF1 β-xylosidase was active across a broad pH range of 4.5-9.5 with maximum activity at pH 6.5. β-xylosidase from *Geobacillus stearothermophilus* T-6 also exhibited maximum activity at 6.5 [26] whereas β-xylosidases from *Geobacillus thermoleovorans* IT-08 [21], *Geobacillus thermodenitrificans* TSAA1 [7] and *G. pallidus* [19] showed their pH optima at 6.0, 7.0 and 8.0, respectively. Activity at broad pH range is a critical feature of applicability of industrial enzymes. β-xylosidases from other reported *Geobacillus* spp. had a narrow pH range except *G. pallidus* [19]. At pH 9.0, less than 10% and 40% relative activity for *Geobacillus thermoleovorans* IT-08 [21] and *Geobacillus thermodenitrificans* TSAA1 [7], respectively compared to 46% from WSUCF1 β-xylosidase whereas *G. pallidus* [19] retained 89% activity.

WSUCF1 β-xylosidase was also active across a broad temperature range with optima at 70°C. Temperature optima was higher than *Geobacillus stearothermophilus* T-6 (65°C, [26]), *Geobacillus thermoleovorans* IT-08 (65°C, [21]), *Geobacillus thermodenitrificans* TSAA1 (60°C, [7]), and *Thermoanaerobacterium saccharolyticum* JW/SL-YS485 (65°C, [5]). At a temperature higher than its optima (i.e. at 75°C), WSUCF1 β-xylosidase retained 77% relative activity which was comparable to relative activities from *Geobacillus thermodenitrificans* TSAA1 [7] and *G. pallidus* [19]. *Geobacillus thermoleovorans* IT-08 [21] β-xylosidase retained less than 40% relative activity at 75°C.

The performance of an enzyme mainly depends upon its operational stability. Enzymatic hydrolysis of lignocellulosic biomass is usually carried out for ≥ 3 days, so there is utmost requirement of enzymes which could retain their stability during this period. Therefore, thermostable enzymes are gaining wide industrial and biotechnological interest due to their ability to withstand the often relatively harsh conditions of industrial processing [27]. WSUCF1 β-xylosidase was found to be highly thermostable when compared with other reported β-xylosidases from thermophiles. WSUCF1 xylosidase retained 93, 83 and 50% activity at 50, 60 and 70°C after the period of 9 days. β-xylosidase from *Geobacillus pallidus* [19] had a half-life of 50 h at 60°C, *Bacillus halodurans* XylBH39 and XylBH43 displayed approximate half-life values of 2.40 and 0.05 h, respectively at 60°C, *Geobacillus thermoleovorans* IT-08 [21] retained more than 70% activity after 1 h, *Thermoanaerobacterium saccharolyticum* JW/SL-YS485 [5] exhibited half-life of 1 h at 67°C, *Geobacillus thermoleovorans* IT-08 [22] retained 50% activity after 35 min at 57.5°C. At 50°C, the WSUCF1 β-xylosidase retained almost 100% relative activity, and therefore this enzyme could be used as a component of commercial enzyme cocktails mainly working optimally at ≤ 50°C. Various studies have also been published to enhance thermal stability of xylanases using genetic engineering [28,29] whereas WSUCF1 β-xylosidase is highly thermostable in its native form. Thus β-xylosidase from *Geobacillus* sp. WSUCF1 appears to be among the most thermostable β-xylosidases from thermophiles having optimum growth temperatures of ≤ 70°C reported to date (Table 2). Higher thermal-stability of WSUCF1 β-xylosidase will allow hydrolysis for extended times to convert lignocellulosic biomass into sugars, leading to decreased amount of enzyme needed for saccharification. It has also been stated that enzymes are the main cost in lignocellulosic conversion process [30]. Due to high thermostability of WSUCF1 β-xylosidase, it could be recycled with its full activity for the subsequent enzymatic hydrolysis. Several studies have been reported on enzyme recycling, which significantly reduce overall cost of the process [31,32]. These discussions showed that thermostability is an essential attribute of an enzyme for its commercial application.

The kinetic properties of WSUCF1 β-xylosidase (K_m - 2.38 mM and V_{max} - 147.0 U/mg) were significantly different from those of the other reported β-xylosidases. Compared to WSUCF1 β-xylosidase, *Thermoanaerobacterium saccharolyticum* JW/SL-YS485 β-xylosidase [5] and *Geobacillus thermodenitrificans* TSAA1 β-xylosidase [7] showed higher K_m of 28 mM and 2.8 mM, respectively, which shows WSUCF1 β-xylosidase has more specificity towards substrate.

β-xylosidase plays a vital role in the conversion of xylobiose and other higher xylo-oligosaccharides to xylose during enzymatic hydrolysis [1]. At commercial scale, performing enzymatic hydrolysis of lignocellulosic biomass at high solid loadings is beneficial as it increases product concentrations and utilize less water. On the other end, this increased product concentrations could lead to enzyme inhibition called end-product inhibition [33]. Therefore for industrial processes, utilization of a xylose tolerant β-xylosidase is very important in order to achieve high concentrations of xylose by avoiding end product inhibition. WSUCF1 β-xylosidase exhibited tolerance to high concentrations of xylose, retaining 80% of initial activity at 100 mM xylose i.e. higher tolerance than β-xylosidase from *Paecilomyces thermophila* which retained only 54% activity in the presence of 100 mM xylose. The other β-xylosidases reported for exhibiting strong tolerance to xylose inhibition are from *Scytalidium thermophilum* ([34] and *Thermatoga thermarum* [11] which were insensitive to inhibition by up to 200 mM xylose.

Complete enzymatic degradation of xylan to xylose is one of the most important and challenging reactions. Endoxylanases are reported to deconstruct the xylan into xylose and different xylo-oligosaccharides. These xylo-oligosaccharides are often found in hydrolysates of lignocellulosic materials as they can't be fermented by ethanalogens leading to lower yields of ethanol [35]. Kont et al. [36] demonstrated that xylo-oligosaccharides are also one of the strong inhibitors of cellulases. Zhang and Viikari [37] and Qing et al. [38] also studied the inhibitory effects of xylo-oligosaccharides on cellobiohydrolase. These studies demonstrate the necessity of enzymes which could breakdown these inhibitory xylo-oligosaccharides to obtain better fermentable sugar yields. Xylo-oligosaccharides are reported to be hydrolyzed to xylose by exo-type xylanolytic β-xylosidase [39]. WSUCF1 β-xylosidase released xylose from all the tested xylo-oligosaccharides (X2-X5) indicating that it is a true β-D-xylosidase [34]. Addition of thermostable WSUCF1 β-xylosidase along with endo-xylanase during birchwood xylan hydrolysis increased final xylose yields to approximately 90% which would further improve the fermentation product yields. Similar observations were observed in β-xylosidases from *Paecilomyces thermophila* [6] and *Thermotoga thermarum* [11], which were active on different xylo-oligosaccharides obtained after xylan hydrolysis.

Conclusions

In this study, a GH39 β-xylosidase from the thermophile *Geobacillus* sp. strain WSUCF1 was cloned and purified. Compared to β-xylosidases from other thermophiles, WSUCF1 β-xylosidase showed more thermostability. It also displayed various other desirable properties including high specific activity, high xylose tolerance, excellent hydrolytic activity on p-nitrophenyl-xylopyranoside and xylo-oligosaccharides, and resistance to detergent. A high degree of synergy was found with respect to the release of reducing sugars from xylans when β-xylosidase was combined with the endo-xylanase. These findings open a path to potential industrial applications of the WSUCF1 β-xylosidase.

Methods

Multiple sequence alignment and phylogenetic analysis

Nucleotide sequence from the whole genome of *Geobacillus* sp. strain WSUCF1 [15] was translated to amino acid sequence using ExPASy-Translate tool (http://web.expasy.org/translate/). Amino acid sequence of *Geobacillus* sp. strain WSUCF1 β-xylosidase was used as a BLAST query for seeking other homologous amino acid sequences. The sequence alignment was created with ClustalW program. Sequences were aligned manually using the Mega 5.2 software. Phylogenetic relationship was inferred using the Neighbor-Joining (NJ) method as described earlier [16].

Cloning and expression of WSUCF1 β-xylosidase

To amplify the 1509 bp β-xylosidase gene sequence from *Geobacillus* sp. strain WSUCF1 (simply referred to as WSUCF1), PCR primers (F 5'- *CGCGAACAGATTG GAGGT* AAGGTTGTAAACGTGCCAAGC-3' and R 5'-*GTGGCGGCCGCTCTATTA*TGAAGAATA CGATGTCA TTTC-3') containing homologous sequences to the ends of the pRham N-His SUMO Kan Vector (Lucigen, Middleton, WI) (italicized) were designed to amplify the β-xylosidase gene sequence. PCR product containing the gene of interest is cloned as a fusion to cleavable SUMO (Small Ubiquitin-like Modifier) tag under the control of the L-rhamnose-inducible rhaPBAD promoter harbored on the pRham™ N-His SUMO Vector. PCR conditions were as follows: initial denaturation at 94°C for 2 min followed by 25 cycles of 94°C for 15 sec, 55°C for 15 sec and 72°C for 1 min 30 sec in 50 μl reaction with final extension of 10 min at 72°C. The amplicon was cloned into the vector according to the manufacturer's instructions, and the resulting plasmid transformed into *E. coli* chemically competent cells (Lucigen, Middleton, WI). After transformation, *E. coli* cells were plated on agar plates containing Terrific broth (TB) and 30 μg/ml kanamycin, incubated for 36 hours at 37°C. The positive clones were screened using colony PCR and confirmed by β-xylosidase activity assay as described below. Cycling conditions for colony PCR were: initial denaturation 94°C, 5 sec; 25 cycles of 94°C for 15 sec, 55°C for 15 sec and 72°C for 1 min; extension 72°C 10 min.

Purification of the recombinant β-xylosidase

A positive clone was grown in 250-mL flask containing 50 ml Terrific broth (TB), kanamycin (30 μg/ml), and the expression of β-xylosidase was induced by addition of 0.2% rhamnose followed by incubation at 37°C for 48 hours. Cells were centrifuged at 10,000 rpm for 15 minutes. Ten ml of clear culture supernatant was applied to a Ni-NTA agarose resin column (1.5 cm × 5 cm) equilibrated with equilibration buffer (50 mM NaH_2PO_4, 300 mM NaCl, 10 mM imidazole, pH 8.0). After binding, the column was washed with wash buffer (50 mM NaH_2PO_4, 300 mM NaCl, 20 mM imidazole, pH 8). The bound enzyme was eluted with elution buffer (50 mM NaH_2PO_4, 300 mM NaCl, 250 mM imidazole, pH 8.0) at a flow rate of 0.5 ml/min collecting 1 ml fractions. Fractions containing β-xylosidase activity were pooled. To remove the His tag from the protein, the protein was cleaved with SUMO specific protease which also contains the His tag (as per manufacturer's protocol). After the cleavage, the protein was separated from the SUMO protease and the His tag by re-performing metal affinity chromatography with Ni-NTA agarose resin column. His tag and SUMO protease bound to the resin while the protein was eluted. The homogeneity of the purified enzyme was checked by sodium dodecyl sulfate-polyacrylamide gel electrophoresis (SDS-PAGE) as described below. Protein concentrations in the samples were determined using 2-D Quant kit (GE Healthcare).

Assay of β-xylosidase activity

For β-xylosidase activity, pNPX, 2.5 mg/mL was used as a substrate. The assay mixture contained 80 μL of substrate, 100 μl 0.1 M phosphate buffer (pH 6.5), and 20 μL of appropriately diluted enzyme. The reaction was carried out at 70°C for 10 min and then stopped by the addition of 100 μL of sodium carbonate (2% w/v). The p-nitrophenol released was determined by measuring the absorbance of each sample at 410 nm. One unit of enzyme activity was defined as the amount of enzyme that released 1 μmol of p-NP per minute under the standard assay conditions.

SDS-PAGE, zymography, and molecular mass determination

SDS-PAGE was performed as described by Laemmli (1970) [40]. The molecular weight standards used were BioRad Precision plus standards. To obtain zymogram of β-xylosidase activity, enzyme (10 μl) was mixed with 10 μl 2X SDS sample buffer. Two different processing conditions were used prior to loading: i) purified enzyme sample was heated to 95°C for 2 min and then loaded to an 8-16% gradient SDS-PAGE gel and ii) enzyme sample was loaded directly to an 8-16% gradient SDS-PAGE gel without any heating. Current was passed through the gel (150 V constant voltage) for about 1 h until the bromophenol blue dye-front migrated to the bottom of the gel. The gel was washed successively with the following for 30 minutes each: 20% isopropanol in phosphate buffer saline (PBS, pH 5.9); 8 M urea in PBS; and with PBS (pH 5.9) 3 times. The gel was incubated with substrate (0.1 mg/ml 4-methylumbelliferyl-β-D-xylopyranoside in PBS, pH 5.9) for 30 min. Finally, the detection of β-xylosidase activity was examined by fluorescence of 4-methylumbelliferone, which is visualized under ultraviolet light. The gel was then stained with Coomassie brilliant blue R-250 (LabSafe GEL Blue, G-Biosciences, St. Louis, MO) to visualize protein.

Characterization of WSUCF1 β-xylosidase

Activity profile of β-xylosidase was determined at different pH 3–10. The enzyme assays were carried in the following discontinuous buffer system under standard assay conditions: 100 mM sodium citrate (pH 3–6), sodium phosphate (pH 6–8), Tris-HCl (pH 7.5-9), or glycine-NaOH (pH 8.5-10). The optimum pH was used to determine the optimum temperature for the β-xylosidase. Experiments were carried out in the temperature range of 40-80°C. In order to assess thermostability of β-xylosidase, 1 ml of enzyme was incubated at 50-80°C for varying time intervals. Subsamples (20 μl) were removed at specific time intervals over the period of incubation. The residual enzyme activity was measured and plotted against time. Initial activity was assumed to be 100% and residual enzyme activity reported as percentages of the initial activity during the incubation period.

Kinetic parameters (K_m and V_{max}) for the purified recombinant β-xylosidase were determined by measuring the enzyme activity using 0–15 mM pNP-X as substrate in 50 mM phosphate buffer (pH 6.5) at 70°C. The data was plotted according to the Lineweaver-Burk method to calculate K_m and V_{max} values.

Substrate specificity of the purified enzyme was investigated under standard assay conditions as described above. Activities towards p-nitrophenyl derivatives were measured by the rate of p-nitrophenol formed during hydrolysis from 2.5 mg/ml of the substrates in 50 mM phosphate buffer (pH 6.5) at 70°C for 10 min, and detected by spectrophotometry at 410 nm. Substrates used were p-nitrophenyl β-D-xylopyranoside, p-nitrophenyl α-L-arabinofuranoside, p-nitrophenyl β-D-cellobioside, p-nitrophenyl β-D-glucopyranoside.

The effect of metal ions on enzyme activity of β-xylosidase was determined by adding chloride salts of Mg^{2+}, Zn^{2+}, Ba^{2+}, Ni^{2+}, Mn^{2+},Ca^{2+}, Co^{2+}, Cu^{2+} and Al^{3+} to the enzyme reaction mixtures at a final concentration of 1 mM. The enzyme reaction mixtures were incubated for 10 min at 70°C, and then activities were measured as described above. The activity of the enzyme without adding any metal ions was considered as 100%.

The inhibitory effect of different concentrations of xylose was determined by incubating 20 μl enzyme solution, 80 μl of 2.5 mg/ml pNPX dissolved in 100 mM phosphate buffer (pH 6.5) and varying amounts of xylose with a final concentration of 0–300 mM at 70°C for 10 min. Residual β-xylosidase activities were measured as described above.

Hydrolysis of xylo-oligosaccharides by the WSUCF1 β-xylosidase

Four μl of 1.5% (w/v) xylo-oligosaccharides was incubated with 0.05 U of the β-xylosidase in 50 mM phosphate buffer (pH 6.5) at 70°C for 1, 2 and 6 h. Aliquots were withdrawn at 0, 1, and 2 h, the reactions were stopped by boiling for 5 min. Products of enzymatic hydrolyses were analyzed by spotting on silica gel plates 60 F-254 (E. Merck, Germany). The plates were developed with butanol-acetic acid-water (2:1:1, v/v/v) solvent system followed by heating for few minutes at 105°C in an oven, after spraying the plates with a methanol-sulfuric acid mixture (9:1, v/v) [11]. A mixture of xylo-oligosaccharides consisting of xylose (X), xylobiose (XB), xylotriose (XTri), and xylotetraose (XTet) and xylopentaose (XP) were used as standards.

Synergistic effect of WSUCF1 endo-xylanase and β-xylosidase on xylan hydrolysis

For enzymatic hydrolysis of xylan, the reaction mixture containing 5 mg purified recombinant endoxylanase/g xylan (purified enzyme from *Geobacillus* sp. WSUCF1, [16]), 2.5 mg recombinant β-xylosidase/g xylan and 20 mg birchwood xylan in 4 ml of 50 mM phosphate buffer (pH 6.5) was incubated at 60°C for 6 to 24 h. Xylose, xylobiose, xylotriose and xylotetraose were determined by high performance liquid chromatography using a 30 mm Micro-Guard Cation H cartridge and a 300 mm Aminex HPX-87H column (Bio-Rad Laboratories, Inc., Hercules, CA) on a 1100 Series HPLC system equipped with a refractive index detector (Agilent Technologies, Santa Clara, CA). Samples (10 μl) were injected onto a heated column (55°C) and eluted at 0.6 ml/min using 5 mM H_2SO_4 as mobile phase.

Competing interests

The authors declare that they have no competing interests.

Authors' contributions

AB designed and performed the experiments and drafted the manuscript. KMB helped to analyze the degradation products and revised the manuscript. RKS directed the over-all study and revised the manuscript. All authors have read and approved the final manuscript.

Acknowledgements

The authors gratefully acknowledge the financial support provided by National Science Foundation – Industry/University Cooperative Research Center (NSF-I/UCRC, Grant #441087). The support from the Department of Chemical and Biological Engineering at the South Dakota School of Mines and Technology is gratefully acknowledged.

Author details

[1]Department of Chemical and Biological Engineering, South Dakota School of Mines and Technology, Rapid City, SD 57701, USA. [2]Renewable Product Technology Research Unit, Agricultural Research Service, National Center for Agricultural Utilization Research, U.S. Department of Agriculture, Peoria, IL 61604, USA. [3]Present address: Great Lakes Bioenergy Research Center, Michigan State University, East Lansing, MI 48824, USA.

References

1. Juturu V, Wu JC: **Microbial xylanases: engineering, production and industrial applications.** *Biotechnol Adv* 2012, **30:**1219–1227.
2. Yeoman CJ, Han Y, Dodd D, Schroeder CM, Mackie RI, Cann IK: **Thermostable enzymes as biocatalysts in the biofuel industry.** *Adv Appl Microbiol* 2010, **70:**1–55.
3. Gao D, Uppugundla N, Chundawat SP, Yu X, Hermanson S, Gowda K, Brumm P, Mead D, Balan V, Dale BE: **Hemicellulases and auxiliary enzymes for improved conversion of lignocellulosic biomass to monosaccharides.** *Biotechnol Biofuels* 2011, **4:**5.
4. Kumar R, Wyman C: *Enzymatic hydrolysis of cellulosic biomass through enhanced removal of oligomers.* 2010. WO2010011957A2.
5. Shao W, Xue Y, Wu A, Kataeva I, Pei J, Wu H, Wiegel J: **Characterization of a novel beta-xylosidase, XylC, from *Thermoanaerobacterium saccharolyticum* JW/SL-YS485.** *Appl Environ Microbiol* 2011, **77:**719–726.
6. Yan QJ, Wang L, Jiang ZQ, Yang SQ, Zhu HF, Li LT: **A xylose-tolerant beta-xylosidase from *Paecilomyces thermophila*: characterization and its co-action with the endogenous xylanase.** *Bioresour Technol* 2008, **99:**5402–5410.
7. Anand A, Kumar V, Satyanarayana T: **Characteristics of thermostable endoxylanase and β-xylosidase of the extremely thermophilic bacterium *Geobacillus thermodenitrificans* TSAA1 and its applicability in generating xylooligosaccharides and xylose from agro-residues.** *Extremophiles* 2013, **17:**357–366.
8. Bravman T, Mechaly A, Shulami S, Belakhov V, Baasov T, Shoham G, Shoham Y: **Glutamic acid 160 is the acid–base catalyst of beta-xylosidase from *Bacillus stearothermophilus* T-6: a family 39 glycoside hydrolase.** *FEBS Lett* 2001, **495:**115–119.
9. Huang Z, Liu X, Zhang S, Liu Z: **GH52 xylosidase from *Geobacillus stearothermophilus*: characterization and introduction of xylanase activity by site-directed mutagenesis of Tyr509.** *J Ind Microbiol Biotechnol* 2014, **41:**65–74.
10. Nanmori T, Watanabe T, Shinke R, Kohno A, Kawamura Y: **Purification and properties of thermostable xylanase and beta-xylosidase produced by a newly isolated *Bacillus stearothermophilus* strain.** *J Bacteriol* 1990, **172:**6669–6672.
11. Shi H, Li X, Gu H, Zhang Y, Huang Y, Wang L, Wang F: **Biochemical properties of a novel thermostable and highly xylose-tolerant β-xylosidase/α-arabinosidase from *Thermotoga thermarum*.** *Biotechnol Biofuels* 2013, **6:**27.
12. Bhalla A, Bansal N, Kumar S, Bischoff KM, Sani RK: **Improved lignocellulose conversion to biofuels with thermophilic bacteria and thermostable enzymes.** *Bioresour Technol* 2013, **128:**751–759.
13. Czjzek M, Ben David A, Bravman T, Shoham G, Henrissat B, Shoham Y: **Enzyme-substrate complex structures of a GH39 beta-xylosidase from *Geobacillus stearothermophilus*.** *J Mol Biol* 2005, **353:**838–846.
14. Rastogi G, Bhalla A, Adhikari A, Bischoff KM, Hughes SR, Christopher LP, Sani RK: **Characterization of thermostable cellulases produced by *Bacillus* and *Geobacillus* strains.** *Bioresour Technol* 2010, **101:**8798–8806.
15. Bhalla A, Kainth AS, Sani RK: **Draft genome sequence of lignocellulose-degrading thermophilic bacterium *Geobacillus* sp. strain WSUCF1.** *Genome Announc* 2013, **1:**4.

16. Bhalla A, Bischoff KM, Uppugundla N, Balan V, Sani RK: **Novel thermostable endo-xylanase cloned and expressed from bacterium *Geobacillus* sp. WSUCF1.** *Bioresour Technol* 2014, **165:**314–318.
17. Viikari L, Alapuranen M, Puranen T, Vehmaanpera J, Siika-Aho M: **Thermostable enzymes in lignocellulose hydrolysis.** *Adv Biochem Eng Biotechnol* 2007, **108:**121–145.
18. Contreras LM, Gómez J, Prieto J, Clemente-Jiménez JM, Las Heras-Vázquez FJ, Rodríguez-Vico F, Blanco FJ, Neira JL: **The family 52 beta-xylosidase from *Geobacillus stearothermophilus* is a dimer: structural and biophysical characterization of a glycoside hydrolase.** *Biochim Biophys Acta* 2008, **1784:**1924–1934.
19. Quintero D, Velasco Z, Hurtado-Gómez E, Neira JL, Contreras LM: **Isolation and characterization of a thermostable beta-xylosidase in the thermophilic bacterium *Geobacillus pallidus*.** *Biochim Biophys Acta* 2007, **1774:**510–518.
20. Baba T, Shinke R, Nanmori T: **Identification and characterization of clustered genes for thermostable xylan-degrading enzymes, beta-xylosidase and xylanase, of *Bacillus stearothermophilus* 21.** *Appl Environ Microbiol* 1994, **60:**2252–2258.
21. Ratnadewi AA, Fanani M, Kurniasih SD, Sakka M, Wasito EB, Sakka K, Nurachman Z, Puspaningsih NN: **β-D-xylosidase from *Geobacillus thermoleovorans* IT-08: biochemical characterization and bioinformatics of the enzyme.** *Appl Biochem Biotechnol* 2013, **170:**1950–1964.
22. Wagschal K, Heng C, Lee CC, Robertson GH, Orts WJ, Wong DW: **Purification and characterization of a glycoside hydrolase family 43 beta-xylosidase from *Geobacillus thermoleovorans* IT-08.** *Appl Biochem Biotechnol* 2009, **155:**304–313.
23. Lama L, Calandrelli V, Gambacorta A, Nicolaus B: **Purification and characterization of thermostable xylanase and beta-xylosidase by the thermophilic bacterium *Bacillus thermantarcticus*.** *Res Microbiol* 2004, **155:**283–289.
24. Eneyskaya EV, Ivanen DR, Bobrov KS, Isaeva-Ivanova LS, Shabalin KA, Savel'ev AN, Golubev AM, Kulminskaya AA: **Biochemical and kinetic analysis of the GH3 family beta-xylosidase from *Aspergillus awamori* X-100.** *Arch Biochem Biophys* 2007, **457:**225–234.
25. Yang JK, Yoon HJ, Ahn HJ, Lee BI, Pedelacq JD, Liong EC, Berendzen J, Laivenieks M, Vieille C, Zeikus GJ, Vocadlo DJ, Withers SG, Suh SW: **Crystal structure of beta-D-xylosidase from *Thermoanaerobacterium saccharolyticum*, a family 39 glycoside hydrolase.** *J Mol Biol* 2004, **335:**155–165.
26. Shallom D, Leon M, Bravman T, Ben-David A, Zaide G, Belakhov V, Shoham G, Schomburg D, Baasov T, Shoham Y: **Biochemical characterization and identification of the catalytic residues of a family 43 β-D-xylosidase from *Geobacillus stearothermophilus* T-6.** *Biochemistry* 2005, **44:**387–397.
27. Turner P, Mamo G, Karlsson EN: **Potential and utilization of thermophiles and thermostable enzymes in biorefining.** *Microb Cell Fact* 2007, **6:**9.
28. Singh SK, Heng C, Braker JD, Chan VJ, Lee CC, Jordan DB, Yuan L, Wagschal K: **Directed evolution of GH43 β-xylosidase XylBH43 thermal stability and L186 saturation mutagenesis.** *J Ind Microbiol Biotechnol* 2014, **41:**489–498.
29. Zhang ZG, Yi ZL, Pei XQ, Wu ZL: **Improving the thermostability of *Geobacillus stearothermophilus* xylanase XT6 by directed evolution and site-directed mutagenesis.** *Bioresour Technol* 2010, **101:**9272–9278.
30. Klein-Marcuschamer D, Oleskowicz-Popiel P, Simmons BA, Blanch HW: **The challenge of enzyme cost in the production of lignocellulosic biofuels.** *Biotechnol Bioeng* 2012, **109:**1083–1087.
31. Tu M, Zhang X, Paice M, MacFarlane P, Saddler JN: **The potential of enzyme recycling during the hydrolysis of a mixed softwood feedstock.** *Bioresour Technol* 2009, **100:**6407–6415.
32. Weiss N, Börjesson J, Pedersen LS, Meyer AS: **Enzymatic lignocellulose hydrolysis: improved cellulase productivity by insoluble solids recycling.** *Biotechnol Biofuels* 2013, **6:**5.
33. Kristensen JB, Felby C, Jørgensen H: **Yield-determining factors in high-solids enzymatic hydrolysis of lignocellulose.** *Biotechnol Biofuels* 2009, **2:**11.
34. Zanoelo FF, PolizeliMdMde L, Terenzi HF, Jorge JA: **Purification and biochemical properties of a thermostable xylose-tolerant beta- D-xylosidase from *Scytalidium thermophilum*.** *J Ind Microbiol Biotechnol* 2004, **31:**170–176.
35. Fujii T, Yu G, Matsushika A, Kurita A, Yano S, Murakami K, Sawayama S: **Ethanol production from xylo-oligosaccharides by xylose-fermenting *Saccharomyces cerevisiae* expressing β-xylosidase.** *Biosci Biotechnol Biochem* 2011, **75:**1140–1146.
36. Kont R, Kurašin M, Teugjas H, Väljamäe P: **Strong cellulase inhibitors from the hydrothermal pretreatment of wheat straw.** *Biotechnol Biofuels* 2013, **6:**135.
37. Zhang J, Viikari L: **Xylo-oligosaccharides are competitive inhibitors of cellobiohydrolase I from *Thermoascus aurantiacus*.** *Bioresour Technol* 2012, **117:**286–291.
38. Qing Q, Yang B, Wyman CE: **Xylooligomers are strong inhibitors of cellulose hydrolysis by enzymes.** *Bioresour Technol* 2010, **101:**9624–9630.
39. Herrmann MC, Vrsanska M, Jurickova M, Hirsch J, Biely P, Kubicek CP: **The β-D-xylosidase of *Trichoderma reesei* is a multifunctional b-D-xylan xylohydrolase.** *Biochem J* 1997, **321:**375–381.
40. Laemmli UK: **Cleavage of structural proteins during the assembly of the head of bacteriophage T4.** *Nature* 1970, **227:**680–685.

11

Ethanol effect on metabolic activity of the ethalogenic fungus *Fusarium oxysporum*

Thomas Paschos[1], Charilaos Xiros[2] and Paul Christakopoulos[3*]

Abstract

Background: *Fusarium oxysporum* is a filamentous fungus which has attracted a lot of scientific interest not only due to its ability to produce a variety of lignocellulolytic enzymes, but also because it is able to ferment both hexoses and pentoses to ethanol. Although this fungus has been studied a lot as a cell factory, regarding applications for the production of bioethanol and other high added value products, no systematic study has been performed concerning its ethanol tolerance levels.

Results: In aerobic conditions it was shown that both the biomass production and the specific growth rate were affected by the presence of ethanol. The maximum allowable ethanol concentration, above which cells could not grow, was predicted to be 72 g/L. Under limited aeration conditions the ethanol-producing capability of the cells was completely inhibited at 50 g/L ethanol. The lignocellulolytic enzymatic activities were affected to a lesser extent by the presence of ethanol, while the ethanol inhibitory effect appears to be more severe at elevated temperatures. Moreover, when the produced ethanol was partially removed from the broth, it led to an increase in fermenting ability of the fungus up to 22.5%. The addition of *F. oxysporum*'s system was shown to increase the fermentation of pretreated wheat straw by 11%, in co-fermentation with *Saccharomyces cerevisiae*.

Conclusions: The assessment of ethanol tolerance levels of *F. oxysporum* on aerobic growth, on lignocellulolytic activities and on fermentative performance confirmed its biotechnological potential for the production of bioethanol. The cellulolytic and xylanolytic enzymes of this fungus could be exploited within the biorefinery concept as their ethanol resistance is similar to that of the commercial enzymes broadly used in large scale fermentations and therefore, may substantially contribute to a rational design of a bioconversion process involving *F. oxysporum*. The SSCF experiments on liquefied wheat straw rich in hemicellulose indicated that the contribution of the metabolic system of *F. oxysporum* in a co-fermentation with *S. cerevisiae* may play a secondary role.

Keywords: Bioethanol, Ethanol inhibition, Ethanol tolerance, Ethanol removal, *Fusarium oxysporum*

Background

Fusarium oxysporum is a filamentous fungus, capable of producing ethanol not only from hexoses but also from pentoses. The presence of many cellulases and hemicellulases in its secretome is reflected to the ability to grow on many lignocellulosic substrates under submerged or solid state conditions [1-3]. The fungus is capable of degrading and fermenting a wide variety of different substrates under anaerobic or limited oxygen conditions in the presence of inhibitory compounds such as furan derivatives, phenolic compounds and weak acids [4]. The low ethanol production rate and in some cases the formation of significant amount of acetic acid as by-product, have been considered as obstacles for its industrial exploitation. However, it has been shown that the efficient lignocellulolytic secretome of this fungus combined with its ability to ferment xylose could significantly improve the ethanol production from lignocellulosic substrates in co-culture with *Saccharomyces serevisiae* [5]. The beneficial properties of *F. oxysporum* can be fully exploited in consolidated bioprocess (CBP) where a microbial system able to degrade lignocellulose and ferment the released sugars in a single reactor is used. CBP is decreasing not only the overall cost but also the environmental impact of the process by minimizing the

* Correspondence: paul.christakopoulos@ltu.se
[3]Biochemical Process Engineering, Division of Chemical Engineering, Department of Civil, Environmental and Natural Resources Engineering, Luleå University of Technology, Luleå SE-971 87, Sweden
Full list of author information is available at the end of the article

addition of commercial lignocellulolytic enzymes. The economics of the process can be further improved by operating at high gravity (HG) conditions (operating at total solids content above 20% w/v). Although HG technology is associated with challenges such as the insufficient mixing of the lignocellulosic slurries, the absence of significant amounts of free water and the presence of high amounts of microbial inhibitors, it has gained much attention as a promising technology due to the possibility to achieve final ethanol concentrations above 4% w/v [6]. The higher the ethanol concentration is, the lower the distillation cost (per g of ethanol produced) is [7]. The tolerance to ethanol of microorganisms used in such processes is therefore crucial for their use in large scale.

Ethanol affects the cellular membranes and influences cell metabolism and macromolecular biosynthesis by inducing the production of heat shock-like proteins, lowering the rate of RNA and protein accumulation [8], enhancing the frequency of petite mutations, altering metabolism, denaturing intracellular proteins and glycolytic enzymes and reducing their activity [9]. These inhibitory effects are reflected to the decreased cell division rate, to the decreased cell volume and to low specific growth rates, while high ethanol concentration reduces cell vitality and increases cell death [10]. Robust fermenting microorganisms like *S cerevisiae,* have developed appropriate mechanisms to deal with several types of damages caused by increased ethanol concentration. The yeast stress response is a transient reprogramming of cellular activities to ensure survival in challenging conditions, protect essential cell components and enable cells to resume their normal metabolic conditions [11].

Although most of the published articles on the field deal with yeasts' ethanol stress and tolerance, there is some evidence concerning ethanol stress and cell response in fungi as well. The addition of ethanol to submerged cultures of *Phanerochaete chrysosporium* affected both the mycelial morphology and the fungal wall permeability and led to decreased pellet diameter and fungal biomass net weight [12]. The presence of ethanol (0.5–2%, v/v) hampered the secretion of cellulases by *Trichoderma reesei* [13]. The authors suggested that ethanol inhibition occurred at a pre-translational level by interfering with either the formation or the stability of cellulase mRNA. Asiimwe et al. reported indications that cytosolic aldehyde dehydrogenases play an important role in ethanol tolerance of the mycorrhizal fungus *Tricholoma vaccinum* [14]. The abundance of certain intracellular metabolites has also been connected to the fungal ethanol stress response: Evidence about the role of sterol glycosides and cerebrosides in the cell response to elevated ethanol levels was provided by the strong increase in the content of sterol glucoside that was observed after the treatment of fungal cells with increased ethanol concentrations [15].

Studying the effects of ethanol not only on the final yields but also on the biomass and ethanol rates is essential in order to evaluate a microorganism with regard to its implementation in biofuels production processes. Furthermore, the possibility of using lignocellulolytic activities in simultaneous saccharification and fermentation (SSF) processes is highly dependent on their stability during such a process. Ethanol is considered to be a non-competitive inhibitor for cellulases and this inhibition probably results from reversible enzyme denaturation [16]. Ethanol, like other organic molecules such as butanol or acetone, binds on the non-catalytic region of the enzyme, causing changes in the shape of the protein molecule, which in turn affects the catalytic activity. However, not all the enzymes are affected by ethanol in the same way. As shown by Chen and Jin the presence of ethanol had a positive effect on β-glucosidase from *Penicillium decumbens*, especially at temperatures above 40°C [17]. Therefore, the study of the effects of ethanol on the cellulolytic activities is of major importance for the evaluation of a cellulolytic system to be used in SSF processes for ethanol production.

Although many scientific reports have been published on both the hydrolytic and the fermentative performance of *F. oxysporum* on lignocellulosic substrates, little study has been published on the ethanol tolerance of this fungus. Early works on the fermentative performance of *F. oxysporum* refer to the effects of ethanol on this fungus claiming a relatively high ethanol tolerance but those statements were only based on the increase in ethanol concentration during the fermentation stage [18-20]. Recently, Hennessy et al., [3] showed that *Agrobacterium tumefaciens*-mediated transformation could be exploited as a tool to generate significant degrees of phenotypic diversity in *F. oxysporum* strain 11C in response to alcohol stress. In the same study, the importance of further investigation of the ethanol tolerance of this fungus is underlined.

The present work is studying for the first time the effects of ethanol on the cellulolytic and xylanolytic activities of *F. oxysporum*, on growth on glucose and xylose under aerobic conditions, and on ethanol production from glucose and xylose during micro-aerobic fermentation. This work complimented with the resent work of Hennessy et al., [3] gives to the scientific community a new prospect for the investigation of the lignocellulose degrading and C5, C6 fermenting fungus *F. oxysporum* as a potential CBP microorganism. Furthermore to overcome the ethanol limits, this study explored an alternative operating condition for fermentation with stepwise partial removal of ethanol. Finally, SSF under high dry matter conditions has been performed in order to explore the contribution of the metabolic system of *F. oxysporum* along with *S. cerevisiae* in ethanol production.

Results

Effect of ethanol on growth (aerobic conditions) and fermentative performance (micro-aerobic conditions) of *F. oxysporum*

Growth stage

The ability of *F. oxysporum* to grow in the presence of ethanol under aerobic conditions was investigated. Various amounts of ethanol (0–60 g/L) were added to the medium prior to inoculation and the fungal growth was monitored over time. The effect of ethanol on growth on glucose was evident even 15 h after inoculation, while it was maximized after 44 h. As shown in Figure 1, the maximum biomass concentration decreased as the initial ethanol concentration increased. In the presence of 30 g/L of ethanol, the biomass produced corresponded to the 77% of the one in the absence of ethanol, while the decrease was dramatic when 40 g/L of ethanol were added to the medium. Almost no growth was observed when the experiments were performed in the presence of 60 g/L ethanol. The inhibitory effect of ethanol appears more severe when *F. oxysporum* was grown on xylose as carbon source. Figure 2 indicates that, the produced biomass was severely diminished when ethanol concentration exceeded 30 g/L and actually no growth was observed over 40 g/L initial ethanol in the broth.

Furthermore, the inhibitory effect of ethanol on fungal growth was clearly reflected not only on biomass yield but also on the growth rate. As shown in Table 1, the specific growth rate decreased dramatically as the ethanol concentration in the broth increased. In order to fully describe the kinetic pattern of ethanol inhibition on cell growth, the kinetic model proposed by Luong [21] was applied:

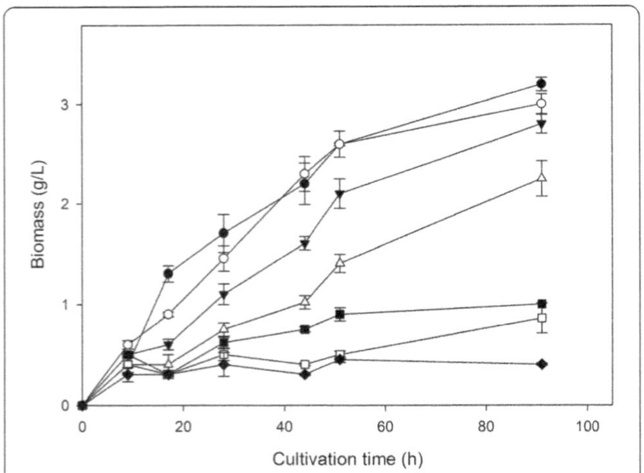

Figure 1 Ethanol effect on *F. oxysporum* growth on glucose. Ethanol was initially added at concentrations up to 6% w/v in the cultivations prior to inoculation, using glucose as carbon source. All experiments were carried out in duplicate, vertical bars indicate the error levels. The symbols stand for: No ethanol added (●), 1% w/v initial ethanol (○), 2% w/v initial ethanol (▼), 3% w/v initial ethanol (Δ), 4% w/v initial ethanol (■),5% w/v initial ethanol (□), 6% w/v initial ethanol (♦).

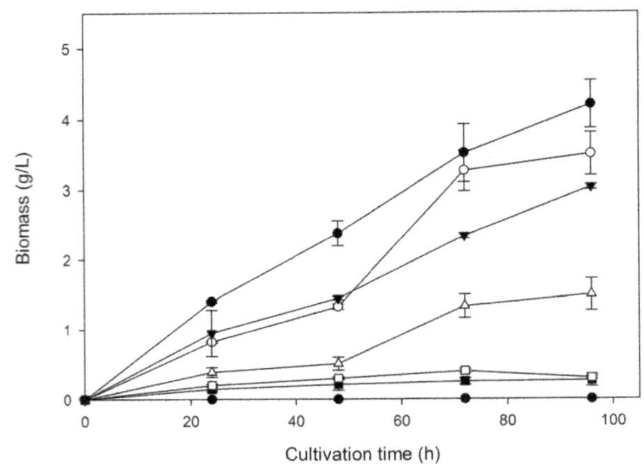

Figure 2 Ethanol effect on *F. oxysporum* growth on xylose. Ethanol was initially added at concentrations up to 6% w/v in the cultivations prior to inoculation, using xylose as carbon source. All experiments were carried out in duplicate, vertical bars indicate the error levels. The symbols stand for: No ethanol added (●), 1% w/v initial ethanol (○), 2% w/v initial ethanol (▼), 3% w/v initial ethanol (Δ), 4% w/v initial ethanol (■),5% w/v initial ethanol (□), 6% w/v initial ethanol (♦).

$$\frac{\mu_i}{\mu_o} = 1\text{-}\left(\frac{P}{P_m}\right)^a \quad (1)$$

Where, μ_0 and μ_i are the maximum specific growth rate and maximum specific growth rate in the presence of ethanol respectively. P_m is the critical ethanol concentration above which cells cannot grow. The relationship between specific growth rate and critical ethanol concentration could be described by the ethanol tolerance index "α" proposed by the same model (a linear relationship for $\alpha = 1$, a hyperbolic relationship when $\alpha > 1$ and a parabolic relationship when $\alpha < 1$) [21]. The average specific growth rates were calculated and plotted against ethanol concentrations to estimate P_m and α of our experimental data, extracting values of 7.2% and 1.05 for P_m and α, respectively when glucose was used as substrate, and $Pm = 3.8\%$ and $\alpha = 2.5$ when the fungus was grown on xylose.

Fermentation stage

In order to investigate the ethanol effect on the ability of the fungus to convert sugars to ethanol, *F. oxysporum* submerged cultivations previously grown under aerobic conditions, were subsequently shifted to micro-aerobic conditions, while various ethanol concentrations (0–6%

Table 1 Growth rates and lag phases of *F.oxysporum's* aerobic growth in the presence and absence of ethanol

Ethanol% (w/v)	0	1	2	3	4	5	6
			Growth on glucose				
$\mu(h^{-1})$	0.045 ± 0,002	0.038 ± 0.001	0.035 ± 0.003	0.033 ± 0.003	0.002 ± 0.003	0.009 ± 0.002	0.009 ± 0.003
Lag phase (h)	9	9	17	28	44	N/A	N/A
			Growth on Xylose				
$\mu(h^{-1})$	0.29 ± 0.003	0.030 ± 0.003	0.026 ± 0.002	0.017 ± 0.002	0.005 ± 0.001	N/A	N/A
Lag phase (h)	9	9	18	48	N/A	N/A	N/A

w/v) were added at the start-up of fermentation stage. Figures 3 and 4 show the net ethanol amounts produced by *F. oxysporum* cells at different initial ethanol concentrations with glucose or xylose as carbon sources respectively. The total ethanol concentration to which the cells are exposed during the fermentation course was the initial amount added to the broth plus the produced ethanol. The experimental results obtained on glucose fermentation showed that both ethanol production rate and final ethanol concentration were affected by the initial ethanol level added in the culture. When glucose used as carbon source, the addition of 2%, 3% and 4% w/v ethanol in the culture media resulted in reduction of 46%, 65% and 74% in the net ethanol production, respectively, while no fermenting activity was observed after the addition of 5% or 6% (w/v) of ethanol. Using xylose as carbon source, the reductions in ethanol yields were 26%, 50% and 68% when 2%, 3% and 4% w/v ethanol was added with no observed ethanol production over 5% (w/v) of initially added ethanol.

Effect of ethanol on the cellulolytic and xylanolytic systems of *F. oxysporum*

To evaluate the potential use of *F. oxysporum* cellulolytic system in consolidated or SSF bioprocesses, the effect of ethanol on the secreted hydrolytic activities was studied by measuring both the FPA and the xylanase activity in the presence of several ethanol levels. To evaluate the obtained data we performed similar studies using commercial enzymatic preparations. The results are shown in Figures 5 and 6. The assay reactions were carried out in two different temperatures, 30°C and 50°C. The first is the most likely temperature for a consolidated or SSF bioprocess, while the second is the standard temperature assay for the most lignocellulolytic enzymatic activities. The cellulolytic activity of *F. oxysporum* at 30°C was not significantly affected in ethanol concentrations up to 2% w/v (Figure 5). Moreover, despite the obvious reduction in higher ethanol concentrations, the residual activity at 6% and 8% (w/v) was measured at 72% and 63% of the initial, respectively. The ethanol inhibition was, as expected,

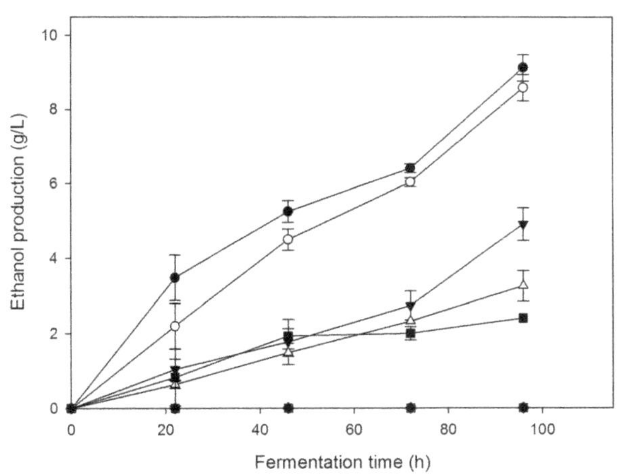

Figure 3 Ethanol effect on *F. oxysporum* fermetative performance on glucose. Ethanol was initially added at concentrations up to 6% w/v in the fermentation broth using glucose as carbon source. All experiments were carried out in duplicate, vertical bars indicate the error levels. The symbols stand for: No ethanol added (●), 1% w/v initial ethanol (○), 2% w/v initial ethanol (▼), 3% w/v initial ethanol (Δ), 4% w/v initial ethanol (■), 5% w/v initial ethanol (□), 6% w/v initial ethanol (♦).

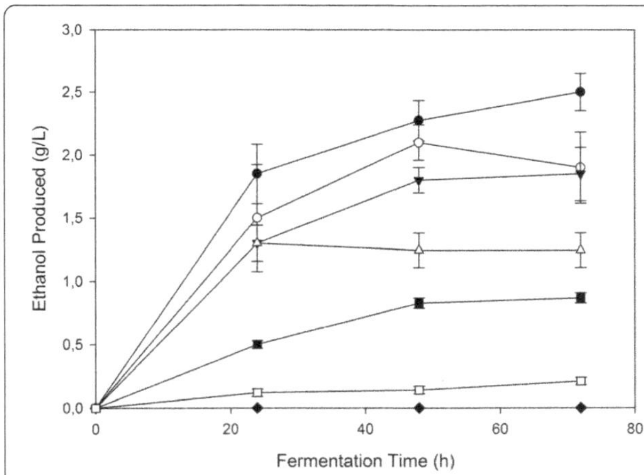

Figure 4 Ethanol effect on *F. oxysporum* fermetative performance on xylose. Ethanol was initially added at concentrations up to 6% w/v in the fermentation broth using glucose as carbon source. All experiments were carried out in duplicate, vertical bars indicate the error levels. The symbols stand for: No ethanol added (●), 1% w/v initial ethanol (○), 2% w/v initial ethanol (▼), 3% w/v initial ethanol (Δ), 4% w/v initial ethanol (■), 5% w/v initial ethanol (□), 6% w/v initial ethanol (♦).

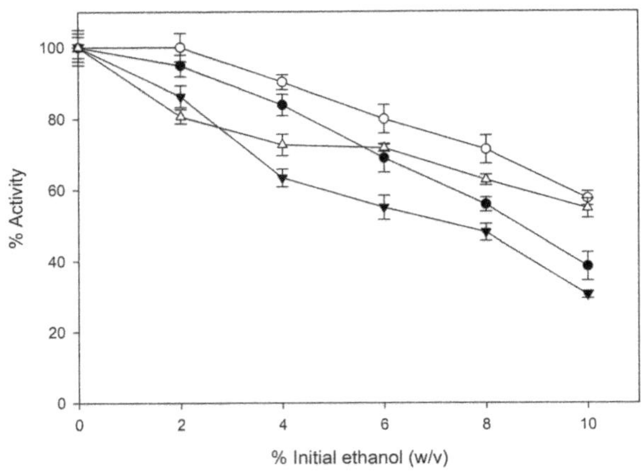

Figure 5 Ethanol effect on cellulolytic enzymatic activities. The activity assays were carried out under the addition of ethanol in the reaction buffer. *F. oxysporum* enzymes compared to mixture of commercial enzymes Celluclast 1,5 L – Novozyme 188. All assays were carried out in duplicate, vertical bars indicate the error levels. (▼) indicates *F. oxysporum* enzyme complex at 50°C and (Δ) at 30°C. (●) indicates commercial enzymes at 50°C and (o) at 30°C.

more severe at 50°C where the enzymatic activity was reduced to 64% and 47% of the initial in the presence of 6% (w/v) and 8% (w/v) of ethanol, respectively. Finally, under the presence of 10% (w/v) ethanol the enzymes exhibited 50% and 70% of their initial activity at 50°C and 30°C, respectively. The decrease in cellulolytic activity showed a similar trend for both the commercial and the *F. oxysporum* enzyme mixtures (Figure 5).

The ethanol effect on xylanolytic activities is shown in Figure 6. Similarly with cellulases the hemicellulolytic activity of *F. oxysporum* was more intensely affected at 50°C. As shown, under most of the ethanol concentrations tested, the commercial mixture retained slightly higher activities (5% to 10%).

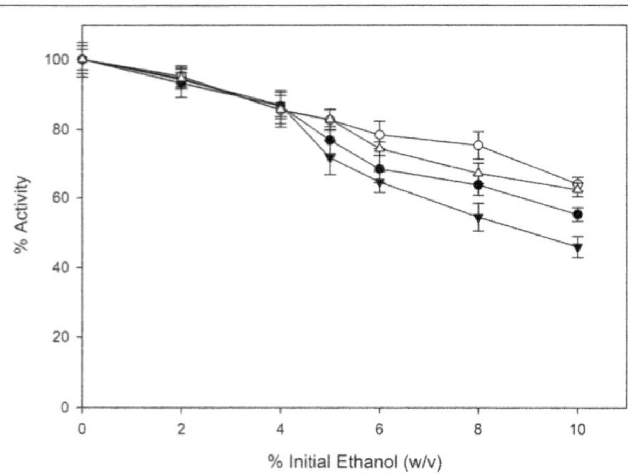

Figure 6 Ethanol effect on hemicellulolytic enzymatic activities. The activity assays were carried out under the addition of ethanol in the reaction buffer. *F. oxysporum* enzymes compared to mixture of commercial enzymes Celluclast 1,5 L – Novozyme 188. All assays were carried out in duplicate, vertical bars indicate the error levels. (▼) indicates *F. oxysporum* enzyme complex at 50°C and (Δ) at 30°C. (●) indicates commercial enzymes at 50°C and (o) at 30°C.

The stability of the lignocellulolytic enzymes in the presence of ethanol is another important parameter regarding their suitability for SSF or CBP processes where they have to withstand elevated ethanol concentrations for many hours. Therefore, ethanol stability tests were performed using *F. oxysporum* enzymes in comparison with commercial enzymes at 30°C in the presence of 25 g/L and 50 g/L ethanol. As shown (Figures 7 and 8), commercial enzymes retained 70% and 65% of their activity at 2.5% and 5% of ethanol over 48 hours of incubation while *F. oxysporum* enzymes retained 52% and 45% of their activity under the same conditions. It is worth to be noticed that the commercially available enzymatic preparations often contain additives in order to increase their stability.

Removal of ethanol during fermentation process

As proved in the present study (Figure 3 and 4), the presence of ethanol had an inhibitory effect on *F. oxysporum*'s fermenting ability. To examine whether the presence of ethanol has irreversible effects on the fungal metabolism, the ethanol was partially removed from the broth every 48 hours, as described in Methods section. Fermentation under vacuum has been performed in the past in order to overcome the inhibitory effects of ethanol [22,23]. Figure 9 shows that the ethanol production of *F. oxysporum* was increased by 22.5% when the ethanol was partially removed from the broth, showing that when the excessive ethanol was removed from the broth, *F. oxysporum* regained its metabolic activity confirming the reversible character of ethanol inhibition on *F. oxysporum* cells under the conditions of the experiment.

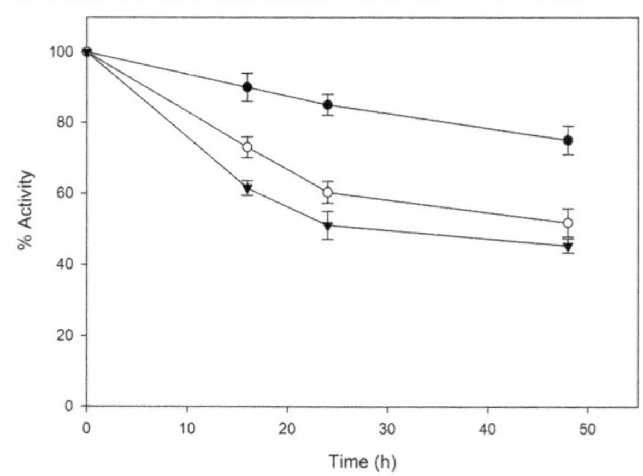

Figure 7 Ethanol effect on lignocellulolytic enzymatic stabilities of *F. oxysporum*. *F. oxysporum* enzymes were incubated at 30°C in the presence of 0% w/v (●), 2,5%w/v (o) and 5% w/v (▼) ethanol.

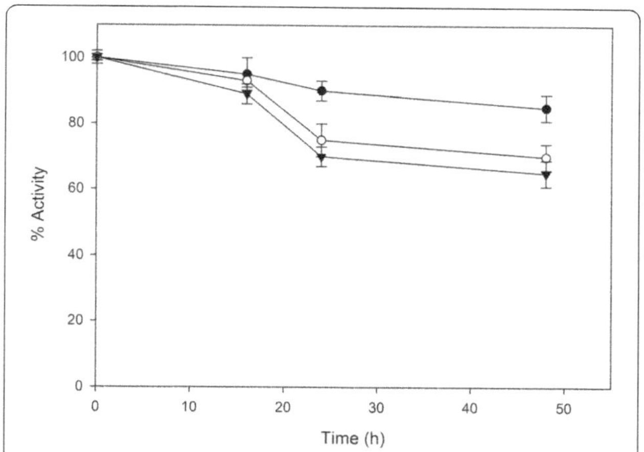

Figure 8 Ethanol effect on lignocellulolytic enzymatic stabilities of commercial enzyme mixture. The commercial enzyme system Celluclast 1,5 L – Novozyme 188 were incubated at 30°C in the presence of 0% w/v (●), 2,5%w/v (○) and 5% w/v (▼) ethanol.

Evaluation of *F. oxysporum* potential during co-fermentation of liquefied wheat straw

Finally, in order to test the hypothesis that *F. oxysporum* can be efficiently used in fermentation processes contributing in xylose fermentation and minimization of the amount of externally added enzymes by exploiting the hydrolytic enzymes produced by the fungus, 50 mg/g DM of *F. oxysporum* cells and/or 5 FPU/g DM enzymes were added along with *S. cerevisiae* in the start-up of fermentation stage. As control experiments, fermentations of the material with (a) *S. cerevisiae* and commercial enzymes, (b) *S. cerevisiae* and no addition of enzymes, (c) *S. cerevisiae* and *F. oxysporum* biomass, (d) *S. cerevisiae* and *F. oxysporum* enzymes were conducted. Results are presented in Table 2. It can be pointed out that the addition of *F. oxysporum* enzyme system can adequately replace the commercial enzyme system reaching 94% of the ethanol yield. Furthermore, when there was no enzyme addition at the startup of the fermentation process, the combined effect of both microorganisms resulted in a slight increase in ethanol production of about 7%, compared to the fermentation only with *S. cerevisiae.* Finally, the addition of *F. oxysporum*'s system in the SSF process increased the ethanol production by about 11%, compared to the fermentation conducted with *S. cerevisiae* and commercial enzymes.

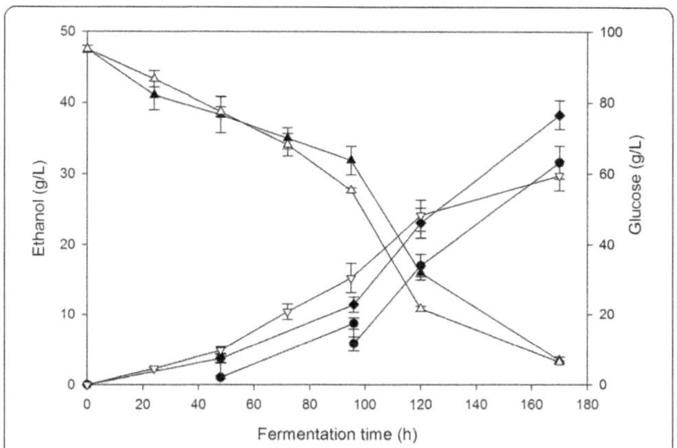

Figure 9 Effect of ethanol removal during the fermentation process. The produced ethanol during fermentation process was removed from the broth every 48 hours. All measurements were carried out in duplicate, vertical bars indicate the error levels. Ethanol present in broth (●), total ethanol produced (ethanol present in broth plus the removed ethanol) (♦), glucose consumption (▲), ethanol production in control experiment (∇) glucose consumption in control experiment (Δ).

Discussion

The results presented here showed that the response of *F. oxysporum* to ethanol can be compared to that of other xylose fermenting microorganisms used in bioethanol production processes. Studies on *Pichia stipites* showed that no growth was observed when the initial ethanol concentration was over 3.5% w/v during cultivation on glucose or xylose [24,25]. Additionally, there was no ethanol production measured on either source over the same ethanol concentration. Similarly, the growth of *Kloeckera africana* on Agave (tequila juice) at ethanol concentration above 2.5% w/v was associated with reduced biomass production [26]. Another study by Bajpai and Margaritis, made on *Kluyveromyces maxianus,* indicated that ethanol caused inhibitory effects on cell growth and fermentation ability [27]. As shown in that study, it was mainly the growth and ethanol production rates that were decreased, while the biomass and ethanol yields were almost not affected by the presence of ethanol. In accordance with the results presented here, a linear negative correlation of the specific growth rate with the initial ethanol concentration and a similar linear relationship between μ_{max} and ethanol concentration was reported [27].

The effects of ethanol on growth of *F. oxysporum* (aerobic conditions) were more severe when xylose was the carbon source in contrast with the ethanol effects in the productive anaerobic stage where the inhibition was stronger on glucose. The difference in the effects on the fermentative performance could be explained by the higher net ethanol production in the case of glucose. Although the net amount produced did not exceed 9 g/L, we must take into account the total ethanol levels. Furthermore, it has been proposed that the toxic effect of the endogenously produced ethanol to the microorganisms is more severe than the effect of the externally added ethanol [28,29]. However, other studies explained similar observations by the cumulative toxic action of ethanol together with other toxic by-products of fermentation, such as organic acids [30,31]. Although organic acids' concentrations were not measured during the

Table 2 Fermentation of liquefied pretreated wheat straw, with and without the addition of *F. oxysporum*'s system

	SSCF	*F. oxysporum* enzymes	*F. oxysporum* cells	Commercial enzymes	No additions in fermentation
% DM	25	25	25	25	25
Enzymes at liquefaction	5 FPU/g DM	5 FPU/g DM	5 FPU/g DM	5 FPU/g DM	5 FPU/g DM
F. oxysporum enzymes added at fermentation	5 FPU/g DM	5 FPU/g DM	None	None	None
Commercial enzymes added at fermentation	None	None	None	5 FPU/g DM	None
F. oxysporum biomass	50 mg/g DM	None	50 mg/g DM	None	None
S. cerevisiae	5 mg/g DM	5 mg/g DM	5 mg/g DM	5 mg/g DM	5 mg/g DM
Ethanol produced (g/L)	**34.7 ± 1.1**	**29.2 ± 0.8**	**21.1 ± 1.0**	**31.2 ± 0.9**	**19.6 ± 0.8**

The addition of *F. oxysporum* biomass and enzyme system where tested for the fermentation of liquefied pretreated wheat straw. *F. oxysporum* enzymes were tested as a substitute of commercial enzymes and *F. oxysporum* biomass for its ability to cooperate with *S. cerevisiae* in ethanol production. SSCF stands for Simultaneous Saccharification and Co-Fermentation.

present study, it has been shown that acetic acid is a major by-product of the metabolism of *F. oxysporum* grown on xylose at similar conditions [32].

During the consolidated and SSF processes the need of an active enzyme system is necessary to maintain high saccharification rates and yields. *F. oxysporum* secrets a multi-enzyme system, rich in cellulases and hemicellulases, which is very efficient in biomass saccharification [33]. *F. oxysporum*'s enzymatic system showed an adequate tolerance to ethanol. Especially at 30°C, the most common temperature for SSF processes, (where *F. oxysporum*'s enzymatic system is more likely to be exploited) the effect of ethanol on the cellulolytic and xylanolytic activities of the enzymatic system was significantly less severe than at 50°C. Different reasons could explain this result. Various researchers have previously shown that ethanol reduces the enzymatic activities of cellulases. Its inhibition effect is temperature depended and increases as the reaction temperature increases [34,35]. Although the inhibitory effects of ethanol on cellulases are in many cases demonstrated studies have reported also destabilizing effects [17,36,37]. It seems that at elevated temperatures, the destabilizing effect of ethanol may accelerate denaturation of the enzyme even if the temperature itself could not cause protein denaturation for such a short period (as in our case, 1 h) [38]. Skovgaard and Jorgensen working on a set of thermo tolerant enzyme mixture found that the effect of ethanol on the activity reduction as the reaction temperature increases is not related to inhibition but rather to destabilization of the enzymes [39].

As shown, the ethanol effect on *F. oxysporum* enzymes is temperature dependent. When the assay is performed at 30°C, *F. oxysporum*'s cellulolytic and xylanolytic activity remains in high levels, in the presence of considerable amounts of ethanol. Even in the presence of 50 g/L of ethanol, *F. oxysporum's* enzymes retain 45% of their activity over 48 hours. This could be beneficial in a consolidated bioprocess where the amount of commercial enzymes added in the hydrolysis step could be reduced by the ability of *F. oxysporum* to produce ethanol tolerant enzymes, able to continue the saccharification during the fermentation stage. Another study indicated that, *F. oxysporum* in mixed culture with *S. cerevisiae* was successfully used to ferment carbohydrates of wet exploded pre-treated wheat straw to ethanol under an SSF mode [5].

Additionally, when the produced ethanol was partially removed from the fermentation system of *F. oxysporum* in a stepwise manner the final ethanol production level was increased by around 23% reaching 38.4 g/L. The increase in ethanol production achieved here is similar to that reported in earlier studies dealing with the *in situ* removal of ethanol from the fermentation broth, where an increase up to 30% in the overall ethanol production could be achieved in most cases [22,40-42]. As shown in Figure 9, the beneficial effect of ethanol removal is clear only when the ethanol present in the medium is above 3% w/v. When ethanol concentration in the broth tends to reach 3%, the fermentation seems to stop. This is consistent with our findings regarding the ethanol tolerance levels of *F. oxysporum*. In cultures where the ethanol was partially removed it is obvious that this removal allowed the total ethanol production to exceed the 3% barrier. These results, suggest that as the ethanol concentration diminishes sufficiently, any metabolic inhibitory effects are reversed. As a result, the fermenting capacity is recovered when ethanol is removed. The ethanol concentration achieved here is very close to the crucial ethanol concentration of 4% (w/v) in the broth which is considered as a minimum prerequisite for a feasible large-scale distillation technology [7].

The straw used at the present work is rich in hemicellulose, and thus suitable to evaluate the ability of *F. oxysporum* not only to hydrolyze it but also to ferment the released xylose. The addition of *F. oxysporum* cells in the fermentation process resulted in 11% increase of the ethanol production (Table 2). Specifically, if ethanol yields are based on the dry material, the mixed microbial culture fermentation led to 138 g ethanol per kg pretreated wheat straw, while fermentation only with *S. cerevisiae* resulted in 128 g ethanol per kg pretreated wheat straw. The results shown in Table 2 indicate that *F. oxysporum*'s system is worthy of further investigation regarding its potential to enhance ethanol production during an SSF process along with *S. cerevisiae* when pentoses are present in the fermentation medium. These results are in accordance with Panagiotou et al. who reported in an earlier study that *F. oxysporum* multienzyme and microbial system had positive effect in ethanol production [5]. Moreno et al. [43] reported ethanol concentration 25 g/L at 20% solid content, which corresponds to 125 g ethanol per kg of pretreated material. Whereas, Jorgensen et al. [35] reported ethanol concentration up to 130 g/kg, applying a similar liquefaction process to that of the present study, in a high hemicellulose content material. At the present work, the liquefaction step took place for 6 hours, and 5 FPU/g DM were used at the liquefaction step supplemented with 5 FPU/g DM at the SSF process. Comparatively the enzyme loadings of Moreno et al., [43] and Jorgensen et al. [35] were 15 FPU/g DM and 7 FPU/g DM respectively. The saccharification times also vary, at the present study 6 hours of liquefaction were followed by 48 h of Simultaneous Saccharification and Co-Fermentation (SSCF), while Moreno et al. [43] conducted 144 h of SSCF and Jorgensen et al. [35] conducted 24 h of liquefaction and 48 h of SSF.

The complexity of SSCF process when dealing with lignocellulosic materials justifies the differences observed by different researchers. Many factors are affecting ethanol yields when operating at high solids fermentations, such as the nutrients deficiency, the presence of other inhibitors, the viscosity, and the inhomogeneity [6]. Moreover, differences between different raw materials and different pretreatments (The SF value of the pretreated wheat straw used in the present study was calculated at 3.58 (same as Jorgensen et al.), while Moreno et al. and presented a process of SF 3.34) make these processes difficult to control and to compare in many cases. Undoubtedly, many factors not being the objectives of this study, are affecting *F. oxysporum* ethanol yields and productivities and no clear conclusions regarding the contribution of *F. oxysporum* metabolic system to xylose fermentation during SSCF can be made.

Considering the above, the results of the present study show that *F. oxysporum*'s addition led to a significant ethanol increase, exploiting the hydrolytic enzymes produced by the fungus and thus reducing the amount of commercial enzymes needed. On the other hand, the ethanol yield, of about 31% of the maximum theoretical based on total carbohydrates content, achieved in the present work indicates that further investigation of the process is required.

Conclusions

The results presented here assess three different aspects of *F. oxysporum* ethanol tolerance (effects on growth, enzymes, fermenting ability) for the use of a microorganism in a bioethanol production process and show that *F. oxysporum* can be used in a biorefinery both as a pentose fermenting and enzyme producing microorganism. Hemicellulose rich streams, produced after pretreatment of the raw material could be used for the production of necessary enzymes for the degradation of lignocellulosic biomass. Although the ethanol tolerance levels of *F.oxysporum* indicated that the fungus could be useful at the early stages of the fermentation, this was not confirmed by the SSCF and SSF results, in those cases *F. oxysporum* biomass did not demonstrate the expected contribution to the ethanol production, probably due to other stress factors not studied during this work (effect of high solids, presence of other inhibitory compounds). Nevertheless, the effect of *F. oxysporum* metabolic system could not be neglected as (even not as high as expected) the addition of *F. oxyporum* biomass increased the produced ethanol in an SSCF experiment along with *F. oxysporum* enzymes and *S. cerevisiae* compared to the one that only enzymes and yeast were added. The satisfactory ethanol tolerance levels of *F. oxysporum* enzymatic system allow the use of these enzymes also in high gravity bioconversion processes. This could lead to the reduction of the overall financial cost of the process, not only by achieving a high final ethanol concentration but also by reducing the amount of commercial enzymes added. To conclude with, this study as well as other studies made on *F. oxysporum* tolerance and adaptability [3,4] indicated that *F. oxysporum* is capable of dealing with ethanol, which makes it a very interesting microorganism for further study on CBP and SSCF processes.

Methods

Microorganism

Fusarium oxysporum F3 isolated from cumin [1], the fungus was grown on potato-dextrose-agar (PDA) slants for 5 days at 30°C. The slants were maintained as a stock culture at 4°C. Commercial dry baker's yeast (Yiotis, Athens, Greece) was used at the pretreated wheat straw fermentations.

Carbon sources – chemicals

Pretreated wheat straw (*Triticum aestivum* L) was used as raw material for ethanol production. The pretreatment of the straw (PWS) was performed in the Inbicon pilot plan at Skærbæk, Denmark. The residence time setpoint in the reactor was 12 min and the reactor temperature was maintained at 185°C by injection of steam. The severity effect (SF) was determined according to Garrote et al. [44] by the following equation:

$$SF = \log(R) = log\left(t \cdot e\frac{T-100}{14.75}\right)$$

Pretreated wheat straw lignocellulose content was (%) cellulose 39.0 ± 1.5 and hemicellulose 21.0 ± 1.2. Structural carbohydrate content (cellulose and hemicellulose) was determined by a method adopted by the NREL protocol [45].

Brewers Grain (BG) and Corn Cobs (CC) were supplied by the Athens Brewery S.A. and the Agricultural University of Athens, respectively. BG composition was analyzed by Xiros et al. [46] and Corn Cobs was analyzed by Katapodis and Christakopoulos [47].

All chemicals and reagents were provided by Sigma-Aldrich (USA).

Commercial enzymes Celluclast 1.5 L and Novozyme 188 were provided by Novozymes (Denmark). The commercial enzymes were used in a 5:1 v/v ratio.

Inoculum

For inoculum production, spores were extracted from the stock slants using 5 ml of sterile distilled water and then cultivated in 250 mL Erlenmeyer flasks containing 100 mL of the following mineral medium:(in g L^{-1}) 1.00 KH_2PO_4, 0.30 $CaCl_2.2H_2O$, 0.30 $MgSO_4.7H_2O$, 10.00 $(NH_4)_2HPO_4$, 6.94 $NaH_2PO_4 \cdot 2H_2O$, 9.52 $Na_2HPO_4 \cdot 2H_2O$ [2], supplemented with either 40 g L^{-1} BG-CC (2/1) or 20 g L^{-1} glucose or xylose, depending on the experimental procedure. The pH was adjusted to 6.0. The flasks were incubated at 30°C for 2 days in an orbital shaker at 200 rpm (Zhicheng ZHWY-211C) for mycelium production.

Aerobic submerged cultivation

Aerobic submerged cultures were carried out in 250 mL Erlenmeyer flasks containing 2 g of glucose or xylose and 100 mL of the above described mineral medium (pH 6.0). Medium and sugars were sterilized separately at 109°C for 40 min. The culture medium was inoculated with 10 ml of 48 h-old inoculum (prepared as described above). The flasks were incubated at 30°C for 7 days at 200 rpm. The flasks were capped with cotton that allowed aerobic conditions and assured sterile conditions. At different time intervals aliquots were aseptically withdrawn and used for biomass and sugar consumption estimation.

For analyzing the potential impact, of the presence of ethanol, on the growth of *F. oxysporum* the following experiment was carried out. In submerged cultures as described above ethanol concentrations up to 6% w/v were exogenously added in the broth. At different time intervals aliquots were aseptically withdrawn and used for biomass and sugar consumption estimation.

Biomass estimation

Samples of the aerobic cultures were filtered using 0.2 μm pure size filter paper, Millipore (USA). The biomass content was measured by weighing the dried samples.

Enzyme production

Crude enzyme extract production was carried out in 3 L Erlenmeyer flasks containing 40 g of carbon sources (BG-CC 2/1) and 1 L of mineral medium. Prior to sterilization, the initial pH of the medium was adjusted to 6.0. The medium was sterilized at 121°C for 20 min and was inoculated with 100 mL of 72 h old inoculum (prepared as described above). The flasks were incubated at 30°C for 5 days at 200 rpm. At the end of the enzyme production stage, the culture was centrifuged (14,000 rpm, 4°C, 40 min) and the clarified supernatant was concentrated using ultra filtration membranes (10,000 kDa) Millipore (USA), there was no buffer addition during ultrafiltration process.

Enzyme assays

One unit (U) of enzyme activity was defined as the amount of enzyme required to liberate 1 μmol of product per minute, at assay temperature.

Filter paper activity (FPA) was determined as described by Wood and Bhat [48]. In a capped 2 ml Eppendorf tube, 1 mL of buffer solution, 0.5 mL of enzyme (with the appropriate dilution so to release less than 1 mg product per mL) and a filter paper (Watman #1 1x6 cm, approximately 50 mg) were added. The mixture was incubated at 50°C for 1 hour at 1000 rpm. The released reducing sugars were measured by the 3,5-dinitrosalicylic acid (DNS) method [49].

The β-1,4-d-endoxylanase activity was determined by incubating the enzyme for 10 minutes at 50°C with 1% birchwood xylan [50]. The released reducing sugars were measured by the 3,5-dinitrosalicylic acid (DNS) method [49].

Ethanol effect on enzymatic activity

For the investigation of the effect of ethanol concentration on the cellulolytic and xylanolytic activity of the enzymes used in this study, assays were carried out in the presence of initial ethanol concentrations up to 10% w/v. The ethanol was added in the reaction buffer solution prior to the enzyme addition. The assay reactions were carried out as described above at either 30°C or 50°C.

Ethanol stability

The ethanol stability of the enzymes was performed at 30°C the SSF temperature at which the enzymes withstand the presence of ethanol for many hours. The enzymes were incubated at 30°C with the addition of two ethanol concentrations, 25 g/L and 50 g/L. At different time intervals aliquots were taken and the cellulolytic activity was measured by the Filter Paper method.

Analytical methods

Reducing sugars concentration was determined according to dinitro-3.5-salicilic acid (DNS) method [47]. The effect of ethanol on the sugar measuring method was determined by control measurements, containing all the ethanol concentrations used and a standard sugar concentration. Ethanol appeared to have no effect on the reducing sugar determining method. Glucose was measured according to commercial enzyme solution of GOD/PAP (glucose oxidase/ peroxidase assay) (Biosis, Greece). Xylose was measured according to commercial D-xylose assay kit (Megazyme, Ireland) Ethanol was analyzed using an HPLC system (Szimadju) equipped with an Aminex HPX-87H column (Bio-Rad, 300 x 7.8 mm, particle size 9 μm) using a Refractive Index (RI) detector. Mobile phase was 5 mM H_2SO_4 in HPLC grade water at 0.6 mL/min flow rate, column temperature was 40°C, injection volume was 50 μl and total runtime was 30 min. All samples were filtered (0,2 μm, Macherey-Nagel) prior to the analysis.

Ethanol production

Ethanol production experiments under micro aerobic conditions were carried out in 250 mL Erlenmeyer flasks containing 2 g of glucose or xylose, sterilized at 109°C for 40 min. 100 mL of culture previously grown on glucose as carbon source for about 4 days (until substrate was fully consumed) at 30°C, were aseptically transferred in the Erlenmeyer flasks provided with needle-pierced rubber stoppers, which ensured micro-aerobic conditions and allowed the release of produced carbon dioxide [51] and the cultures were incubated at 30°C for 7 days and 80 rpm (Zhicheng ZHWY-211C). At different time intervals, aliquots were aseptically withdrawn and used for ethanol production and substrate consumption estimation.

For analyzing the potential impact, of the presence of ethanol, on the fermenting ability of *F. oxysporum,* micro aerobic cultivations of *F. oxysporum* were carried out, as mention before, in the presence of extracellular, exogenously added ethanol. The initial ethanol concentrations were up to 6% w/v. At different time intervals aliquots were aseptically withdrawn and used for ethanol and substrate consumption estimation.

All ethanol production experiments were carried out as described by Xiros et al. [4], Dogaris et al. [51] and Matsakas and Christakopoulos [52]. After each sampling, nitrogen (0.1 vvm) was flushed in the cultures for 10 minutes to assure the anaerobic conditions.

Partial ethanol removal

Ethanol production experiments under micro aerobic conditions were carried out in 250 mL Erlenmeyer flasks containing 9.5 g of glucose sterilized at 109°C for 40 min. 100 mL of culture previously grown with glucose as carbon source for about 4 days (until glucose was fully consumed) at 30°C, were aseptically transferred in the Erlenmeyer flasks and the cultures were incubated at 30°C for 7 days at 80 rpm (Zhicheng ZHWY-211C). Every two days the broth was aseptically centrifuged and transferred to a rotary evaporator (Rotavapor RE 11, Buchi (Switzerland)) where the contained ethanol was partially removed. The vacuum process was carried out at 75°C for approximately 10 minutes. Consequently, the broth was returned in the Erlenmeyer flasks along with the centrifuged fungal cells. At this stage, nitrogen was flushed in the cultures to assure the anaerobic conditions. For avoiding broth condensation, the loss of water during the evaporation process was estimated by weighting. Equal amount of sterile water (less than 3 mL each time) was added in the culture.

Liquefaction of hydrothermaly treated wheat straw

For the liquefaction and saccharification of the pretreated wheat straw a reactor was manufactured in house. The reactor consists of a cylinder drum 25 cm wide and 60 cm diameter, and a rotating shaft for mixing the material. A 0,55 kW motor was used for the rotation of the shaft. The rotation speed could be controlled from 0 up to 20 rpm, and the motor can be programmed to shift its rotation clockwise and anti-clockwise. An external jacket filled with oil was used for temperature control up to 90°C. The experiments were executed at 50°C, using 30% DM of straw supplemented with buffer (phosphate-citrate, 50 mM, pH 5,5) and an enzyme mixture of 5 FPU per g DM, consisting of Celluclast 1.5 L and Novozyme 188 (both from Novozymes A/S, Bagsværd, Denmark) in ratio of 5:1 v/v. The activity of this mixture was measured at 81 FPU/ml by the filter paper assay. The mixing speed was 7 rpm shifting clockwise and anti-clockwise twice a minute. The duration of the liquefaction was 6 h.

Fermentation of liquefied wheat straw

The liquefied at 30% DM material was used as carbon source for ethanol production. *S. cerevisiae* (Dry Baker's Yeast) and *F. oxysporum* (previously grown on glucose for 4 days under submerged cultivation) were used as the fermentative microorganisms. 5 mg/g DM dry yeast and 50 mg/g DM *F. oxysporum* centrifuged cells were added. Moreover 5 FPU/g DM of either commercial enzymes (Celluclast 1.5 L and Novozym 188) or *F. oxysporum's* enzymes were added to enhance the hydrolysis. The final DM in the mixed fermentation was 25% w/w due to the addition of the enzymes and the fungal biomass. All experiments were carried out, in duplicate, under submerged cultivation at pH 5.5 and T = 30°C in 250 mL Erlenmeyer flasks.

Competing interests

The authors declare that they have no competing interests.

Authors' contributions

PC conceived the study. PC and CX designed the experiments. TP carried on the experimental work. PC, TP and CX drafted the manuscript. PC and CX revised the manuscript. All authors read and approved the final manuscript.

Acknowledgements

The research leading to these results has received funding from the European Community's 7th Framework Program (FP7/2007-2013) under grant agreement 213139 – the HYPE project.

Thomas Paschos would like to thank the Hellenic Scholarship Foundation for a grand.

Paul Christakopoulos would like to thank Bio4Energy, a strategic research environment appointed by the Swedish government, for supporting this work.

Author details

[1]Biotechnology Laboratory, School of Chemical Engineering, National Technical University of Athens, 9 Iroon Polytechniou Str, Zografou Campus, Athens 5780, Greece. [2]Industrial Biotechnology, Department of Chemical and Biological Engineering, Chalmers University of Technology, Kemivägen 10, Gothenburg 41296, Sweden. [3]Biochemical Process Engineering, Division of Chemical Engineering, Department of Civil, Environmental and Natural Resources Engineering, Luleå University of Technology, Luleå SE–971 87, Sweden.

References

1. Christakopoulos P, Macris BJ, Kekos D. Direct fermentation of cellulose to ethanol by *Fusarium oxysporum*. Enzym Microb Technol. 1989;11:236–9.
2. Xiros C, Christakopoulos P. Enhanced ethanol production from brewer's spent grain by a *Fusarium oxysporum* consolidated system. Biotechnol Biofuels. 2009;2:4.
3. Hennesey RC, Doohan F, Mullins E. Generating phenotypic diversity in a fungal biocatalyst to investigate alcohol stress tolerance encountered during microbial cellulosic biofuel production. Plos One. 2013;8(10):e77501.
4. Xiros C, Vafiadi C, Paschos T, Christakopoulos P. Toxicity tolerance of *Fusarium oxysporum* towards inhibitory compounds formed during pretreatment of lignocellulosic materials. J Chem Technol Biotechnol. 2011;86(2):223–30.
5. Panagiotou G, Topakas E, Moukouli M, Christakopoulos P, Olsson L. Studying the ability of *Fusarium oxysporum* and recombinant *Saccharomyces cerevisiae* to efficiently cooperate in decomposition and ethanolic fermentation of wheat straw. Biomass Bioenergy. 2011;35(8):3727–32.
6. Koppram R, Tomás-Pejó E, Xiros C, Olsson L. Lignocellulosic ethanol production at high-gravity: challenges and perspectives. Trends Biotechnol. 2014;32(1):46–53.
7. Galbe M, Sassner P, Wingren A, Zacchi G. Process engineering economics of bioethanol production. Adv Biochem Eng/Biotechnol. 2007;108:303.
8. Chandler M, Stanley GA, Rogers P, Chambers P. A genomic approach to defining the ethanol stress response in the yeast *Saccharomyces cerevisiae*. Ann Microbiol. 2004;54:427–54.
9. Hu XH, Wang MH, Tan T, Li JR, Yang H, Leach L. Genetic dissection of ethanol tolerance in the budding yeast *Saccharomyces cerevisiae*. Genetics. 2007;175:1479–87.
10. Birch RM, Walker GM. Influence of magnesium ions on heat shock and ethanol stress responses of *Saccharomyces cerevisiae*. Enzym Microb Technol. 2000;26:678–87.
11. Stanley D, Bandara A, Fraser S, Chambers PJ, Stanley GA. The ethanol stress response and ethanol tolerance of *Saccharomyces cerevisiae*. J Appl Microbiol. 2010;109:13–24.
12. Gomma EO. Ethanol induced response in *Phanerochaete chrysosporium* and its role in the decolorization of triarylmethane dye. Ann Microbiol. 2011;62:1403–9.
13. Haab D, Gassner B, Kubicek CP. Protein hypersecretory *Trichoderma-Reesei* mutant rut C-30 displays increased ethanol and polyene resistance. J Biotechnol. 1993;29:97–108.
14. Asiimwe T, Krause K, Schlunk I, Kothe E. Modulation of ethanol stress tolerance by aldehyde dehydrogenase in the mycorrhizal fungus *Tricholoma vaccinum*. Mycorrhiza. 2012;22(6):471–84.
15. Sakaki T, Zahringer U, Warnecke DC, Fahl A, Knogge W, Heinz E. Sterol glycosides and cerebrosides accumulate in *Pichia pastoris*, rhynchosporium secalis and other fungi under normal conditions or under heat shock and ethanol stress. Yeast. 2001;18:679–95.
16. Holtzapple M, Cognata M, Shu Y, Hendrikson C. Inhibition of *Trichoderma reesei* cellulase by sugars and solvents. Biotechnol Bioeng. 1990;36:275–87.
17. Chen H, Jin S. Effect of ethanol and yeast on cellulase activity and hydrolysis of crystalline cellulose. Enzym Microb Technol. 2006;39(7):1430–2.
18. Suihko ML. The fermentation of different carbon sources by *Fusarium oxysporum*. Biotechnol Lett. 1983;5:721–4.
19. Enari TM, Suihko ML. Ethanol production by fermentation of pentoses and hexoses from cellulosic materials. Crit Rev Biotechnol. 1983;1(3):229–40.
20. Singh A, Kumar PK. *Fusarium oxysporum*: status in bioethanol production. Crit Rev Biotechnol. 1991;11(2):129–47.
21. Luong JHT. Kinetics of ethanol inhibition in alcohol fermentation. Biotechnol Bioeng. 1985;27:280–5.
22. Nguyen VD, Auresenia J, Kosuge H, Tan RR, Brondial Y. Vacuum fermentation integrated with separation process for ethanol production. Biochem Eng. 2011;55(3):208–14.
23. Lee JH, Woodard JC, Pagan RJ, Rogers PL. Vacuum fermentation for ethanol production using strains of *Zymomonas Mobilis*. Biotechnol Lett. 1981;3(4):177–82.
24. Lee TY, Kim MD, Kim KY, Park K, Ryu YW, Seo JH. A parametric study on ethanol production from xylose by *Pichia stipitis*. Biotechnol Bioprocess Eng. 2000;5:27–31.
25. Meyrial V, Delgenes JP, Romieu C, Moletta R, Gounot AM. Ethanol tolerance and activity of plasma membrane ATPase in *Pichia stipitis* grown on D-xylose or on D-glucose. Enzym Microb Technol. 1995;17:535–40.
26. Diaz-Montano DM, Favela-Torres E, Cordova J. Improvement of growth, fermentative efficiency and ethanol tolerance of *Kloeckera africana* during the fermentation of Agave tequilana juice by addition of yeast extract. J Sci Food Agric. 2010;90(2):321–8.
27. Bajpai P, Margaritis A. Ethanol inhibition kinetics of *Kluyveromyces marxianus* grown on Jerusalem artichoke juice. Appl Environ Microbiol. 1982;44 (6):1325–9.
28. Du Preez JC, Bosch M, Prior BA. Temperature profiles of growth and ethanol tolerance of the xylose-fermenting yeasts *Candida shehatae* and *Pichia stipites*. Appl Microbiol Biotechnol. 1987;25:521–5.
29. Casey GP, Ingledew WM. Reevaluation of alcohol synthesis and tolerance in brewer's yeast. ASBC J. 1985;43:75–83.
30. Dombek KM, Ingram LO. Determination of the intracellular concentration of ethanol in *Saccharomyces cerevisiae* during fermentation. Appl Environ Microbiol. 1986;51:197–200.
31. Woo JM, Yang KM, Kim SU, Blank LM, Park JB. High temperature stimulates acetic acid accumulation and enhances the growth inhibition and ethanol production by *Saccharomyces cerevisiae* under fermenting conditions. Appl Microbiol Biotechnol. 2014;98(13):6085–94.
32. Panagiotou G, Christakopoulos P, Olsson L. The influence of different cultivation conditions on the metabolome of *Fusarium oxysporum*. J Biotechnol. 2005;118:304–15.
33. Xiros C, Katapodis P, Christakopoulos P. Evaluation of *Fusarium oxysporum* cellulolytic system for an efficient hydrolysis of hydrothermally treated wheat straw. Bioresour Technol. 2009;100:5362–5.
34. Wu Z, Lee YY. Inhibition of the enzymatic hydrolysis of cellulose by ethanol. Biotechnol Lett. 1997;19:977–9.
35. Jorgensen H, Vibe-Pedersen J, Larsen J, Felby C. Liquefaction of lignocellulose at high-solids concentrations. Biotechnol Bioeng. 2007;96:862–70.

36. Ferchak JD, Pye EK. Effect of cellobiose, glucose, ethanol, and metal ions on the cellulase enzyme complex of *Thermomonospora fusca*. Biotechnol Bioeng. 1983;25:2865–72.
37. Jing X, Zhang X, Bao J. Inhibition performance of lignocellulose degradation products on industrial cellulase enzymes during cellulose hydrolysis. Appl Biochem Biotechnol. 2009;159:696–707.
38. Yoshikawa H, Hirano A, Arakawa T, Shiraki K. Mechanistic insights into protein precipitation by alcohol. Int J Biol Macromol. 2012;50(3):865–71.
39. Skovgaard PA, Jorgensen H. Influence of high temperature and ethanol on thermostable lignocellulolytic enzymes. J Ind Microbiol Biotechnol. 2013;40(5):447–56.
40. Cysewski GR, Wilke CR. Rapid ethanol fermentations using vacuum and cell recycle. Biotechnol Bioeng. 1977;19:1125–43.
41. Roffler SR, Blanch HW, Wilke CR. *In situ* recovery of fermentation products. Trends Biotechnol. 1984;2:129–36.
42. Taylor F, Marquez MA, Johnston DB, Goldberg NM, Hicks KB. Continuous high-solids corn liquefaction and fermentation with stripping of ethanol. Bioresour Technol. 2010;101:4403–8.
43. Moreno AD, Tomas-Pejo E, Ibarra D, Ballesteros M, Olsson L. *In situ* laccase treatment enhances the fermentability of steam-exploded wheat straw in SSCF processes at high dry matter consistencies. Bioresour Technol. 2013;143:337–43.
44. Garrote G, Domínguez H, Parajó JC. Hydrothermal processing of lignocellulosic materials. Holz Roh-Werkst. 1999;57:191–202.
45. Sluiter A, Hames B, Ruiz R, Scarlata C, Sluiter J, Templeton D. Determination of structural carbohydrates and lignin biomass. Laboratory analytical protocol. Golden: National Renewable Energy Laboratory; 2005.
46. Xiros C, Topakas E, Katapodis P, Christakopoulos P. Hydrolysis and fermentation of brewer's spent grain by *Neurospora crassa*. Bioresour Technol. 2008;99:5427–35.
47. Katapodis P, Christakopoulos P. Enzymic production of feruloyl xylo-oligosaccharides from corn cobs by a family 10 xylanase from *Thermoascus aurantiacus*. LWT-Food Sci Technol. 2007;41(7):1239–43.
48. Wood TM, Bhat KM. Methods for measuring cellulase activities. Methods Enzymo. 1988;160:87–112.
49. Miller GL. Use of dinitrosalicylic acid reagent for determination of reducing sugar. Anal Chem. 1959;31:426–8.
50. Bailey MJ, Biely P, Poutanen K. Interlaboratory testing of methods for assay of xylanase activity. J Biotechnol. 1992;23:257–70.
51. Dogaris I, Gkounta O, Mamma D, Kekos D. Bioconversion of dilute-acid pretreated sorghum bagasse to ethanol by *Neurospora crassa*. Appl Microbiol Biotechnol. 2012;95:541–50.
52. Matsakas L, Christakopoulos P. Fermentation of liquefacted hydrothermally pretreated sweet sorghum bagasse to ethanol at high-solids content. Bioresour Technol. 2013;127:202–8.

Expression of the *Thermobifida fusca* xylanase Xyn11A in *Pichia pastoris* and its characterization

Longmei Zhao, Jiang Geng, Yaoqi Guo, Xiudong Liao, Xuhui Liu, Rujuan Wu, Zhaojun Zheng and Rijun Zhang*

Abstract

Background: Xylan is a major component of plant cells and the most abundant hemicellulose. Xylanases degrade xylan into monomers by randomly cleaving β-1,4-glycosidic bonds in the xylan backbone, and have widespread potential applications in various industries. The purpose of our study was to clone and express the endoxylanase gene *xynA* of *Thermobifida fusca* YX in its native form and with a C-terminal histidine (His) tag in *Pichia pastoris* X-33. We analyzed and compared these two forms of the protein and examined their potential applications in various industries.

Results: The *xynA* gene from *T. fusca* YX was successfully cloned and expressed using *P. pastoris* X-33. We produced a recombinant native form of the protein (rXyn11A) and a C-terminal His-tagged form of the desired protein (rXyn11A-$(His)_6$). The specific activities of rXyn11A and rXyn11A-$(His)_6$ in culture supernatants approached 149.4 and 133.4 U/mg, respectively. These activities were approximately 4- and 3.5-fold higher than those for the non-recombinant wild-type Xyn11A (29.3 U/mg). Following purification, the specific activities of rXyn11A and rXyn11A-$(His)_6$ were 557.35 and 515.84 U/mg, respectively. The specific activity of rXyn11A was 8% higher than that of rXyn11A-$(His)_6$. Both recombinant xylanases were optimally active at 80°C and pH 8.0, and exhibited greater than 60% activity between pH 6–9 and 60–80°C. They exhibited similar pH stability, while rXyn11A exhibited better thermostability; N-glycosylation enhanced the thermostability of both recombinant xylanases. The products of beechwood xylan hydrolyzed by both xylanases included xylobiose, xylotriose, xylotetraose and xylopentaose.

Conclusions: The C-terminal His tag had adverse effects when added to the Xyn11A protein. The thermostability of both recombinant xylanases was enhanced by N-glycosylation. Their stabilities at a high pH and temperature indicate their potential for application in various industries.

Keywords: *Thermobifida fusca* Xyn11A, *Pichia pastoris*, Expression, C-terminal His-tag, Characterization, Glycosylation

Background

Xylan, the major component in plant cells and the most abundant hemicellulose, is composed of β-1,4-linked xylopyranosyl residues [1,2]. Recently, lignocellulose, the most plentiful renewable biomass produced from photosynthesis, has attracted worldwide attention as a raw material for bioconversion [3]. Degrading all components of lignocellulose requires the synergistic activity of a large variety of enzymes with different specificities, such as exo-1,4-β-glucanases, endo-1,4-β-glucanases, β-glucosidases, endo-xylanases, pectin methyl esterases, and laccase [4].

Among these, xylanases (β-1,4-Endoxylanases, EC 3.2.1.8) are the core enzymes responsible for the degradation of xylan into monomers through the random cleavage of β-1,4-glycosidic bonds in the xylan backbone [4,5]. Xylanases have widespread potential applications in the textile, feed, beverage, and biofuel industries [2,6-10]; moreover, they form a major group of industrial enzymes used in the paper and pulp industry. The release of lignin from paper pulp can be facilitated by the hydrolysis of xylan [2]. Xylanases can facilitate the hydrolysis of lignocellulosic substrates as xylanases ameliorate the "blocking effect" of xylan, one of the major mechanisms that limits the accessibility of cellulose to cellulase enzymes. Xylanases also interact synergistically with cellulases to improve cellulose accessibility by increasing fiber swelling and porosity [11].

According to amino acid sequences, structural folds, and catalytic mechanisms, xylanases are classified into glycoside hydrolases (GH) families 5, 7, 8, 10, 11, and 43 [12-14]. An industrially suitable xylanase should have

* Correspondence: feedbiotech@yahoo.com
Laboratory of Feed Biotechnology, State Key Laboratory of Animal Nutrition, College of Animal Science and Technology, China Agricultural University, Beijing 100193, China

certain specific properties; for example, high activity and stability at high temperatures (>70°C) or at high pH (>8.0) [15]. *Thermobifida fusca* is a thermophilic actinomycete and a major degrader of plant cell walls in heated organic materials such as compost piles and rotting hay [16]. The xylanase Xyn11A, from *T. fusca*, belongs to the G (11) family and contains a substrate-binding domain. Xyn11A has thermostable properties, unique among family G xylanases, that are especially useful for industrial purposes [17].

The yeast *Pichia pastoris* is an excellent and efficient system for the expression of secreted and intracellular proteins [18]. A number of xylanases from a variety of organisms have been cloned and expressed in *P. pastoris* [19-25]. Some xylanases have been cloned and expressed in the bacterium *Escherichia coli* [20,26-28]. A comparison of expression levels and the activities of products in *P. pastoris* and *E. coli* revealed that expression levels of active proteins in *P. pastoris* were much higher than those in *E. coli* [29]. Recombinant proteins produced in *P. pastoris* were very easily N-glycosylated *via* post-translational modification, although the properties of glycosylated proteins were not always distinct from unglycosylated proteins. The thermostability of recombinant proteins can be decreased or increased after glycosylation [29,30].

We selected the *xynA* gene encoding Xyn11A from *T. fusca* YX for cloning and expression in *P. pastoris*. To facilitate the purification of recombinant proteins, some of the proteins were fused to a polyhistidine tag (His) at the C-terminus [31]; however, recombinant proteins containing a His-tag can differ from their wild-type counterparts with respect to dimerization or oligomerization properties [32]. We therefore also attempted to express *xynA* in its native form in a *P. pastoris* expression system and compared the properties of the native form with those of the His-tagged form.

Results and discussion

Analysis of *xynA* gene from *T. fusca* genomic DNA

We amplified an 891-bp DNA fragment encoding the mature Xyn11A peptide (296 aa) by polymerase chain reaction (PCR) from the genomic DNA of *T. fusca* YX using primers (*xyA-f, xyA-r, xyA-his-r*). The fragment was then successfully cloned into the pMD19-T vector (TaKaRa, Dalian, China). According to the Xyn11A protein sequence (GenBank Accession Number U01242), the mature Xyn11A peptide comprises a 189-aa catalytic domain (CD), an 86-aa xylan binding domain (XBD), and a 21-aa Gly-Pro-rich region that connects the CD and XBD [17]. The nucleotide and corresponding aa sequences of Xyn11A are shown in Figure 1. Based on previous studies, there were seven possible N-glycosylation sites (Asn-X-Ser/Thr) [33], with very little O-linked glycosylation observed in *P. pastoris* [34]. Supposing that all of the possible N-glycosylations occurred and the average glycosylated side chain was $Man_{10}GlcNAc$ [34], then the molecular mass of the target protein would increase by about 12.9 kDa compared with the native wild-type Xyn11A (31.9 kDa).

Expression of recombinant Xyn11A and Xyn11A-$(His)_6$ in *P. pastoris*

The pPICZα-*xynA* and pPICZα-*xynA*-$(His)_6$ expression plasmids were successfully generated as described in Methods. The coding sequences for Xyn11A and Xyn11A-$(His)_6$ were cloned in frame with the α-factor signal sequence in the *P. pastoris-E. coli* shuttle vector pPICZα-A. After confirmation by sequencing, the recombinant plasmids were linearized using restriction endonuclease *Pme*I and electroporated into *P. pastoris* X-33. The pPICZα-A vector was used as a negative control and transformed into *P. pastoris* X-33. Transformants harboring pPICZα-*xynA* or pPICZα-*xynA*-$(His)_6$ were grown on YPDS agar plates containing 100 μg/mL Zeocin and selected for high resistance to Zeocin (2000 μg/mL); successful transformants were obtained.

The pPICZα-*xynA* or pPICZα-*xynA*-$(His)_6$ plasmids carrying the *xynA* gene were regulated by an alcohol oxidase gene promoter that could be induced by 0.5% (v/v) methanol. The maximum activities of recombinant Xyn11A (rXyn11A) and recombinant Xyn11A-$(His)_6$ (rXyn11A-$(His)_6$) in BMMY medium reached 1157.63 and 964.97 U/mL, respectively, after 96 h of cultivation. The specific activity of a crude preparation of rXyn11A (149.4 U/mg) was about 12% higher than that of rXyn11A-$(His)_6$ (133.4 U/mg). The C-terminal His tag was located behind the XBD which was not essential for efficient hydrolysis of soluble xylan[17], therefore the C-terminal His tag affected the activity of rXyn11A-$(His)_6$ through its dimerization or oligomerization properties [32].

Several protein bands corresponding to a molecular mass of about 45 kDa were observed as the major protein in supernatants from *P. pastoris*/pPICZα-*xynA* and *P. pastoris*/pPICZα-*xynA*-$(His)_6$ cultures. This protein was not detected in the supernatant of *P. pastoris* cultures transformed with pPICZα-A (Figure 2). These 45-kDa proteins accumulated in supernatants obtained at different culture times (Figure 2). The molecular mass of native wild-type Xyn11A should be 31.9 kDa [17]; however, activity assays revealed that these 45-kDa proteins possessed xylanase activity (Figure 3), and thus we assign them as the recombinant xylanases with some glycosylated side chains. Xylanase activity was not detected in the supernatant of *P. pastoris* transformed with pPICZα-A. These results were consistent with those from previous studies, indicating the generation of several glycosylated proteins [15,34]. A separate protein band (Figure 3B, arrows) was thought to be the proteolytic degradation product of recombinant xylanase including the CD. The target proteins rXyn11A and rXyn11A-$(His)_6$

```
  1 gccgtgacctccaacgagaccgggtaccacgacgggtacttctac
    A  V  T  S  N  E  T  G  Y  H  D  G  Y  F  Y
 46 tcgttctggaccgacgcgcctggaacggtctccatggagctgggc
    S  F  W  T  D  A  P  G  T  V  S  M  E  L  G
 91 cctggcggaaactacagcacctcctggcggaacaccgggaacttc
    P  G  G  N  Y  S  T  S  W  R  N  T  G  N  F
136 gtcgccggtaagggatgggccaccggtggccgccggaccgtgacc
    V  A  G  K  G  W  A  T  G  G  R  R  T  V  T
181 tactccgccagcttcaacccgtcgggtaacgcctacctgaccctc
    Y  S  A  S  F  N  P  S  G  N  A  Y  L  T  L
226 tacgggtggacgcggaacccgctcgtggagtactacatcgtcgaa
    Y  G  W  T  R  N  P  L  V  E  Y  Y  I  V  E
271 agctggggcacctaccggcccaccggtacctacatgggcacggtg
    S  W  G  T  Y  R  P  T  G  T  Y  M  G  T  V
316 accaccgacggtggtacctacgacatctacaagaccacgcggtac
    T  T  D  G  G  T  Y  D  I  Y  K  T  T  R  Y
361 aacgcgccctccatcgaaggcacccggaccttcgaccagtactgg
    N  A  P  S  I  E  G  T  R  T  F  D  Q  Y  W
406 agcgtccgccagtccaagcggaccagcggtaccatcaccgcgggg
    S  V  R  Q  S  K  R  T  S  G  T  I  T  A  G
451 aaccacttcgacgcgtgggcccgccacggtatgcacctcggaacc
    N  H  F  D  A  W  A  R  H  G  M  H  L  G  T
496 cacgactacatgatcatggcgaccgagggctaccagagcagcgga
    H  D  Y  M  I  M  A  T  E  G  Y  Q  S  S  G
541 tcctccaacgtgacgttgggcaccagcggcggtggaaaccccggt
    S  S  N  V  T  L  G  T  S  G  G  G  N  P  G
586 gggggcaacccccccggtggcggcaacccccccggtggcggtggc
    G  G  N  P  P  G  G  G  N  P  P  G  G  G  G
631 tgcacggcgacgctgtccgcgggccagcagtggaacgaccgctac
    C  T  A  T  L  S  A  G  Q  Q  W  N  D  R  Y
676 aacctcaacgtcaacgtcagcggctccaacaactggaccgtgacc
    N  L  N  V  N  V  S  G  S  N  N  W  T  V  T
721 gtgaacgttccgtggccggcgaggatcatcgccacctggaacatc
    V  N  V  P  W  P  A  R  I  I  A  T  W  N  I
766 cacgccagctacccggactcccagaccttggttgcccggcctaac
    H  A  S  Y  P  D  S  Q  T  L  V  A  R  P  N
811 ggcaacggcaacaactggggcatgacgatcatgcacaacggcaac
    G  N  G  N  N  W  G  M  T  I  M  H  N  G  N
856 tggacgtggcccacggtgtcctgcagcgccaac[tag]
    W  T  W  P  T  V  S  C  S  A  N  *
```

Figure 1 Nucleotide sequence and corresponding amino acid sequence of the Xyn11A from *T. fusca* YX. The catalytic domain of Xyn11A is underlined. The xylan binding domain of Xyn11A is indicated by a dotted underline. The Gly-Pro-rich linker region is double underlined. Seven putative N-glycosylation sites are highlighted in grey. The stop codon is boxed and marked by asterisk (*).

constituted up to 63.7 and 62.3%, respectively, of all soluble proteins in supernatants as determined by Bandscan 5.0 software (Glyko, Novato, CA, USA).

Purification of recombinant xylanases

Purification of the recombinant proteins was performed as described in Methods, and a summary is presented in Table 1. The rXyn11A and rXyn11A-$(His)_6$ were purified 3.7-fold and 3.9-fold to homogeneity with recovery yields of 8.5% and 19.8%, respectively. The purification fold was not high because the recombinant xylanases constituted so much of the total protein in the supernatants (Figure 2). The specific activity of rXyn11A was 557.35 U/mg while that of rXyn11A-$(His)_6$ was 515.84 U/mg. Purified fractions of rXyn11A and rXyn11A-$(His)_6$ were subjected to 15% sodium dodecyl sulfate polyacrylamide gel electrophoresis (SDS-PAGE) and revealed several protein bands with molecular masses of about 45 kDa, which were distinct from the molecular mass of wild-type Xyn11A (31.9 kDa) and the calculated molecular mass of rXyn11A-$(His)_6$ (32.7 kDa), respectively (Figure 4).

Although rXyn11A-$(His)_6$ could be purified to homogeneity with a simpler procedure than rXyn11A, the latter exhibited higher specific activity than the former, consistent with results from a previous study where it was shown that the recombinant native form of cellulase was more active and stable than a C-terminal His-tagged cellulase produced in *E. coli* [35].

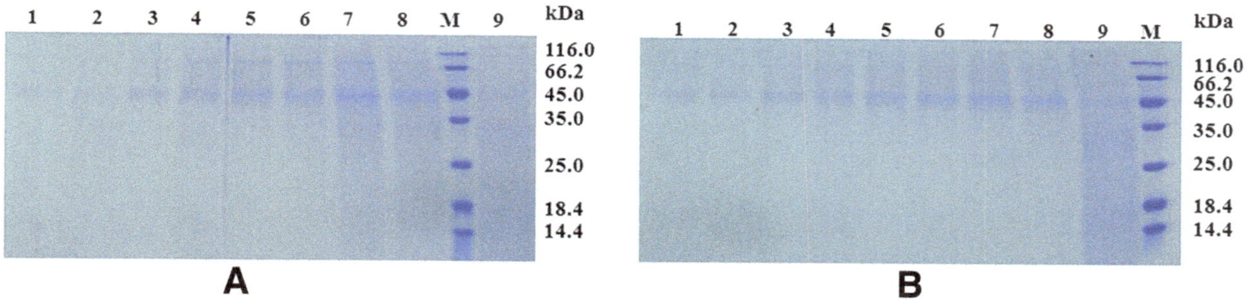

Figure 2 Analysis of the accumulation of recombinant xylanases expressed in *P. pastoris X-33*. Lane M: the molecular weight marker; **A**: SDS-PAGE of rXyn11A, Lane 1–8: supernatants of transformant harboring pPICZα-*xynA* obtained every 12 hours (12 h, 24 h, 36 h, 48 h, 60 h, 72 h, 84 h, 96 h), Lane 9: supernatant of transformant harboring pPICZα-A; **B**: SDS-PAGE of rXyn11A-(His)$_6$, Lane 1–8: supernatants of transformant harboring pPICZα-*xynA*-(His)$_6$ obtained every 12 hours (12 h, 24 h, 36 h, 48 h, 60 h, 72 h, 84 h, 96 h), Lane 9: supernatant of transformant harboring pPICZα-A. All samples containing 16 μL supernatant were loaded onto a 15% polyacrylamide gel.

Characterization of the recombinant xylanases

Both rXyn11A and rXyn11A-(His)$_6$ were optimally active at pH 8.0, and exhibited >60% activity between pH 6.0–9.0 (Figure 5A). Optimal activity was seen at 80°C, with >60% activity between 60 and 80°C (Figure 5B). Activities increased as the temperature rose, resulting in maximum activity at 80°C, and then decreased rapidly when the temperature increased above 80°C. The residual activities of both recombinant xylanases were >60% over a broad pH range (3.0–10.0), indicating their stability and that they would be of great practical use in various industrial applications (Figure 5C). The rXyn11A xylanase exhibited >80% residual activity between 30 and 80°C, while rXyn11A-(His)$_6$ exhibited 57% residual activity at 80°C (Figure 5D). After incubation at 90°C for 30 min, rXyn11A and rXyn11A-(His)$_6$ retained more than 10% of their initial activities (Figure 5D). We observed >20% residual activity for rXyn11A and rXyn11A-(His)$_6$ after treatment at 90°C for 10 min (Figure 5E and F). Both of the recombinant xylanases showed better thermal stabilities than the native wild-type Xyn11A from *T. fusca* [17]. Compared with results in a previous study regarding the same xylanase that was not glycosylated in *P. pastoris* [36], rXyn11A and rXyn11A-(His)$_6$ exhibited greater thermostability, possibly because of glycosylation of the recombinant xylanases.

Analysis of deglycosylation

Analysis of rXyn11A and rXyn11A-(His)$_6$ by SDS-PAGE revealed that their molecular weights (about 45 kDa) were increased by about 13 kDa compared with the native wild-type Xyn11A (31.9 kDa; Figure 4). Following treatment with endo-H to remove carbohydrate moieties, a protein band was observed at about 32 kDa by SDS-PAGE (Figure 4). These findings suggested that N-glycosylation of rXyn11A and rXyn11A-(His)$_6$ accounted for about 13 kDa.

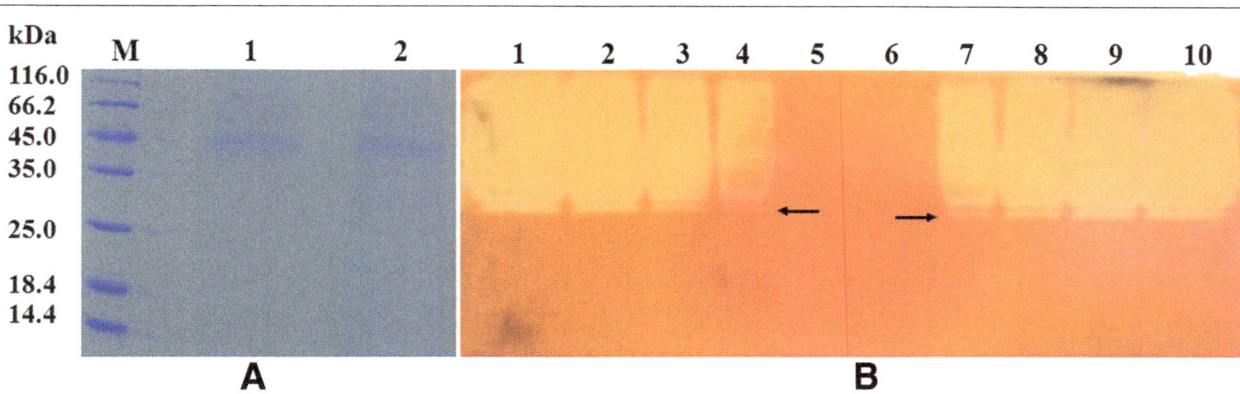

Figure 3 Identification of the recombinant xylanases about 45 kDa. Lane M: the molecular weight marker; **A**: SDS-PAGE of recombinant xylanases stained with Coomassie Brilliant Blue R-250 (samples containing 16 μL supernatant were loaded onto a 15% polyacrylamide gel), Lane 1: supernatant of transformant harboring pPICZα-*xynA*-(His)$_6$, Lane 2: supernatant of transformant harboring pPICZα-*xynA*; **B**: SDS-PAGE activity staining of recombinant xylanases using Congo Red (samples were loaded onto a 15% polyacrylamide gel), Lane 1–4: supernatants of transformant harboring pPICZα-*xynA*-(His)$_6$ with different loading volumes (15 μL, 10 μL, 5 μL, 2.5 μL), Lane 5–6: supernatant of transformant harboring pPICZα-A, Lane 7–10: supernatants of transformant harboring pPICZα-*xynA* with different loading volumes (2.5 μL, 5 μL, 10 μL, 15 μL).

Table 1 Summary of the purification of rXyn11A and rXyn11A-(His)$_6$

Purification steps	Total activity (U)	Total protein (mg)	Specific activity (U · mg^{-1})	Purification (fold)	Yield (%)
P. pastoris X-33/ pPICZα-*xynA*					
The supernatant	11576.3	77.5	149.4	1	100
Sephadex G-25	9205.5	36.1	254.9	1.7	79.5
Mono Q 5/50 GL	2462.3	6.3	390.2	2.6	21.3
Ultrafiltration	979.3	1.8	557.4	3.7	8.5
P. pastoris X-33/pPICZα-*xynA*-(His)$_6$					
The supernatant	9649.7	72.4	133.4	1	100
Ni-NTA agarose	3057.0	15.8	193.6	1.5	31.7
Ultrafiltration	1908.6	3.7	515.8	3.9	19.8

Both the deglycosylated rXyn11A and rXyn11A-(His)$_6$ showed optimal activities at 70°C and pH 7.0 (Figure 6A and B), which were decreased compared with the N-glycosylated recombinant xylanases (80°C and pH 8.0). The deglycosylated xylanases showed similar pH stability to the N-glycosylated recombinant xylanases (Figure 6C), but the thermostability was decreased. After incubation at 80°C and 90°C for 30 min, deglycosylated rXyn11A and rXyn11A-(His)$_6$ retained less than 30% and 5% of their initial activities (Figure 6D), respectively. SDS-PAGE of deglycosylated rXyn11A and rXyn11A-(His)$_6$ without denaturation showed that rXyn11A-(His)$_6$ was not thoroughly deglycosylated (Figure 7), thus the change in thermostability of rXyn11A-(His)$_6$ was less than that for rXyn11A. The C-terminal His tag did not have a major effect on the activity and stability of recombinant xylanase after deglycosylation (Figure 6).

The recombinant xylanases had higher optimum temperatures and exhibited better thermostability, possibly because of N-glycosylation. Our results corresponded with those in a previous study that showed that the N-glycosylation of recombinant enzymes expressed in *P. pastoris* enhanced their thermal stabilities [29,30,37].

Based on the amino acid sequence of mature Xyn11A (Figure 1), there appear to be seven potential N-glycosylation sites (N-X-S/T: NET 5–7, NYS 34–36, NPS 66–68, NVT 183–185, NVS 230–232, NWT 236–238, NWT 285–287); these sites were predicted by NetNGlyc 1.0Server (http://www.cbs.dtu.dk/services/NetNGlyc/). Among these N-glycosylation sites, the first four were located within the CD and the last three sites were within the XBD. There were six sites (NET 5–7, NYS 34–36, NVT 183–185, NVS 230–232, NWT 236–238, NWT 285–287) with >54% probability of being glycosylated. A possible explanation for the higher thermostability caused by glycosylation may be that the long outer chains added during glycosylation can affect the folding or function of a foreign protein in *P. pastoris* [33]. In this study, the optimum pH was altered from 7.0 to 8.0 and the optimum temperature increased from 65°C to 80°C. These changes may be attributable to the four N-glycosylation sites located within the CD. To investigate the relationship

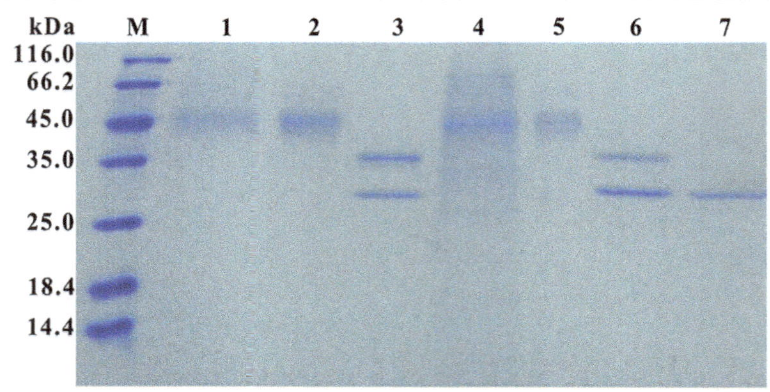

Figure 4 SDS-PAGE analyses of the purification and deglycosylation of recombinant xylanases expressed in *P.pastoris X-33*. Lane M: the molecular weight marker; Lane 1 and 4: supernatant of transformant harboring pPICZα-*xynA*-(His)$_6$ and pPICZα-*xynA*, respectively; Lane 2 and 5: the purified rXyn11A-(His)$_6$ and rXyn11A, respectively; Lane 3 and 6: the purified rXyn11A-(His)$_6$ and rXyn11A deglycosylated with Endo H, respectively; Lane 7: Endo H.

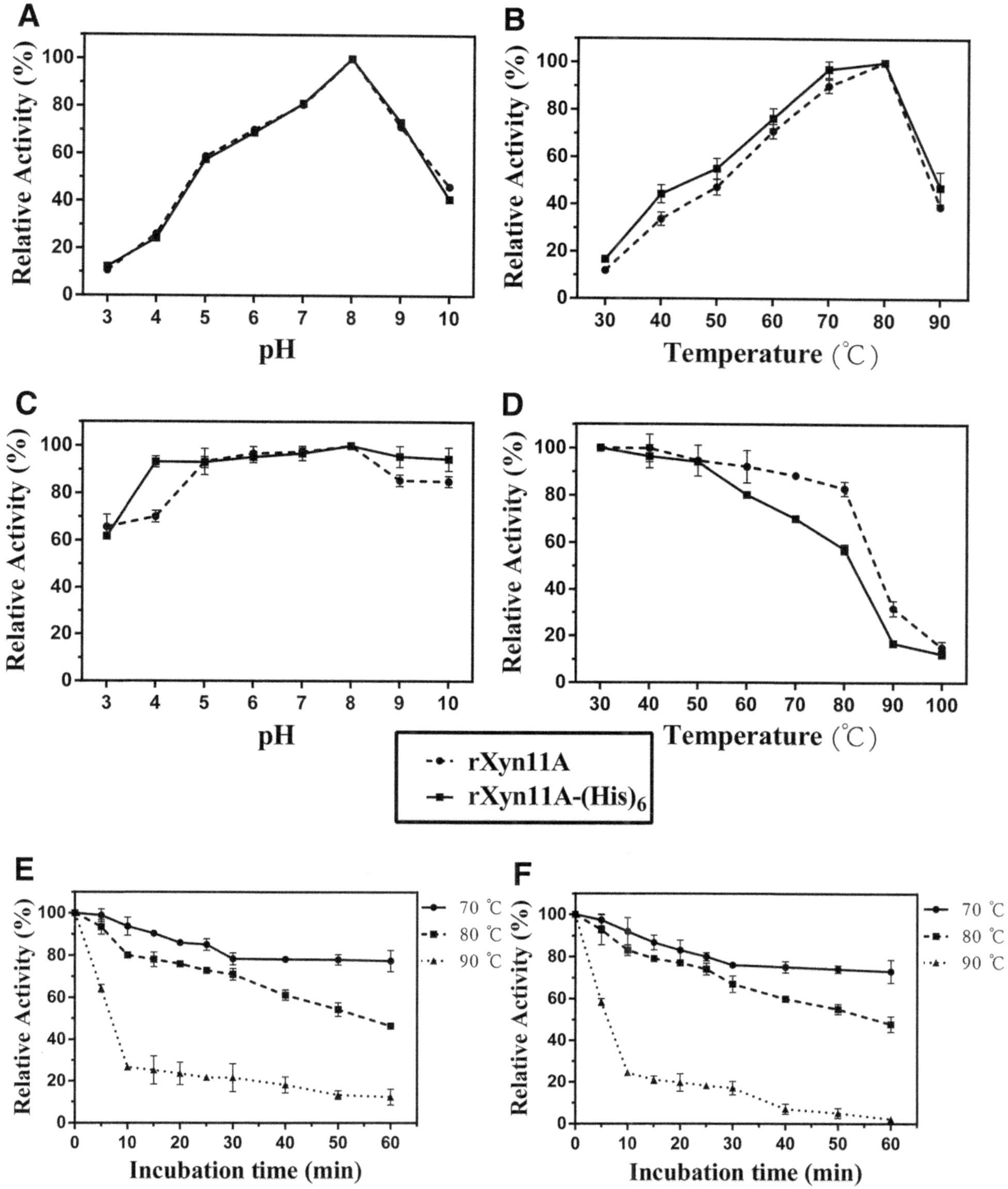

Figure 5 Effect of pH and temperature on activity and stability of rXyn11A and rXyn11A-(His)$_6$. **A**: Optimum pH; **B**: optimum temperature; **C**: pH stability; **D**: thermal stability; **E**: thermostability of rXyn11A (incubated at 70°C, 80°C, 90°C for different time); **F**: thermostability of rXyn11A-(His)$_6$ (incubated at 70°C, 80°C, 90°C for different time).

between N-glycosylation and xylanase thermostability, we used Swiss Model to perform homology-modelling of Xyn11A (1–187 AA, the catalytic domain). There are one α-helix and several β-sheets in this domain (Figure 8A), and the β-sheets form a groove (Figure 8B) which houses the active site [38]. Several potential N-glycosylation sites

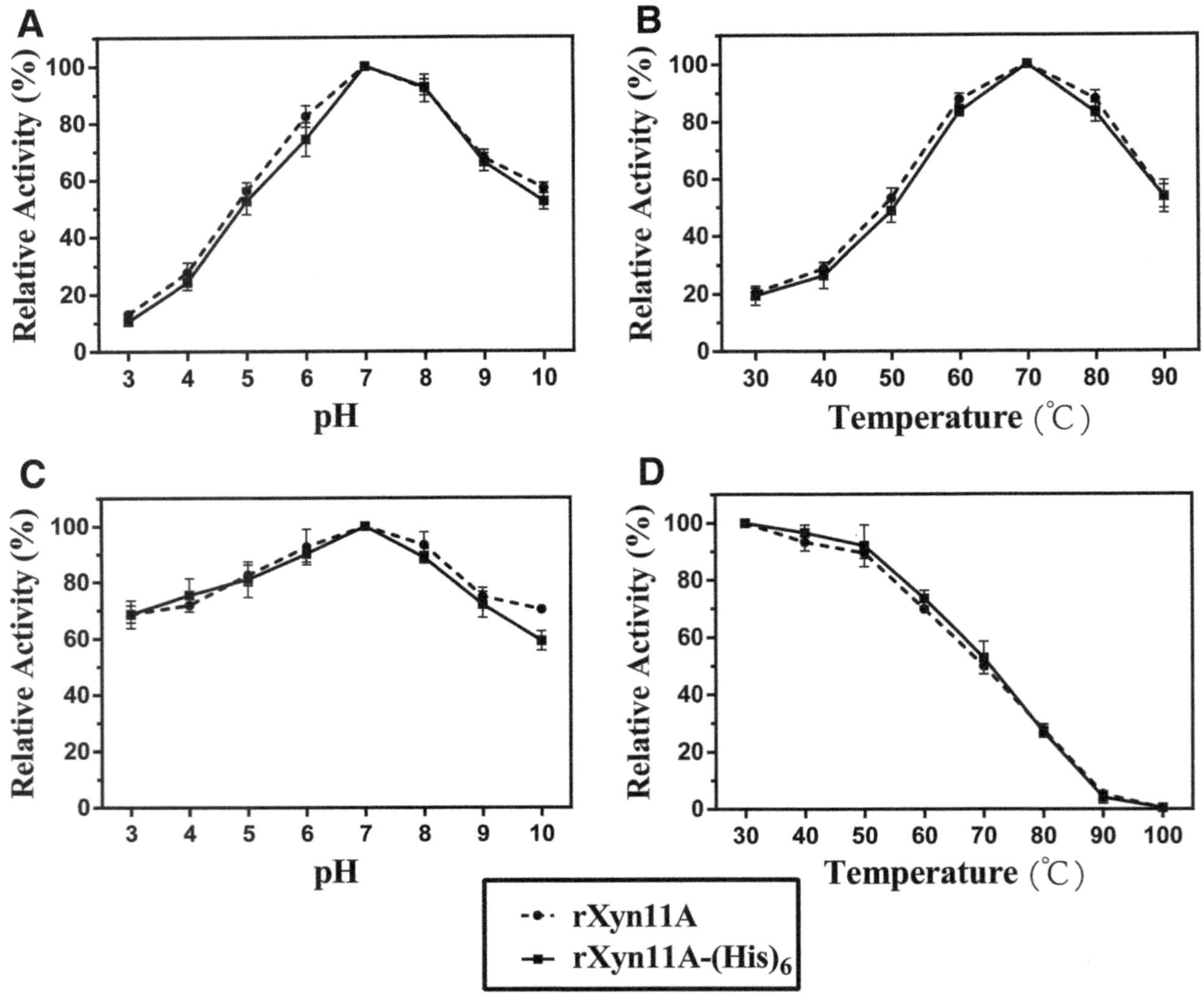

Figure 6 Effect of pH and temperature on activity and stability of deglycosylated rXyn11A and rXyn11A-(His)$_6$. A: Optimum pH; **B**: optimum temperature; **C**: pH stability; **D**: thermal stability.

were just included in the catalytic domain. Furthermore, it is believed that thermostable enzymes usually contain a large number of surface polar residues which generate more water-mediated networks surrounding the enzyme [38]. Therefore, the thermostability of rXyn11A and rXyn11A-(His)$_6$ may be increased by N-glycosylation. The glycosylation of recombinant proteins alters some properties of the protein but not to a great extent [15,39]; in certain cases glycosylation can adversely affect thermostability in *P. pastoris* [29]. In a study examining the same xylanase, it was seen that when the linearized pPIC9K-tfx plasmid was transformed into *P. pastoris* GS115 after the xylanase gene was cloned into the vector, the expressed recombinant xylanase was not glycosylated and was the same size as the native xylanase [36]. These results showed that the difference between plasmids, as well as protein properties, may affect the glycosylation of recombinant proteins expressed in *P. pastoris*.

Analysis of hydrolytic products

The products of beechwood xylan hydrolyzed by rXyn11A and rXyn11A-(His)$_6$ were analyzed using thin-layer chromatography (TLC) (Figure 9) and high-performance ion chromatography (HPIC) (Table 2). For TLC, a series of oligosaccharides with a degree of polymerization of two and above were observed on the silica gel plate. The main hydrolytic products were X2–X5, while many higher oligomers were still present in the digest. During hydrolysis, the concentrations of xylooligosaccharides increased for 4 h and then decreased as the xylooligosaccharides were hydrolyzed by the recombinant xylanases (Figure 9). The HPIC results revealed that rXyn11A is better at hydrolyzing beechwood xylan than rXyn11A-(His)$_6$. These findings

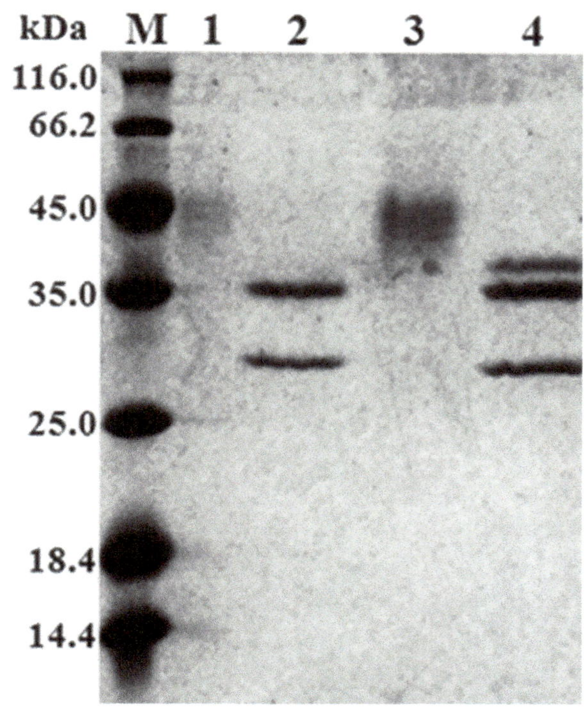

Figure 7 SDS-PAGE analyses of deglycosylated rXyn11A and rXyn11A-(His)$_6$ without denaturation. Lane M: the molecular weight marker; Lane 1 and 3: the purified rXyn11A and rXyn11A-(His)$_6$, respectively; Lane 2 and 4: the purified rXyn11A and rXyn11A-(His)$_6$ deglycosylated with Endo H without denaturation, respectively.

imply that the C-terminal His tag probably has adverse effects on xylanase activity. The C-terminal His tag may alter the structure of recombinant xylanases. The TLC and HPIC analyses showed that the main hydrolytic products were higher oligomers than xylose. Even after 24 h, there was little xylose released *via* hydrolysis. Therefore, our results indicate that rXyn11A and rXyn11A-(His)$_6$ hydrolyzed xylan by cleaving the intramolecular glycosidic bonds of xylooligosaccharides, as they are both endoxylanases [17]. The distribution pattern of hydrolytic products was identical for rXyn11A and rXyn11A-(His)$_6$.

Conclusions

We have described the cloning and expression of a xylanase from *T. fusca*, and compared the characteristics of the recombinant wild-type xylanase, and the xylanase with a C-terminal His tag, expressed in *P. pastoris*. The characteristics of rXyn11A and rXyn11A-(His)$_6$ were similar with respect to thermostability and pH tolerance. The activity of rXyn11A was higher than that of rXyn11A-(His)$_6$, with both their activities higher than that of native Xyn11A. Therefore, we conclude that the C-terminal His tag had adverse effects on enzyme activity. N-glycosylation enhanced the thermostability of the recombinant xylanases, suggesting they could be applied to the degradation of cellulosic materials and in other industries.

Methods

Chemicals and enzymes

Restriction endonucleases, T4 DNA Ligase and Endoglycosidase H (Endo H) were purchased from New England Biolabs (Ipswich, MA, USA). Yeast extract and peptone were obtained from OXOID (Basingstoke, Hampshire, UK), yeast nitrogen base (YNB) without amino acids was purchased from BD (Sparks, MD, USA), biotin was obtained from Amersco (Solon, OH, USA), and Zeocin was purchased from Invitrogen (San Diego, CA, USA). Ni–NTA agarose and silica gel plates 60 F 254 was purchased from Merck (Darmstadt, Germany). Beechwood xylan was purchased from Sigma Chemical Company (St. Louis, MO, USA). Xylose (X1) was from Wako (Osaka,

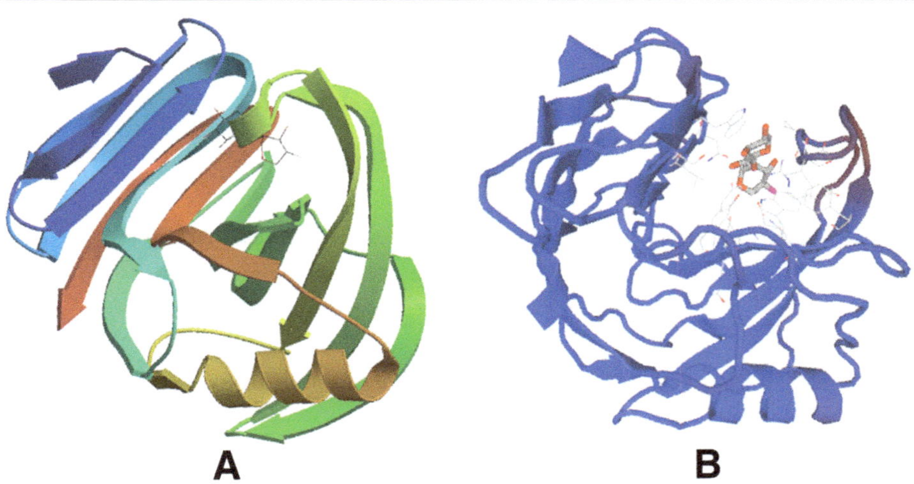

Figure 8 Homology-modelling of Xyn11A. A: Model of Xyn11A (the catalytic domain); **B**: Xyn11A model (the catalytic domain) showing the groove (housing the active site).

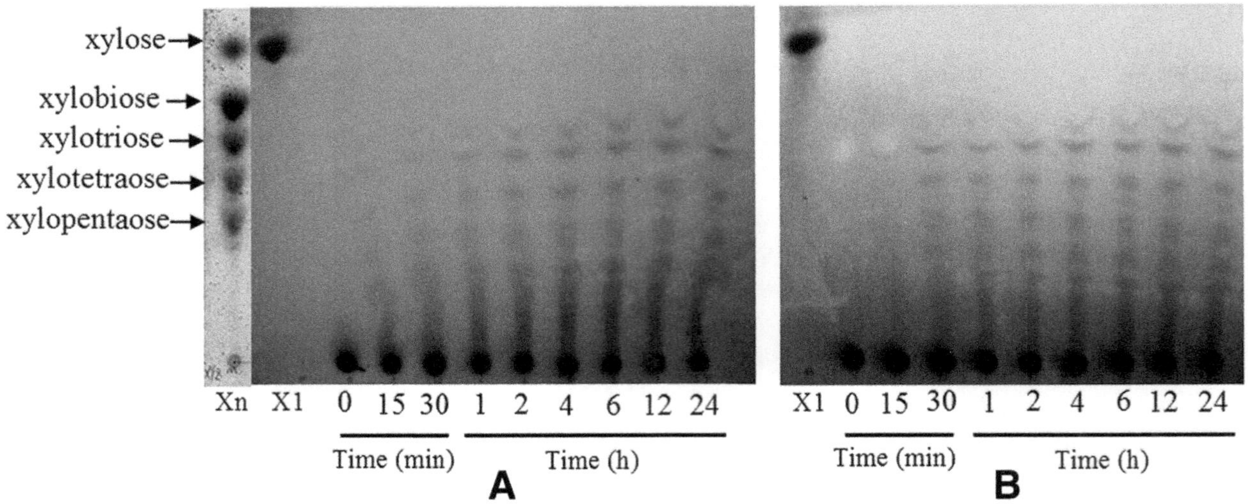

Figure 9 Thin-layer chromatogram of hydrolytic products of beechwood xylan by recombinant xylanases. **A**: Degradation of beechwood xylan by rXyn11A. **B**: Degradation of beechwood xylan by rXyn11A-(His)$_6$.

Japan). All other chemicals used were of reagent grade obtained from standard sources.

Strains and plasmids

Genomic DNA from *T. fusca* YX used in this study was purchased from ATCC (catalog BAA-629D-5). For cloning, pMD19-T vector (TaKaRa, Dalian, China) and E. coli DH5α (Tiangen, Beijing, China) were used according to the manufacturers' recommendation. *P. pastoris* X-33 and pPICZα-A vector from EasySelect Pichia Expression Kit were purchased from Invitrogen (San Diego, CA, USA).

Construction of the plasmid for Xyn11A expression

In order to obtain the target gene *xynA* encoding the mature region of Xyn11A from *T. fusca* YX genome, three primers (*xyA-f, xyA-r, xyA-his-r*) were designed. The mature *xynA* gene was PCR amplified from genomic DNA of *T. fusca* YX with *xyA-f* (5′-CGGAATTCGCCGTGACC TCCAACGAGA-3′) as forward primer and *xyA-r* (5′-GCTCTAGATTACTA GTTGGCGCTGCAGGACA-3′) as reverse primer (*EcoR*I and *Xba*I restriction sites are denoted by the underline). For the PCR of His-tagged *xynA* gene, the reverse primer *xyA-his-r* (5′-GCTCTAGATTA CTA ATGATGATGATGATGATGGTTGGCGCTGCA GGACA-3′) was used (*Xba*I restriction site and (His)$_6$-tag are denoted by the underline). Both PCR amplifications were carried out in a Thermal cycler (Bio-Rad, Hercules, CA, USA) as the following conditions: an initial denaturation at 95°C for 5 min, 30 cycles of 95°C for 30 s, 55°C for 30 s and 72°C for 1 min, followed by one cycle of 72°C for 10 min. Afterwards, both PCR products were cloned into pMD19-T vector, yielding plasmids pMD19-*xynA* and pMD19-*xynA*-(His)$_6$, respectively.

To obtain the secretive expression of rXyn11A and rXyn11A-(His)$_6$, the plasmid pMD19-*xynA*/pMD19-*xynA*-(His)$_6$ and *P. pastoris-E. coli* shuttle vector pPICZα-A were both digested with *EcoR*I and *Xba*I and mutually ligated. After being transformed into *E. coli* DH5α, transformants were selected at 37°C on the low salt LB agar plates (5 g/L yeast extract, 10 g/L tryptone, 5 g/L NaCl, 15 g/L agar, and adjusted pH to 7.5) containing 25 μg/mL Zeocin. The recombinant plasmids designated as pPICZ*α*-*xynA* and pPICZ*α*-*xynA*-(His)$_6$ were obtained and sequenced by Invitrogen (Shanghai, China).

Table 2 Hydrolytic products concentration and percentage content of beechwood xylan by rXyn11A and rXyn11A-(His)$_6$

Enzyme	Reaction time (h)	Hydrolytic products concentrations (μg/mL) and percentage content				
		Xylose	Xylobiose	Xylotriose	Xylotetraose	xylopentaose
rXyn11A	6	12.0 (0.4%)	178.6 (6.0%)	630.8 (21.3%)	1007.9(34.1%)	1130.2(38.2%)
	12	16.9 (0.5%)	179.7 (5.4%)	661.5 (19.9%)	1113.1 (33.6%)	1347.6 (40.6%)
	24	24.1 (0.7%)	169.1 (5.2%)	649.6 (20.0%)	1152.6 (35.5%)	1255.3 (38.6%)
rXyn11A-(His)$_6$	6	12.5 (0.8%)	69.8 (4.2%)	295.3 (18.0%)	593.1 (36.1%)	672.2 (40.9%)
	12	12.8 (0.7%)	90.2 (5.3%)	317.8 (18.5%)	589.0 (34.4%)	704.0 (41.1%)
	24	17.4 (1.1%)	85.5 (5.3%)	298.9 (18.5%)	552.5 (34.1%)	664.4 (41.0%)

Transformation of *P. pastoris* and screening of multi-copy transformants

Both the recombinant plasmids pPICα-*xynA* and pPICZα-*xynA*-$(His)_6$ were linearized by restriction digestion with *Pme*I. Afterwards linearized DNA were transformed into *P. pastoris* X-33 by electroporation using MicroPulser (Bio-Rad, Hercules, CA, USA) under the PIC setting. Then the *P. pastoris* transformants were selected on the YPDS agar plates (10 g/L yeast extract, 20 g/L peptone, 20 g/L dextrose, 20 g/L agar and 1 M sorbitol) containing 100 μg/mL Zeocin at 30°C. After that, the transformants obtained were checked for the target gene integration using the extracted *P. pastoris* genomic DNA by PCR.

In order to get multi-copy transformants, the recombinant clones obtained were dotted onto the YPDS agar plates with higher Zeocin concentration (for example, 1000 μg/mL, 2000 μg/mL) and incubated at 30°C for about 2 days.

Expression of recombinant xylanases in *P. pastoris*

For the expression of rXyn11A and rXyn11A-$(His)_6$ in recombinant *P. pastoris*, a single colony was inoculated into 25 mL BMGY medium (10 g/L yeast extract, 20 g/L peptone, 100 mM potassium phosphate (pH 6.0), 13.4 g/L YNB, 4×10^{-4} g/L biotin, 1% (v/v) glycerol) in a 250 mL baffled flask and shaken (220 rpm) at 29°C for about 18 hours. Then the cells were harvested by centrifuging at 3000 × g for 5 min, and resuspended to an OD600 of 1.0 in BMMY medium (10 g/L yeast extract, 20 g/L peptone, 100 mM potassium phosphate (pH 6.0), 13.4 g/L YNB, 4×10^{-4} g/L biotin, 0.5% (v/v) methanol) about 100 mL in a 1 L baffled flask, shaken at 29°C for 4 days. To maintain induction, 100% methanol was added to the culture to a final concentration of 0.5% every 24 h during the induction phase.

Purification of recombinant xylanases

For the purification of rXyn11A, the culture supernatant was collected by centrifuging at 12000 rpm for 5 min. The ÄKTA avant 25 chromatography system (GE Healthcare, Little Chalfont, UK) was used for the purification. The supernatant was loaded on a Sephadex G-25 (GE Healthcare, Little Chalfont, UK) column which was pre-equilibrated with 20 mM Tris–HCl buffer (pH 8.2). Binding protein was eluted by the same buffer, the eluted protein was pooled and loaded on an anion-exchange Mono Q column which was equilibrated with 20 mM Tris–HCl buffer (pH 8.2). Then the Mono Q column was eluted by 20 mM Tris–HCl buffer (pH 8.2) containing 0.9 M NaCl at 2 ml/min. Afterwards the purified protein was pooled and concentrated by the ultrafiltration with a 10 kDa molecular weight cut-off membrane in the Amicon Ultra-15 centrifugal filter unit (Merck Millipore, MA, USA). The concentrated protein was stored at 4°C.

For the purification of rXyn11A-$(His)_6$, the harvested culture supernatant was filtrated with 0.22 μm Millex-GP Filter Unit (Merck Millipore, MA, USA) after centrifuging the culture at 12000 rpm for 5 min, and then the filtrated supernatant was directly applied to Ni-NTA column (Merck, Darmstadt, Germany) which was equilibrated with 50 mM PBS buffer (pH 7.4) at 1 mL/min. After the column was washed by 50 mM PBS buffer (pH 7.4) at 2 mL/min, the enzyme was eluted successively with 50 mM PBS buffer (pH 7.4) containing 50 mM and 100 mM imidazole at 2 ml/min until A280 of the effluent kept constant. The eluted enzyme was pooled and concentrated by the ultrafiltration with a 10 kDa molecular weight cut-off membrane in the Amicon Ultra-15 centrifugal filter unit (Merck Millipore, MA, USA). The purified enzyme was stored at 4°C for later use.

Recombinant xylanase activity assays and protein determinations

The culture collected at 96 h was centrifuged and the xylanases activities of culture supernatants were measured [40] according to the increase in reducing sugar given by the dinitrosalicylic acid (DNS) method [41] using xylose as a standard. The reaction mixture containing 900 μL of 1% (w/v) beechwood xylan prepared in 0.2 M glycine-NaOH buffer (pH 8.0) and 100 μL of appropriately diluted xylanase enzyme solution was incubated at 80°C for 10 min. After the reaction was stopped by adding 1000 μL DNS solution (1% 3, 5-dinitrosalicyclic acid, 20% potassium sodium tartrate, 1% NaOH, 0.2% phenol, 0.05% Na_2SO_3) to the mixtures, the samples were vortexed and boiled for 10 min, followed by measuring the absorbance of 200 μL sample at 540 nm using iMark Microplate Reader (Bio-Rad, Hercules, CA, USA). One unit of xylanase activity was defined as the amount of xylanase required to catalyze the release of 1 μmol of xylose equivalent in 1 minute under the assay condition. All of the xylanase activity values presented were averages obtained from triplicate measurements. The protein concentrations of culture supernatant were measured by Bio-Rad Protein Assay kit (Hercules, CA, USA) in triplicate.

SDS-PAGE and xylanase activity staining

SDS-PAGE was performed to detect the rXyn11A and rXyn11A-$(His)_6$ with 5% stacking gel and 15% separating gel, and the protein bands were stained by Coomassie brilliant blue R-250.

Activity staining of recombinant xylanases was performed to display the visible xylanase activity bands on the gel [42] with slight modification. A 15% separating gel with 0.1% xylan in it was prepared for separating proteins. The protein samples prepared for the activity staining were heated for 10 min at 60°C in the presence of 2% SDS, but without 2-mercaptoethanol. After electrophoresis, the gel was immediately washed by five successive 30-min washes in 0.2 M cold glycine-NaOH buffer (pH 8.0) with first two washes contained 25% (v/v) isopropanol to remove the SDS. Then the gel was submersed in 0.2 M glycine-NaOH buffer (pH 8.0) and incubated at 80°C for 1 h. Subsequently, the gel was stained in 0.2% Congo Red for 30 min and destained with 1 M NaCl. Visible and clear activity bands against deep red background indicated the degradation of xylan.

Deglycosylation of recombinant xylanases

Both the purified rXyn11A and purified rXyn11A-$(His)_6$ were denatured by heating at 100°C for 10 min, and then the endoglycosidase H was added and the reaction mixtures were incubated at 37°C for 2 hour. And all manipulations were according to the instructions of manufacturer. Subsequently, SDS-PAGE was performed to detect the separation of reaction products.

For the activity assay of deglycosylated recombinant xylanases, the endo H digestion reactions were slightly adapted and conducted as follows, the reaction mixtures containing endo H and purified rXyn11A or rXyn11A-$(His)_6$ were incubated at 37°C for 4 hour without denaturation process. Subsequently, SDS-PAGE was performed to detect the separation of reaction products.

Characterization of recombinant xylanases in pH and temperature

To determine the optimal pH of rXyn11A and rXyn11A-$(His)_6$ as well as deglycosylated rXyn11A and rXyn11A-$(His)_6$, the xylanase activities were assayed in 0.2 M various buffers such as Macllvaine's buffer (pH 3.0-7.0) and glycine-NaOH buffer (pH 8.0-10.0) according to the standard assay method. For the determination of xylanases stabilities at various pH values, enzymes were pre-incubated in different pH buffers for 60 min at 55°C. After cooling, the residual activities were measured as the standard method.

To measure the temperature optima of rXyn11A and rXyn11A-$(His)_6$ as well as deglycosylated rXyn11A and rXyn11A-$(His)_6$, the xylanase activities were determined among the temperature range from 30°C to 90°C by the standard assay method. For the determination of xylanases thermostability, enzymes were pre-incubated at different temperatures (30–100°C) in 0.2 M glycine-NaOH buffer (pH 8.0) for 30 min, and the remained activities were measured as the standard method after cooling.

The thermostability of rXyn11A and rXyn11A-$(His)_6$ was further investigated under different incubated time at 70, 80, 90°C. Aliquots were withdrawn at different time intervals (0 min, 5 min, 10 min, 15 min, 20 min, 25 min, 30 min, 40 min, 50 min, 60 min). After cooling, the remnant activities were assayed according to the standard method.

Structural modelling of Xyn11A

The homology-modelling of Xyn11A was performed by Swiss-Model (http://swissmodel.expasy.org/) using the template (PDB entry 3zse), and the amino acid sequence identity was 99.47% [43,44].

Analysis of hydrolysis of beechwood xylan by the recombinant xylanases

To analyze the hydrolytic products of beechwood xylan, 1.0% beechwood xylan in 2 ml of 0.2 M glycine-NaOH buffer was mixed with 10 μL of rXyn11A and 10 μL of rXyn11A-$(His)_6$, respectively. The reaction mixtures were incubated at 80°C for 24 h. The aliquots at different time intervals were withdrawn and analyzed by both thin-layer chromatography (TLC) and high-performance ion chromatography (HPIC). For analysis of TLC, 10 μL of the samples were spotted on the silica gel plates 60 F 254 (Merck, Darmstadt, Germany). A mixture containing xylose (X1), xylobiose (X2), xylotriose (X3), xylotetraose (X4) and xylopentaose (X5) dissolved in water was used as the standard. Chromatography was developed in butanol-acetic acid-water (2:1:1, v/v) solvent system. After heating at 130°C for a couple of minutes in an oven, the plate was colorated by spraying with methanol-sulfuric acid mixture (95:5, v/v) and heated in an oven at 130°C for 5 min. The manipulations were modified from the method [45]. For analysis of HPIC, the samples at different time intervals (6 h, 12 h, 24 h) and standard xylooligosaccharides were analyzed with CarboPac^{TM} PA200 anion-exchange column (3 mm × 250 mm; Dionex, CA, USA), pure water as mobile phase (0.45 mL/min), and injection volumes were 10 μL. Sugar peaks were screened by ICS 5000 Electrochemical Detector (Dionex, CA, USA).

Abbreviations

rXyn11A: Recombinant Xyn11A; rXyn11A-(His)6: Recombinant Xyn11A with C-terminal His-tagged; CD: Catalytic domain; XBD: Xylan binding domain; X1: Xylose; X2: Xylobiose; X3: Xylotriose; X4: Xylotetraose; X5: Xylopentaose; PCR: Polymerase chain reaction; YPDS: Yeast extract peptone dextrose medium with 1 M sorbitol; BMGY: Buffered glycerol-complex medium; BMMY: Buffered methanol-complex medium; PBS: Phosphate buffer saline; SDS–PAGE: Sodium dodecyl sulphate polyacrylamide gel electrophoresis; TLC: Thin-layer chromatography; HPIC: High-performance ion chromatography.

Competing interests

The authors declare that they have no competing interests.

Authors' contributions

LZ conceived and designed the study, performed the experiments, and drafted the manuscript. JG participated in the design of the study. RZ conceived of the study, participated in its design and helped to draft the manuscript. LZ, JG, YG, XDL, XHL, RW, ZZ and RZ reviewed and edited the manuscript. All authors read and approved the final manuscript.

Acknowledgements

This research was supported by the 'Five-twelfth' National Science and Technology Support Program of China (2011BAD26B0403).

References

1. Beg QK, Kapoor M, Mahajan L, Hoondal GS. Microbial xylanases and their industrial applications: a review. Appl Microbiol Biotechnol. 2001;56:326–38.
2. Subramaniyan S, Prema P. Biotechnology of microbial xylanases: enzymology, molecular biology, and application. Crit Rev Biotechnol. 2002;22:33–64.
3. Li H, Kim NJ, Jiang M, Kang JW, Chang HN. Simultaneous saccharification and fermentation of lignocellulosic residues pretreated with phosphoric acid-acetone for bioethanol production. Bioresour Technol. 2009;100:3245–51.
4. Van Dyk JS, Pletschke BI. A review of lignocellulose bioconversion using enzymatic hydrolysis and synergistic cooperation between enzymes-Factors affecting enzymes, conversion and synergy. Biotechnol Adv. 2012;30:1458–80.
5. Wong KKY, Tan LUL, Saddler JN. Multiplicity of beta-1,4-xylanase in microorganisms-functions and applications. Microbiol Rev. 1988;52:305–17.
6. Beg QK, Bhushan B, Kapoor M, Hoondal GS. Enhanced production of a thermostable xylanase from *Streptomyces* sp. QG-11-3 and its application in biobleaching of eucalyptus kraft pulp. Enzyme Microb Tech. 2000;27:459–66.
7. Subramaniyan S, Prema P. Cellulase-free xylanases from *Bacillus* and other microorganisms. FEMS Microbiol Lett. 2000;183:1–7.
8. Techapun C, Poosaran N, Watanabe M, Sasaki K. Thermostable and alkaline-tolerant microbial cellulase-free xylanases produced from agricultural wastes and the properties required for use in pulp bleaching bioprocesses: a review. Process Biochem. 2003;38:1327–40.
9. Verma D, Satyanarayana T. Cloning, expression and applicability of thermo-alkali-stable xylanase of *Geobacillus thermoleovorans* in generating xylooligosaccharides from agro-residues. Bioresour Technol. 2012;107:333–8.
10. Bae HJ, Kim HJ, Kim YS. Production of a recombinant xylanase in plants and its potential for pulp biobleaching applications. Bioresour Technol. 2008;99:3513–9.
11. Hu J, Arantes V, Saddler JN. The enhancement of enzymatic hydrolysis of lignocellulosic substrates by the addition of accessory enzymes such as xylanase: is it an additive or synergistic effect? Biotechnol Biofuels. 2011;4:36.
12. Henrissat B, Davies G. Structural and sequence-based classification of glycoside hydrolases. Curr Opin Struct Biol. 1997;7:637–44.
13. Cantarel BL, Coutinho PM, Rancurel C, Bernard T, Lombard V, Henrissat B. The Carbohydrate-Active EnZymes database (CAZy): an expert resource for Glycogenomics. Nucleic Acids Res. 2009;37:D233–8.
14. Goncalves TA, Damasio ARL, Segato F, Alvarez TM, Bragatto J, Brenelli LB, et al. Functional characterization and synergic action of fungal xylanase and arabinofuranosidase for production of xylooligosaccharides. Bioresour Technol. 2012;119:293–9.
15. Cheng YF, Yang CH, Liu WH. Cloning and expression of *Thermobifida* xylanase gene in the methylotrophic yeast *Pichia pastoris*. Enzyme Microb Tech. 2005;37:541–6.
16. Bachmann SL, McCarthy AJ. Purification and cooperative activity of enzymes constituting the xylan-degrading system of *Thermomonospora fusca*. Appl Environ Microb. 1991;57:2121–30.
17. Irwin D, Jung ED, Wilson DB. Characterization and sequence of a *Thermomonospora fusca* xylanase. Appl Environ Microb. 1994;60:763–70.
18. Romanos MA, Scorer CA, Clare JJ. Foreign gene expression in yeast: a review. Yeast. 1992;8:423–88.
19. Ruanglek V, Sriprang R, Ratanaphan N, Tirawongsaroj P, Chantasigh D, Tanapongpipat S, et al. Cloning, expression, characterization, and high cell-density production of recombinant endo-1,4-beta-xylanase from *Aspergillus niger* in *Pichia pastoris*. Enzyme Microb Tech. 2007;41:19–25.
20. Wang Y, Zhang H, He Y, Luo H, Yao B. Characterization, gene cloning, and expression of a novel xylanase XYNB from *Streptomyces olivaceoviridis* A1. Aquaculture. 2007;267:328–34.
21. Jeya M, Thiagarajan S, Lee JK, Gunasekaran P. Cloning and expression of GH11 xylanase gene from *Aspergillus fumigatus* MKU1 in *Pichia pastoris*. J Biosci Bioeng. 2009;108:22–9.
22. Krainer F, Dietzsch C, Hajek T, Herwig C, Spadiut O, Glieder A. Recombinant protein expression in Pichia pastoris strains with an engineered methanol utilization pathway. Microb Cell Fact. 2012;11:22.
23. Vanz AL, Lunsdorf H, Adnan A, Nimtz M, Gurramkonda C, Khanna N, et al. Physiological response of Pichia pastoris GS115 to methanol-induced high level production of the Hepatitis B surface antigen: catabolic adaptation, stress responses, and autophagic processes. Microb Cell Fact. 2012;11:103.
24. Varnai A, Tang C, Bengtsson O, Atterton A, Mathiesen G, Eijsink V. Expression of endoglucanases in Pichia pastoris under control of the GAP promoter. Microb Cell Fact. 2014;13:57.
25. He J, Yu B, Zhang KY, Ding XM, Chen DW. Expression of endo-1, 4-beta-xylanase from Trichoderma reesei in Pichia pastoris and functional characterization of the produced enzyme. BMC Biotechnol. 2009;9:56–66.
26. Huang JL, Wang GX, Xiao L. Cloning, sequencing and expression of the xylanase gene from a *Bacillus subtilis* strain B10 in *Escherichia coli*. Bioresour Technol. 2006;97:802–8.
27. Baez A, Majdalani N, Shiloach J. Production of recombinant protein by a novel oxygen-induced system in Escherichia coli. Microb Cell Fact. 2014;13:50.
28. Songsiriritthigul C, Buranabanyat B, Haltrich D, Yamabhai M. Efficient recombinant expression and secretion of a thermostable GH26 mannan endo-1,4-beta-mannosidase from Bacillus licheniformis in Escherichia coli. Microb Cell Fact. 2010;9:20.
29. Li PZ, Anumanthan A, Gao XG, Ilangovan K, Suzara VV, Duezguenes N, et al. Expression of recombinant proteins in *Pichia Pastoris*. Appl Biochem Biotechnol. 2007;142:105–24.
30. Zhao XH, Wang W, Wang FQ, Wei DZ. A comparative study of beta-1, 4-endoglucanase (possessing beta-1, 4-exoglucanase activity) from *Bacillus subtilis* LH expressed in *Pichia pastoris* GS115 and *Escherichia coli* Rosetta (DE3). Bioresour Technol. 2012;110:539–45.
31. Zhang F, Chen JJ, Ren WZ, Nie GX, Ming H, Tang SK, et al. Cloning, expression and characterization of an alkaline thermostable GH9 endoglucanase from *Thermobifida halotolerans* YIM 90462(T). Bioresour Technol. 2011;102:10143–6.
32. Wu JW, Filutowicz M. Hexahistidine (His_6)-tag dependent protein dimerization: A cautionary tale. Acta Biochim Pol. 1999;46:591–9.
33. Cereghino JL, Cregg JM. Heterologous protein expression in the methylotrophic yeast *Pichia pastoris*. FEMS Microbiol Rev. 2000;24:45–66.
34. Grinna LS, Tschopp JF. Size distribution and general structural features of N-linked oligosaccharides from the methylotrophic yeast, *Pichia pastoris*. Yeast. 1989;5:107–15.
35. Zhang D, Lax AR, Raina AK, Bland JM. Differential cellulolytic activity of native-form and C-terminal tagged-form cellulase derived from *Coptotermes formosanus* and expressed in *E. coli*. Insect Biochem Mol Biol. 2009;39:516–22.
36. Sun JY, Liu MQ, Weng XY, Qian LC, Gu SH. Expression of recombinant *Thermomonospora fusca* xylanase A in *Pichia pastoris* and xylooligosaccharides released from xylans by it. Food Chem. 2007;104:1055–64.
37. Sevo M, Degrassi G, Skoko N, Venturi V, Ljubijankic G. Production of glycosylated thermostable Providencia rettgeri penicillin G amidase in Pichia pastoris. FEMS Yeast Res. 2002;1:271–7.
38. van Bueren AL, Otani S, Friis EP, Wilson KS, Davies GJ. Three-dimensional structure of a thermophilic family GH11 xylanase from Thermobifida fusca. Acta Crystallogr F. 2012;68:141–4.
39. Van Wyk N, Den Haan R, Van Zyl WH. Heterologous co-production of *Thermobifida fusca* Cel9A with other cellulases in *Saccharomyces cerevisiae*. Appl Microbiol Biotechnol. 2010;87:1813–20.
40. Bailey MJ, Biely P, Poutanen K. Interlaboratory testing of methods for assay of xylanase activity. J Biotechnol. 1992;23:257–70.
41. Miller GL. Use of Dinitrosalicylic acid reagent for determination of reducing sugar. Anal Chem. 1959;31:426–8.
42. Chen P, Buller CS. Activity staining of xylanases in polyacrylamide gels containing xylan. Anal Biochem. 1995;226:186–8.
43. Arnold K, Bordoli L, Kopp J, Schwede T. The SWISS-MODEL workspace: a web-based environment for protein structure homology modelling. Bioinformatics. 2006;22:195–201.
44. Biasini M, Bienert S, Waterhouse A, Arnold K, Studer G, Schmidt T, et al. SWISS-MODEL: modelling protein tertiary and quaternary structure using evolutionary information. Nucleic Acids Res. 2014;42:W252–8.
45. Zhang M, Jiang Z, Yang S, Hua C, Li L. Cloning and expression of a *Paecilomyces thermophila* xylanase gene in *E. coli* and characterization of the recombinant xylanase. Bioresour Technol. 2010;101:688–95.

13

Overexpression of *STARCH BRANCHING ENZYME II* increases short-chain branching of amylopectin and alters the physicochemical properties of starch from potato tuber

David A Brummell[1*], Lyn M Watson[1], Jun Zhou[1], Marian J McKenzie[1], Ian C Hallett[2], Lyall Simmons[3], Margaret Carpenter[3] and Gail M Timmerman-Vaughan[3]

Abstract

Background: Starch is biosynthesised by a complex of enzymes including various starch synthases and starch branching and debranching enzymes, amongst others. The role of all these enzymes has been investigated using gene silencing or genetic knockouts, but there are few examples of overexpression due to the problems of either cloning large genomic fragments or the toxicity of functional cDNAs to bacteria during cloning. The aim of this study was to investigate the function of potato *STARCH BRANCHING ENZYME II* (*SBEII*) using overexpression in potato tubers.

Results: A hybrid *SBEII* intragene consisting of potato cDNA containing a fragment of potato genomic DNA that included a single intron was used in order to prevent bacterial translation during cloning. A population of 20 transgenic potato plants exhibiting *SBEII* overexpression was generated. Compared with wild-type, starch from these tubers possessed an increased degree of amylopectin branching, with more short chains of degree of polymerisation (DP) 6–12 and particularly of DP6. Transgenic lines expressing a *GRANULE-BOUND STARCH SYNTHASE* (*GBSS*) RNAi construct were also generated for comparison and exhibited post-transcriptional gene silencing of *GBSS* and reduced amylose content in the starch. Both transgenic modifications did not affect granule morphology but reduced starch peak viscosity. In starch from *SBEII*-overexpressing lines, the increased ratio of short to long amylopectin branches facilitated gelatinisation, which occurred at a reduced temperature (by up to 3°C) or lower urea concentration. In contrast, silencing of *GBSS* increased the gelatinisation temperature by 4°C, and starch required a higher urea concentration for gelatinisation. In lines with a range of *SBEII* overexpression, the magnitude of the increase in SBEII activity, reduction in onset of gelatinisation temperature and increase in starch swollen pellet volume were highly correlated, consistent with reports that starch swelling is greatly dependent upon the amylopectin branching pattern.

Conclusion: This work reports the first time that overexpression of *SBEII* has been achieved in a non-cereal plant. The data show that overexpression of *SBEII* using a simple single-intron hybrid intragene is an effective way to modify potato starch physicochemical properties, and indicate that an increased ratio of short to long amylopectin branches produces commercially beneficial changes in starch properties such as reduced gelatinisation temperature, reduced viscosity and increased swelling volume.

Keywords: Amylopectin branching, *GBSS* silencing, *SBEII* overexpression, Starch gelatinisation, Starch properties, *Solanum tuberosum*

* Correspondence: david.brummell@plantandfood.co.nz
[1]The New Zealand Institute for Plant & Food Research Limited, Food Industry Science Centre, Private Bag 11600, Palmerston North 4442, New Zealand
Full list of author information is available at the end of the article

Background

Starch serves as the most important carbohydrate storage reserve for plants, and provides the foundation for human and animal nutrition as well as a raw material for numerous food and industrial applications [1]. Starch consists of two molecules, long essentially linear chains of α-(1 → 4)-linked glucopyranosyl residues (amylose), and a much larger, highly branched molecule (amylopectin) consisting of thousands of short α-(1 → 4)-linked glucan chains of degree of polymerisation (DP) 6 to >100 that are attached by α-(1 → 6)-linked branch points [2].

Amylose and amylopectin are synthesised in the plastids, where they assemble into a semicrystalline granule. Amylopectin is the major component, making up 70–80% of starch in most species. Amylose is synthesised by granule-bound starch synthase (GBSS), whereas a large complex of enzymes is required to synthesise amylopectin. This complex consists of four soluble starch synthases (SSI, SSII, SSIII, SSIV) and two types of starch branching enzyme (SBEI, SBEII), with various debranching enzymes, kinases and other enzymes also involved [2-4]. The availability of various mutants lacking particular enzyme activities has shown how the properties of starch from different species are altered by changes to composition or structure. For example, starch from the *waxy* mutant of maize (*Zea mays* L.) (deficient in GBSS and virtually amylose-free) had a lower gelatinisation temperature, whereas that from wheat (*Triticum aestivum* L.) had an unaltered gelatinisation temperature but altered pasting properties [5]. Starch from potato (*Solanum tuberosum* L.) where *GBSS* had been down-regulated produced gels with improved stability and clarity [6].

Differences in amylopectin branching affect granule crystallinity, which together with differences between species in granule size and shape result in altered thermal, pasting and biophysical properties [7,8]. Amylopectin branching is carried out by SBEI and SBEII, enzymes that create branch points by cleaving α-(1 → 4)-linkages and reattaching the glucan chain via an α-(1 → 6)-linkage. Alterations in SBE activity affect the number and size distribution of amylopectin branches. In cereals, there are three isoforms of SBE: SBEIa, SBEIIa and SBEIIb, and in the *amylose extender* mutant of maize that lacks SBEIIb, amylopectin had fewer branches and a higher proportion of longer chains [7] that are essentially amylose. Similarly in rice (*Oryza sativa* L.), down-regulation of *SBEIIb* decreased the proportion of short branches and increased the proportion of intermediate and long branches [9]. In barley (*Hordeum vulgare* L.), only when both *SBEIIa* and *SBEIIb* were down-regulated was a significant reduction in the frequency of amylopectin branches observed [10]. Indeed, down-regulation of all three isoforms of SBE in barley produced starch that was entirely amylose [11]. *In vitro* experiments with rice enzymes found that SBEIIb transferred chains of DP6 and DP7, SBEIIa transferred chains of DP6–15, and SBEIa produced a range of short and intermediate chains of DP ≤ 40 [12].

In potato only two isoforms of SBE are involved, SBEI (also known as SBE B) and SBEII (also known as SBE A) [13,14]. SBEI is the major isoform, but its down-regulation had little effect on starch structure [15,16]. Knockdown of the minor SBEII isoform, however, increased the average chain length of amylopectin, which, as in the maize *amylose extender* mutant [17], resulted in an increase in apparent amylose content due to the increased proportion of longer linear amylopectin chains [14]. Knockdown of both *SBEI* and *SBEII* produced potato starch that was >70% apparent amylose and had a very high phosphorus content [18].

Overexpression of SBE activity to modify starch structure is a much more difficult proposition than knockdown, due to the enzyme's deleterious effects on bacteria during cloning of the constructs [14]. Overexpression of bacterial glycogen synthase or glycogen branching enzyme produced starch with more branches and more short chains [19-21], but only one study has successfully overexpressed a plant SBE enzyme. This involved the cloning of an 18 kb genomic fragment containing the native promoter and coding region, including 21 introns, of the rice *SBEIIb* gene, and expressing it in a rice *SBEIIb*-deficient background [22]. We aimed to examine the role of potato SBEII in starch branching and resultant effects on potato starch properties. Potato is now the third most important food crop in the world and the most important non-cereal [23], whose starch also has many uses in the food and manufacturing industries [1] due to its advantages over cereal starches (see Discussion). To do this we employed a simple hybrid cDNA/gDNA intragene construct containing a single intron to prevent bacterial translation, expressed under the control of the strong tuber-specific potato *GBSS* promoter. An additional way to increase the proportion of branched amylopectin in the granule is to down-regulate *GBSS* and reduce amylose content, so knockdown *GBSS* lines were also generated for comparison.

Results

Characterisation of transgenic potato plants

Initial experiments confirmed an earlier report that a complete cDNA of potato *SBEII* was unable to be propagated in *Escherichia coli* [14]. Despite the use of several different bacterial strains, very few colonies were produced and all were subsequently discovered to contain frame-shift mutations (data not shown). Therefore, a complete but hybrid *SBEII* intragene containing a single intron to prevent bacterial translation was assembled from cDNA and genomic DNA fragments (See Methods and Additional file 1: Figure S1). A population of 28 plants transformed with the *SBEII-intron* intragene under the control of the *GBSS* promoter was generated (Additional file 1: Figure S2). Of these, 20 plantlets showing good root development were grown to maturity, and tubers exhibited a range of *SBEII* mRNA abundances (Figure 1A). *SBEII* mRNA accumulation in these lines ranged from less than wild-type controls (line 1047–3), moderate overexpression (lines 1047–1, 2, 4, 5, 6, 7, 8, 9, 11, 14, 18, 20, 22, 24) to high overexpression (lines 1047–12, 15, 17, 23). Line 1047–19 exhibited approximately wild-type mRNA abundance of *SBEII* but with an additional anomalous larger *SBEII* transcript presumably derived from an aberrant insertion event. To confirm that mRNA accumulation of the structurally unrelated starch synthase enzymes SSII and SSIII was not affected by manipulation of *SBEII* expression, the same gel blot was stripped and re-hybridised. Sequential re-hybridisation with labelled probes for *SSII* and *SSIII* showed that the mRNA abundances of these two genes were essentially indistinguishable from wild-type and empty vector controls (Additional file 1: Figure S3).

Of a population of 24 plants putatively transformed with a *GBSS* inverted repeat transgene, only 12 were positive for the presence of both the transgene and the selectable marker gene (Additional file 1: Figure S2). Of these 12 lines, 11 showed a strong reduction in *GBSS* mRNA accumulation (Figure 1B). One line, 1041–4, did not show any apparent posttranscriptional gene silencing, and this line was retained as an additional control for some tests since it was transgenic but *GBSS* mRNA accumulation was not reduced relative to wild-type.

To assess the effect of overexpression of a *SBEII* intragene on SBE enzyme activity, SBEII was separated from the major and predominant SBEI by separating proteins using native polyacrylamide gel electrophoresis (PAGE), followed by an in-gel activity reaction and staining for SBE reaction products (Figure 2A). The two isoforms of SBE were identified based on the migration of rice and wheat enzymes [17,22,24] and the relative activities of potato SBEI and SBEII [14]. SBEII activity approximately reflected the abundance of *SBEII* mRNA, being similar to wild-type in line 1047–3, and with a range of activities greater than wild-type in a selection of other lines (WT < 1047–5 < 1047–23 = 1047–12 ≤ 1047–17 ≤ 1047–15). The greatest enhancement was in lines 1047–15 and 17, which possessed SBEII activity almost 3-fold that of wild-type (Figure 2B).

Effects of altered *GBSS* and *SBEII* expression on the structure of starch

The effectiveness of silencing of *GBSS* was determined by measuring the percentage of amylose present in the starch from these plants (Table 1). Wild-type and empty vector control starches were approximately 19–26% amylose, but this was reduced to 3.1–5.6% in lines 1041–1, 3 and 16. In line 1041–4, which showed no reduction in *GBSS* mRNA abundance relative to controls (Figure 1B), amylose comprised 21.3% of the total starch, a content within the range found in controls. The percentage of amylose in the 1047 lines overexpressing *SBEII* ranged from 17.0 to 23.4%, and was similar to controls.

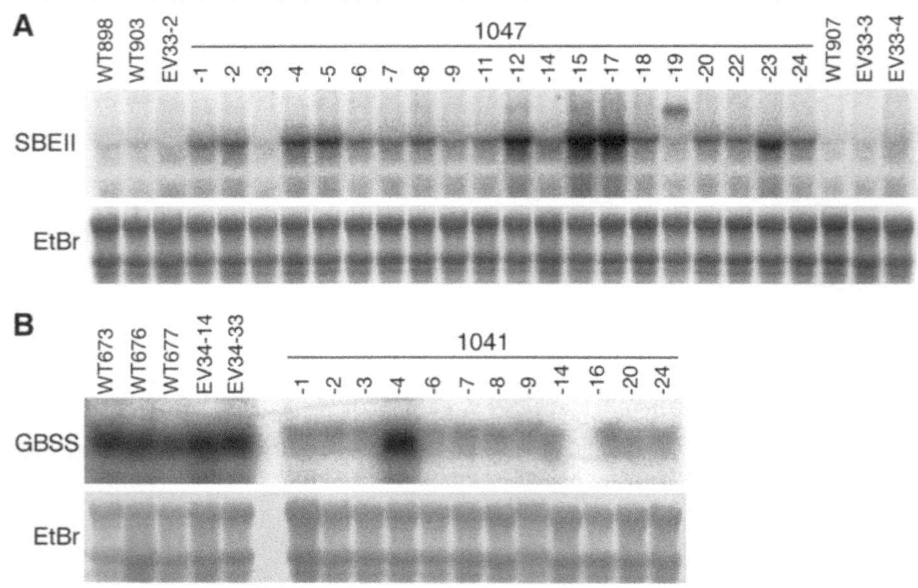

Figure 1 RNA gel blots showing alterations to *SBEII* and *GBSS* mRNA abundance in transgenic plants. **(A)** *SBEII* mRNA abundance in tubers from plants transformed with a *SBEII* overexpression construct (1047 lines). **(B)** *GBSS* mRNA abundance in tubers from plants transformed with a *GBSS* RNAi construct (1041 lines). Wild-type (WT) and empty vector transformants (EV) are shown for comparison. Ethidium bromide (EtBr) staining of RNA (with contrast reversed) was used to confirm the equivalency of loading.

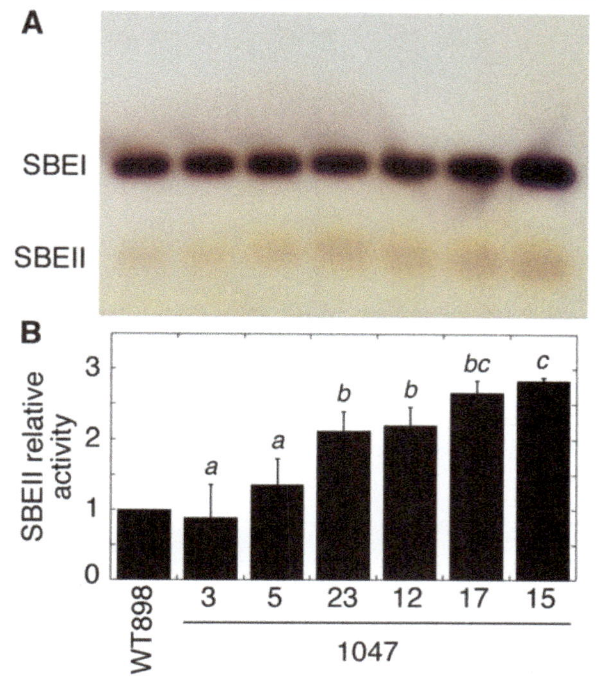

Figure 2 SBE activity in tubers from a series of lines transformed with a *SBEII* overexpression construct. **(A)** Native PAGE activity gel with SBE activity visualized by iodine staining. **(B)** Mean relative SBEII activity ± SD as determined from the density of the spots. Data are from technical replicate protein preparations from the same pooled tissue as Figure 1 run on different gels on different days (n = 3). Means not sharing a common letter are significantly different between groups at $P = 0.05$ as determined by LSD after a one-way ANOVA test.

Table 1 Amylose content in starch from potato tubers of a range of lines exhibiting silencing of *GBSS* (1041 lines) or overexpressing *SBEII* (1047 lines)

Line	% Amylose
WT676	26.3 ± 0.3
WT898	21.4 ± 0.3
WT903	21.5 ± 0.9
EV33-3	20.6 ± 0.2
EV33-5	18.8 ± 0.8
EV34-33	22.2 ± 0.2
1041-1	4.9 ± 0.3
1041-3	3.1 ± 1.0
1041-4	21.3 ± 0.2
1041-16	5.6 ± 0.3
1047-3	19.2 ± 0.3
1047-5	17.0 ± 0.3
1047-12	17.9 ± 0.6
1047-15	23.4 ± 0.8
1047-17	19.4 ± 0.3
1047-23	22.2 ± 0.8

Starch was also prepared and assayed from wild-type (WT) and empty vector (EV) control plants grown alongside the transgenic lines. Data are means ± SD (n = 4 technical replicates).

Size exclusion chromatography was used to separate amylose from amylopectin (Figure 3). Elution profiles consisted of two main regions, a sharp peak eluting at 9–11 min representing high molecular weight amylopectin, and a broader more disperse lower molecular weight peak eluting at 12–17 min representing amylose [25-27]. The small peaks eluting at 18–19 min were probably due to free sugars. The relative areas of the amylopectin and amylose peaks showed that suppression of *GBSS* in line 1041–3 substantially reduced the content of amylose and that overexpression of *SBEII* in line 1047–17 slightly reduced the content of amylose, consistent with the biochemical assays (Table 1). The amylopectin peaks for both transgenic lines eluted at slightly higher molecular weight than wild-type.

A modified fluorophore-assisted carbohydrate electrophoresis (FACE) method was used to examine the effects of altered *SBEII* and *GBSS* expression on the chain length distribution of debranched polyglucans (Figure 4). Polyglucans from wild-type showed a moderate number of side chains of DP6, few side-chains of DP7-9, and a predominant peak of side chains spanning the DP ~ 10-24 range (Figure 4A, C). The number of side chains of DP ≥ 37 was small. Overexpression of *SBEII* in various lines had little effect on the chain length distribution of moderately sized chains of DP13-18, slightly reduced the abundance of larger chains of DP ≥ 20 and increased the number of short chains of DP ≤ 12 (Figure 4B). The greatest effect was on the number of chains of DP6, with the increase in lines 1047–5, 12, 15, 17 and 23 ranging from 19 to 56% greater than wild-type (Figure 4A). However, there was no obvious correlation between the increase in DP6 chain abundance and relative SBEII activity. The only

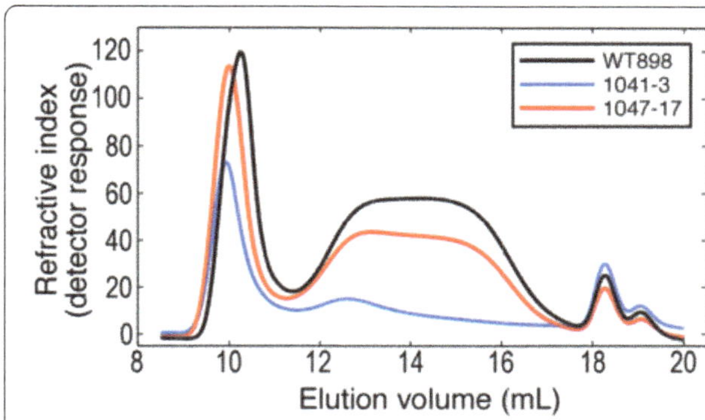

Figure 3 Size separation of tuber starch polyglucans. Starch was solubilised in NaOH and gelatinised by heat before size exclusion chromatography on a Waters Ultrahydrogel Linear column. Curves are the means of measurements on three independent biological replicates per line. Carbohydrate in the eluate was measured by refractive index.

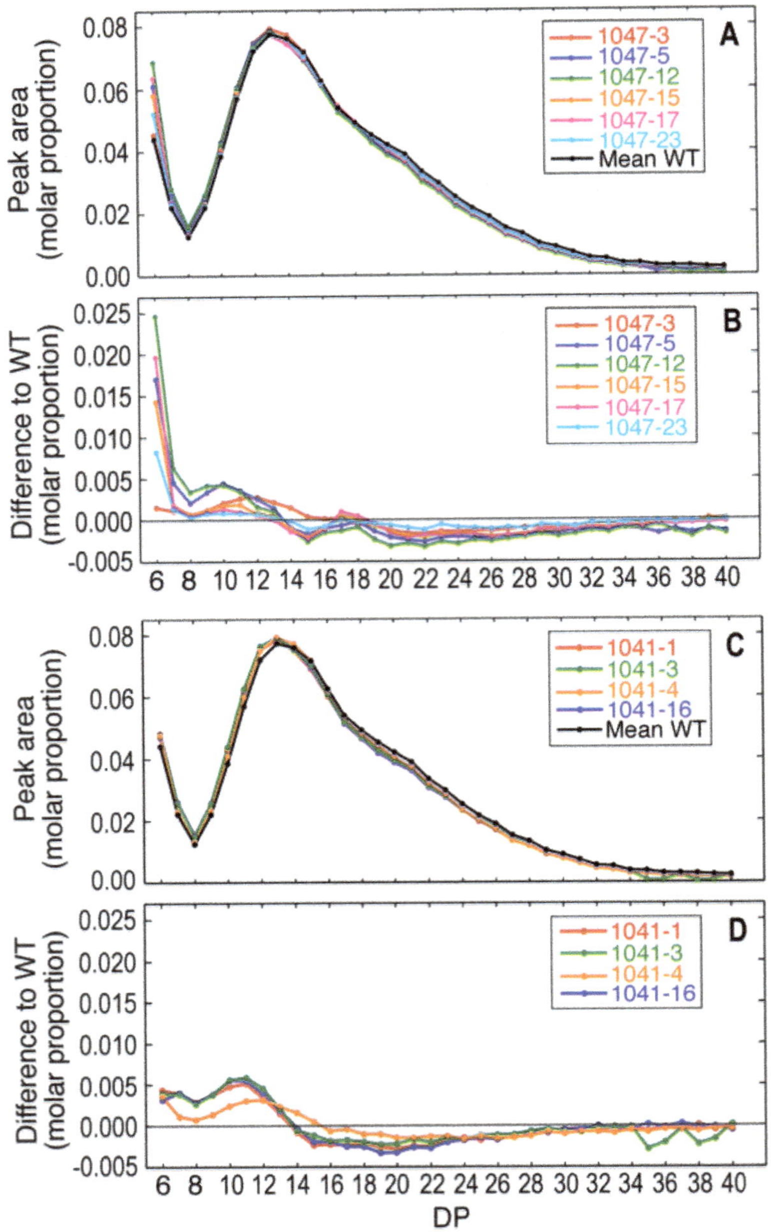

Figure 4 Chain length distribution of amylopectin unit chains. **(A)** Comparison of size distribution of amylopectin unit chains in starch from tubers with altered *SBEII* expression. **(B)** Difference plot showing changes in chain length distribution relative to wild-type. **(C)** Comparison of size distribution of amylopectin unit chains in starch from tubers with altered *GBSS* expression. **(D)** Difference plot showing changes in chain length distribution relative to wild-type.

line in which an increase in the abundance of DP6 side chains was not observed was line 1047–3 (Figure 4B), which was the line in which SBEII mRNA abundance and activity were not increased (Figures 1 and 2).

In transgenic lines with a higher proportion of amylopectin in the granule due to reduced amylose content (lines 1041–1, 3, 16), the chain length distribution of polyglucans was very similar to wild-type (Figure 4C). There was a small increase in the number of side chains of DP6-13, and a small decrease in those of DP15-28 (Figure 4D), changes that were much less pronounced in line 1041–4 that did not exhibit silencing of *GBSS* (Figure 1). This suggests that the reduced activity of GBSS not only increased the proportion of amylopectin in starch (by reducing the biosynthesis of amylose), but also had some small effects on its structure.

Amylopectin chain length distribution and the phosphorus content of the starch are related, with phosphorylation predominantly occurring on longer amylopectin chains [28]. Many studies of starch phosphorylation underestimate the true abundance of phosphate groups by measuring only C6-linked phosphates, and ignoring the more acid-labile C3 linkage. We have used an improved method that allows determination of phosphate groups with both linkages. The ratio of C6 phosphate content to C3 phosphate content was ~4.5:1 in all lines tested (Table 2). Total phosphate content was ~40 nmol per mg starch in a wild-type line, an empty vector transformant and two lines with knockdown of *GBSS* expression. However, total phosphate content was reduced by ~25% in two lines in which *SBEII* expression was increased, with both C6- and C3-linked phosphate content being reduced.

Effects of altered *GBSS* and *SBEII* expression on the morphology of starch granules

To test whether the morphology of the starch granules was altered by the transgenic modifications, a range of microscopy techniques was employed (Figure 5). Brightfield microscopy showed a wide range of granule sizes in wild-type starch (Figure 5A), and neither knockdown of *GBSS* nor overexpression of *SBEII* had any obvious effect on the size range distribution of the granules. Granules from wild-type and *SBEII* overexpressors stained dark blue with iodine (Figure 5B), whereas granules from *GBSS* knockdown plants showed the pale purple colour typical of a low amylose content but with visible dark blue staining at the hilum core. Examination under polarised light detected birefringence visible as a 'Maltese cross' shape and blue and yellow sectors (Figure 5C). Birefringence was higher in the *GBSS* knockdown line, presumably since these lines are higher in amylopectin content, which has a more ordered orientation (crystallinity) than amylose. An increased birefringence in the *GBSS* knockdown line was also observed using differential interference contrast microscopy (Figure 5D). Variable pressure scanning electron microscopy showed that the external morphology of the starch granules was not affected by *GBSS* knockdown or by *SBEII* overexpression (Figure 5E).

Table 2 Phosphate content of tuber starches from lines exhibiting silencing of *GBSS* (1041 lines) or overexpressing *SBEII* (1047 lines)

Line	C6 PO_4	C3 PO_4	Total
WT898	32.8 ± 1.8	7.5 ± 0.6	40.3
EV33-5	29.0 ± 2.7	6.4 ± 0.6	35.4
1041-3	35.2 ± 6.1	8.8 ± 1.9	44.0
1041-16	31.7 ± 1.6	7.1 ± 0.5	38.8
1047-15	24.1 ± 2.5	5.1 ± 0.3	29.1
1047-17	24.5 ± 2.1	5.8 ± 0.1	30.2

WT: wild-type; EV: empty vector. Data are mean nmol Glc6P per mg starch ± SD and mean nmol Glc3P per mg starch ± SD (n = 3 biological replicates). Total nmol phosphate per mg starch was calculated as the sum of C6 plus C3 phosphates.

Details of the internal structure of the granules were examined by fluorescent labelling with 8-amino-1,3,6-pyrenetrisulfonic acid (APTS) combined with confocal laser scanning microscopy (Figure 5F). APTS labels the single reducing end of each molecule with high efficiency, and since amylose molecules are much smaller than the highly branched amylopectin molecules, they exhibit a higher fluorescence (on a fluorescently-labelled reducing end per total Glc residue basis). A small intensely labelled region at the hilum core was visible in all the lines, indicating a high concentration of amylose at this site. This was observed even in the low amylose *GBSS* knockdown line, and is consistent with the iodine staining of this site seen under brightfield microscopy. Growth rings were more visible in the wild-type, suggesting that the deposition of amorphous and crystalline growth rings may have been altered in the transgenic lines. However, overall fluorescence was lower in both of the transgenic lines, and was too low for clear visualisation of growth rings in the *SBEII* overexpressors.

Thermal and gelatinisation properties of transgenically modified starch

The different contents of amylose, short-branched amylopectin and phosphorus in the starch would be expected to result in altered thermal and gelatinisation properties. Both knockdown of *GBSS* expression and overexpression of *SBEII* markedly altered the onset of gelatinisation temperature (T_o), but in opposite ways (Table 3). In wild-type and empty vector controls, T_o ranged from 54.2 to 55.4°C, whereas in three lines with reduced *GBSS* expression and lowered amylose content (1041–1, 3 and 16), T_o was increased by ~4°C, to 59.3 to 60.0°C. In the transgenic line 1041–4, with unaltered *GBSS* expression, T_o was not increased relative to controls. In contrast, overexpression of *SBEII* in various 1047 lines reduced T_o by up to 3°C (Table 3). In these lines, a trend towards lowered T_o (WT > 1047–5 = 1047–23 > 1047–12 ≥ 1047–15 ≥ 1047–17) correlated strongly ($r = -0.931$) with the trend of increasing SBEII activity (Figure 2). In line 1047–3, which showed SBEII activity similar to wild-type, T_o was not reduced and was within the range of variability seen in control lines. The temperature of gelatinisation peak (T_p), temperature of gelatinisation conclusion (T_c) and gelatinisation enthalpy (ΔH) showed similar trends to T_o in all of the lines.

An examination of starch pasting properties using a rapid visco analyser (RVA) detected differences between the lines (Figure 6), including the lowered T_o of starch from line 1047–17 and the increased T_o of starch from line 1041–3. Other than the onset temperature, pasting profiles were similar in shape, but with a reduction in peak viscosity of ~10% for line 1047–17 and ~24% for line 1041–3. Final viscosities were also reduced in both transgenic lines.

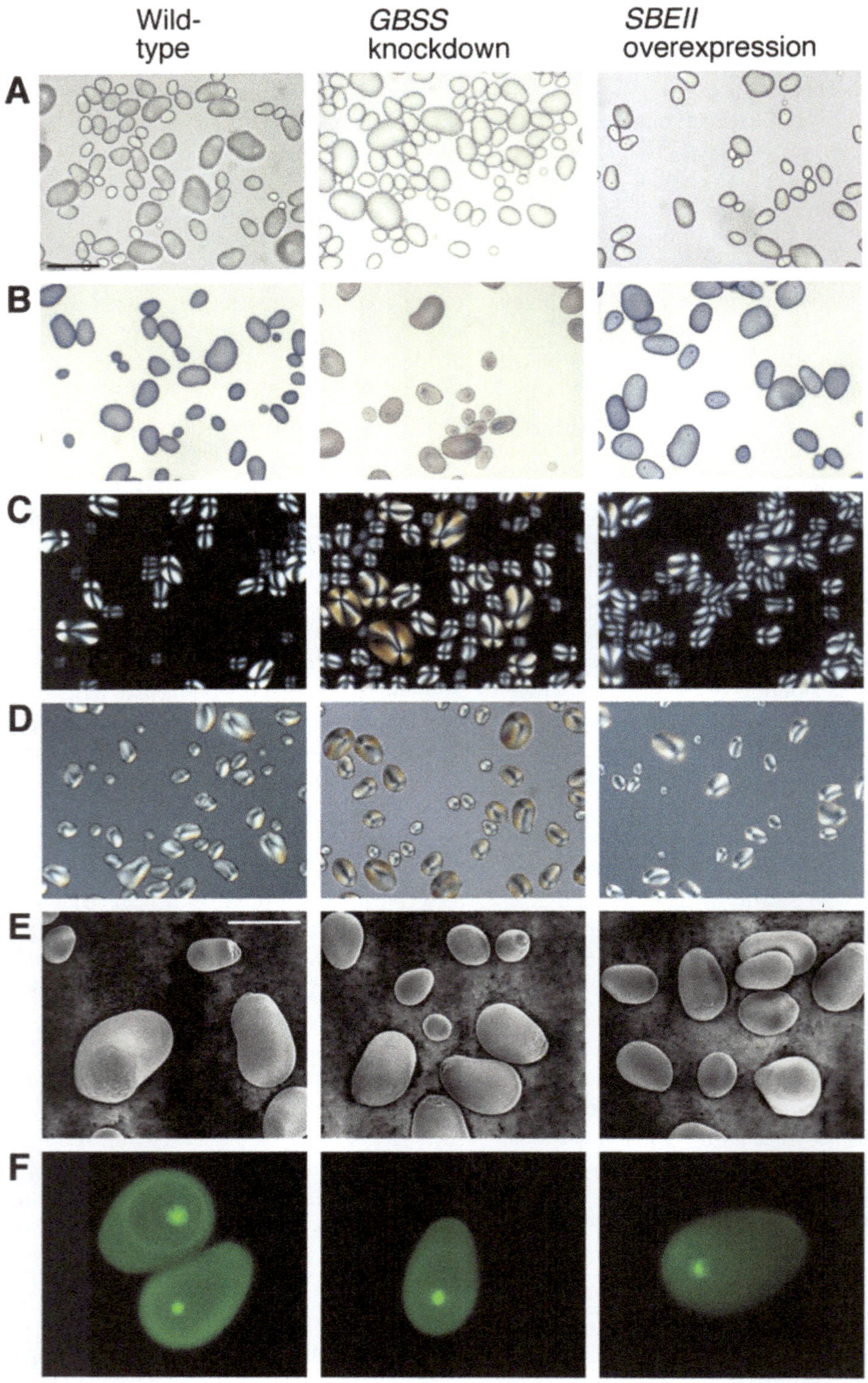

Figure 5 Images of tuber starch granules from wild-type and transgenically modified potato tubers. **(A)** Brightfield. **(B)** Brightfield, starch stained with I_2/KI. **(C)** Polarised light. **(D)** Differential interference contrast. **(E)** Variable pressure scanning electron microscopy. **(F)** Optical section of starch fluorescently labelled with APTS taken using confocal scanning laser microscopy. Lines used were WT898, 1041–3 and 1047–17. The scale bar (in panel **A**) for light microscopy pictures **(A–D)** represents 100 μm, and for the scanning electron microscopy **(E)** represents 40 μm.

Table 3 Thermal properties of tuber starch granules from plants exhibiting silencing of *GBSS* (1041 lines) or overexpressing *SBEII* (1047 lines)

Line	T_o (°C)	T_p (°C)	T_c (°C)	ΔH (J g^{-1})
WT676	54.7 ± 0.7^{e}	58.1 ± 0.4^{ef}	61.6 ± 0.0^{d}	15.4 ± 0.4^{bcde}
WT898	55.3 ± 0.2^{e}	58.5 ± 0.1^{ef}	62.1 ± 0.1^{d}	15.9 ± 0.2^{bcde}
WT903	55.4 ± 0.6^{e}	58.6 ± 0.3^{ef}	61.6 ± 0.0^{d}	15.7 ± 0.1^{bcde}
EV33-3	54.2 ± 0.7^{cd}	57.7 ± 0.4^{d}	61.6 ± 0.5^{d}	15.8 ± 2.7^{bcde}
EV33-5	55.4 ± 0.6^{e}	58.7 ± 0.4^{fg}	62.1 ± 0.1^{d}	15.0 ± 0.5^{abcd}
EV34-33	54.9 ± 0.2^{de}	58.2 ± 0.1^{de}	61.7 ± 0.0^{d}	14.9 ± 0.5^{abc}
1041-1	59.3 ± 0.1^{f}	63.1 ± 0.2^{h}	67.5 ± 0.1^{f}	18.2 ± 0.1^{f}
1041-3	59.6 ± 0.6^{f}	63.8 ± 0.4^{i}	68.7 ± 0.3^{g}	17.2 ± 1.5^{ef}
1041-4	53.8 ± 0.5^{c}	57.8 ± 0.2^{d}	61.9 ± 0.6^{d}	16.8 ± 1.7^{def}
1041-16	60.0 ± 0.4^{f}	63.4 ± 0.3^{hi}	67.9 ± 0.3^{f}	17.9 ± 0.4^{f}
1047-3	55.2 ± 0.3^{e}	59.1 ± 0.5^{g}	64.9 ± 1.2^{e}	16.6 ± 1.5^{cdef}
1047-5	53.6 ± 0.8^{c}	57.0 ± 0.7^{c}	60.9 ± 0.1^{c}	14.2 ± 1.6^{ab}
1047-12	52.8 ± 0.9^{b}	55.9 ± 0.5^{b}	59.1 ± 0.2^{ab}	13.7 ± 0.5^{a}
1047-15	52.3 ± 0.4^{ab}	55.6 ± 0.2^{b}	59.2 ± 0.2^{b}	14.5 ± 0.5^{ab}
1047-17	52.0 ± 0.3^{a}	55.0 ± 0.1^{a}	58.5 ± 0.2^{a}	14.6 ± 0.4^{ab}
1047-23	53.7 ± 0.2^{c}	56.7 ± 0.2^{c}	59.7 ± 0.3^{b}	15.0 ± 1.3^{abcd}

WT: wild-type; EV: empty vector. Data are means ± SD (n = 4 biological replicates). Means in the same column not sharing a common letter are significantly different between groups at $P = 0.05$ as determined by LSD after a one-way ANOVA test.
T_o = temperature of onset of gelatinisation peak.
T_p = temperature of gelatinisation peak.
T_c = temperature of gelatinisation conclusion.
ΔH = gelatinisation enthalpy of starch.

The greatest difference in physicochemical properties was seen in the gelatinisation of starch granules in urea. In 4 M urea, starch granules from line 1041–3 showed very little gelatinisation, with a swollen pellet volume only one-third that of wild-type starch granules (Figure 7A). Starch from lines 1041–1 and 1041–16 produced similar results (data not shown). In contrast, starch granules from line 1047–17 exhibited a huge amount of gelatinisation, producing a large and diffuse pellet with a swollen volume 3-fold greater than that of wild-type. Starch from line 1047–15 produced similar results (data not shown). Examination of a range of urea concentrations showed that gelatinisation was initiated at a lower urea concentration in starch granules from 1047–17, but followed a similar curve to wild-type and achieved a similar maximum swollen volume (Figure 7B). Gelatinisation followed a very different pattern in starch granules from 1041–3, beginning at a higher urea concentration and showing a remarkably steep increase in swollen pellet volume over a change in urea concentration from 4 M to 6 M. The maximum swelling volume achieved by 1041–3 was up to 25% greater than wild-type, and at urea concentrations above 7 M occupied the entire 1 mL volume of the original liquid provided.

The extent of *SBEII* overexpression had a clear effect on the swollen volume of the starch granules in 4 M

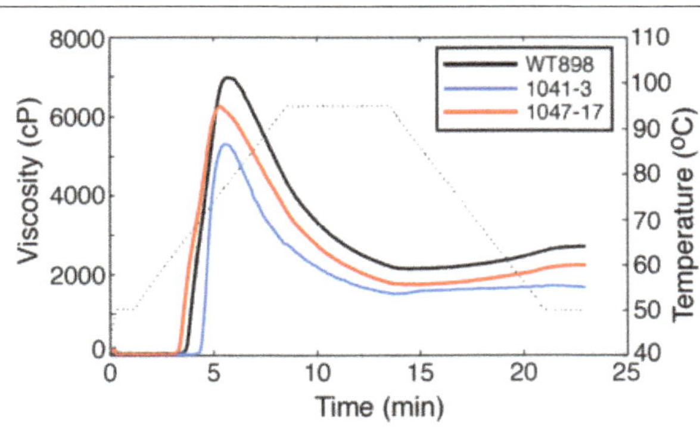

Figure 6 Comparison of pasting properties of tuber starch from wild-type and transgenic lines. Pasting properties were assessed in starch from lines exhibiting *GBSS* knockdown (1041–3) or *SBEII* overexpression (1047–17). Changes in viscosity of 7% (w/w) starch suspensions during heating and subsequent cooling were measured by RVA. Curves are the means of measurements on three independent biological replicates per line. The thin dotted line shows the change in temperature during measurement.

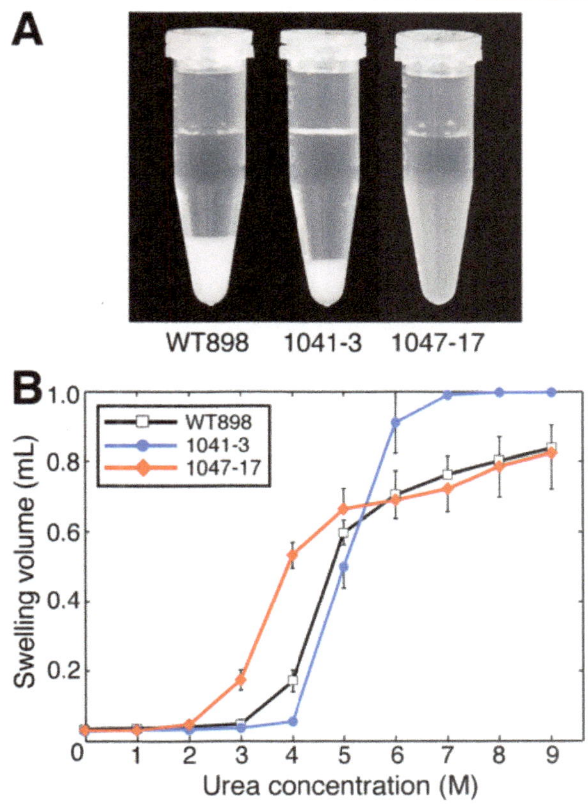

Figure 7 Changes in gelatinisation of tuber starch granules from transgenic potato lines. **(A)** Effect of *GBSS* knockdown (1041–3) or *SBEII* overexpression (1047–17) on starch granule swelling in 4 M urea. **(B)** Comparison of swelling of starch granules in a range of urea concentrations. Gelatinisation was assessed by measuring the volume of the swollen starch pellet. Data are mean pellet volumes ± SD (n = 4 biological replicates).

urea (Figure 8). In line 1047–3, swollen pellet volume was less than wild-type. However, the other lines formed a series of increasing swollen pellet volume, in the order WT < 1047–5 = 1047–23 < 1047–12 ≤ 1047–17 ≤ 1047–15. This order reflected the trend in increasing SBEII activity (Figure 2), confirming that SBEII activity and starch swelling power are positively correlated ($r = 0.939$). In contrast, the increase in swollen pellet volume was strongly but negatively correlated ($r = -0.976$) to the reduction of T_o (Table 3).

Discussion

The manipulation of starch properties is important commercially for various industrial applications, as well as for the production of food ingredients. Several loss-of-function mutants and RNAi knockdown studies have helped elucidate the role of particular enzymes in starch biosynthesis [3], but overexpression of starch biosynthetic enzymes is more difficult to accomplish and few examples exist. In one study, overexpression of *SBEIIb* in rice produced starch with increased branching, reduced T_o, increased solubility and increased swelling power [22]. However, potato starch has several advantages over cereal starch for commercial applications. Rice starch granules are small (around 5 μm) and polygonal with sharp edges [29], a high lipid content and a high T_o, whereas potato starch granules are of a large oval shape of average size 40 μm and are less crystalline than rice starch [8,30,31]. Starch from potato is also highly phosphorylated relative to cereal starch [32]. The large granule size, low crystallinity, low lipid content and high phosphate content give potato starch a low temperature for onset of gelatinisation, and a high swelling power and viscosity that are useful for both culinary and industrial applications.

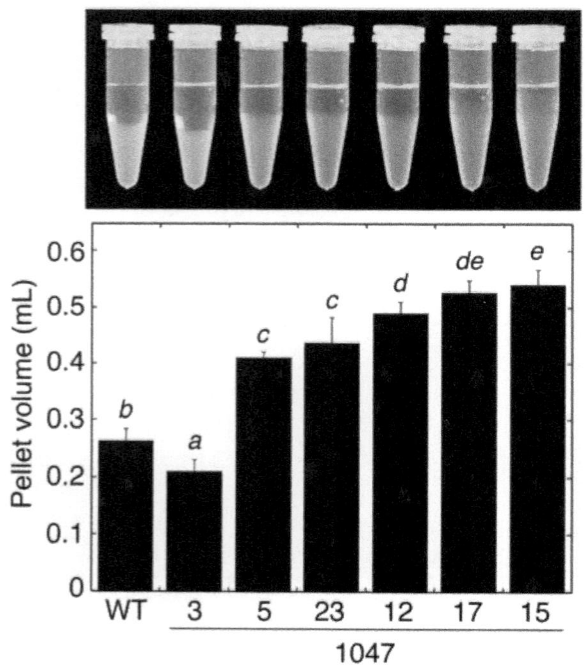

Figure 8 Effects of altered *SBEII* expression on the gelatinisation of tuber starch granules in 4 M urea. Data are mean pellet volumes ± SD (n = 3 biological replicates). Means not sharing a common letter are significantly different between groups at $P = 0.05$ as determined by LSD after a one-way ANOVA test. A photograph of one of the replicates is shown above the histogram.

Amylopectin chains have a polymodal size distribution, and models of amylopectin structure propose a long C chain that contains the single reducing terminal residue, long inner B chains that carry additional branches, and shorter, outer A chains that do not carry additional branches and are linked to the B chains [33]. Polymodal chain size distributions of DP 6–12, 13–24, 25–36 and ≥37 are believed to correspond to A chains (DP6–12), and various classes of longer B chains [34]. The abundance of short A chains of DP 6–12 probably determines the starch crystalline allomorph [31]. A range of *SBEII* overexpression levels increased the number of amylopectin chains of DP ≤ 12 and caused a particularly large increase in those of DP6 (Figure 4B). This did not result in any obvious difference in the shape or size of the granules or in their iodine staining properties (Figure 5), presumably because amylose content was not substantially affected (Table 1). Iodine staining is largely due to the content of amylose, which has a 20-fold stronger binding capacity for iodine than does amylopectin [26]. In contrast, knockdown of *GBSS* resulted in a pale purple colour upon iodine staining (Figure 5B), indicating a low amylose content, but with dark blue staining and intense fluorescence at the hilum (Figure 5B,F). A high concentration of amylose at the hilum core even in *GBSS*-silenced lines is consistent with previous observations [6,27]. Knockdown of *GBSS* did not bring about an alteration in the appearance of the granules (Figure 5), [27].

Starch gelatinisation, caused by heating starch in excess water (or exposure to chaotropic agents such as urea), is due to disruption of the molecular order in the crystalline regions and results in irreversible granular swelling and starch solubilisation. The temperature required for starch gelatinisation, and the swelling volume achieved, are affected by granule size and shape, the amylose:amylopectin ratio, amylopectin branching, phosphorus content and other factors. Overexpression of *SBEII* created amylopectin with a high abundance of short DP6 side chains and facilitated gelatinisation, enabling gelatinisation to occur at a lower temperature (Table 3) and at a lower concentration of urea (Figure 7). Granule size and shape were not affected by *GBSS* knockdown or *SBEII* overexpression, but amylose content, amylopectin branching and phosphorus content were altered. Starch phosphorylation occurs mainly on the longer branches of amylopectin, with only a minor proportion on short amylopectin chains, and is essentially absent on amylose [28,35]. Thus, knockdown of *GBSS* expression and the resulting low amylose content had little effect on the degree of starch phosphorylation (Table 2). In *SBEII* overexpression lines, the increased number of short branches of DP ≤ 12 and reduced number of longer branches of DP ≥ 20 should reduce total starch phosphorus content to some extent, and a 25% reduction was observed.

Studies on cereal starch have suggested that swelling power is exclusively due to the properties of the amylopectin, with amylose and lipids inhibiting swelling [36]. In *SBEII*-overexpressing line 1047–17, the swelling volume in 4 M urea was increased to approximately 65% of the maximal volume achieved (in 9 M urea), compared with only 20% for wild-type (Figure 7). A very low amylose content in line 1041–3 changed the gelatinisation response to urea concentration, and increased swelling power at >6 M urea (Figure 7), [17,37], but amylose content in the *SBEII* overexpressing lines was hardly different from wild-type (Table 1). This suggests that amylopectin branching pattern was the major factor influencing starch swelling power. Previous studies have found that T_o decreased when there was an increased frequency of short amylopectin branches, due either to knockdown of *isoamylase1* or overexpression of *SBEIIb* in rice [22,29], or knockdown of *SSII/SSIII* or overexpression of bacterial glycogen branching enzyme in potato [21,38]. The converse finding, that an increased T_p results from a reduction in the frequency of short branches due to knockdown of *SBEII* genes in rice and barley [9,10] or the *amylose extender* (knockout of *SBEIIb*) mutant of maize [7], confirms the influence of starch branching frequency on thermal properties.

In lines with a range of *SBEII* expression, there was a correlation between SBEII activity, the degree of reduction in T_o and the degree of increase in the swelling volume in 4 M urea. In other studies, comparisons between transgenic potato lines found that swelling power was decreased when either amylopectin branching was reduced (knockdown of *SBEI/SBEII*), or phosphorus content was reduced (knockdown of *GWD* (glucan, water dikinase)) [39]. A knockdown of *GWD* that reduced starch C6-linked phosphorus content by 85% had little effect on T_o orT_p, although swelling power was considerably reduced [39,40]. We observed increased swelling power in starch from our *SBEII* overexpressing lines, despite a 25% reduction in phosphorus content. Thus, increased amylopectin branching rather than small changes in amylose or phosphorus content most likely produced both the lowered T_o and increased swelling power. Consistent with the data presented here, increased branching in rice starch was also reported to increase swelling volume [22], although differences were not quantified and the swelling power of rice starch is very small relative to that of potato starch [41].

The present study shows that overexpression of potato *SBEII* increases the proportion of short chains of DP6-12, consistent with *in vitro* experiments that found potato SBEII transfers a high number of chains of DP6 [42] and knockdown of *SBEII* that caused an increase in average chain length [14]. We propose that this results in an abnormal amylopectin structure with reduced crystallinity, thus altering gelatinisation properties. In contrast, the increase in T_o found in the *GBSS* knockdown lines was most likely due to the reduced amylose content, since the degree of amylopectin branching and the phosphorus content were not substantially altered (Figure 4D, Table 2). An increase in the gelatinisation temperature of low-amylose potato starch has been noted previously [37,39].

Both genetic manipulations caused a lowered peak viscosity as determined by RVA, which measures changes in viscosity of the starch as it is heated and cooled under constant stirring (Figure 6). A rapid rise in viscosity upon heating caused by gelatinisation is followed by a decline as the amylopectin matrix structure is broken down by mechanical shearing. The two amylose-containing starches (WT898 and 1047–17) showed a higher peak viscosity and a slightly higher setback viscosity after cooling than line 1041–3. The reduced peak and final viscosities of low-amylose starches, such as that from 1041–3, are consistent with previous studies [15,43]. Starch from a line overexpressing *SBEII* also had a reduced viscosity, presumably due to the increased ratio of short to long amylopectin branches [21]. This is useful commercially since a reduction in the viscosity of potato starch is highly desirable in the production of manufactured products, in order to facilitate handling [37].

Conclusions

The data presented here show that overexpression of the potato *SBEII* gene can be achieved in potato, and that a relatively small single-intron construct obviates the difficulties of either manipulating a very large genomic fragment or cloning a functional *SBEII* cDNA gene in bacteria. The use of potato rather than cereals is advantageous since transformation is more facile and starch yields are high. The abundance of short amylopectin branches and thermal properties are highly correlated, and elsewhere highly branched starch with many short chains has been shown to have a lower gelatinisation temperature and improved rheological properties [44]. The T_o of native potato starch is up to 10°C lower than that of rice starch [41], and in the work presented here the increased abundance of short amylopectin branches has lowered it by another 3°C and reduced viscosity while increasing swelling power. Such starches may be expected to find many uses in the food and manufacturing industries.

Additionally, although modifying the amylopectin branching pattern *in planta* requires the use of genetic modification, starch produced in this way may replace chemically modified starch in food applications and thus remove environmental concerns and food additive labelling requirements. Previously, studies of overexpression of starch branching enzymes in potato have used bacterial enzymes [19-21]. Since the transgene used for overexpression in the current work consisted of a modified potato *SBEII* gene that did not contain any non-potato elements and was controlled by a potato promoter and terminator (both from the *GBSS* gene), the approach could be used to create entirely intragenic plants where all the transferred DNA (including T-DNA borders) is derived from the potato genome [45]. The use of intragenic technologies with all-potato DNA may ultimately remove the need for genetic modification labelling [45].

Methods

Construct assembly and transgenic plants

It did not prove possible to propagate a complete cDNA of potato *SBEII* in *Escherichia coli*, so an alternative strategy was employed. A suitable short intron to prevent translation of a functional SBEII protein in bacteria was identified at amino acid 439, and primers (Additional file 1: Table S1) were designed to incorporate existing *Afl*II and *Sac*I sites in the potato sequence that flanked the putative intron. PCR amplification from potato (cv 'Karaka') gDNA using primers FW3 and RW10 produced a 374 bp fragment that contained an intron of 231 bp.

The upstream portion of potato *SBEII* cDNA was PCR-amplified using primers SBE-A and RW11 (incorporating a *Afl*II site) to give cDNA1, and the downstream portion using primers F4 (incorporating a *Sac*I site) and SBE-B to give cDNA2. All PCR products were verified by DNA sequencing. cDNA1 and the genomic fragment containing the intron were ligated together using the *Afl*II site, then the cDNA1:gDNA fragment (blunt-*Sac*I) and the cDNA2 fragment (*Sac*I-blunt) were ligated in a three-way ligation into expression vector GBSSPTpUC57 (a gift from Dr A. J. Conner), between the *GBSS* promoter and the *GBSS* terminator. This produced a 4.46 kb construct pDB1047 consisting of $GBSS_{pro}$:cDNA1:gDNA: cDNA2:$GBSS_{term}$ (Additional file 1: Figure S1).

To produce an inverted repeat construct for silencing of *GBSS*, two fragments of potato *GBSS* cDNA were amplified by PCR using primer pairs GBSS-A and GBSS-B, and GBSS-C and GBSS-D (Additional file 1: Table S1). Primers GBSS-A and GBSS-B amplified a fragment of 1364 bp spanning amino acids 132–585 of the 607-amino acid protein, adding an initiating methionine codon and restriction sites *Xho*I upstream and *Pst*I downstream. Primers GBSS-C and GBSS-D amplified a fragment of 987 bp spanning amino acids 135–463, adding restriction sites *Xba*I and *Stu*I upstream and *Pst*I downstream. The GBSS-A/GBSS-B PCR product was ligated in the sense orientation to the GBSS-C/GBSS-D PCR product in the antisense orientation, using the *Pst*I site. This DNA was ligated into expression vector GBSSPTpUC57 between the *GBSS* promoter and the *GBSS* terminator to produce construct pJZ1041.

Construct pDB1047 was transferred into binary vector pMOA33, conferring resistance to kanamycin, and construct pJZ1041 was transferred into binary vector pMOA34, conferring resistance to hygromycin [46]. These binary vectors were electroporated into *Agrobacterium tumefaciens* LBA4404, which was used to transform *Solanum tuberosum* cv 'Karaka', a cultivar common in New Zealand, as described [47]. Plantlets were confirmed as transgenic by screening for the presence of the transgene and the selectable marker gene using PCR (Additional file 1: Figure S2). Empty vector transformants were produced similarly, and non-transformed wild-type (WT) material was regenerated through the tissue culture process (lacking the selectable antibiotic) as controls. Primary transformants were grown and screened as described below.

Plant growth and molecular analysis

Plantlets were potted in plastic bags (16 L) of potting mix and grown to maturity in a containment glasshouse in Palmerston North (40° 21′ S, 175° 36′ E) during the southern summer and autumn months (January to May). Glasshouse conditions were approximately 15 h daylight per day, with supplementary cooling and heating providing a maximum daytime temperature of 27°C and a minimum night-time temperature of 16°C. The plant aerial portions were allowed to senesce naturally. Three tubers per plant were peeled, chopped into small cubes and frozen in liquid nitrogen before being stored at −80°C for RNA or enzyme assay. RNA was prepared using the hot borate method [48], and 10 µg per lane was separated by electrophoresis in 1.2% agarose denaturing formaldehyde gels. The RNA was blotted to Hybond-XL nylon membrane (GE Healthcare, Little Chalfont, UK) and immobilized by ultraviolet irradiation. Labelled probes were produced using PCR products SBEII cDNA2 and GBSS-A/GBSS-B as templates with [α-^{32}P]dATP and the Klenow fragment of DNA polymerase I. RNA gel blots were hybridised overnight with the labelled probes in Church and Gilbert [49] buffer at 65°C, washed in 1x SSC at 65°C and exposed to X-ray film. For analysis of starch physicochemical properties, independent primary transformants were multiplied by excising axillary buds of plantlets in tissue culture and regenerating them to whole plants and tubers in subsequent summers. Biological replicates consisted of pooled tissue of three tubers from three or four different but genetically identical plants.

Native PAGE determination of SBEII activity

Tuber tissue was powdered in liquid nitrogen, and 100 mg was extracted with 100 μL of extraction buffer (50 mM HEPES-NaOH, pH 7.4, 2 mM $MgCl_2$, 50 mM 2-mercaptoethanol, 12.5% (v/v) glycerol), incubated on ice for 5 min then centrifuged twice at 16,000 × *g* for 15 min [17]. Soluble proteins in the supernatant were used as the crude enzyme preparation. Proteins (150 μg per lane) were separated on native 5% (w/v) polyacrylamide gels containing substrates for SBE activity as described by Tetlow *et al.* [24]. After electrophoresis, gels were rinsed, incubated in reaction buffer at 30°C for 2 h then stained with I_2/KI as described by Nishi *et al.* [17]. Gels were photographed immediately after staining. Gel pictures were converted to greyscale, and relative density in the SBEII spots quantified using MultiGauge software (Fujifilm, Tokyo, Japan).

Starch preparation

Starch was prepared by chopping peeled fresh tubers into small pieces, adding an equal amount of ice-cold water and crushing the tissue with a potato masher. Starch granules were washed through a 200 μm mesh sieve, and the suspension allowed to settle overnight. The supernatant and any discoloured layer above the white starch pellet were discarded. The starch was resuspended in a large volume of ice-cold water and the settling step repeated. The pellet of starch granules was freeze-dried and used for analysis.

Microscopy

Aqueous, unstained suspensions of starch grains were observed by brightfield, differential interference contrast (DIC) and simple polarising microscopy (crossed polars but no phase plate) using an Olympus Vanox AHT3 microscope (Olympus Optical Co Ltd., Tokyo, Japan). Brightfield microscopy was also used to view starch stained with iodine using a stain solution of 0.2% I_2 in a 2% aqueous solution of KI at a dilution ratio of 1:100 stain:starch suspension. For scanning electron microscopy (SEM), dry starch powder was lightly sprinkled onto carbon adhesive tabs attached to aluminium stubs and was viewed uncoated in a FEI Quanta 250 SEM (FEI Company, Hillsboro, OR, USA) using variable pressure mode to reduce charging (chamber pressure 60 Pa, accelerating voltage 5 kV).

The reducing ends of starch molecules were fluorescently labelled within the starch granules using APTS as described by Blennow *et al.* [27]. Optical sections were taken under a Leica SP5 DM6000B laser scanning confocal microscope, using an excitation wavelength of 488 nm and detection at the interval from 500 to 535 nm.

Size exclusion chromatography

Aliquots of starch were suspended in 2 M NaOH, gelatinised at 100°C then dissolved in water as described by Blennow *et al.* [26]. The solubilised starch was applied to a Waters Ultrahydrogel Linear column (300 × 7.8 mm) with an exclusion limit of 7×10^6 Da, and eluted with 10 mM NaOH at 50°C. Carbohydrate content in the effluent was recorded on-line by measuring refractive index (Waters 2414 Refractive Index detector (Waters Corp, MA, USA)). A large peak at the end of the elution profiles (due to NaOH) was omitted.

Starch composition

The percentage of amylose present in the starch was measured using the Megazyme K_AMYL Amylose kit (Megazyme, Bray, Ireland). Starch branching patterns were assessed using a micro-scale FACE procedure to measure chain length size distribution [50]. The degree of C6 and C3 starch phosphorylation was determined using a mass spectrometric method [51].

Starch properties

Thermal analysis of gelatinisation properties was performed using a Q2000 differential scanning calorimeter (DSC) (TA Instruments, New Castle, DE, USA). Starch (~5 mg) was accurately weighed into Tzero aluminium hermetic pans (TA Instruments), and 2.33 μL water mg^{-1} starch added to give a suspension containing 70% water. The pans were hermetically sealed and left at room temperature (20°C) for at least 2 h. DSC conditions were: equilibration at 10°C for 5 min, then heating at 5°C min^{-1} to 100°C. The heat flow curves were analysed using TA Universal Analysis 2000 software version 4.5A build 4.5.0.5 using the "integrate peak sigmoid horizontal" option over the range 50–80°C. Variables calculated were peak onset temperature (T_o), peak maximum temperature (T_p), peak conclusion temperature (T_c) and peak area (ΔH), which provides an estimate of enthalpy in J g^{-1}.

Starch pasting properties were evaluated using a rapid visco analyser (RVA) (Newport Scientific, Sydney, Australia). Approximately 1 g of starch was accurately weighed and dried at 130°C to constant weight. A starch suspension was prepared at 7% (w/w) on a dry weight basis in distilled water, with a total weight of 28 g. The suspension was equilibrated at 50°C for 1 min, heated to 95°C at 6°C min^{-1}, maintained at 95°C for 5 min, then cooled to 50°C at 6°C min^{-1}, with constant stirring at 160 rpm.

The swelling volume of starch was determined using a urea procedure [17]. Freeze-dried starch (20 mg) was mixed with 1 mL of urea solution (0 to 9 M). The mixtures were allowed to stand at room temperature (23°C) for 24 h, centrifuged at 8000 × *g* for 20 min, then allowed to stand at room temperature for 1 h. The volumes of the swollen pellet were determined by subtracting the volume of the supernatant from the original 1 mL volume.

Statistical analysis

Means were compared using one-way ANOVA. Least significant differences (LSDs) were used to establish which means differed significantly ($P = 0.05$). Since for SBEII activity the data was relative and was scaled so that wild-type was 1, the wild-type data was not included in the ANOVA; pooled standard errors from the ANOVA were used in one-sample t-tests to test whether each mean was significantly different from 1. Correlation coefficients were calculated between means. Analysis was carried out using Genstat (version 17, 2014, VSNI Ltd, Hemel Hempstead, UK).

Additional file

Additional file 1: Table S1. Primers used for PCR amplification. **Figure S1.** Sequence of the hybrid cDNA/gDNA SBEII-intron construct. **Figure S2.** PCR screening of transgenic plants for the presence of the transgene and the selectable marker gene. **Figure S3.** RNA gel blot analysis of *SSII* and *SSIII* mRNA abundance in tubers from plants overexpressing *SBEII*.

Abbreviations

APTS: 8-amino-1,3,6-pyrenetrisulfonic acid; DP: Degree of polymerisation; DSC: Differential scanning calorimeter; FACE: Fluorophore-assisted carbohydrate electrophoresis; GBSS: Granule-bound starch synthase; PAGE: Polyacrylamide gel electrophoresis; RVA: Rapid visco analyser; SBE: Starch branching enzyme; SS: Starch synthase.

Competing interests

The authors declare that they have no competing interests.

Authors' contributions

DAB conceived and designed the study, DAB, LWM, JZ, MJM, ICH, LS, MC and GMT-V performed experiments, DAB wrote the paper. All authors discussed the results and commented on the manuscript. All authors read and approved the final manuscript.

Acknowledgements

We thank Tony Conner for the generous gift of GBSSPTpUC57 plasmid, Jianyu Chen (Manawatu Microscopy and Imaging Centre, Massey University) for assistance with confocal microscopy, Kerry Sullivan and Ronan Chen for assistance with RNA preparation, Tonya Frew and Rebecca Cooper for running the FACE analysis, Nigel Joyce for mass spectrometer operations, Ian King for care of the plants, Duncan Hedderley for statistical analysis and Nigel Larsen and Paul Johnston for comments on the manuscript. This work was funded by contract number C02X0805 from the Foundation for Research, Science and Technology of New Zealand.

Author details

[1]The New Zealand Institute for Plant & Food Research Limited, Food Industry Science Centre, Private Bag 11600, Palmerston North 4442, New Zealand. [2]The New Zealand Institute for Plant & Food Research Limited, Mount Albert Research Centre, Private Bag 92169, Auckland 1142, New Zealand. [3]The New Zealand Institute for Plant & Food Research Limited, Canterbury Agriculture & Science Centre, Private Bag 4704, Christchurch 8140, New Zealand.

References

1. Jobling S. Improving starch for food and industrial applications. Curr Opinion Plant Biol. 2004;7:210–8.
2. Zeeman SC, Kossmann J, Smith AM. Starch: its metabolism, evolution, and biotechnological modification in plants. Annu Rev Plant Biol. 2010;61:209–34.
3. Ball SG, Morell MK. From bacterial glycogen to starch: understanding the biogenesis of the plant starch granule. Annu Rev Plant Biol. 2003;54:207–33.
4. Kötting O, Kossmann J, Zeeman SC, Lloyd JR. Regulation of starch metabolism: the age of enlightenment? Curr Opinion Plant Biol. 2010;13:321–9.
5. Yoo SH, Jane J. Structural and physical characteristics of waxy and other wheat starches. Carbohydr Poly. 2002;49:297–305.
6. Visser RGF, Suurs LCJM, Bruinenberg PM, Bleeker I, Jacobsen E. Comparison between amylose-free and amylose containing potatoes. Starch-Stärke. 1997;49:438–43.
7. Yuan RC, Thompson DB, Boyer CD. Fine structure of amylopectin in relation to gelatinization and retrogradation behavior of maize starches from three *wx*-containing genotypes in two inbred lines. Cereal Chem. 1993;70:81–9.
8. Srichuwong S, Sunarti TC, Mishima T, Isono N, Hisamatsu M. Starches from different botanical sources I: Contribution of amylopectin fine structure to thermal properties and enzyme digestibility. Carbohydr Poly. 2005;60:529–38.
9. Butardo VM, Fitzgerald MA, Bird AR, Gidley MJ, Flanagan BM, Larroque O. et al. Impact of down-regulation of *starch branching enzyme IIb* in rice by artificial microRNA- and hairpin RNA-mediated RNA silencing. J Exp Bot. 2011;62:4927–41.
10. Regina A, Kosar-Hashemi B, Ling S, Li Z, Rahman S, Morell M. Control of starch branching in barley defined through differential RNAi suppression of starch branching enzyme IIa and IIb. J Exp Bot. 2010;61:1469–82.
11. Carciofi M, Blennow A, Jensen SL, Shaik SS, Henriksen A, Buléon A, et al. Concerted suppression of all starch branching enzyme genes in barley produces amylose-only starch granules. BMC Plant Biol. 2012;12:223.
12. Nakamura Y, Utsumi Y, Sawada T, Aihara S, Utsumi C, Yoshida M, et al. Characterization of the reactions of starch branching enzymes from rice endosperm. Plant Cell Physiol. 2010;51:776–94.
13. Larsson CT, Khoshnoodi J, Ek B, Rask L, Larsson H. Molecular cloning and characterization of starch-branching enzyme II from potato. Plant Mol Biol. 1998;37:505–11.
14. Jobling SA, Schwall GP, Westcott RJ, Sidebottom CM, Debet M, Gidley MJ, et al. A minor form of starch branching enzyme in potato (*Solanum tuberosum* L.) tubers has a major effect on starch structure: cloning and characterisation of multiple forms of *SBE A*. Plant J. 1999;18:163–71.
15. Flipse E, Suurs LCJM, Keetels CJAM, Kossmann J, Jacobsen E, Visser RGF. Introduction of sense and antisense cDNA for branching enzyme in the amylose-free potato mutant leads to physico-chemical changes in the starch. Planta. 1996;198:340–7.
16. Safford R, Jobling SA, Sidebottom CM, Westcott RJ, Cooke D, Tober KJ, et al. Consequences of antisense RNA inhibition of starch branching enzyme activity on properties of potato starch. Carbohydr Poly. 1998;35:155–68.
17. Nishi A, Nakamura Y, Tanaka N, Satoh H. Biochemical and genetic analysis of the effects of *Amylose-Extender* mutation in rice endosperm. Plant Physiol. 2001;127:459–72.
18. Schwall GP, Safford R, Westcott RJ, Jeffcoat R, Tayal A, Shi YC, et al. Production of very-high-amylose potato starch by inhibition of SBE A and B. Nature Biotechnol. 2000;18:551–4.
19. Shewmaker CK, Boyer CD, Wiesenborn DP, Thompson DB, Boersig MR, Oakes JV, et al. Expression of *Escherichia coli* glycogen synthase in the tubers of transgenic potatoes (*Solanum tuberosum*) results in a highly branched starch. Plant Physiol. 1994;104:1159–66.
20. Kortstee AJ, Vermeesch AMS, de Vries BJ, Jacobsen E, Visser RGF. Expression of *Escherichia coli* branching enzyme in tubers of amylose-free transgenic potato leads to an increased branching degree of the amylopectin. Plant J. 1996;10:83–90.
21. Kortstee AJ, Suurs LCJM, Vermeesch AMG, Keetels CJAM, Jacobsen E, Visser RGF. The influence of an increased degree of branching on the physico-chemical properties of starch from genetically modified potato. Carbohydr Poly. 1998;37:173–84.
22. Tanaka N, Fujita N, Nishi A, Satoh H, Hosaka Y, Ugaki M, et al. The structure of starch can be manipulated by changing the expression levels of starch branching enzyme IIb in rice endosperm. Plant Biotechnol J. 2004;2:507–16.
23. FAOSTAT. Food and Agriculture Organization of the United Nations Statistics Division. 2013. http://faostat3.fao.org. Accessed 19 February 2015.
24. Tetlow IJ, Wait R, Lu Z, Akkasaeng R, Bowsher CG, Esposito S, et al. Protein phosphorylation in amyloplasts regulates starch branching enzyme activity and protein–protein interactions. Plant Cell. 2004;16:694–708.

25. Kubo A, Fujita N, Harada K, Matsuda T, Satoh H, Nakamura Y. The starch-debranching enzymes isoamylase and pullulanase are both involved in amylopectin biosynthesis in rice endosperm. Plant Physiol. 1999;121:399–409.
26. Blennow A, Bay-Smidt AM, Bauer R. Amylopectin aggregation as a function of starch phosphate content studied by size exclusion chromatography and on-line refractive index and light scattering. Int J Biol Macromol. 2001;28:409–20.
27. Blennow A, Hansen M, Schulz A, Jørgensen K, Donald AM, Sanderson J. The molecular deposition of transgenically modified starch in the starch granule as imaged by functional microscopy. J Struct Biol. 2003;143:229–41.
28. Blennow A, Bay-Smidt AM, Wischmann B, Olsen CE, Møller BL. The degree of starch phosphorylation is related to the chain length distribution of the neutral and the phosphorylated chains of amylopectin. Carbohydr Res. 1998;307:45–54.
29. Fujita N, Kubo A, Suh DS, Wong KS, Jane JL, Ozawa K, et al. Antisense inhibition of isoamylase alters the structure of amylopectin and the physicochemical properties of starch in rice endosperm. Plant Cell Physiol. 2003;44:607–18.
30. Hizukuri S, Kaneko T, Takeda Y. Measurement of the chain length of amylopectin and its relevance to the origin of crystalline polymorphism of starch granules. Biochim Biophys Acta. 1983;760:188–91.
31. Buléon A, Colonna P, Planchot V, Ball S. Starch granules: structure and biosynthesis. Int J Biol Macromol. 1998;23:85–112.
32. Lim ST, Kasemsuwan T, Jane JL. Characterization of phosphorus in starch by ^{31}P-nuclear magnetic resonance spectroscopy. Cereal Chem. 1994;71:488–93.
33. Peat S, Whelan WJ, Thomas GJ. Evidence of multiple branching in waxy maize starch. J Chem Soc. 1952;1952:4546–8.
34. Hanashiro I, Abe JI, Hizukuri S. A periodic distribution of the chain length of amylopectin as revealed by high-performance anion-exchange chromatography. Carbohydr Res. 1996;283:151–9.
35. Takeda Y, Hizukuri S. Location of phosphate groups in potato amylopectin. Carbohydr Res. 1982;102:321–7.
36. Tester RF, Morrison WR. Swelling and gelatinization of cereal starches. I. Effects of amylopectin, amylose, and lipids. Cereal Chem. 1990;67:551–7.
37. Visser RGF, Suurs LCJM, Steeneken PAM, Jacobsen E. Some physicochemical properties of amylose-free potato starch. Starch-Stärke. 1997;49:443–8.
38. Edwards A, Fulton DC, Hylton CM, Jobling SA, Gidley M, Rössner U, et al. A combined reduction in activity of starch synthases II and III of potato has novel effects on the starch of tubers. Plant J. 1999;17:251–61.
39. Wickramasinghe HAM, Blennow A, Noda T. Physico-chemical and degradative properties of *in-planta* re-structured potato starch. Carbohydr Poly. 2009;77:118–24.
40. Kozlov SS, Blennow A, Krivandin AV, Yuryev VP. Structural and thermodynamic properties of starches extracted from GBSS and GWD suppressed potato lines. Int J Biol Macromol. 2007;40:449–60.
41. Singh N, Singh J, Kaur L, Sodhi NS, Gill BS. Morphological, thermal and rheological properties of starches from different botanical sources. Food Chem. 2003;81:219–31.
42. Rydberg U, Andersson L, Andersson R, Aman P, Larsson H. Comparison of starch branching enzyme I and II from potato. Eur J Biochem. 2001;268:6140–5.
43. Jobling SA, Westcott RJ, Tayal A, Jeffcoat R, Schwall GP. Production of a freeze-thaw-stable potato starch by antisense inhibition of three starch synthase genes. Nature Biotechnol. 2002;20:295–9.
44. Jane J, Chen YY, Lee LF, McPherson AE, Wong KS, Radosavljevic M, et al. Effects of amylopectin branch chain length and amylose content on the gelatinization and pasting properties of starch. Cereal Chem. 1999;76:629–37.
45. Conner AJ, Barrell PJ, Baldwin SJ, Lokerse AS, Cooper PA, Erasmuson AK, et al. Intragenic vectors for gene transfer without foreign DNA. Euphytica. 2007;154:341–53.
46. Barrell PJ, Conner AJ. Minimal T-DNA vectors suitable for agricultural deployment of transgenic plants. BioTechniques. 2006;41:708–10.
47. Barrell PJ, Shang YJ, Cooper PA, Conner AJ. Alternative selectable markers for potato transformation using minimal T-DNA vectors. Plant Cell Tiss Organ Cult. 2002;70:61–8.
48. Wan CY, Wilkins TA. A modified hot borate method significantly enhances the yield of high-quality RNA from cotton (*Gossypium hirsutum* L.). Anal Biochem. 1994;223:7–12.
49. Church GM, Gilbert W. Genomic sequencing. Proc Natl Acad Sci U S A. 1984;81:1991–5.
50. Murray S, McKenzie M, Butler R, Baldwin S, Sutton K, Batey I, et al. Quantitative, small-scale, fluorophore-assisted carbohydrate electrophoresis implemented on a capillary electrophoresis-based DNA sequence analyzer. Anal Biochem. 2011;413:104–13.
51. Carpenter M, Joyce N, Butler R, Genet R, Timmerman-Vaughan G. A mass spectrometric method for quantifying C3 and C6 phosphorylation of starch. Anal Biochem. 2012;431:115–9.

14

Rapid high-yield expression of a candidate influenza vaccine based on the ectodomain of M2 protein linked to flagellin in plants using viral vectors

Eugenia S. Mardanova[1], Roman Y. Kotlyarov[1], Victor V. Kuprianov[1], Liudmila A. Stepanova[2], Liudmila M. Tsybalova[2], George P. Lomonosoff[3] and Nikolai V. Ravin[1*]

Abstract

Background: The extracellular domain of matrix protein 2 (M2e) of influenza A virus is a promising target for the development of a universal vaccine against influenza because M2e sequences are highly conserved among human influenza A strains. However, native M2e is poorly immunogenic, but its immunogenicity can be increased by delivery in combination with adjuvants or carrier particles. It was previously shown that fusion of M2e to bacterial flagellin, the ligand for Toll-like receptor (TLR) 5 and powerful mucosal adjuvant, significantly increases the immunogenicity and protective capacity of M2e.

Results: In this study, we report for the first time the transient expression in plants of a recombinant protein Flg-4M comprising flagellin of *Salmonella typhimurium* fused to four tandem copies of the M2e peptide. The chimeric construct was expressed in *Nicotiana benthamiana* plants using either the self-replicating potato virus X (PVX) based vector, pA7248AMV-GFP, or the cowpea mosaic virus (CPMV)-derived expression vector, pEAQ-*HT*. The highest expression level up to 30 % of total soluble protein (about 1 mg/g of fresh leaf tissue) was achieved with the PVX-based expression system. Intranasal immunization of mice with purified Flg-4M protein induced high levels of M2e-specific serum antibodies and provided protection against lethal challenge with influenza virus.

Conclusions: This study confirms the usefulness of flagellin as a carrier of M2e and its relevance for the production of M2e-based candidate influenza vaccines in plants.

Keywords: Viral vector, Potato virus X, M2 protein, Flagellin, Influenza, Vaccine

Background

Influenza is a widely distributed viral infection of humans and animals. The high variability of the surface viral proteins, hemagglutinin and neuraminidase, results in the appearance of a new epidemic strain every one to two years [1]; this requires the production of new vaccines at the same frequency. A promising solution to this problem is the development of recombinant vaccines that can be rapidly produced in standard expression systems. One of the most promising conserved antigens of influenza virus is the extracellular domain of transmembrane protein M2, M2e [2]. The length of M2e is only 24 amino acid residues (including the initiator methionine). Its sequence is virtually unchanged in all human isolates since 1933 [3–5], and in strains of avian influenza A it differs in only a few amino acids. By itself M2e possesses low immunogenicity [6] but this can be increased by delivery in a multimeric form or in combination with complex adjuvants or delivery systems [3, 5, 7–9].

Toll-like receptors (TLRs) have been shown to play a critical role in controlling the adaptive immune response, by properly arming dendritic and other antigen-presenting cells, triggering important costimulatory and regulatory mechanisms, and promoting antigen presentation [10, 11].

* Correspondence: nravin@biengi.ac.ru
[1]Centre 'Bioengineering', Russian Academy of Sciences, Prosp. 60-letya Oktyabrya, bld 7-1, 117312 Moscow, Russia
Full list of author information is available at the end of the article

It has been previously shown that genetically fusing an antigen of interest to bacterial flagellin, the ligand for TLR5, significantly increases the immunogenicity and protective capacity of the antigen [12, 13]. Notably flagellin is particularly active as mucosal adjuvant, opening the possibility of non-invasive intranasal administration of vaccines. The ability of flagellin to be both the platform and the adjuvant for different vaccines was demonstrated in various models of infection including influenza [13–19].

Recombinant proteins, including those for medical purposes, can be produced in a variety of expression systems, including bacteria, yeast, plants, and mammalian cells. The advantages of using plants for protein production include the low final cost and inherent safety of products due to the absence of pathogens common for plants and animals. Recombinant proteins in plants may be obtained by stable genetic transformation (either nuclear or plastid) or through transient expression [20–22]. Creation of lines of stably transformed plants and, subsequently, production of a protein from such plants requires considerable time and cost. The level of expression of target proteins by transgenic plants is usually relatively low, resulting in a high cost of products due to the difficulty of their purification [23, 24]. The use of self-replicating viral vectors or efficient non-replicating transient expression vectors is an alternative approach providing sufficient amounts of the desired protein (up to 1–2 g/kg of plant biomass) [20] to conduct immunological studies. To create such vectors, sequences from tobacco mosaic virus (TMV), cowpea mosaic virus (CPMV) or potato virus X (PVX) have frequently been used. The genomes of these viruses are small and consist of positive-strand RNA. The vectors are introduced into plant cells by infiltration of the plant leaf tissue with *Agrobacterium* suspensions carrying the vector within the T-DNA portion of a binary plasmid resulting in transfer of vector DNA into the recipient plant cells during the agroinfection [25, 26]. Such expression systems have been used for production of a variety of pharmaceutical proteins, including antibodies and candidate vaccines [22, 27, 28]. The latter have included influenza virus hemagglutinin [29–31] and HBc virus-like particles carrying M2e peptide [9].

In this work four tandem copies of M2e sequences based on human and avian influenza virus isolates were genetically fused to *Salmonella typhimurium* flagellin and produced in *Nicotiana benthamiana* plants using transient expression using either a replicating PVX vector [32] or the non-replicating CPMV-*HT* system [33, 34]. Intranasal immunization of mice with the plant-produced hybrid protein induced the desired immune response and conferred protective immunity against a lethal influenza virus challenge.

Results

Viral vectors for expression of the hybrid protein consisting of flagellin linked to four copies of influenza M2e peptide in N. benthamiana plants

The highly conserved extracellular domain of the influenza virus M2 protein, M2e, is a promising target for the development of recombinant "universal" vaccines against influenza [5]. However, M2e is poorly immunogenic and needs to be presented in a multimeric form or linked to a carrier molecule such as virus-like particles [3]. Here we use flagellin of *Salmonella typhimurium* as such carrier and simultaneously as an adjuvant. In order to make a broader range candidate vaccine, a hybrid protein Flg-4M containing two copies of human consensus M2e sequence and two copies of the M2e peptide of avian influenza virus strain A/Chicken/Kurgan/05/2005 fused to the C-terminus of *S. typhimurium* FljB (arranged as Flg-M2eh-M2ek-M2eh-M2ek) was designed. In order to facilitate folding of hybrid protein, copies of M2e peptides were separated from each other by flexible glycine-rich linker (Fig. 1a). The empty flagellin (lacking M2e) was used as a control. Both proteins, Flg-4M and Flg, were designed with N-terminal 6-histidine tag to facilitate their purification.

Two plant virus-based transient expression systems were used for production of recombinant Flg-4M protein in *N. benthamiana*. The first vector, pA7248AMV-GFP [32], is a replicating system based on the PVX genome. This vector includes the 5'-UTR of the PVX genome, the polymerase gene, the first promoter of subgenomic RNA, the translation enhancer represented by the sequence of the 5'-untranslated region of RNA 4 of the alfalfa mosaic virus (AMV), the GFP gene flanked by unique restriction sites *Asc*I and *Sma*I, the last 60 nucleotides of the coat protein gene, and the 3'-untranslated region of the PVX genome. This construction is nserted between the 35S promoter and 35S terminator and cloned in the binary vector pBIN19 [32]. To create the viral vector allowing production of the Flg-4M protein, a sequence of the hybrid gene was cloned in pA7248AMV-GFP at the *Asc*I and *Sma*I sites to replace GFP, which resulted in construction of a recombinant vector pA7248AMV-Flg-4M (Fig. 1b). Recombinant vector pA7248AMV-Flg for expression of Flg in plants was constructed in a similar way.

The second expression system, pEAQ, is based on CPMV. This is a non-replicating system which uses production of a highly translatable mRNA to achieve high level expression. The recombinant vector pEAQ-Flg-4M contains 5'-UTR and 3'-UTR from CPMV RNA-2 upstream and downstream, respectively, of the hybrid gene Flg-4M (Fig. 1b).

Expression of the hybrid protein Flg-4M in N. benthamiana plants

To express the protein Flg-4M, vectors were inserted into *A. tumefaciens* strain GV3101, which was used for

(A)

MHHHHHHGSAQVINTNSLSLLTQNNLNKSQSALGTAIERLSSGLRINSAKDDAAGQAIANRF
TANIKGLTQASRNANDGISIAQTTEGALNEINNNLQRVRELAVQSANSTNSQSDLDSIQAEI
TQRLNEIDRVSGQTQFNGVKVLAQDNTLTIQVGANDGETIDIDLKQINSQTLGLDSLNVQKA
YDVKDTAVTTKAYANNGTTLDVSGLDDAAIKAATGGTNGTASVTGGAVKFDADNNKYFVTIG
GFTGADAAKNGDYEVNVATDGTVTLAAGATKTTMPAGATTKTEVQELKDTPAVVSADAKNAL
IAGGVDATDANGAELVKMSYTDKNGKTIEGGYALKAGDKYYAADYDEATGAIKAKTTSYTAA
DGTTKTAANQLGGVDGKTEVVTIDGKTYNASKAAGHDFKAQPELAEAAAKTTENPLQKIDAA
LAQVDALRSDLGAVQNRFNSAITNLGNTVNNLSEARSRIEDSDYATEVSNMSRAQILQQAGT
SVLAQANQVPQNVLSLLRGS*SLLTEVETPIRNEWGCRCNDSSD*LQGGGSG**SLLTEVETPTRN**
EWECRCSDSSDGGGSG*SLLTEVETPIRNEWGCRCNDSSD*ELGGGSG**SLLTEVETPTRNEWEC**
RCSDSSD

(B)

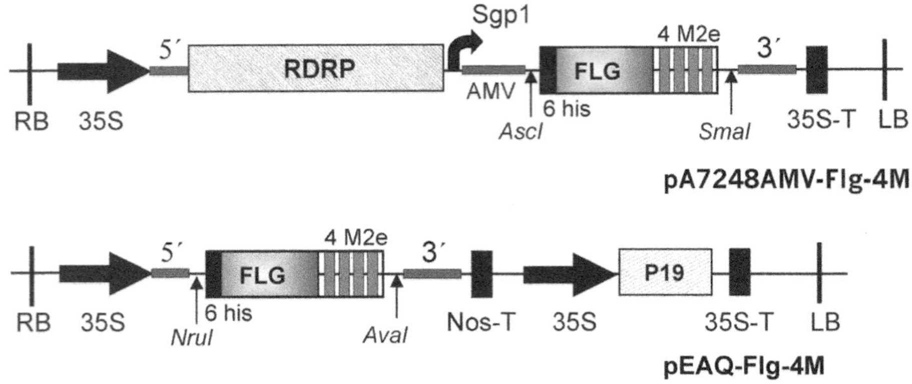

Fig. 1 Structure of the expression constructs used in this study. **a** Amino acid sequence of Flg-4M protein. The sequences of human "consensus" M2e peptides, M2eh, are italicized; the sequences of M2e peptide of avian influenza virus strain A/Chicken/Kurgan/05/2005, M2ek, are shown in bold; the linker sequences are underlined. Flagellin sequence is shaded in grey. **b** Structures of viral vectors pA7248AMV-Flg-4M and pEAQ-Flg-4M. The tDNA regions of the binary vectors are shown. RDRP, RNA dependent RNA polymerase of PVX; Sgp1, subgenomic promoter of the PVX protein 25 K; AMV, the leader sequence of RNA 4 of the alfalfa mosaic virus; 6his, 6 histidine tag; FLG, flagellin of *S. typhimurium*; 4M2e, four copies of M2e peptide (arranged as M2eh-M2ek-M2eh-M2ek); 35S, promoter; 35S-T, terminator of the cauliflower mosaic virus RNA; NosT, terminator of nopaline synthase gene; LB and RB are the left and right borders of tDNA; p19, gene of silencing suppressor P19 from the tomato bushy stunt virus; 5′ and 3′, the 5′-UTR and the 3′-UTR from PVX (in pA7248AMV-Flg-4M) and CPMV (in pEAQ-Flg-4M)

agroinfiltration of leaves of *N. benthamiana* plants. Agroinfiltration zones for two vectors were located within one leaf. Protein samples were isolated in 3, 4, 5, 6, and 7 days after infection. For every day three leaves were taken for analysis. Protein samples were analyzed using SDS-PAGE and western-blotting (Fig. 2). Maximal expression was observed at fourth day. Summarizing the expression results for 15 leaves we found out that pEAQ-Flg-4M construct provides expression level of about 30 % in comparison to pA7248AMV-Flg-4M.

Given the higher levels of expression obtained, we used vector pA7248AMV-Flg-4M for the scaled-up production of protein Flg-4M in *N. benthamiana* plants. The total soluble protein was isolated from leaves at 4th day after agroinfiltration and analyzed by SDS-PAGE. The data presented in Fig. 3a show that Flg-4M protein is highly expressed and accounted for about 30 % of the soluble protein fraction (~1 mg/g of fresh leaf tissue). The recombinant protein was specifically revealed in western blot analysis with the monoclonal antibodies specific to M2e (Fig. 3b). The recombinant fusion protein carrying an N-terminal 6-histidine tag was purified to a purity of greater than 90 % (Fig. 3a) using standard affinity chromatography on Ni-NTA resin and used for animal experiments. The recombinant Flg protein was expressed in *N. benthamiana* using vector pA7248AMV-Flg and purified in a similar way.

Antigenicity of plant-produced protein Flg-4M

The display of the M2e peptide in an accessible manner when it is fused with flagellin in the Flg-4M protein was examined by ELISA. To this end plates were coated with serial dilutions of Flg-4M protein and empty Flg, then probed with the M2e-specific and the 6-his tag-specific monoclonal antibodies (Fig. 4). While both Flg-4M and Flg reacted with the anti-his antibodies, only Flg-4M reacted with the M2e-specific antibodies. These results confirm the presence of M2e in the purified Flg-4M protein and its accessibility for antibodies.

The M2e peptides naturally exposed on the surface of infected cells as tetramers tends to aggregate due to the

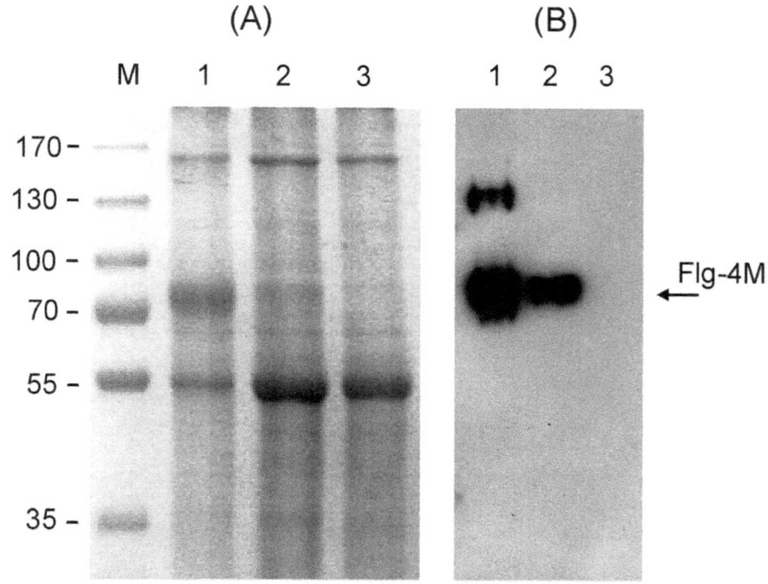

Fig. 2 Efficiency of PVX- and CPMV-based vectors. Coomassie brilliant blue stained gel (**a**) and western blot (**b**) of proteins isolated form *N. benthamiana* plants and separated by SDS-PAGE. M, – molecular weight marker (kD); 1 and 2, – total soluble proteins isolated at fourth day after agroinfiltration from leaves infiltrated with vector pA7248AMV-Flg-4M and pEAQ-Flg-4M, respectively; 3, – total soluble proteins isolated from empty leaves. Antibodies against M2e were used in western blotting

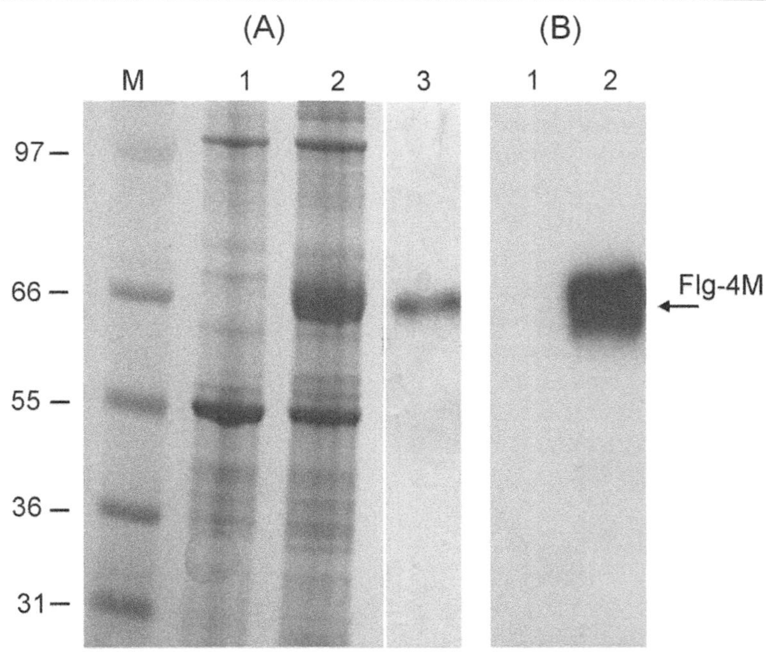

Fig. 3 Expression and purification of Flg-4M protein. Coomassie brilliant blue stained gel (**a**) and western blot (**b**) of proteins isolated form *N. benthamiana* plants and separated by SDS-PAGE. M, – molecular weight marker (kD); 1, – total soluble proteins isolated from empty leaves; 2, – total soluble proteins isolated at fourth day after agroinfiltration from leaves infiltrated with vector pA7248AMV-Flg-4M; 3, – purified Flg-4M protein. Antibodies against M2e were used in western blotting

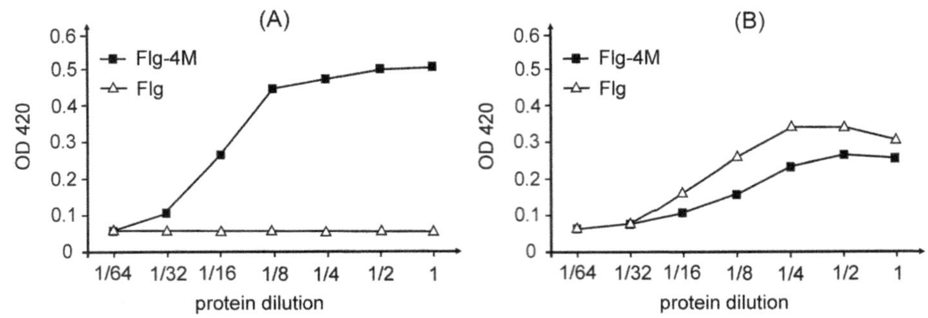

Fig. 4 Antigenicity of Flg-4M protein. Two-fold dilutions of recombinant Flg-4M or Flg proteins (starting from 30 μg/ml) were coated on ELISA plates which were then probed with monoclonal antibodies specific for M2e (**a**) or 6-his tag present both in Flg-4M and Flg (**b**)

presence of two cysteine residues per M2e copy that lead to disulfide bond formation and intra- and interprotein cross-linking. In order to control the lack of aggregation of Flg-4M protein we also performed western-blot analysis of purified protein under reducing and non-reducing conditions. The recombinant protein migrated at expected position and was revealed by the monoclonal antibodies specific to M2e, suggesting the lack of aggregation (see Additional file 1). No high-molecular weight aggregates or particulate structures were detected by the electron microscopy of purified protein samples.

Immunogenicity and protectivity of plant-produced protein Flg-4M

To characterize the immunogenicity and protective action, we performed two sets of animal experiments. In the first case mice were immunized with the plant-produced preparation of Flg-4M protein, in the second experiment an additional group of mice was immunised with plant-produced flagellin lacking M2e. Mice were vaccinated intranasally three times at two-week intervals with 10 μg protein without additional adjuvants. Blood samples were taken 2 weeks after the second and/or third immunizations. Serum was analysed by ELISA to identify IgG antibodies directed against M2e, using plates coated with synthetic peptides G-37 and G-50, whose sequences corresponded to the human "consensus" M2e sequence and the M2e of influenza strain A/Chicken/Kurgan/05/2005, respectively. A strong immune response was activated after the second immunisation and further increased after the third (Fig. 5). Induced antibodies efficiently bind to both synthetic M2e peptides G-50 and G-37, which are present in the candidate vaccine, although the immune response towards G37, the human M2e sequence, was stronger.

In the first experiment the immunised and control mice were challenged with $10LD_{50}$ of mouse-adapted human influenza strain A/PR/8/34 (H1N1) and with $5LD_{50}$ of avian influenza strain A/Chicken/Kurgan/05/2005 (H5N1). As shown in Fig. 6a, 75 % mice immunized with Flg-4M, survived the lethal challenge of A/Chicken/Kurgan/05/2005 (H5N1), while the rate of survival among the control mice was 14 %. The challenge of immunized mice with $10LD_{50}$ of A/PR/8/34 (H1N1) led to a 50 % survival, while all control mice died upon infection (Fig. 6b).

In the second experiment $5LD_{50}$ of influenza strain A/PR/8/34 (H1N1) was used for challenge. Figure 6c shows that mice immunised with Flg-4M were completely protected from lethal virus challenge, while mice immunised with Flg or PBS were not protected. The morbidity of the disease was monitored by measuring the weight of the animals. The data presented in Fig. 6d reveal that immunisation did not eliminate morbidity but reduced it relative to the control.

Discussion

The purpose of the present work was to develop a plant-produced candidate influenza vaccine based on M2e

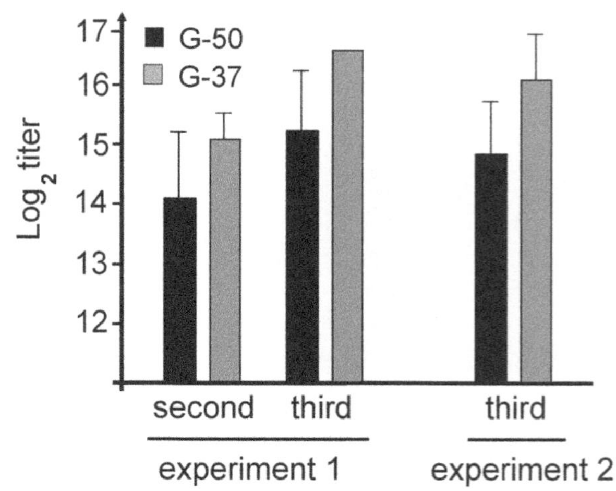

Fig. 5 Titers of IgG antibodies to synthetic M2e peptides G-50 and G-37 in sera of immunized mice. The data obtained in two independent experiments after the second and the third immunizations are shown. The results are expressed as the mean titer ± standard deviation for each group of five mice. Titers lower than 200 were detected for mice immunized with PBS or Flg

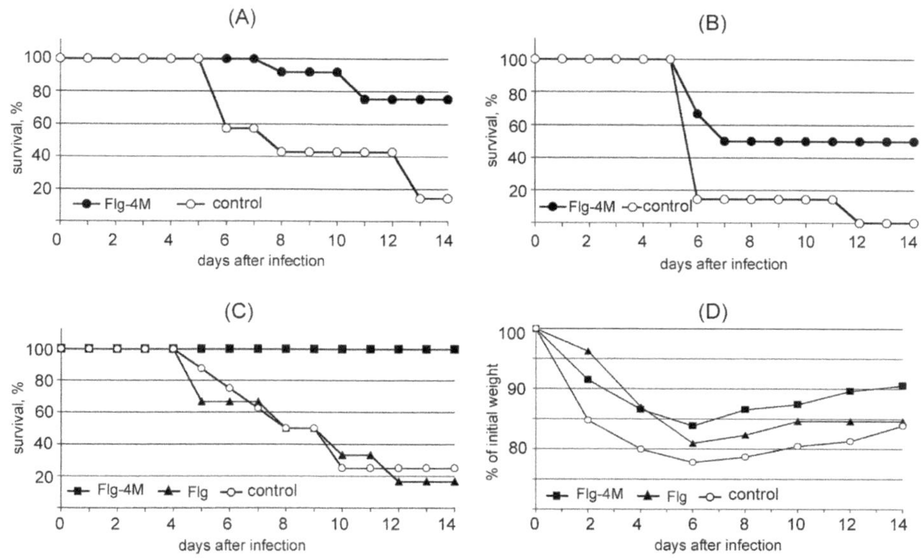

Fig. 6 Protective efficiency of the plant-produced Flg-4M protein. Experiment 1: mice were challenged with $5LD_{50}$ of A/Chicken/Kurgan/05/2005 (**a**) or with $10LD_{50}$ of A/PR/8/34 (**b**). Experiment 2: mice were challenged with $5LD_{50}$ of influenza strain A/PR/8/34 (**c, d**). Survival (**a, b, c**) and weight (**d**) of immunized and control mice were monitored for 14 days post-challenge. Control, mice immunized with PBS

peptide linked to bacterial flagellin acting as an adjuvant for mucosal immunization. It was previously demonstrated [13] that immunization of mice with *E. coli*-produced flagellin linked to four tandem copies of human "consensus" M2e, provides protection against lethal challenge with human influenza A strain. Although the M2e sequence is well conserved in almost all influenza strains isolated from humans, the sequences of M2e peptides of newly emerging highly pathogenic strains of animal origin, such as avian influenza, differ from the human consensus, suggesting that a specific M2e-based vaccine against these strains would be required [35]. Combination of several M2e peptides of different origin in one vaccine protein could allow development of candidate vaccine efficient against the corresponding strains [36]. Therefore in addition to human consensus M2e peptide we included M2e peptides of avian influenza virus into the hybrid protein based on bacterial flagellin.

Although plants for a long time have been recognized as a promising host for expression of recombinant proteins, including vaccines, there are no universal approaches ensuring high level expression for any protein of interest. Here we used two plant viral-based expression system for production of Flg-4M protein in *N. benthamiana*. The first is replicating vector pA7248AMV-GFP based on PVX genome, and the second vector pEAQ comprising fragments of CPMV genome is unable to replicate in plant cell. In this instance the replicating PVX-based vector gave higher expression of the Flg-4M protein than pEAQ. The results obtained here are in contrast to those found when comparing the expression of the Human Papillomavirus 8 L1 protein using a replicating TMV-based vector and the pEAQ-*HT* system where the latter gave a 15-fold higher level [37]. There are several possible explanations for this difference in expression levels between the two systems: it may reflect an intrinsic higher level of expression from the replicating versus the non-replicating system or it may relate to the strain of *A. tumefaciens* (GV3101 vs LBA4404) used for infiltration. The target protein was expressed in *N. benthamiana* cells at very high level up to 30 % of total soluble protein for pA7248AMV-GFP (~1 mg/g of fresh leaf tissue). It is much higher than 280 μg/g of fresh weight leaves reported for expression of the flagellin alone using non-replicating transient expression system [38]. Analysis of the plant-expressed Flg-4M protein by transmission electron microscopy did not reveal the presence of aggregates. This coupled with observation that the protein remained in the supernatant after ultracentrifugation suggests that the protein remains in an essentially unaggregated state.

Intranasal immunization of mice with the plant-produced Flg-4M protein induced an immune response against M2e and was responsible for development of protective immunity against the lethal influenza infection. Complete protection was observed in an experiment where immunized mice were challenged with $5LD_{50}$ of human influenza strain A/PR/8/34. Notably the protection strongly depended on the presence of M2e in the hybrid protein, since immunisation with plant-produced empty Flg confer no extra protection relative to the negative control. In other experiments the protection was, however, not 100 % complete that could be related to low immunization dose (only 10 μg) and, in case of experiment with $10LD_{50}$ of A/PR/8/34 strain, by the use of too high infection dose, as evidenced by complete mortality in the control group.

The lack of complete protection of mice was observed by Stepanova *et al.* [39] in case of Flg-4M protein produced in *E. coli*: 90 % survival upon challenge with $5LD_{50}$ of A/Chicken/Kurgan/05/2005 and 80 % survival upon challenge with $5LD_{50}$ of A/PR/8/34 strain, although a higher dose of vaccine protein, 50 μg, was used for each of three immunisation.

Three other examples of the plant-based expression of the M2e epitope have been reported. Ravin *et al.* [9], using the PVX-based vector employed in this work, produced HBc virus-like particles carrying the M2e peptide with the yield about 1-2 % of total soluble protein; the plant-produced particles induced a protective immune response (90 % survival) against $1LD_{50}$ of homologous influenza A infection. Virus-like particles formed by L1 protein of human papillomavirus fused to M2e were produced in *N. benthamiana* with the pEAQ-*HT* expression system with the yield in the range from 45 to 120 μg/g of fresh weight [40]. Another approach is construction of recombinant tobacco mosaic virus carrying M2e epitope as a fusion with the coat protein; such viruses are capable of systemic infection of *N. benthamiana* plants and induced a protective immune response against challenge with $5LD_{50}$ of homologous influenza A virus [41]. Altogether these results indicated that production of M2e-based candidate influenza vaccines in plants is feasible. The particular advantage of flagellin as a mucosal adjuvant suggests that expression of flagellin-M2e hybrid protein in plants by agroinfiltration with a recombinant viral vectors is a promising approach for production of recombinant vaccines against influenza.

Conclusions

In this study, we expressed the chimeric protein comprising flagellin of *Salmonella typhimurium* fused to four tandem copies of the influenza M2e peptide in *N. benthamiana* plants. The use of self-replicating viral vector based on PVX genome allowed to achieve very high level of expression up to 1 mg/g of fresh leaf tissue. Intranasal immunization of mice with purified vaccine protein induced high levels of M2e-specific serum antibodies and provided protection against lethal challenge with influenza virus. This work shows that expression of fusion proteins based on flagellin and M2e in plants is feasible and could became a promising approach for manufacturing of "universal" recombinant vaccines against influenza.

Methods

Media, reagents and synthetic gene

Bacteria were grown in LB broth or on plates with LB agar at 37 °C (*Escherichia coli*) or at 28 °C (*Agrobacterium tumefaciens*). If necessary, the media were supplemented with antibiotics: kanamycin (50 μg/ml), rifampicin (50 μg/ml), or gentamycin (25 μg/ml). Monoclonal antibodies to the M2e (clone M2A10) were generously provided by Dr. Peter Sveshnikov from the Russian Research Center of Molecular Diagnostics and Therapy.

Previously we constructed an artificial gene Flg-4M encoding a hybrid protein carrying an N-terminal histidine tag and consisting of flagellin of *Salmonella typhimurium* (Flg) with four copies of M2e peptide linked to its C terminus [39]. The M2e sequences corresponded to a human "consensus" M2e peptide (SLLTEVETPIRNEWGCRCNDSS DP, M2eh), and the M2e peptide of avian influenza virus strain A/Chicken/Kurgan/05/2005 (SLLTEVETPTRNEWEC RCSDSSD, M2ek). Individual copies of the M2e peptides were separated from each other by flexible glycine-rich linkers. The hybrid gene, arranged as Flg-M2eh-M2ek-M2eh-M2ek, was cloned in expression vector pQE30 resulting in plasmid pQE30_Flg2M2eh2M2ek. The amino acid sequence of the hybrid protein Flg-4M is shown in Fig. 1a.

Vectors for expression of target proteins in plants

For expression of the hybrid protein Flg-4M in plants the corresponding gene was cloned into vectors pA7248AMV-GFP (based on the PVX genome [32]) and pEAQspecialK-GFP-HT (a non-replicating CPMV-based vector [34]). These are the binary vectors and can be maintained in both *E. coli* and *A. tumefaciens*. The sequence encoding Flg-4M was amplified by PCR using primers AscI-Flg-F (ATGGCG CGCCATGCATCACCATCACCATCACGGA) and M2K-SmaI-R (ATCCCGGGAATTAAGCTGGGGTACCCTAAT CG) or NruI-Flg-F (ATTCGCGAATGCATCACCATCAC CATCACGGA) and M2K-AvaI-R (ATCTCGAGAATTAA GCTGGGGTACCCTAATCG) using pQE30_Flg2M2eh2-M2ek as a template for PCR-amplification.

For expression of flagellin lacking M2e (Flg) the corresponding gene was amplified by PCR using primers AscI-Flg-F (ATGGCGCGCCATGCATCACCATCACCA TCACGGA) and Flg-SmaI-R (ATCCCGGGCTAACG-TAACAGAGACAGCACGTTC) and pQE30_Flg2M2eh 2M2ek as a template.

The PCR fragments encoding Flg-4M were cloned in pA7248AMV-GFP at the *AscI*/*SmaI* sites or in pEA QspecialK-GFP-HT at the *NruI*/*AvaI* sites. The resulting recombinant vectors, pA7248AMV-Flg-4M and pEAQ-Flg-4M (Fig. 1b), were transferred from *E. coli* into the *A. tumefaciens* strain GV3101 using electroporation. Likewise the PCR fragment encoding Flg was cloned in pA7248AMV-GFP at the *AscI*/*SmaI* sites resulting in recombinant vector pA7248AMV-Flg for expression of 6-his tagged flagellin in plants.

Agroinfiltration of plants

Agrobacteria carrying recombinant binary vectors were grown overnight with shaking at 28 °C. The cells (1.5 ml) were harvested by centrifugation for 5 min at 4000 g, and the pellet was resuspended in 1.5 ml of buffer containing 10 mM MES (pH 5.5) and 10 mM $MgSO_4$. Leaves of *N. benthamiana* plants were infiltrated with suspension of agrobacteria (optical absorption of the solution $OD_{600} \sim 0.2$) using a syringe without a needle.

For expression of pA7248AMV-Flg-4M, the plant leaves were concurrently infiltrated with agrobacteria carrying binary vector driving expression of silencing suppressor 24-kDa protein (p24) from grapevine leafroll-associated virus-2 [42] to suppress virus-induced gene silencing. The same protocol was used for expression of Flg.

Isolation of proteins from plant tissues

The plant-produced Flg-4M protein carrying 6-histidine tag was isolated under denaturing conditions on Ni-NTA Resin (Promega, USA). Four to five days after infiltration, the *N. benthamiana* leaves were powdered to homogeneity in liquid nitrogen. A solution containing 6 M Guanidine HCI, 0.1 M NaH_2PO_4, 0.01 M Tris (pH 8.0) was added to the powdered plant tissue (5 mL per 1 g). The resulting mixture was centrifuged at 14,000 g for 15 min, and the supernatant was applied to Ni-NTA-resin equilibrated with the same buffer and incubated for 60 min. Then the resin was sequentially washed with PBS (50 mM NaH_2PO_4, 300 mM NaCl, pH 8.0) buffer containing 10 mM and 20 mM imidazole, respectively. The recombinant protein Flg-4M was eluted with PBS containing 0.5 M imidazole. After elution the protein was dialyzed against PBS using Slide-A-Lyzer Mini dialysis devices (Thermo Scientific, USA). Dialyzed protein preparation was centrifuged at 100,000 g to remove Rubisco aggregates and then supernatant was stored at –20 °C in PBS. Total soluble proteins were measured by a Bradford assay (Bio-Rad) following manufacturer's instructions. The same protocol was used for purification of Flg.

SDS-PAGE and western-blotting of protein preparations

Protein samples were diluted in SDS-PAGE sample buffer including β-mercaptoethanol. The samples were boiled for 5 min and subjected to SDS-PAGE in a 10 % (w/v) gel. After electrophoresis, the gel was either stained with Coomassie brilliant blue or the proteins were transferred from the gel onto a Hybond-P membrane (GE Healthcare, USA) by electroblotting. To prevent nonspecific binding of antibodies with the membrane, it was treated with 5 % (w/v) solution of dry milk in TBS-T (150 mM NaCl, 20 mM Tris, 0.1 % Tween 20, pH 8.0) buffer. The membrane was probed with mouse monoclonal antibodies specific for the M2e peptide and then incubated with rabbit anti-mouse antibodies conjugated with peroxidase. Specific protein complexes were detected using a Western Blot ECL Plus kit (GE Healthcare, USA). The intensity of the bands in stained gels and blots was determined using Nonlinear. Dynamics. TotalLab. TL120.v2009-NULL software.

Western blot analysis of purified Flg-4M under non-reducing conditions was performed as above but that β-mercaptoethanol was not included in the sample buffer and the sample was not boiled before loading onto a gel.

Protein ELISA

To confirm identity and accessibility of M2e to antibodies, ELISA plates were coated overnight at 4 °C with serial dilutions of Flg-4M or Flg proteins in sodium bicarbonate buffer pH 8.5. Plates were treated with a blocking buffer supplemented with 0.1 % (w/v) BSA for 2 h at 37 °C and washed three times using PBS with Tween. Plates were probed with monoclonal antibody specific for M2e or 6-His tag for 1 h at 37 °C. As a conjugate, rat monoclonal anti-mouse IgG (Imtek, Russia) labelled with a horseradish peroxidase was used at a 1:10,000 dilution. After adding tetramethylbenzidine substrate (BD Bioscience) and monitoring color development, the reaction was stopped by H_2SO_4, and OD at 450 nm was measured on a microplate spectrophotometer.

Synthetic peptides

The following synthetic peptides were tested as a positive control in the ELISA experiments:

G-50 SLLTEVETPTRNEWECRCSDSSD (M2ek, A/Chicken/Kurgan/05/2005)

G-37 SLLTEVETPIRNEWGCRCNDSSD (M2eh, human consensus). Residues that differ between the sequences are underlined.

Analysis of production of antibodies against M2e

Female Balb/c mice weighing 18–20 g were immunized intranasally thrice at two week intervals. The dose of Flg-4M and Flg proteins was 10 μg/mouse and no additional adjuvant was used. The mouse sera were studied two weeks after the second and third immunizations. The sera of non-immunized mice were used as a negative control.

For ELISA, 96-well plates with a high adsorption capacity (Greiner, Germany) were covered with synthetic peptides G-50 (M2ek) and G-37 (M2eh) at a concentration of 5 mg/ml (w/v) in phosphate buffer, pH 7.2–7.4 and kept overnight at 4 °C. Plates were treated with a blocking buffer (0.01 M PBS, pH 7.2–7.4) supplemented with 5 % (v/v) fetal calf serum for 1 h at room temperature and washed three times using PBS with Tween. A series of 1:2 dilutions of the different serum samples, starting with a 1:400 dilution, were loaded onto the antigen-coated plates in the blocking buffer and then incubated for 1 h at room temperature. As a conjugate, rat monoclonal anti-mouse IgG (Invitrogen) labelled with a horseradish peroxidase was used at a 1:2000 dilution. Tetramethylbenzidine (BD Bioscience) was used as a substrate. The reaction was monitored by measuring the OD at 450 nm. The endpoint titres are defined as the highest dilution producing an OD value twice that of the background (serum of non-immunised mice).

Influenza viruses and challenge

The following mouse-adapted influenza viruses were used to challenge animals immunised by the candidate vaccines: A/PR/8/34 (H1N1) at the dose of $10LD_{50}$ or $5LD_{50}$, and A/Chicken/Kurgan/05/2005 (H5N1) at the dose of $5LD_{50}$. The virus was administered intranasally in a total volume of 100 μl to mice anaesthetised by ether. The mice were monitored for weight loss and survival rate daily following the viral challenge for a period of 2 weeks. Survival rate was determined by death or a cut-off of 25 % in body weight loss at which point animals were euthanized.

Ethics statement

The study was carried out in accordance with the Russian Guidelines for the Care and Use of Laboratory Animals. The protocol was approved by the Committee for Ethics of Animal Experimentation of the Research Institute of Influenza (Permit Number: 01213). All efforts were made to minimize animal suffering.

Additional file

Additional file 1: Figure S1. Western- blot analysis of purified Flg-4M protein under reducing and nonreducing conditions.

Abbreviations

AMV: 5′-untranslated region of RNA 4 of the alfalfa mosaic virus; CPMV: Cowpea mosaic virus; M2e: The extracellular domain of matrix protein 2 of influenza A virus; PVX: Potato virus X; TLR: Toll-like receptor; TMV: Tobacco mosaic virus.

Competing interests

The authors declare that they have no competing interests.

Authors' contributions

EM constructed recombinant viral vectors, carried out the plant expression and drafted the manuscript. RK and VK designed and constructed the genes encoding the fusion Flg-4M protein. LS and LT performed the animal experiments and analysed the data. GPL participated in the design of the study and helped to draft the manuscript. NR conceived of the study, and participated in its design and coordination and helped to draft the manuscript. All authors read and approved the final manuscript.

Acknowledgements

This work was supported by the joint grant of Royal Society (UK) International Exchanges Scheme (grant IE130251) and Russian Foundation for Basic Research (project 13-04-92621) and by the fellowship of the President of Russia for young scientists awarded to R.Y. Kotlyarov. The work was also supported by the UK Biotechnological and Biological Sciences Research Council (BBSRC) Institute Strategic Programme Grant "Understanding and Exploiting Plant and Microbial Secondary Metabolism" (BB/J004596/1) and the John Innes Foundation.

Author details

[1]Centre 'Bioengineering', Russian Academy of Sciences, Prosp. 60-letya Oktyabrya, bld 7-1, 117312 Moscow, Russia. [2]Research Institute of Influenza, Russian Federation Ministry of Health, 15/17 Prof. Popova str., 197376 St. Petersburg, Russia. [3]Department of Biological Chemistry, John Innes Centre, Norwich Research Park, NR4 7UH Norwich, UK.

References

1. Webster RG, Bean WJ, Gorman OT, Chambers TM, Kawaoka Y. Evolution and ecology of influenza A viruses. Microbiol Rev. 1992;56:152–79.
2. Fiers W, De Filette M, El Bakkouri K, Schepens B, Roose K, Schotsaert M, et al. M2e-based universal influenza A vaccine. Vaccine. 2009;27:6280–3.
3. Fiers W, De Filette M, Birkett A, Neirynck S, Min Jou W. A "universal" human influenza A vaccine. Virus Res. 2004;103:173–6.
4. Ito T, Gorman OT, Kawaoka Y, Bean WJ, Webster RG. Evolutionary analysis of the influenza A virus M gene with comparison of the M1 and M2 proteins. J Virol. 1991;65:5491–8.
5. Neirynck S, Deroo T, Saelens X, Vanlandschoot P, Jou WM, Fiers W. A universal influenza A vaccine based on the extracellular domain of the M2 protein. Nat Med. 1999;5:1157–63.
6. Feng J, Zhang M, Mozdzanowska K, Zharikova D, Hoff H, Wunner W, et al. Influenza A virus infection engenders a poor antibody response against the ectodomain of matrix protein 2. Virol J. 2006;3:102.
7. Liu W, Peng Z, Liu Z, Lu Y, Ding J, Chen YH. High-epitope density in a single recombinant protein molecule of the extracellular domain of influenza A virus M2 protein significantly enhances protective immunity. Vaccine. 2004;2:366–71.
8. Mozdzanowska K, Feng J, Eid M, Kragol G, Cudic M, Otvos L, et al. Induction of influenza type A virus-specific resistance by immunization of mice with a synthetic multiple antigenic peptide vaccine that contains ectodomains of matrix protein 2. Vaccine. 2003;21(19–20):2616–26.
9. Ravin NV, Kotlyarov RY, Mardanova ES, Kuprianov VV, Migunov AI, Stepanova LA, et al. Plant-produced recombinant influenza vaccine based on virus-like HBc particles carrying an extracellular domain of M2 protein. Biochemistry (Mosc). 2012;77:33–40.
10. Iwasaki A, Medzhitov R. Toll-like receptor control of the adaptive immune responses. Nat Immunol. 2004;5:987–95.
11. Takeda K, Kaisho T, Akira S. Toll-like receptors. Annu Rev Immunol. 2003;21:335–76.
12. McDonald WF, Huleatt JW, Foellmer HG, Hewitt D, Tang J, Desai P, et al. A West Nile virus recombinant protein vaccine that co-activates innate and adaptive immunity. J Infect Dis. 2007;195:1607–17.
13. Huleatt JW, Nakaar V, Desai P, Huang Y, Hewitt D, Jacobs A, et al. Potent immunogenicity and efficacy of a universal influenza vaccine candidate comprising a recombinant fusion protein linking influenza M2e to the TLR5 ligand flagellin. Vaccine. 2008;26:201–14.
14. Jeon SH, Ben-Yedidia T, Arnon R. Intranasal immunization with synthetic recombinant vaccine containing multiple epitopes of influenza virus. Vaccine. 2002;20:2772–80.
15. Hong SH, Byun Y-H, Nguyen CT, Kim SY, Seong BL, Park S, et al. Intranasal administration of flagellin-adjuvanted inactivated influenza vaccine enhances mucosal immune responses to protect mice against lethal infection. Vaccine. 2012;30:466–74.
16. Liu G, Tarbet B, Song L, Reiserova L, Weaver B, Chen Y, et al. Immunogenicity and efficacy of flagellin-fused vaccine candidates targeting 2009 pandemic H1N1 influenza in mice. PLoS One. 2011;6:e20928.
17. Liu G, Song L, Reiserova L, Trivedi U, Li H, Liu X, et al. Flagellin-HA vaccines protect ferrets and mice against H5N1 highly pathogenic avian influenza virus (HPAIV) infections. Vaccine. 2012;30:6833–8.
18. Song L, Zhang Y, Yun NE, Poussard AL, Smith JN, Smith JK, et al. Superior efficacy of a recombinant flagellin: H5N1 HA globular head vaccine is determined by the placement of the globular head within flagellin. Vaccine. 2009;27:5875–84.
19. Wang B-Z, Xu R, Quan F-S, Kang SM, Wang L, Compans RW. Intranasal immunization with influenza VLPs incorporating membrane-anchored flagellin induces strong heterosubtypic protection. PLoS One. 2010;5:e13972.
20. Gleba Y, Klimyuk V, Marillonnet S. Viral vectors for the expression of proteins in plants. Curr Opin Biotechnol. 2007;18:134–41.
21. Lico C, Chen Q, Santi L. Vectors for production of recombinant proteins in plants. J Cell Physiol. 2008;216:366–77.

22. Thuenemann EC, Lenzi P, Love AJ, Taliansky M, Becares M, Zuniga S, et al. The use of transient expression systems for the rapid production of virus-like particles in plants. Curr Pharm Des. 2013;19:5564–73.
23. Edelbaum O, Stein D, Holland N, Gafni Y, Livneh O, Novick D, et al. Expression of active human interferon-β in transgenic plants. J Interferon Res. 1992;12:449–53.
24. Kusnadi AR, Nikolov ZL, Howard JA. Production of recombinant proteins in transgenic plants: practical considerations. Biotechnol Bioeng. 1997;56:473–84.
25. Kapila J, De Rycke R, van Montagu M, Angenon G. An Agrobacterium-mediated transient gene expression system for intact leaves. Plant Sci. 1997;122:101–8.
26. Marillonnet S, Thoeringer C, Kandzia R, Klimyuk V, Gleba Y. Systemic *Agrobacterium tumefaciens*-mediated transfection of viral replicons for efficient transient expression in plants. Nat Biotechnol. 2005;23:718–23.
27. Cañizares MC, Nicholson L, Lomonossoff GP. Use of viral vectors for vaccine production in plants. Immunol Cell Biol. 2005;83:263–70.
28. Chichester JA, Yusibov V. Plants as alternative systems for production of vaccines. Hum Vaccines. 2007;3:146–9.
29. D'Aoust MA, Couture MM, Charland N, Trepanier S, Landry N, Ors F, et al. The production of hemagglutinin-based virus-like particles in plants: a rapid, efficient and safe response to pandemic influenza. Plant Biotechnol J. 2010;8:607–19.
30. Mett V, Musiychuk K, Bi H, Farrance CE, Horsey A, Ugulava N, et al. A plant-produced influenza subunit vaccine protects ferrets against virus challenge. Viruses. 2008;2:33–40.
31. Shoji Y, Chichester JA, Jones M, Manceva SD, Damon E, Mett V. Plant-based rapid production of recombinant subunit hemagglutinin vaccines targeting H1N1 and H5N1 influenza. Hum Vaccines. 2011;7:41–50.
32. Mardanova ES, Kotliarov RIu, Ravin NV. Increased efficiency of recombinant proteins production in plants due to optimized translation of RNA of viral vector. Mol Biol (Mosk). 2009;43(3):568-71.
33. Sainsbury F, Lomonossoff GP. Extremely high-level and rapid transient protein production in plants without the use of viral replication. Plant Physiol. 2008;148:1212–8.
34. Sainsbury F, Thuenemann EC, Lomonossoff GP. pEAQ: versatile expression vectors for easy and quick transient expression of heterologous proteins in plants. Plant Biotechnol J. 2009;7:682–93.
35. De Filette M, Fiers W, Martens W, Birkett A, Ramne A, Löwenadler B, et al. Improved design and intranasal delivery of an M2e-based human influenza A vaccine. Vaccine. 2006;24:6597–601.
36. Blokhina EA, Kuprianov VV, Stepanova LA, Tsybalova LM, Kiselev OI, Ravin NV, et al. A molecular assembly system for presentation of antigens on the surface of HBc virus-like particles. Virology. 2013;435:293–300.
37. Matić S, Masenga V, Poli A, Rinaldi R, Milne RG, Vecchiati M, et al. Comparative analysis of recombinant Human Papillomavirus 8 L1 production in plants by a variety of expression systems and purification methods. Plant Biotechnol J. 2012;10:410–21.
38. Girard A, Saron W, Bergeron-Sandoval LP, Sarhan F, Archambault D. Flagellin produced in plants is a potent adjuvant for oral immunization. Vaccine. 2011;29:6695–703.
39. Stepanova L, Kotlyarov R, Kovaleva A, Potapchuk M, Korotkov A, Sergeeva M, et al. Protection against multiple influenza A virus strains induced by candidate recombinant vaccine based on heterologous M2e peptides linked to flagellin. PLoS One. 2015;10(3):e0119520.
40. Matić S, Rinaldi R, Masenga V, Noris E. Efficient production of chimeric human papillomavirus 16 L1 protein bearing the M2e influenza epitope in Nicotiana benthamiana plants. BMC Biotechnol. 2011;11:106.
41. Petukhova NV, Gasanova TV, Stepanova LA, Rusova OA, Potapchuk MV, Korotkov AV, et al. Immunogenicity and protective efficacy of candidate universal influenza A nanovaccines produced in plants by tobacco mosaic virus-based vectors. Curr Pharm Des. 2013;19:5587–600.
42. Chiba M, Reed JC, Prokhnevsky AI, Chapman EJ, Mawassi M, Koonin EV, et al. Diverse suppressors of RNA silencing enhance agroinfection by a viral replicon. Virology. 2006;346:7–14.

15

Molecular characterization of two A-type P450s, *WsCYP98A* and *WsCYP76A* from *Withania somnifera* (L.) Dunal: expression analysis and withanolide accumulation in response to exogenous elicitations

Satiander Rana[1], Wajid Waheed Bhat[1], Niha Dhar[1], Shahzad A Pandith[1], Sumeer Razdan[1], Ram Vishwakarma[2] and Surrinder K Lattoo[1*]

Abstract

Background: Pharmacological investigations position withanolides as important bioactive molecules demanding their enhanced production. Therefore, one of the pivotal aims has been to gain knowledge about complete biosynthesis of withanolides in terms of enzymatic and regulatory genes of the pathway. However, the pathway remains elusive at the molecular level. P450s monooxygenases play a crucial role in secondary metabolism and predominantly help in functionalizing molecule core structures including withanolides.

Results: In an endeavor towards identification and characterization of different P450s, we here describe molecular cloning, characterization and expression analysis of two A-type P450s, *WsCYP98A* and *WsCYP76A* from *Withania somnifera*. Full length cDNAs of *WsCYP98A* and *WsCYP76A* have open reading frames of 1536 and 1545 bp encoding 511 (58.0 kDa) and 515 (58.7 kDa) amino acid residues, respectively. Entire coding sequences of *WsCYP98A* and *WsCYP76A* cDNAs were expressed in *Escherichia coli* BL21 (DE3) using pGEX4T-2 expression vector. Quantitative real-time PCR analysis indicated that both genes express widely in leaves, stalks, roots, flowers and berries with higher expression levels of *WsCYP98A* in stalks while *WsCYP76A* transcript levels were more obvious in roots. Further, transcript profiling after methyl jasmonate, salicylic acid, and gibberellic acid elicitations displayed differential transcriptional regulation of *WsCYP98A* and *WsCYP76A*. Copious transcript levels of both P450s correlated positively with the higher production of withanolides.

Conclusions: Two A-types P450 *WsCYP98A* and *WsCYP76A* were isolated, sequenced and heterologously expressed in *E. coli*. Both P450s are spatially regulated at transcript level showing differential tissue specificity. Exogenous elicitors acted as both positive and negative regulators of mRNA transcripts. Methyl jasmonate and salicylic acid resulted in copious expression of *WsCYP98A* and *WsCYP76A*. Enhanced mRNA levels also corroborated well with the increased accumulation of withanolides in response to elicitations. The empirical findings suggest that elicitors possibly incite defence or stress responses of the plant by triggering higher accumulation of withanolides.

Keywords: *Withania somnifera*, Withanolides, Cytochrome P450 monooxygenase, mRNA, Phytohormones, *E.coli*

* Correspondence: sklattoo@iiim.ac.in
[1]Plant Biotechnology Division, CSIR-Indian Institute of Integrative Medicine, Canal Road, Jammu Tawi-180001, India
Full list of author information is available at the end of the article

Background

Cytochrome P450s form a huge superfamily of heme-containing monooxygenases present in all domains of life. These are pivotal in detoxification of xenobiotics, drug metabolism, assimilation of carbon sources and formation of secondary metabolites. Presently, there are more than 18500 P450 genes that have been identified across all the kingdoms of life [1,2]. The preponderance of P450s is more in plants per species than in animals and their content in angiosperm genomes can reach up to 2% [3,4]. Plant cellular and metabolic processes are dependent on P450s to drive the reactions such as hydroxylation, oxidative demethylation, desaturation, epoxidation, oxidative rearrangement of the carbon skeleton and oxidative C–C bond cleavage [5]. P450s are endoplasmic reticulum (ER) localised, requiring auxiliary reductases for the activation of molecular oxygen for different reactions. These reductases transfer two electrons in a single step from cofactor NAD(P)H to heme centre of P450s [6]. P450 monooxygenases constitute a highly regio and stereo-specific class of definite substrate-specific enzymes. By gene annotations approximately, 1% of the total genes in *Arabidopsis thaliana* and other model plants are found to be cytochrome P450s. *Arabidopsis* genome contains 244 genes and 28 pseudo-genes representing cytochrome P450 [3]. Involvement of P450s in diverse physiological processes *in vivo,* leads to biosynthesis and catabolism of huge array of complex metabolites like lignin, pigments, defence compounds, fatty acids, hormones and signaling molecules [7-9].

Owing to their regio and stereo-specific catalysing versatility, these are potential targets for industrial biocatalysis. P450s have been applied in industry for the investigation of new drugs, medicine or xenobiotics [10-12]. Remarkable variety of chemical reactions catalysed and enormous number of substrates attacked, have earned P450s the reputation of "the most versatile biological catalysts in nature" [13-15]. Thus, emphasizing identification and characterization of P450s for elucidation of various biosynthetic pathways. Plant P450s are presumed to be polyphyletic in their origin and generally classified into two main clades, A-type and non-A-type. A-type P450s are involved in 'plant specific metabolism' and are responsible for the biosynthesis of diverse natural products [7].

Withania somnifera is a medicinal plant of immense repute known to synthesize a suite of low molecular weight natural products such as flavonoids, alkaloids, terpenoids, tannins, resins and sterols through secondary metabolism [16-18]. These metabolites are derived from distinct metabolic pathways. The main constituents reported for significant pharmacological activities are withanolides, a group of naturally occurring triterpenoids. Withanolides are synthesized *via* both mevalonate (MVA) and non-mevalonate (MEP/DOXP) pathways [19]. Recently, we have elucidated the regulatory role of three branch point oxidosqualene cyclases (β-amyrin synthase, lupeol synthase and cycloartenol synthase) which showed that repression of β-amyrin synthase and lupeol synthase led to diversion of common precursor flux towards cycloartenol synthase resulting in enhanced accumulation of withanolides [20]. We have also characterized two paralogs of cytochrome P450 reductase (*WsCPR1* and *WsCPR2*) wherein *WsCPR2* expression was inducible in response to exogenous elicitors [21]. Two more genes namely squalene synthase and squalene epoxidase were also characterized from same plant species [22]. The regulation pattern of squalene synthase, squalene epoxidase, cycloartenol synthase and *WsCPR1* and *WsCPR2* in leaves and roots at various developmental stages in relation to dynamics of withanolides accumulation, indicated towards their tissue-specific biosynthesis [22]. In our earlier findings, we have reported that exogenous application of methyl jasmonate (MeJA) and salicylic acid (SA) increases withanolide content due to the presence of stress regulatory elements within the upstream promoter regions of their genes [20,23].

We, in present study report isolation, heterologous expression and transcript profiling of two A-type P450 monooxygenases viz. *WsCYP98A* (GenBank: HM585369) and *WsCYP76A* (GenBank: KC008573). These are supposed to be involved mainly in the shikmate and phenylpropanoid pathway biosynthesizing the various defence molecules during biotic and abiotic stress. CYP98A (p-coumarate 3-hydroxylase) catalyzes the meta-hydroxylation of *p*-coumarate derivatives for the biosynthesis of lignins and chlorogenic acid, which constitutes a crucial bottleneck in phenylpropanoid pathway [24]. The members of CYP76 gene family characterized from many plant species are known to catalyse various oxidative reactions. For instance, *Catharanthus roseus* CYP76A26 converts both iridodial and iridotrial into 7-deoxyloganetic acid. Whereas, *C. roseus* CYP76B6 and CYP76C1 from *A. thaliana* execute hydroxylation of geraniol [25,26].

In present investigation, full length genes of *WsCYP98A* and *WsCYP76A* were cloned and heterologously expressed in *Escherichia coli. WsCYP98A* shares maximum homology with reported *CYP98A* of *Capsicum annum* (95%) and *Solanum tuberosum* (94%) whereas, *WsCYP76A* was similar to *CYP76A2* of *S. tuberosum* (83%) and *S. melongena* (83%). The phylogenetic analysis demonstrated that both genes belong to their respective families and are close homologs to other *CYP98A* and *CYP76A* genes of Solanaceae family. The prediction of three dimensional structures, ligand binding site and conserved amino acid residues of both proteins were done using homology modelling analysis. The transcript profiling of *WsCYP98A* and *WsCYP76A* in different tissues of *W. somnifera* was performed using quantitative real time PCR (qRT-PCR). Expression profiles in response to exogenous elicitors namely MeJA, SA and gibberellic acid (GA_3) were also studied for both the P450s. Empirical findings reveal their possible role in defence mechanism *via* enhanced accumulation of withanolides. Purview of literature suggests that this is the only report of type-A *WsCYP98A* and *WsCYP76A* from *W. Somnifera*.

Methods

Plant material

A WS-3 rich genetic stock of *W. somnifera* designated as WS-Y-08 grown at IIIM experimental farm (CSIR-Indian Institute of Integrative Medicine, Jammu, India, 32°44′N longitude, 74°55′E latitude; 305 m in altitude) was used as a source material. *In vitro* plants were raised through induction of axillary buds. These cultures were used for MeJA, SA and GA_3 treatments. Samples were collected, frozen immediately in liquid nitrogen, and stored at −80°C.

RNA isolation and cDNA synthesis

Total RNA was extracted using Trizol reagent (Sigma, St. Louis, USA) as per manufacturer's protocol. Concentration of isolated RNAs was estimated by measuring the absorbance at 260 nm in a spectrophotometer (AstraAuriga, Cambridge, UK). Further, quality of RNA was assessed by determining the ratio of absorbance at 260 and 280 nm ($A_{260/280}$) and formaldehyde-denatured agarose gel electrophoresis. Total RNA (5 μg) was incubated with RNase free DNase (Fermentas, Burlington, Canada) at 37°C for 30 min. First strand cDNA synthesized using the RevertAid cDNA synthesis kit (Fermentas, Burlington, Canada) in a total volume of 20 μl containing 3 μg total RNA, 10 mM dNTPs, 10 μM oligo (dT) primer, 1 μl M-MuLV reverse transcriptase (200 U/μl) and 1X first strand buffer (250 mM Tris–HCl, pH 8.3, 250 mM KCl, 20 mM $MgCl_2$, 50 mM DTT). The reaction was incubated for 60 min at 42°C followed by 5 min at 70°C to inactivate the reverse transcriptase.

Amplification of *WsCYP98A* and *WsCYP76A*

Degenerate primers (Table 1) based on the conserved regions of P450s were designed by multiple sequence alignment of different P450 monooxygenases sequences retrieved from the GenBank database at National Center for Biotechnology Information (NCBI). Optimization of polymerase chain reaction conditions allowed amplification of cDNA fragments of *WsCYP98A* and *WsCYP76A* under following cycling conditions: one cycle of 94°C for 3 min, 35 cycles of 94°C for 30 s, 55–60°C for 30 sec and 72°C for 1 min followed by a final extension of 72°C for 10 min in a thermal cycler (Bio-Rad Laboratories, Hercules, USA). Amplicons examined by agarose gel electrophoresis were cloned into pTZ57R/T vector (Fermentas, Burlington, Canada), and then transformed into *E. coli* DH5α host strain. Cloned amplicons were sequenced using an automated DNA sequencer (ABI Prism 3130XL; Applied Biosystems, Foster City, USA). The nucleotide sequences obtained were analysed using the similarity search BLAST program and subsequently used for designing gene specific primers (GSPs).

5′ and 3′ RACE PCR

Remaining 5 & 3 cDNA ends of putative *WsCYP98A* and *WsCYP76A* were obtained using RLM-RACE kit according to the product manual (Ambion, Austin, TX, USA). cDNAs obtained were subjected to nested PCR using GSPs as listed in Table 1. Two rounds of PCR were performed, first round with 5′ RACE-OUT corresponding to the 5′ RACE adapter sequence and 5′ CYP98-OUT and 5′ CYP76-OUT as specific outer primers, followed by a second round of PCR with 5′ RACE-IN as inner adapter specific primer and 5′ CYP98-IN and 5′ CYP76-IN as inner GSPs. In both rounds, PCR reactions of 50 μl containing 1.0 μl cDNA as template (except for second round where amplified products of 1st round were used as template), 2 μl of 10 μM 5′ CYP98-OUT and 5′ CYPP76-OUT , 2 μl of 5′ RACE-OUT, 45.0 μl master Mix (34.5 μl PCR-grade water, 10 mM Tris HCl; pH 9.0, 50 mM KCl, 2.5 mM $MgCl_2$, 200 μM dNTPs, 2.5 U Taq DNA polymerase) were subjected to following cycling conditions: One cycle of 94°C for 3 min and 35 cycles of 94°C for 30 s, 60°C for 30 s, 72°C for 2 min with a final step at 72°C of 10 min. Similarly for 3′ RACE, first strand cDNA was synthesized from total RNA using supplied 3′ RACE adapter primer. cDNA was subjected to nested PCR using outer and inner primers specific to 3′ RACE adapter along with 3′ GSPs with same reaction volume and cycling conditions as described for 5′ RACE. The nested amplification products of both 5′ RACE PCR and 3′ RACE PCR were purified and cloned into pTZ57R/T cloning vector (Fermentas, Burlington, Canada). The ligation mixtures were transformed into *E. coli* cells (New England Biolabs, Herts, UK). The clones were picked individually and amplified in 10 mL of Luria-Bertani (LB) medium at 37°C overnight. The plasmid DNA from each clone was extracted using a DNA plasmid Miniprep Kit (Promega, Madison, USA) and sequenced using M13 primers.

Full-length cloning of *WsCYP98A* and *WsCYP76A*

By comparing and aligning the sequences of core fragments, 5′ RACE and 3′ RACE products, the full-length cDNAs of *WsCYP98A* and *WsCYP76A* were generated and subsequently amplified with primers FullCYP98F, FullCYP98R and FullCYP76F, FullCYP76R (Table 1). A high fidelity proof-reading DNA polymerase (New England Biolabs, Herts, UK) was employed for amplification under PCR conditions; One cycle of 94°C for 3 min , 35 cycles of 94°C for 30 s, 60°C for 30 s, 72°C for 1 min. The final extension was at 72°C for 10 min. PCR products were

Table 1 List of primers used in the study

Primers	Sequence (5′ – 3′)	Direction
Degenerate		
CYP450-1 F.	CCN(A/C/G/T)TAY(C/T)GGH(A/C/T)GD(G/A/T)TACTGGAGR(A/G)CAG	Forward
CYP450-1R.	AGTCD(G/A/T)GD(G/A/T)CAW(A/T)D(G/A/T)GCCCATTCD(G/A/T)AG	Reverse
CYP450-2 F.	ATATGGGCY(C/T)GATTATGGV(G/A/C)CCTCA	Forward
CYP450-2R.	AGCATH(A/C/T)AGH(A/C/T)GGAGTTGGAGGY(C/T)TG	Reverse
5′ and 3′ RACE		
5′ Adapter*	GCUGAUGGCGAUGAAUGAACACUGCGUUUGCUGGCUUUGAUGAAA	
5′ RACE-OUT*	GCTGATGGCGATGAATGAACACTG	Forward
5′ RACE-IN*	CGCGGATCCGAACACTGCGTTTGCTGGCTTTGATG	Forward
5′ CYP98-OUT	TGTTATCAGAATTATTGCAATCTCTGT	Reverse
5′ CYP98-IN	AGCCGTAACCTCATCTTCACGAATGG	Reverse
5′ CYP76-OUT	CGTCTCTGTACCAGCTAAAAACATTTCCAC	Reverse
5′ CYP76-IN	GATCTCATGTTCTGATAACTTGGCTGGTT	Reverse
3′ Adapter*	GCGAGCACAGAATTAATACGACTCACTATAGGT_{12}V(G/A/C)N(A/C/T/G)	
3′ RACE-OUT*	GCGAGCACAGAATTAATACGACT	Reverse
3′ RACE- IN*	CGCGGATCCGAATTAATACGACTCACTATAGG	Reverse
3′ CYP98-OUT	TGATCGGCTATGAGCGCGTGATGAA	Forward
3′ CYP98-IN	AAGCGCTAAGGTTGCAGCCTCCAA	Forward
3′ CYP76-OUT	AGAGATCCTGAATGTTGGGATGACCCTAT	Forward
3′ CYP76-IN	AGAGATCCTGAATGTTGGGATGACCCTAT	Forward
Full length		
FullCYP98F	ATGGCAATTCCCTTAGCTGCTGCAATCCCTC	Forward
FullCYP98R	TTATATGTCCACTGCAATTCGTTTATACAACTCAGC	Reverse
FullCYP76F	ATGGAATGGGAATGGAGCTATTTGT	Forward
FullCYP76R	CTAGATAAGATTGATGAGTGTCTCA	Reverse
Expression (*E. coli*)		
CYP98*BamHI*F	CG**GGATCC**ATGGCAATTCCCTTAGCTGCTGCAA	Forward
CYP98*SalI*R	GCTC**GTCGAC**TTATATGTCCACTGCAATTCGT	Reverse
CYP76*BamHI*F	CG**GGATCC**ATGGAATGGGAATGGAGCTATTTGT	Forward
CYP76*SalI*R	GCTC**GTCGAC**CTAGATAAGATTGATGAGTGTCTCA	Reverse
Real-Time analysis		
RTCYP98F	GTGATGGATGAACAAGGAAACGA	Forward
RTCYP98R	CCGTGCTTAGAAAATGCATCCTC	Reverse
RTCYP76 F.P	TAGAATGGGCACTAGCAGAGCTTTTGCG	Forward
RTCYP76 R.P	CTTGCATATAAGGGAGATTGTCGATGTCGTT	Reverse
ActinF	GAGAGTTTTGATGTCCCTGCCATG	Forward
ActinR	CAACGTCGCATTTCATGATGGAGT	Reverse

*Primers marked with star were provided with the kits. Added restriction sites for cloning into pGEX4T-2 vector are bold.

analysed on 1.2% agarose gels and visualized under UV light. The resulted amplified products were ligated in pJET vector and transformed into *E.coli* DH5α.

In silico analysis

BLAST (http://blast.ncbi.nlm.nih.gov/Blast.cgi) was used to find similarity of amplified P450s in GenBank database. Translate tool (http://www.expasy.ch/tools/dna.html) was used to predict the open reading frame (ORF). The properties of deduced amino acid sequences of *WsCYP98A* and *WsCYP76A* were estimated by using ProtParam (http://www.expasy.ch/tools/protparam.html), SPLIT v.4.0 (http://split.pmfst.hr/split/4/) and TMHMM (http://www.cbs.dtu.dk/services/TMHMM/) programs.

Protein sequences were retrieved from the GenBank through BLASTp algorithm at the National Center for Biotechnology Information (NCBI) using *WsCYP98A* and *WsCYP76A* sequences as query and several homologous sequences of different plant species were selected. Multiple sequence alignment was performed using ClustalX program with default parameters [27]. For phylogenetic analysis, sequences were aligned employing ClustalW program (http://www.ebi.ac.uk/Tools/msa/clustalw2/) and phylogenetic tree was constructed by Neighbor-joining method using MEGA 6 software [28]. Bootstrap analysis with 100 replicates was also conducted in order to obtain confidence levels for the branches. Protein secondary structures were determined by SOPMA program [29].

Prediction of three-dimensional structures of *WsCYP98A* and *WsCYP76A*

Three-dimensional structures of *WsCYP98A* and *WsCYP76A* were predicted using Phyre^2 server (Protein Homology/analogY Recognition Engine V 2.0) [30] using crystal structure of cytochrome P450 1a2 (PDB: 2hi4) as a template. Ligand binding sites were predicted using 3DLigandSite. Ramachandran plot analysis of both *WsCYP98A* and *WsCYP76A* was performed online using RAMPAGE web server [31]. Structurally, evolutionary and functionally important regions were identified in deduced protein sequences by ConSurf [32].

Recombinant protein expression in *E. coli*

Full length coding sequences of *WsCYP98A* and *WsCYP76A* genes were modified by adding restriction sites (Table 1). Full length primers were modified by adding *BamHI/SalI* restriction sites for both genes for directional cloning in pGEX4T-2 vector and heterologous expression in *E. coli.* ORFs containing modified restriction sites were excised from pJET cloning vector and ligated into pre-digested, purified pGEX4T-2. Further, ligated mixtures containing pGEX-*WsCYP98A* and pGEX-*WsCYP76A* expression cassettes were transformed into *E. coli* BL21 (DE3). Individual positive colony harbouring transformed expression cassettes were inoculated separately into 100 ml of LB containing 100 µg/ml of ampicillin and incubated overnight at 37°C. 1% culture was transferred into 100 ml of LB media containing the corresponding antibiotic and incubated at 37°C, until optical density ($A_{600\ nm}$) reached 0.4-0.5. Protein expression was induced by adding Iso-propyl β-D-1-thiogalactopyranoside (IPTG; Fermentas, Berligton, Canada) into the cultures at the concentration of 0.2 mM to 1 mM. The cultures were constantly incubated at 25°C for 8 h. The induced bacterial cells were harvested at an interval of 2 h by centrifugation and resuspended in 6X sodium dodecyl sulphate–polyacrylamide gel electrophoresis sample buffer (SDS-PAGE; 0.375 M Tris pH 6.8, 12% SDS, 60% glycerol, 0.6 M DTT, 0.06% bromophenol blue). The expression of target proteins was analysed on 10% SDS-PAGE.

Tissue-specific gene expression analysis using qRT-PCR

To study tissue-specific expression, total RNA was extracted from leaves, stalks, roots, flowers and berries respectively. Total RNA (5 µg) was incubated with RNase free DNase (Fermentas, Burlington, Canada) at 37°C for 30 min. cDNA was synthesized from 3 µg of RNA using RevertAid cDNA synthesis kit according to manufacturer's instruction. qRT-PCR amplification was performed using the Step One Real-time PCR System (Applied Biosystems, Foster City, USA). Briefly, the standard 20 µl of reaction included 0.2 µL cDNA template, 200 nM of each primers, and 10 µL SYBR Premix Ex (Takara, Otsu, Japan) under following cycling conditions: one cycle of 94°C for 1 min, 40 cycle of 94°C for 10 s, 60°C for 15 s and 72°C for 15 s. Two primers, Actin F and Actin R were used to amplify housekeeping actin gene as control. All samples were analysed in triplicate and the specificity of each primer pair was validated by a dissociation curve (a single peak was observed for each primer pair). The real-time PCR amplification data were exported into Microsoft Excel and gene expression levels were calculated based on the comparative C_t method [33].

Effect of elicitors on the expression of *WSCYP98A* and *WsCYP76A*

For elicitor treatment the micro-propagated plantlets were pre-cultured in Murashige and Skoog (MS) liquid medium supplemented with 3% sucrose, inositol (100 mg l^{-1}) and incubated at 25 ± 1°C under 16 h photoperiod with light intensity of 30 µmol m^{-2} s^{-1} provided by cool, white fluorescent tubes of 40 W (Philips, Calcutta, India) for 2 weeks. Relative humidity was maintained at 50-60%. The plantlets growing in liquid MS medium were supplemented with MeJA (0.1 mM), SA (0.1 mM), and GA_3 (0.1 mM) dissolved in ethanol and untreated were kept as control with same amount of ethanol. The tissue from each treated sample was harvested after 6, 12, 24 and 48 h for RNA isolation. cDNA was obtained from each treated sample including control, using the same RNA isolation and cDNA synthesis protocols as described above. Effect of elicitor treatment on expression profile was studied using qRT- PCR as described above.

Withanolide analysis using HPLC

Withanolides were extracted and quantified as described by Dhar *et al.* [34]. Concisely, samples of *W. somnifera* were powdered and extracted with ethanol-water (50:50; *v/v*) with magnetic stirring at room temperature (25 ± 2°C). Extracts were filtered and the solvent was removed under vacuum. The extracts (20 mg/ml) obtained from each sample of the plant material were prepared in HPLC-grade methanol–water (50:50; *v/v*) for quantitative analysis. 1.2 mg per 2 ml of the standards of withanolide A (WS-1), withanone (WS-2) and withaferine A (WS-3) were prepared in HPLC-grade methanol. HPLC analysis was performed with Shimadzu HPLC system (Shimadzu, Tokyo, Japan) equipped with 515 quaternary gradient pump, 717 Rheodyne injector, 2996 PDA detector and CLASS-VP software v 6.14. All samples were filtered through 0.45 μM filters (Millipore, Bedford, USA). Extracts of *W. somnifera* samples were separated on a RP-18e (4.6 × 100 mm, 5 μm) (Merck, Bangalore, India) column. The mobile phase consisted of methanol–water (60:40; *v/v*) delivered at a flow rate of 0.5 ml/min. The samples were analyzed at 30°C to provide efficiency to the peaks. The UV chromatograms were recorded at 237 nm.

Results and discussion

Molecular cloning of *WsCYP98A* and *WsCYP76A*

P450 superfamily contain genes of diverse functions involved both in primary as well as secondary metabolism. These are heme-thiolate proteins containing iron atom coordinated to a proximal cysteine and receive electrons from NAD(P)H *via* a FAD-domain of auxiliary reductases. In the present investigation, two monooxygenases from *W. somnifera, WsCYP98A* and *WsCYP76A* have been isolated and expressed in *E. coli* BL21 (DE3). Quantitative assessment of gene expression pattern in different plant organs and relative transcript abundance in response to exogenous elicitors was also evaluated and corroborated with withanolide production to extrapolate and infer their possible role during biotic and abiotic stresses.

Full length cDNAs of *WsCYP98A* and *WsCYP76A* genes were obtained from the leaf tissue of WS-3 rich chemo-variant by degenerate PCR and RACE methods (*WsCYP98A*: HM585369 and *WsCYP76A*: KC008573). Degenerate primers based on conserved regions were designed by employing multiple sequence alignment strategy using ClustalW. A 550 bp and 520 bp core fragments were obtained from initial RT-PCR reactions. The amplicons were confirmed as members of CYP98 and CYP76 families by sequencing and similarity searches using BLASTn. Further, extension of both amplified core fragments toward 5′ and 3′ ends by using RACE-PCR succeeded in amplification of remaining cDNA portions. Using GSPs designed from start codons and stop codons, 1536 bp and 1548 bp open reading frames (ORFs) were amplified from the pool of leaf cDNA designated as *WsCYP98A* and *WsCYP76A* and theses encoded 511 and 515 amino acid residues respectively (Figure 1A and B). The first methionine as per First-AUG rule was considered as initiator codon. *WsCYP98A* contains downstream untranslated region (UTR; 230 bp) at 3′ ends whereas *WsCYP76A* comprises of 3′-UTR of 180 bp. Similarity search showed *Ws*CYP98A shares 65-95% homology with protein sequences of many other species while *Ws*CYP76A shares 39-89% homology.

In silico characterization of deduced *Ws*CYP98A and *Ws*CYP76A

*Ws*CYP98A and *Ws*CYP76A have predicted isoelectric points 8.13 and 7.99 and molecular masses of 58.0 kDa and 58.7 kDa respectively. Multiple sequence alignment with P450s of taxonomically diverse species showed presence of cysteine heme ligand signature which is characteristic signature of P450 sequences This motif was detected at positions 434–443 aa (FGAGRRVCPG) and 444–453 aa (FGAGRRMCVG) for *Ws*CYP98A and *Ws*CYP76A respectively (Figure 2A and B). Like all eukaryotic monooxygenases, a string of amino acid residues which helps in anchoring to ER membrane was detected at the N-terminal end of each monooxygenase. This membrane anchor region is also essential for normal interaction between P450s and their redox partners. Their presence was also predicted by TMHMM and SPLIT 4.0 bioinformatics programs (Additional file 1: Figure S1). To ascertain the degree of evolutionary relatedness, Neighbor-joining phylogenetic tree was constructed with MEGA 6.0 software from the ClustalW alignment of *Ws*CYP98A and *Ws*CYP76A with a number of homologous P450 sequences of different plants retrieved from the NCBI GenBank database. *Ws*CYP98A and *Ws*CYP76A corresponded to two separate phylogenetic clans in accordance with the amino acid similarity among their proteins (Figure 3). Prediction of secondary structure of *Ws*CYP98A and *Ws*CYP76A proteins by SOPMA program revealed that they consist of α-helixes (52.84%; 50.68%), β-turns (6.07%; 4.47%) joined by extended strands (8.61%; 10.49%), and random coils (32.49%; 34.37%). Three dimensional structural models for *Ws*CYP98A and *Ws*CYP76A were generated using Phyre^2 with >90% accuracy on the basis of homology based modelling using cytochrome P450 1a2 (PDB: 2HI4) as template (Figure 4A and C). The amino acid residues involved in ligand binding were also predicted using the 3DLigandSite tool as depicted in Figure 4B and D. Analysis of the evolutionary conservation of *Ws*CYP98A and *Ws*CYP76A amino acids were performed using ConSurf program. Several residues with high scores were found to be functional and structural residues of the proteins by ConSeq servers (Additional file 2: Figure S2). Ramachandran plot analysis of *Ws*CYP98A showed 461 aa (90.6%) residues in the most favourable region, 32 aa (6.3%) residues in the additional allowed region and 16 (3.1%) in the outlier region. Similarly, in case of *Ws*CYP76A, 480 aa (93.6%) residues were in the most favourable region, 20 aa (3.9%) residues in the additional allowed region and 13 (2.5%) in the outlier region (Additional file 3: Figure S3).

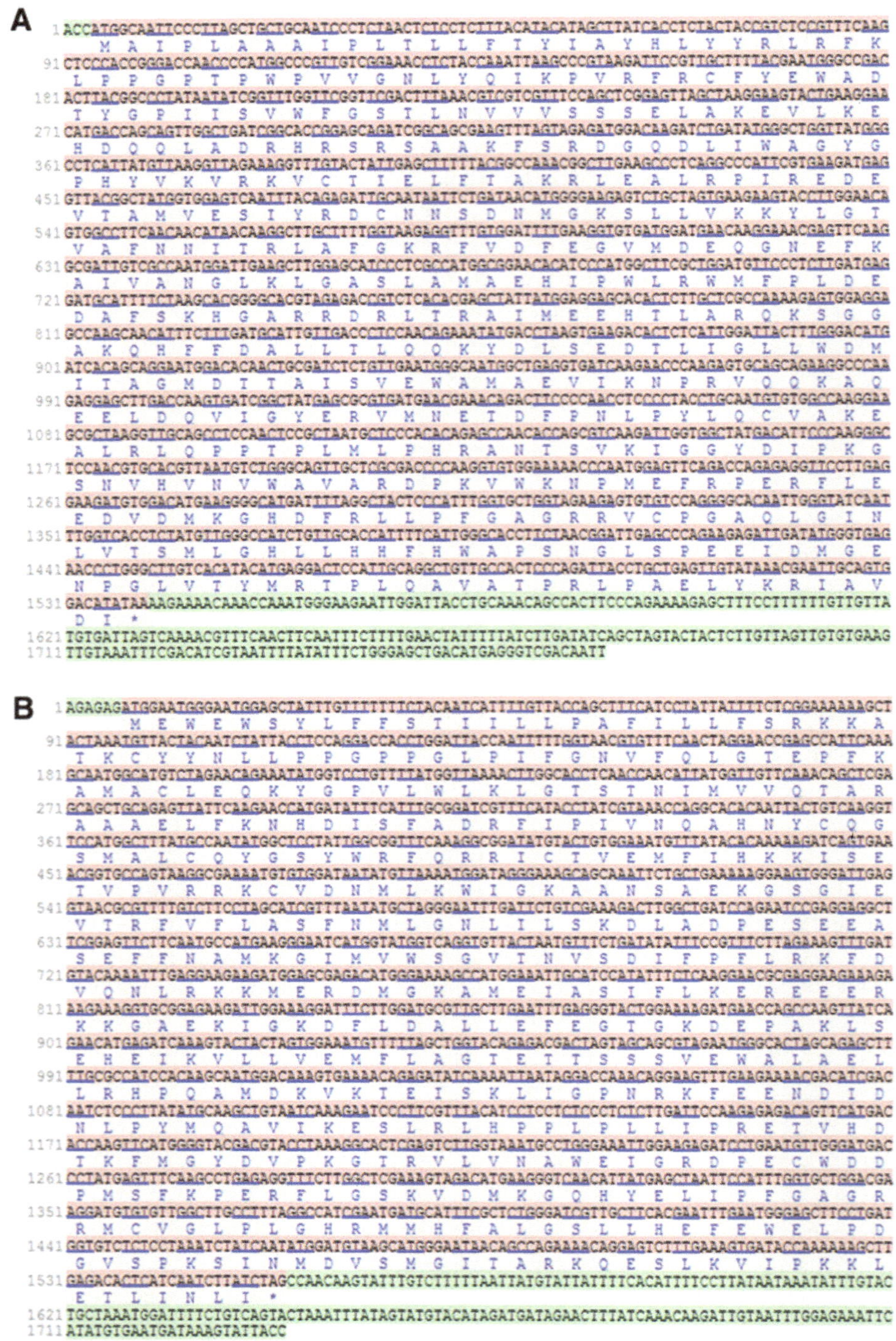

Figure 1 Nucleotide and the deduced amino acid sequence of *WsCYP98A* (A) and *WsCYP76A* (B) from *Withania somnifera*. The start codon (ATG) present at 4th and 7th positions whereas stop codons at 1552 and 1537 bp, respectively.

Heterologous expression in *E. coli*

For heterologous expression, entire coding sequences of *WsCYP98A* and *WsCYP76A* cDNAs were expressed in *E. coli* using pGEX 4 T-2 expression vector system. The ORFs were released from pJET-*WsCYP98A* and pJET-*WsCYP76A* using *BamH*I/*Sal*I restriction enzymes, and inserted into vector pGEX4T-2. The recombinant expression vectors with the inserted *WsCYP98A* and *WsCYP76A* constructs were identified by PCR analysis and restriction digestion using *BamH*I/*Sal*I. Heterologous expression of

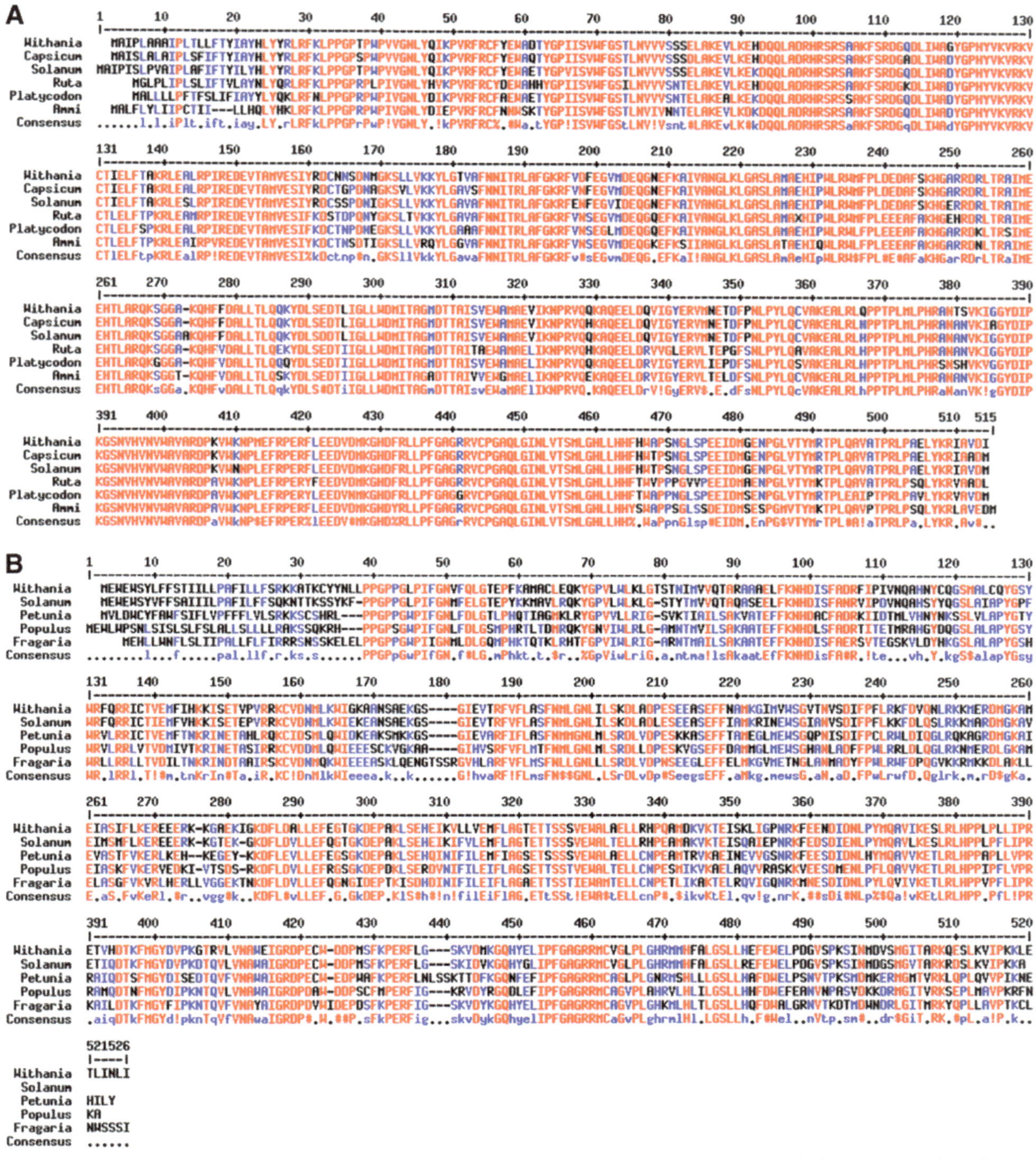

Figure 2 Multiple sequence alignment of deduced amino acid sequences of *WsCYP98A* and *Ws*CYP76A with their respective homologs using Multalign tool. For *WsCYP98A* **(A)**, sequences used for alignment were from *Withania somnifera* (*WsCYP98A*: HM036710), *Solanum lycopersicum* (ACF17644.1) *Capsicum annuum* (ACF17644.1), *Ruta graveolens (AEG19446.1), Platycodon grandiflorus* (AEM63674.1), *Ammi majus* (AAT06912.1). Consensus residues present were shown below. For *WsCYP76A* **(B)**, sequences used for alignment were from *Withania somnifera* (*WsCYP98A*: HM036710), *Solanum lycopersicum* (XP_004249463), *Populus trichocarpa* (XP_002309107.1), *Fragaria vesca* (XP_004305369.1), *Petunia x hybrida* (BAC53892.1). Consensus residues present were shown below.

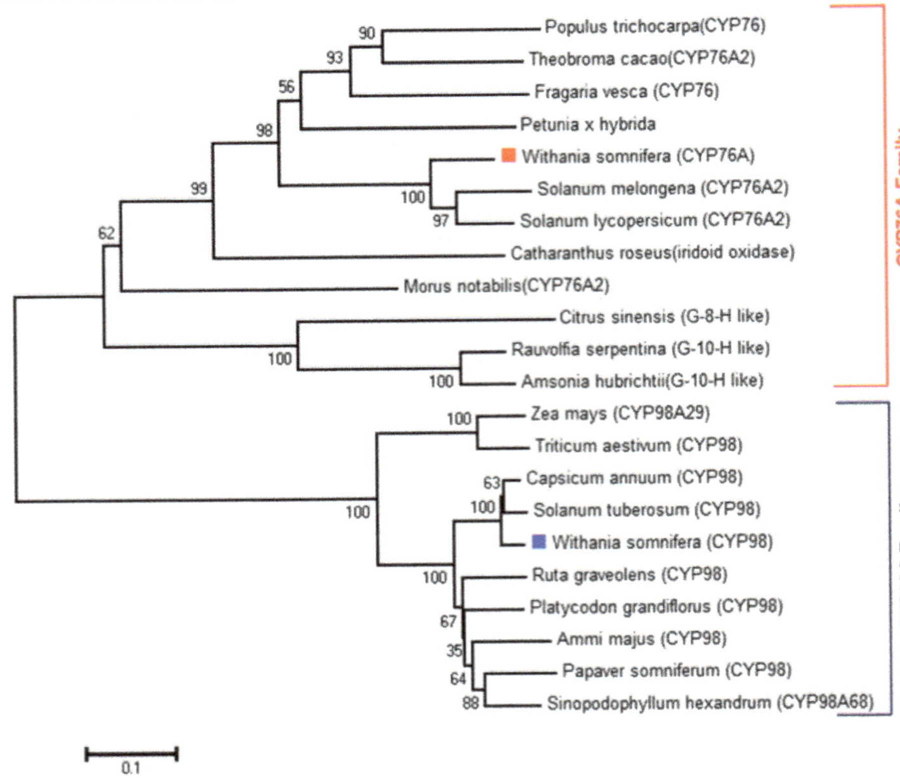

Figure 3 Phylogenetic analysis of deduced amino acid sequences of *WsCYP98A* and *WsCYP76A* was inferred using the Neighbour-joining method employing MEGA 6.0 software. For *WsCYP98A* total of 10 sequences and for *WsCYP76A*, 12 sequences including *Withania somnifera* were used for analysis. The sequences used were from following plant specieis: *Withania somnifera* (*WsCYP98A*: HM036710; *WsCYP76A*: GU808569), *Solanum lycopersicum* (ACF17644.1) *Capsicum annuum* (ACF17644.1) *Solanum tuberosum (XP_006347680.1), Ruta graveolens (AEG19446.1), Platycodon grandiflorus* (AEM63674.1), *Ammi majus* (AAT06912.1), *Zea mays* (ACG25686.1), *Triticum aestivum* (CAE47490.1), *Papaver somniferum* (AFK73720.1), *Sinopodophyllum hexandrum* (AGC29945.1), *Solanum melongena* (CAA50648.1), *Solanumlycopersicum* (XP_004249463), *Populus trichocarpa* (XP_002309107.1), *Fragaria vesca* (XP_004305369.1), *Petunia x hybrida* (BAC53892.1), *Theobroma cacao* (XP_007027589.1), *Catharanthus roseus (iridoid oxidase;* AHK60833.1*)*, *Morus notabilis* (EXC33897.1), *Rauvolfia serpentina* (G-10-H like; AGX93053.1), *Amsonia hubrichtii* (G-10-H like; AGX93052.1, *Citrus sinensis* (G-8-H like; XP_006477331.1). Both P450 confined to separated clan and exhibited maximum similarity with their respective members of CYP98 and CYP76 families.

proteins was induced with different concentrations of IPTG. SDS-PAGE analysis demonstrated that optimum expression of proteins was observed at 25°C using 0.8 mM IPTG after 6–8 h of induction. The fusion protein having molecular weight of ~84.06 kDa and ~84.7 kDa appeared in the lysate of recombinant *E. coli* transformed with the expression cassettes pGEX-*WsCYP98A* and pGEX-*WsCYP76A*, respectively (Figure 5). The apparent molecular masses of these recombinant proteins were in good agreement with molecular mass of GST (26 kDa) and those calculated from the deduced primary structures of *Ws*CYP98A and *Ws*CYP76A proteins. Previously, we were able to optimize the functional expression of *WsCPRs* in soluble fraction with its membrane anchor [21]. Nevertheless, *Ws*CYP98A and *Ws*CYP76A were not expressed in soluble fraction in ample amount. Maximum amount of recombinant protein of two P450s were localized to inclusion bodies which hindered their purification.

Tissue-specific gene expression analysis

To study *WsCYP98A* and *WsCYP76A* gene expression pattern in different tissues of *W. somnifera*, cDNA libraries were prepared separately from RNA samples extracted from leaves, stalks, roots, flowers and berries (unripen) of four month old plant. Tissue-specific cDNAs were used as templates for qRT-PCR. The results showed highest expression of *WsCYP98A* in stalk (Figure 6A). Similar findings have been reported for coumarate 3-hydroxylase in *A. thaliana* where the distribution of transcripts was more preponderant in stem. High frequencies of *CYP98A* transcripts have also been observed in poplar and pine xylem, soybean hypocotyl and stem, as well as cotton fibres indicating towards its higher expression in lignin synthesizing tissues [35-37]. However, two closely related homologs *CYP98A8* and *CYP98A9* expressed predominantly in floral tissues which is very distinct from other CYP98 members. The higher expression in floral tissue probably helps in

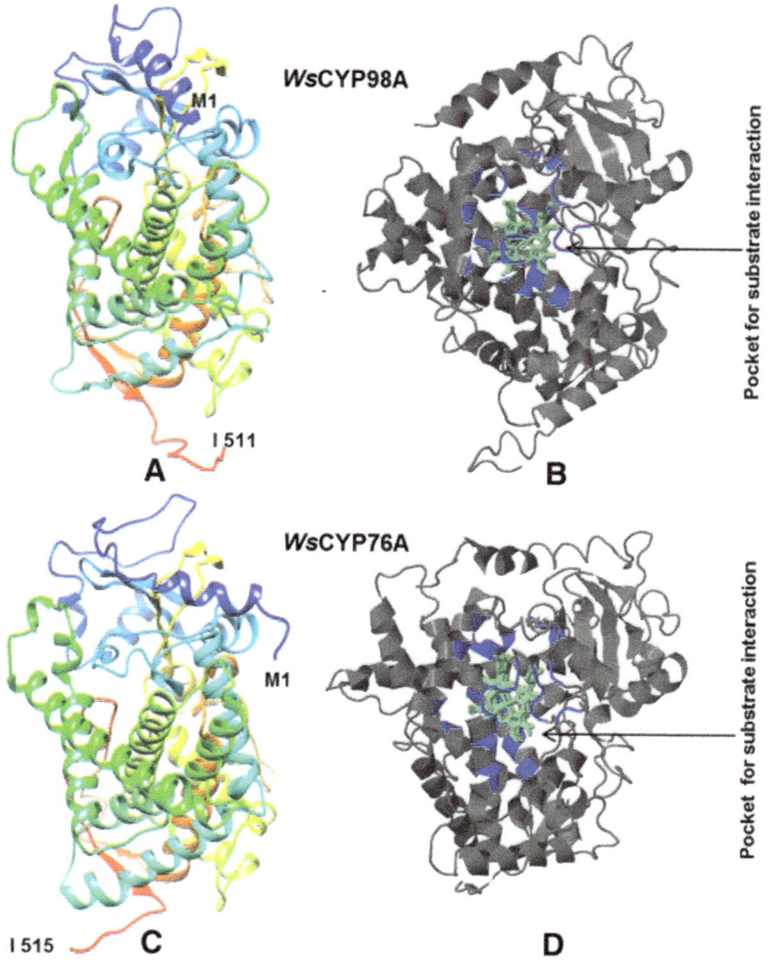

Figure 4 Prediction of three dimensional structures of *Ws*CYP98A and *Ws*CYP76A using Phyre² based on homology modelling approach. **A** & **C**: Cartoon display of the 3-D structures of *Ws*CYP98A and *Ws*CYP76A as predicted by Phyre² using crystal structure of cytochrome P450 1a2 (PDB: 2HI4) as template. **B** & **D**: Predicted ligand (shown in green) binding sites as predicted by 3DLigandSite Web Server. The residues involved in substrate binding and active site are shown in the center core of structure.

pollen coat development [38]. *WsCYP76A* transcripts were found predominantly expressing in roots of *W. somnifera* (Figure 6B). Likewise, expression of *CYP76A3* was reported higher in petunia roots. Higher expression of *WsCYP76A* in roots possibly protects plant against attack by plant pathogens [39]. The comparative analysis of transcriptomes in *Salvia miltiorrhiza* showed abundant expression of *CYP76AH1* in roots rather than other plant tissues and found to be involved in tanshinone biosynthetic pathway [40]. The least expression of both monooxygenases was observed in berries.

Effect of elicitors on *WsCYP98A* and *WsCYP76A*

In vitro cultures established *via* micro-propagation were used for MeJA, SA and GA_3 treatments. Micro-shoots were grown for 2 weeks in MS liquid medium. Subsequently, micro-shoots were supplemented with MeJA (0.1 mM), SA (0. 1 mM) and GA_3 (0.1 mM) for 48 h. The untreated were kept as control. Samples were harvested after 6, 12, 24, and 48 h of interval. Equal amounts of DNase-treated RNA (3 μg) of control and treated samples were converted into cDNA. The effect of MeJA, SA and GA_3 on expression profile of *WsCYP98A* and *WsCYP76A* was studied using qRT-PCR.

One of the important endogenous plant hormones MeJA is known to modulate biosynthesis of many of the secondary metabolites that play crucial role in the adaptation of plants to various biotic and biotic stresses [41]. MeJA has been reported to stimulate the biosynthesis of alkaloids by inducing the expression of several genes from the monoterpenoid branch of the monoterpene indole alkaloid biosynthetic pathway [42]. In present study, MeJA induced expression of both P450s post 48 h of treatment. *WsCYP98A* showed 2-fold increase in expression after

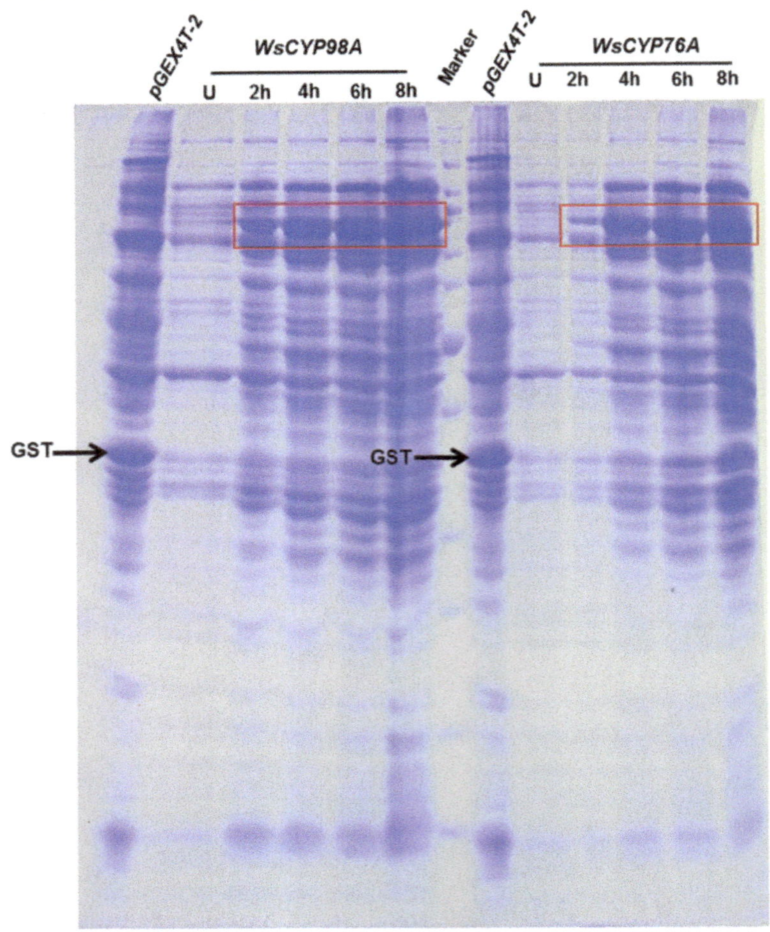

Figure 5 Sodium dodecyl sulphate–polyacrylamide gel electrophoresis (SDS-PAGE: 10%) pattern of proteins obtained from *E.coli* BL21 (DE3) transformed with pGEX-*WsCYP98A* and pGEX-*WsCYP76A*. Lane 1 & 8; Cell lysate of *E. coli* BL21 (DE3) cells containing the empty vector pGEX-4 T2 obtained at 4 h post-induction with 0.8 mM IPTG. Lane 2 & 9; Whole cell lysate of uninduced *Ws*CYP98A and *Ws*CYP76A, Lane 7; Standard protein marker Lane 3–6 & 9–12; Cell lysate of *Ws*CYP98A and *Ws*CYP76A induced with 0.8 mM IPTG harvested after 2 h, 4 h, 6 h and 8 h respectively.

48 h of application of MeJA while *WsCYP76A* showed 6-fold increase in the expression after 48 h of MeJA application (Figure 7A).

SA has been studied in detail as stress signaling molecule. It is implicated in signal transduction of numerous processes including the biosynthesis of some important secondary metabolites in plants [43-45]. SA has been demonstrated to induce the accumulation of triterpenoids, ginsenosides in *Panx ginseng* and glycyrrhizin in *Glycyrrhizaglabra*, respectively [44,46]. Production of sesquiterpenoids, such as bilobalide in *Ginkgo biloba* and artemisinin in *Artemisia annua* are also stimulated by exogenous application of SA [47,48]. SA treated plantlets showed upto 2.5-3 fold increase in accumulation of *WsCYP98A* and *WsCYP76A* after time points of 24 h and 48 h, respectively (Figure 7B). The results obtained are fully in agreement with the previous studies wherein microarray analysis revealed up-regulation of several P450s from *A. thaliana* in response to SA treatment [49]. Likewise, enhanced expression of *CYP98A5* gene has been noticed in *Physilus vulgaris* when leaves were treated with derivatives of salicylic acids [50]. The induction of *LjC3H* in *Lonicera japonium* has been induced by the treatment of MeJA and UV radiation [51,52]. Up-regulation of transcript levels of *WsCYP98A* and *WsCYP76A* in response to elicitors may be due to their possible implication in defence mechanism.

CYP98A is key gene of phenylpropanoid pathway which leads to formation of flavonoids, phenolic and lignin. During pathogens attack and wound healing, deposition of lignin and suberin synthesis occurs in the cell wall [53]. Thus, the abundance of transcripts of *WsCYP98A* in treated plantlets appears to be a defence response due to elicitor triggered stimulus. The expression of *WsCYP76A* have also enhanced in response to MeJA and SA treatment. Members of same family like *CYP76B6* is strongly induced

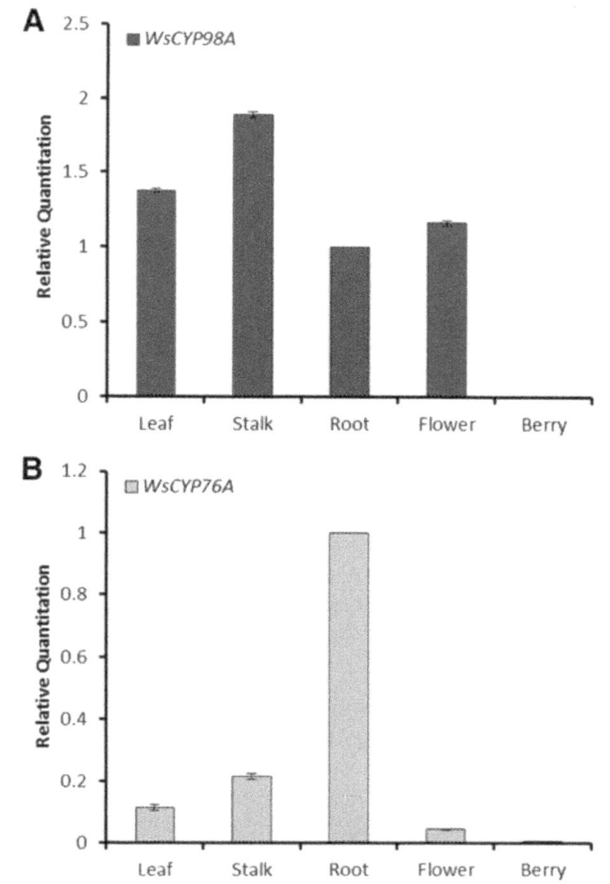

Figure 6 Quantitative assessment of the expression of (A) *WsCYP98A* and (B) *WsCYP76A* in different tissues of *Withania somnifera*. Data were compared and analysed with analysis of variance (*ANOVA*). Values are means, with standard errors indicated by bars, representing three independent biological samples, each with three technical replicates.

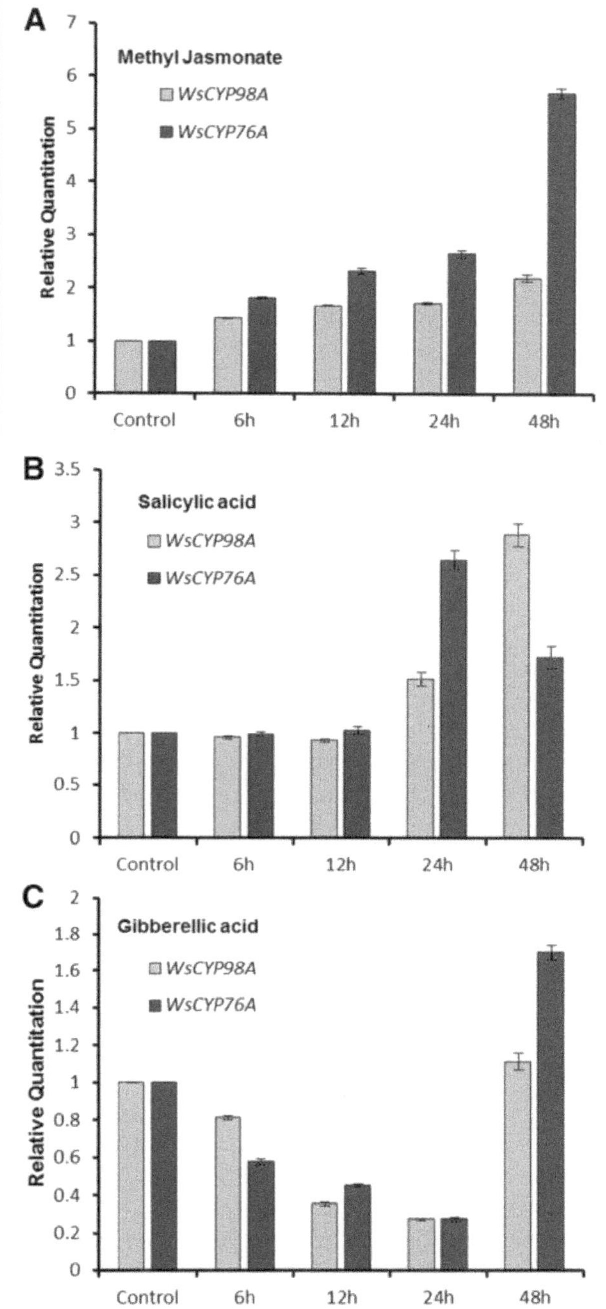

Figure 7 Quantitative real-time analysis of *WsCYP98A* and *WsCYP76A* expression in micro-propagated *Withania somnifera* induced by (A) methyl jasmonate (MeJA; 0.1 mM), (B) salicylic acid (SA; 0.1 mM) and (C) gibberellic acid (GA_3; 0.1 mM) treatments. MeJA, SA and GA_3 were added to the micro-propagated plantlets pre-cultured for 2 weeks on MS liquid medium. The samples were harvested at time points of 6, 12, 24, and 48 h. Actin was kept as internal control. Data were compared and analysed with analysis of variance (*ANOVA*). Experiments were performed in triplicate with similar results with standard errors indicated by bars, representing three independent biological samples, each with three technical replicates.

by MeJA treatment along with other terpenoid indole alkaloid biosynthesis genes in *C. roseus* cell culture. After 48 h of MeJA treatment the expression of *WsCYP76A* was about 6-fold higher than control. Closely related homologs of CYP76 family are mainly involved in biosynthesis of phytoalexin and are mostly multifunctional oxidases such as CYP76M7, which seem to be involved in the production of antifungal phytocassanes [54]. *CYP76B6* has been shown to be a highly specialized multifunctional enzyme catalyzing two sequential oxidation steps leading to the formation of 8-oxogeraniol from geraniol [55].

GA_3 is a diterpene plant growth regulator, controlling various growth and developmental process from seed germination to flower and fruit development. Broadly, GAs have been implicated in various cross talk mechanisms among different signaling pathways triggered by other plant growth regulators [56]. It impinges on a common transcription module of other phytohormones thus promoting many similar developmental responses in plants [57]. It has also been contended that GA_3 orchestrates the metabolite fluxes between the primary and secondary metabolism influencing the production of isoprenoids and carotenoids [58,59]. Exogenous application of GA_3 resulted in decrease in the expression of *WsCYP98A* as well as *WsCYP76A* (Figure 7C). Lowest transcript levels were observed after 24 h of treatment which indicated that GA_3 acts as negative regulator of these two genes. After 48 h of treatment the transcript accumulation of *WsCYP98A* was comparable to control while as *WsCYP76A* showed 0.5-fold higher expression in comparison to control.

Effect of elicitors on accumulation of withanolides

Elicitor treatment is an efficient strategy to induce *de novo* synthesis of secondary metabolites. Endogenous signal molecules like MeJA and SA act as key signaling compounds in various stress responses and plants subjected to various vagaries often result in higher accumulation of secondary metabolites such as alkaloids, terpenoids, flavonoids, phenolic compounds and phytoalexins [60,61]. To establish a correlation between expression profiles and metabolite flux, withanolides extracted from the treated samples were subjected to HPLC analysis (Additional file 4: Figure S4). There was appreciable increase in WS-1 and WS-3 over a period of time in response to elicitor treatments. In MeJA treated samples, there was a significant increase in WS-1 ($53.583 \pm 1.02 - 137.530 \pm 1.76$ μg g^{-1} of dry weight) and WS-3 ($584.012 \pm 2.67 - 2504.937 \pm 3.52$ μg g^{-1} of dry weight) (Figure 8A). SA treated samples showed marked increase in both WS-1 ($129.934 \pm 1.04 - 230.864 \pm 1.54$ μg g^{-1} of dry weight) and WS-3($448.257 \pm 2.59 - 1467.235 \pm 2.58$ μg g^{-1} of dry weight) with no traces of WS-2 (Figure 8B). Estimation of withanolide accumulation in GA_3 treated samples demonstrated a gradual decline in WS-3 ($397.326 \pm 3.54 - 295.1673 \pm 3.18$ μg g^{-1} of dry weight) while as WS-1 increased moderately up to 24 h ($58.475 \pm 2.2 - 112.667 \pm 2$ μg g^{-1} of dry weight) but after 48 h there was a slight decrease in WS-1 (84.214 ± 2.01 μg g^{-1} of dry weight). WS-2 level also peaked at 6 h (44.323 ± 2.49 μg g^{-1} of dry weight) in GA_3 treated plantlets followed by a decrease at 24 h (10.867 ± 0.73 μg g^{-1} of dry weight) and un-detectable concentration after 48 h (Figure 8C). Absence or low production of WS-2 is possibly linked to inherently lower levels of WS-2 accumulation even in cultivated accessions of *W. somnifera* [37]. Application of exogenously applied phytohormones has been shown to impinge on physiological and metabolic processes in many plant organs and cell cultures. In some of the reports accumulation of anthocyanins, glutathione and flavonoids have shown varied response to GA_3 treatments [62].

In our previous reports, MeJA and SA up-regulated the expression of squalene synthase, squalene epoxidase and *WsCPR2* genes which play an important regulatory role in the phytosterol biosynthesis [63]. These elicitors are important components of signal transduction cascades activating plant's defence response against pathogen attack and often result in increased metabolite accumulation.

Conclusion

The increased demand of secondary metabolites for medicinal purposes coupled with low yield and supply has spurred the interest in enhanced production of bioactive metabolites. Recent developments and upsurge in synthetic biology approach in combination with rapid advances in systems biology and metabolic engineering have enabled the manipulation of genome of microorganisms to produce heterologous molecules in a manner that was previously much more demanding in terms of integrating the metabolic circuitries in heterologous host [64]. P450s play a critical role in heterologous and homologous systems as they are able to catalyse regio- and stereospecific hydroxylation reactions that are extremely difficult to carry out using chemical methods [65]. These P450-catalysed reactions are pivotal steps in the biosynthesis of variety of compounds and optimization of P450 enzyme activities are key targets in yield improvement efforts to render such approaches economically feasible [66]. The characterization of P450s and their role in biosynthesis of plant secondary metabolites is a fascinating area of investigation. Present endeavour was aimed towards identification and characterization of different P450 monooxygenases from *W. somnifera*. We were able to clone two important P450 monoxygenases from *Withania* which seem to play an active role in defence mechanism by coping up with biotic or abiotic stresses. It also gets reflected in higher accumulation of withanolides. The full length genes were successfully transformed in *E. coli* for heterologous expression. Time-course study revealed that IPTG had positive influence on protein expression in *E. coli*. Expression pattern for *WsCYP98A* and *WsCYP76A* in different plant tissues was studied using qRT-PCR which showed highest expression of *WsCYP98A* in stalk but the expression of *WsCYP76A* was more obvious in roots. In present study, MeJA and SA acted as positive regulators whereas GA_3 behaved as a negative regulator for *WsCYP98A* and *WsCYP76A*. To establish a correlation between enhanced transcript levels in response to activation of defence mechanism we also analysed withanolide accumulation. As the elicitors play key role as signaling molecules during stress conditions and enhance production of secondary metabolites, we observed discernible changes in withanolide concentrations. MeJA elicitation significantly increased the WS-3 accumulation over a period of 48 h. These results were in conformity with our earlier studies where MeJA and SA mediated induction of squalene synthase, squalene

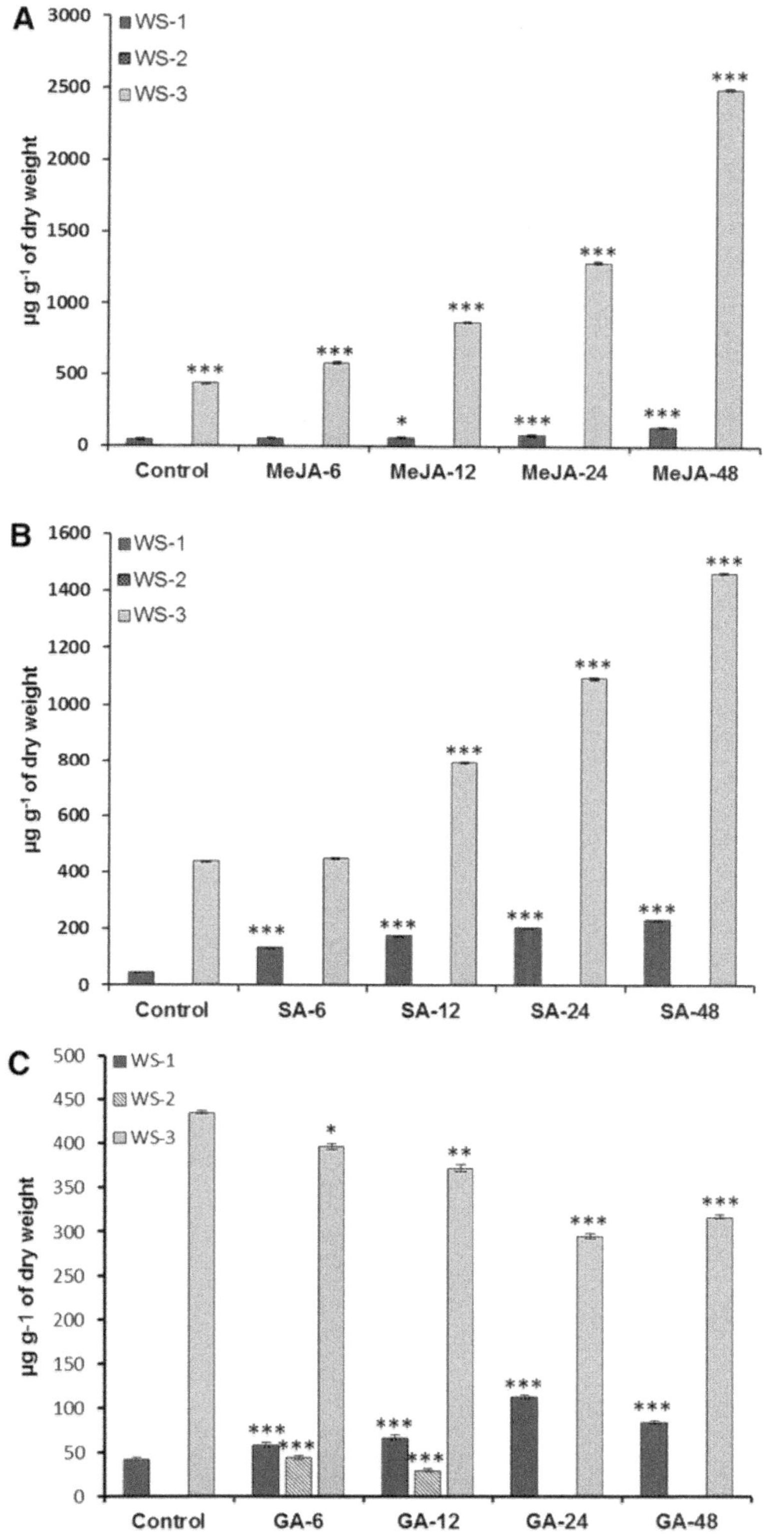

Figure 8 Time-course effect of elicitor treatments on withanolides accumulation in response to (A) methyl jasmonate (0.1 mM), (B) salicylic acid (0.1 mM) and (C) gibberellic acid (0.1 mM) at different time points. Variation in three key withanolides viz. withanolide A (WS-1), withanone (WS-2) and withaferine A (WS-3) was confirmed by HPLC analysis at 6, 12, 24 and 48 h. All values obtained were means of triplicate with standard errors. Time course accumulation of WS-1, WS-2 and WS-3 was statistically significant at *$p < 0.05$, **$p < 0.01$ and ***$p < 0.001$ levels.

epoxidase and *Ws*CPR2 also led to enhanced withanolide content [63,67]. GA_3 was found to be a negative regulator and resulted in low levels of mRNA transcripts. The accumulation of withanolides in GA_3 treated samples showed remarkable decrease in WS-3 and WS-1 accumulation. These results are positively corroborating with some of our earlier findings where the exogenous application of GA_3 resulted in decrease in the cycloartenol synthase at translational level resulting in the decline of withanolide content [20]. Possibly, GA_3 application diverts the flux towards the formation of primary metabolites for cell elongation and development because it has been reported to be associated with the promotion of germination, growth, and flowering [62]. For homologous modulation of withanolide biosynthetic pathway, we have successfully incited the induction of transformed hairy roots using *Agrobacterium rhizogenes*. Hairy roots provide an excellent experimental system to understand the regulatory role of different P450s with their redox partners for enhanced production of withanolides (Unpublished).

Additional files

Additional file 1: Figure S1. Transmembrane domain prediction of **(A)** *WsCYP98A* and **(B)** *WsCYP76A* using TMHMM web server.

Additional file 2: Figure S2. Evolutionary conserved key residue analysis of **(A)** *Ws*CYP98A and **(B)** *Ws*CYP76A identified using Consurf, an empirical Bayesian inference based web server. Residue conservation from variable to conserve is shown in blue (1) to violet (9). The residues involved in binding of the donor moieties are shown in the centre of the structures.

Additional file 3: Figure S3. Ramachandran plot of *Ws*CYP98A *and* *Ws*CYP76A 3D models of *Withania somnifera* using RAMPAGE server. Most favoured regions are coloured red, additional allowed, generously allowed, and disallowed regions are indicated as yellow, light yellow and white fields, respectively.

Additional file 4: Figure S4. HPLC analysis of withanolide A (WS-1), withanone (WS-2) and withaferin A (WS-3) at 6 h and 48 h from methyl jasmonate (0.1 mM), salicylic acid (0.1 mM) and gibberellic acid (0.1 mM) treated micro-shoots of *Withania somnifera*.

Abbreviations

ER: Endoplasmic reticulum; GSPs: Gene specific primers; RACE: Rapid amplification of cDNA ends; ORF: Open reading frame; UTR: Untranslated region; *E. coli*: *Escherichia coli*; IPTG: Isopropyl- β-D-thiogalactopyranoside; SDS-PAGE: Sodium dodecyl sulphate polyacrylamide gel electrophoresis; qRT-PCR: Quantitative real-time polymerase chain reaction; MeJA: Methyl jasmonate; SA: Salicylic acid; GA_3: Gibberellic acid; HPLC: High performance liquid chromatography.

Competing interests

The authors declare that they have no competing interests.

Authors' contributions

Conceived and designed the experiments: SKL, RV, SR. Performed the experiments: SR, WWB, ND, SAP, SR. Analyzed the data: SR, SKL. Contributed reagents/materials/analysis tools: RV, SKL. Wrote the paper: SR, SKL. All authors read and approved the final manuscript.

Acknowledgements

This work was supported by a grant from the Council of Scientific and Industrial Research (CSIR), Government of India, New Delhi under Network Project BSC-0108. S. Rana, W.W.B, N.D. and S. Razdan are thankful to CSIR, Government of India, New Delhi for Senior Research Fellowship (CSIR-SRF). S A. Pandith is grateful to University Grants Commission, Government of India, New Delhi for Senior Research Fellowship (UGC-SRF).

Author details

[1]Plant Biotechnology Division, CSIR-Indian Institute of Integrative Medicine, Canal Road, Jammu Tawi-180001, India. [2]Medicinal Chemistry Division, CSIR-Indian Institute of Integrative Medicine, Canal Road, Jammu Tawi-180001, India.

References

1. Cheng Q, Guengerich FP: **Identification of Endogenous Substrates of Orphan Cytochrome P450 Enzymes Through the Use of Untargeted Metabolomics Approaches.** In *Cytochrome P450 Protocols*, Volume 987. Edited by Phillips IR, Shephard EA, de Montellano PR O. New York: Humana Press; 2013:71–77.
2. Nelson DR, Nebert DW: **Cytochrome P450 (CYP) Gene Superfamily.** In *eLS*. West Sussex: John Wiley & Sons, Ltd; 2011.
3. Nelson D, Ming R, Alam M, Schuler M: **Comparison of cytochrome P450 genes from six plant genomes.** *Trop Plant Biol* 2008, **1**(3–4):216–235.
4. Nelson D, Werck-Reichhart D: **A P450-centric view of plant evolution.** *Plant J* 2011, **66**(1):194–211.
5. Schuler MA, Werck-Reichhart D: **Functional genomics of P450s.** *Annu Rev Plant Biol* 2003, **54:**629–667.
6. de Montellano PRO: *Cytochrome P450: Structure, Mechanism, and Biochemistry*. New York: Springer; 2005.
7. Bak S, Beisson F, Bishop G, Hamberger B, Hofer R, Paquette S, Werck-Reichhart D: **Cytochromes P450.** *Arabidopsis Book* 2011, **9**(10):6.
8. Werck-Reichhart D, Feyereisen R: **Cytochromes P450: a success story.** *Genome Biol* 2000, **1**(6):8.
9. Schuler MA: **The role of cytochrome P450 monooxygenases in plant-insect interactions.** *Plant Physiol* 1996, **112**(4):1411–1419.
10. Miners JO: **Evolution of drug metabolism: hitchhiking the technology bandwagon.** *Clin Exp Pharmacol Physiol* 2002, **29**(11):1040–1044.
11. Guengerich FP: **Cytochrome P450 enzymes in the generation of commercial products.** *Nat Rev Drug Discov* 2002, **1**(5):359–366.
12. Guengerich FP: **Special issue: P450 catalysis mechanisms introduction.** *Arch Biochem Biophys* 2011, **507**(1):1–2.
13. Sono M, Roach MP, Coulter ED, Dawson JH: **Heme-containing oxygenases.** *Chem Rev* 1996, **96**(7):2841–2888.
14. Guengerich FP: **Uncommon P450-catalyzed reactions.** *Curr Drug Metab* 2001, **2**(2):93–115.
15. Coon M: **Cytochrome P450: Nature's most versatile biological catalyst.** *Annu Rev Pharmacol Toxicol* 2005, **45:**1–25.
16. Kirson I, Glotter E, Lavie D, Abraham A: **Constitutents of *Withania somnifera* Dun. Part XII The withanolides of an indian chemotype.** *J Chem Soc C: Organic* 1971, 2032–2044.
17. Lattoo SK, Dhar RS, Khan S, Bamotra S, Dhar AK: **Temporal sexual maturation and incremental staminal movement encourages mixed mating in Withania somnifera - An insurance for reproductive success.** *Curr Sci* 2007, **92**(10):1390–1399.
18. Rana S, Dhar N, Bhat WW, Razdan S, Khan S, Dhar RS, Dutt P, Lattoo SK: **A 12-deoxywithastramonolide-rich somaclonal variant in *Withania somnifera* (L.) Dunal-molecular cytogenetic analysis and significance as a chemotypic resource.** *In Vitro Cell Dev Biol-Plant* 2012, **48**(5):546–554.
19. Chaurasiya ND, Sangwan NS, Sabir F, Misra L, Sangwan RS: **Withanolide biosynthesis recruits both mevalonate and DOXP pathways of isoprenogenesis in Ashwagandha *Withania somnifera* L. (Dunal).** *Plant Cell Rep* 2012, **31**(10):1889–1897.
20. Dhar N, Rana S, Razdan S, Bhat WW, Hussain A, Dhar RS, Vaishnavi S, Hamid A, Vishwakarma R, Lattoo SK: **Cloning and functional characterization of three branch point oxidosqualene cyclases from *Withania somnifera* (L.) dunal.** *J Biol Chem* 2014, **289**(24):17249–17267.

21. Rana S, Lattoo SK, Dhar N, Razdan S, Bhat WW, Dhar RS, Vishwakarma R: **NADPH-cytochrome P450 reductase: molecular cloning and functional characterization of Two paralogs from *Withania somnifera* (L.) dunal.** *PLoS One* 2013, **8**(2):e57068.
22. Dhar N, Rana S, Bhat WW, Razdan S, Pandith SA, Khan S, Dutt P, Dhar RS, Vaishnavi S, Vishwakarma R, Lattoo SK: **Dynamics of withanolide biosynthesis in relation to temporal expression pattern of metabolic genes in *Withania somnifera* (L.) Dunal: a comparative study in two morpho-chemovariants.** *Mol Biol Rep* 2013, **40**(12):7007–7016.
23. Bhat WW, Dhar N, Razdan S, Rana S, Mehra R, Nargotra A, Dhar RS, Ashraf N, Vishwakarma R, Lattoo SK: **Molecular characterization of UGT94F2 and UGT86C4, Two glycosyltransferases from *picrorhiza kurrooa*: comparative structural insight and evaluation of substrate recognition.** *PLoS One* 2013, **8**(9):e73804.
24. Karamat F, Olry A, Doerper S, Vialart G, Ullmann P, Werck-Reichhart D, Bourgaud F, Hehn A: **CYP98A22, a phenolic ester 3′-hydroxylase specialized in the synthesis of chlorogenic acid, as a new tool for enhancing the furanocoumarin concentration in *Ruta graveolens*.** *BMC Plant Biol* 2012, **12**(1):152.
25. Miettinen K, Dong L, Navrot N, Schneider T, Burlat V, Pollier J, Woittiez L, van der Krol S, Lugan R, Ilc T, Verpoorte R, Oksman-Caldentey KM, Martinoia E, Bouwmeester H, Goossens A, Memelink J, Werck-Reichhart D: **The seco-iridoid pathway from *Catharanthus roseus*.** *Nat Commun* 2014, **5**. doi:10.1038/ncomms4606.
26. Hofer R, Dong L, Andre F, Ginglinger JF, Lugan R, Gavira C, Grec S, Lang G, Memelink J, Van der Krol S, Bouwmeester H, Werck-Reichhart D: **Geraniol hydroxylase and hydroxygeraniol oxidase activities of the CYP76 family of cytochrome P450 enzymes and potential for engineering the early steps of the (seco)iridoid pathway.** *Metab Eng* 2013, **20**:221–232.
27. Larkin MA, Blackshields G, Brown NP, Chenna R, McGettigan PA, McWilliam H, Valentin F, Wallace IM, Wilm A, Lopez R, Thompson JD, Gibson TJ, Higgins DG: **Clustal W and ClustalX version 2.0.** *Bioinformatics* 2007, **23**(21):2947–2948.
28. Tamura K, Stecher G, Peterson D, Filipski A, Kumar S: **MEGA6: molecular evolutionary genetics analysis version 6.0.** *Mol Biol Evol* 2013, **30**(12):2725–2729.
29. Geourjon C, Deleage G: **SOPMA: significant improvements in protein secondary structure prediction by consensus prediction from multiple alignments.** *Comput Appl Biosci* 1995, **11**(6):681–684.
30. Kelley LA, Sternberg MJE: **Protein structure prediction on the Web: a case study using the Phyre server.** *Nat Protoc* 2009, **4**(3):363–371.
31. Lovell SC, Davis IW, Arendall WB 3rd, de Bakker PI, Word JM, Prisant MG, Richardson JS, Richardson DC: **Structure validation by Calpha geometry: phi, psi and Cbeta deviation.** *Proteins* 2003, **50**(3):437–450.
32. Celniker G, Nimrod G, Ashkenazy H, Glaser F, Martz E, Mayrose I, Pupko T, Ben-Tal N: **ConSurf: using evolutionary data to raise testable hypotheses about protein function.** *Isr J Chem* 2013, **53**(3–4):199–206.
33. Livak KJ, Schmittgen TD: **Analysis of relative gene expression data using real-time quantitative PCR and the $2^{-\Delta\Delta CT}$ method.** *Methods* 2001, **25**(4):402–408.
34. Dhar RS, Verma V, Suri KA, Sangwan RS, Satti NK, Kumar A, Tuli R, Qazi GN: **Phytochemical and genetic analysis in selected chemotypes of *Withania somnifera*.** *Phytochemistry* 2006, **67**(20):2269–2276.
35. Sterky F, Regan S, Karlsson J, Hertzberg M, Rohde A, Holmberg A, Amini B, Bhalerao R, Larsson M, Villarroel R, Van Montagu M, Sandberg G, Olsson O, Teeri TT, Boerjan W, Gustafsson P, Uhlen M, Sundberg B, Lundeberg J: **Gene discovery in the wood-forming tissues of poplar: analysis of 5, 692 expressed sequence tags.** *Proc Natl Acad Sci U S A* 1998, **95**(22):13330–13335.
36. Allona I, Quinn M, Shoop E, Swope K, St Cyr S, Carlis J, Riedl J, Retzel E, Campbell MM, Sederoff R, Whetten RW: **Analysis of xylem formation in pine by cDNA sequencing.** *Proc Natl Acad Sci U S A* 1998, **95**(16):9693–9698.
37. Schoch G, Goepfert S, Morant M, Hehn A, Meyer D, Ullmann P, Werck-Reichhart D: **CYP98A3 from *Arabidopsis thaliana* is a 3 ′-hydroxylase of phenolic esters, a missing link in the phenylpropanoid pathway.** *J Biol Chem* 2001, **276**(39):36566–36574.
38. Matsuno M, Compagnon V, Schoch GA, Schmitt M, Debayle D, Bassard JE, Pollet B, Hehn A, Heintz D, Ullmann P, Lapierre C, Bernier F, Ehlting J, Werck-Reichhart D: **Evolution of a novel phenolic pathway for pollen development.** *Science* 2009, **325**(5948):1688–1692.
39. Imaishi H, Ishitobi U: **Molecular cloning of CYP76A3, a novel cytochrome P450 from Petunia hybrida catalyzing the omega-hydroxylation of myristic acid.** *Biologia Plantarum* 2008, **52**(2):242–250.
40. Yang L, Ding G, Lin H, Cheng H, Kong Y, Wei Y, Fang X, Liu R, Wang L, Chen X, Chen X, Yang C: **Transcriptome analysis of medicinal plant *Salvia miltiorrhiza* and identification of genes related to tanshinone biosynthesis.** *PLoS One* 2013, **8**(11):e80464.
41. Martin D, Tholl D, Gershenzon J, Bohlmann J: **Methyl jasmonate induces traumatic resin ducts, terpenoid resin biosynthesis, and terpenoid accumulation in developing xylem of Norway spruce stems.** *Plant Physiol* 2002, **129**(3):1003–1018.
42. Oudin A, Mahroug S, Courdavault V, Hervouet N, Zelwer C, Rodriguez-Concepcion M, St-Pierre B, Burlat V: **Spatial distribution and hormonal regulation of gene products from methyl erythritol phosphate and monoterpene-secoiridoid pathways in *Catharanthus roseus*.** *Plant Mol Biol* 2007, **65**(1–2):13–30.
43. Draper J: **Salicylate, superoxide synthesis and cell suicide in plant defence.** *Trends Plant Sci* 1997, **2**(5):162–165.
44. Ali MB, Yu KW, Hahn EJ, Paek KY: **Methyl jasmonate and salicylic acid elicitation induces ginsenosides accumulation, enzymatic and non-enzymatic antioxidant in suspension culture *Panax ginseng* roots in bioreactors.** *Plant Cell Rep* 2006, **25**(6):613–620.
45. Hayat Q, Hayat S, Irfan M, Ahmad A: **Effect of exogenous salicylic acid under changing environment: a review.** *Environ Exp Bot* 2010, **68**(1):14–25.
46. Shabani L, Ehsanpour AA, Asghari G, Emami J: **Glycyrrhizin production by *in vitro* cultured *Glycyrrhiza glabra* elicited by methyl jasmonate and salicylic acid.** *Russ J Plant Physiol* 2009, **56**(5):621–626.
47. Kang SM, Min JY, Kim YD, Kang YM, Park DJ, Jung HN, Kim SW, Choi MS: **Effects of methyl jasmonate and salicylic acid on the production of bilobalide and ginkgolides in cell cultures of *Ginkgo biloba*.** *In Vitro Cell Dev Biol-Plant* 2006, **42**(1):44–49.
48. Pu GB, Ma DM, Chen JL, Ma LQ, Wang H, Li GF, Ye HC, Liu BY: **Salicylic acid activates artemisinin biosynthesis in *Artemisia annua* L.** *Plant Cell Rep* 2009, **28**(7):1127–1135.
49. Narusaka Y, Narusaka M, Seki M, Umezawa T, Ishida J, Nakajima M, Enju A, Shinozaki K: **Crosstalk in the responses to abiotic and biotic stresses in *Arabidopsis*: analysis of gene expression in cytochrome P450 gene superfamily by cDNA microarray.** *Plant Mol Biol* 2004, **55**(3):327–342.
50. Basson AE, Dubery IA: **Identification of a cytochrome P450 cDNA (CYP98A5) from *Phaseolus vulgaris*, inducible by 3,5-dichlorosalicylic acid and 2,6-dichloro isonicotinic acid.** *J Plant Physiol* 2007, **164**(4):421–428.
51. Pu G, Wang P, Zhou B, Liu Z, Xiang F: **Cloning and characterization of *Lonicera japonica* p-coumaroyl ester 3-hydroxylase which is involved in the biosynthesis of chlorogenic acid.** *Biosci Biotechnol Biochem* 2013, **77**(7):1403–1409.
52. Peng X, Li W, Wang W, Bai G: **Cloning and characterization of a cDNA coding a hydroxycinnamoyl-CoA quinate hydroxycinnamoyl transferase involved in chlorogenic acid biosynthesis in *Lonicera japonica*.** *Planta Med* 2010, **76**(16):1921–1926.
53. Nawrath C: **The biopolymers cutin and suberin.** *The Arabidopsis book/Am Soc Plant Biol* 2002, **1**:e0021.
54. Swaminathan S, Morrone D, Wang Q, Fulton DB, Peters RJ: **CYP76M7 is an ent-cassadiene C11 alpha-hydroxylase defining a second multifunctional diterpenoid biosynthetic gene gluster in rice.** *Plant Cell* 2009, **21**(10):3315–3325.
55. Ehlting J, Sauveplane V, Olry A, Ginglinger JF, Provart NJ, Werck-Reichhart D: **An extensive (co-)expression analysis tool for the cytochrome P450 superfamily in *Arabidopsis thaliana*.** *BMC Plant Biol* 2008, **8**:47.
56. Olszewski N, Sun TP, Gubler F: **Gibberellin signaling: biosynthesis, catabolism, and response pathways.** *Plant Cell* 2002, **14**(Suppl):S61–S80.
57. Bai MY, Shang JX, Oh E, Fan M, Bai Y, Zentella R, Sun TP, Wang ZY: **Brassinosteroid, gibberellin and phytochrome impinge on a common transcription module in *Arabidopsis*.** *Nat Cell Biol* 2012, **14**(8):810–817.
58. Mansouri H, Asrar Z, Mehrabani M: **Effects of gibberellic acid on primary terpenoids and delta-tetrahydrocannabinol in *Cannabis sativa* at flowering stage.** *J Integr Plant Biol* 2009, **51**(6):553–561.
59. Jones AMP, Saxena PK, Murch SJ: **Elicitation of secondary metabolism in *Echinacea purpurea* L. by gibberellic acid and triazoles.** *Eng Life Sci* 2009, **9**(3):205–210.
60. Ebel J, Mithofer A: **Early events in the elicitation of plant defence.** *Planta* 1998, **206**(3):335–348.

61. van der Fits L, Memelink J: **ORCA3, a jasmonate-responsive transcriptional regulator of plant primary and secondary metabolism.** *Science* 2000, **289**(5477):295–297.
62. Weiss D, Ori N: **Mechanisms of cross talk between gibberellin and other hormones.** *Plant Physiol* 2007, **144**(3):1240–1246.
63. Bhat WW, Lattoo SK, Razda S, Dhar N, Rana S, Dhar RS, Khan S, Vishwakarma RA: **Molecular cloning, bacterial expression and promoter analysis of squalene synthase from *Withania somnifera* (L.) Dunal.** *Gene* 2012, **499**(1):25–36.
64. Paddon CJ, Keasling JD: **Semi-synthetic artemisinin: a model for the use of synthetic biology in pharmaceutical development.** *Nat Rev Microbiol* 2014, **12**(5):355–367.
65. Morant M, Bak S, Moller BL, Werck-Reichhart D: **Plant cytochromes P450: tools for pharmacology, plant protection and phytoremediation.** *Curr Opin Biotechnol* 2003, **14**(2):151–162.
66. Keasling JD: **Synthetic biology and the development of tools for metabolic engineering.** *Metab Eng* 2012, **14**(3):189–195.
67. Razdan S, Bhat WW, Rana S, Dhar N, Lattoo SK, Dhar RS, Vishwakarma RA: **Molecular characterization and promoter analysis of squalene epoxidase gene from *Withania somnifera* (L.) Dunal.** *Mol Biol Rep* 2013, **40**(2):905–916.

16

Heterologous expression and biochemical characterization of a highly active and stable chloroplastic CuZn-superoxide dismutase from *Pisum sativum*

Narendra Tuteja[1*], Panchanand Mishra[2†], Sandep Yadav[1†], Marjan Tajrishi[1], Sudhir Baral[2] and Surendra Chandra Sabat[2*]

Abstract

Background: CuZn-Superoxide dismutase (SOD) is a unique enzyme, which can catalyzes the dismutation of inevitable metabolic product i.e.; superoxide anion into molecular oxygen and hydrogen peroxide. The enzyme has gained wide interest in pharmaceutical industries due to its highly acclaimed antioxidative properties. The recombinant expression of this protein in its enzymatically active and stable form is highly desired and hence optimization of culture conditions and characterization of the related biochemical properties are essential to explore the significance of the enzyme in physiological, therapeutic, structural and transgenic research.

Results: High-level expression of the chloroplastic isoform of *Pisum sativum* CuZn-SOD was achieved at 18°C, upon isopropyl β-D-1-thiogalactopyranoside induction and the process was optimized for maximum recovery of the protein in its soluble (enzymatically active) form. Both crude and purified protein fractions display significant increase in activity following supplementation of defined concentration Cu ($CuSO_4$) and Zn ($ZnSO_4$). Yield of the purified recombinant protein was ~ 4 mg L^{-1} of culture volume and the bacterial biomass was ~ 4.5 g L^{-1}. The recombinant pea chloroplastic SOD was found to possess nearly 6 fold higher superoxide dismutase activity and the peroxidase activity was also 5 fold higher as compared to commercially available CuZn-superoxide dismutase. The computational, spectroscopic and biochemical characterization reveals that the protein harbors all the characteristics features of this class of enzyme. The enzyme was found to be exceptionally stable as evident from pH and temperature incubation studies and maintenance of SOD activity upon prolonged storage.

Conclusions: Overexpression and purification strategy presented here describes an efficient protocol for the production of a highly active and stable CuZn-superoxide dismutase in its recombinant form in *E. coli* system. The strategy can be utilized for the large-scale preparation of active CuZn-superoxide dismutase and thus it has wide application in pharmaceutical industries and also for elucidating the potential of this protein endowed with exceptional stability and activity.

Keywords: Heterologous expression, CD spectroscopy, Homology modelling, Ni-NTA purification, Peroxidase activity, Superoxide dismutase

* Correspondence: narendra@icgeb.res.in; surendrachandra@gmail.com
†Equal contributors
[1]Plant Molecular Biology Group, International Centre for Genetic Engineering and Biotechnology (ICGEB), Aruna Asaf Ali Marg, New Delhi 110067, India
[2]Stress Biology Laboratory, Gene Function and Regulation, Institute of Life Sciences, Bhubaneswar 751023, Odisha, India

Background

Superoxide dismutases (SOD, EC 1.15.1.1) are ubiquitous enzyme present in plants, animals and microbes and acts as the first line of defense against the toxic effects of superoxide ($O_2^{\cdot -}$) [1].These metallo-enzymes catalyze the dismutation of $O_2^{\cdot -}$ radicals to yield molecular oxygen (O_2) and hydrogen peroxide (H_2O_2) [1,2]. Based on the metal ion cofactor identified in their active site, three types of SODs (CuZn-SOD, Fe-SOD and Mn-SOD) are reported to be present in the plant cell. Amongst these, CuZn-SODs are the most prominent form, and its isoforms have been reported to be present in cytosol, chloroplast, peroxisomes, and in extracellular location also [3,4]. In chloroplast, the CuZn -SOD serves as an important component of water–water cycle and hence prevents plant from photo-oxidative injury [5].

The chloroplastic CuZn-SOD display significant differences with cytosolic SODs in terms of their amino acid composition, spectral characteristics, sensitivity towards H_2O_2, and immunological properties [6]. The two isozymes are believed to evolved independently and diverged at very early stage of its acquisition [7]. In plant species like *Arabidopsis thaliana*, mustard, bean and citrus Fe-SOD is a major SOD isoform in the leaf, indicating that Fe-SOD alone could remove superoxide from photosynthesis process. Yet, in pea, the level of Fe-SOD is much lower as compared to chloroplastic CuZn-SOD, which could indicate that in these group of plants, chloroplastic CuZn-SOD is primarily involved in protecting chloroplasts from $O_2^{\cdot -}$ mediated damage of photosynthetic machinery [8].

Except few cases, eukaryotic CuZn-SODs are generally homodimeric, consisting of two identical subunits with one copper (Cu^{2+}) and one zinc (Zn^{2+}) ion per subunit [9]. The two subunits are held together by hydrophobic interaction and the arrangement of amino acids is conserved at the dimer interface [10]. The three dimensional structure is characterized by a β-barrel structure composed of eight β-sheets with three loops, which are connected through disulfide bridge. The content of α-helical structure accounts for 5% only, exclusive to loop region. The active copper ion is solvent exposed and serves as redox partner for $O_2^{\cdot -}$, whereas the zinc ion is buried inside the β-barrel and is crucial for stabilization of its structure [11]. CuZn-SODs are believed to be one of the most stable globular protein families studied so far. They retain the native structure in 8 M urea, 4% SDS, has a melting temperature as high as 80°C and are also resistant to proteolytic degradation [12]. Several factors contribute to the enzyme's exceptional stability, including the close packing of hydrophobic interfaces in the dimeric eukaryotic enzymes, the tight packing of the eight stranded β -barrel scaffold, the presence of an intra-chain disulfide bridge and the stabilizing effect of metal cofactors [13,14]. In spite of the stringent conservation of the three dimensional structure displayed by all the eukaryotic CuZn-SODs, enzymes isolated from various natural sources often exhibit difference in their kinetic properties and stability [12]. The study of the molecular determinants that are responsible for those differences could provide hints to understand factors governing protein stability and activity.

The heterologous expression of CuZn-SODs in bacteria and yeast often results in the impairment of recombinant protein activity [15]. This may be due to formation of apo-protein and/or improper folding of the protein in its recombinant form. Nevertheless, the overexpression studies are usually accompanied by supplementation of copper and zinc in growth medium [16,17]. In yeast, copper regulates the expression of cytosolic CuZn-SOD at both transcriptional and post-translational levels, while there is no copper-mediated transcriptional regulation of CuZn-SOD in human cells and higher plants, instead there is a translational and/or post-translational mechanism for CuZn-SOD activation [18].

SOD is a unique enzyme which can eliminate $O_2^{\cdot -}$ and therefore the enzyme is widely used in parenteral administrative in case of various diseases principally related to oxidative stress like musculoskeletal inflammation, Alzheimer disease, Crohon's disease, intestinal fat absorption, and as an anti-inflammatory agent [19]. It is also used as an oral administration for radiation therapy, treating ulcers in the eye cornea and for reducing aging symptoms such as wrinkles etc. [19]. Thus, importance of SOD has been acknowledged worldwide in pharmaceutical industries and there are various efforts to develop more efficient and stable enzyme and improvement of *in vivo* pharmacological activity of SOD by conjugating with polysaccharides, hemoglobin and lecithin [20,21].

In the present study, we describe the recombinant expression and biochemical characterization of a highly active chloroplastic CuZn-SOD from *Pisum sativum* (PschSOD) in *E. coli* expression system. The bacterial growth conditions were optimized for the recovery of maximum protein in the enzymatically active (soluble) form. The recombinantly purified protein was further examined for its spectral and biochemical characteristics. The protein displays very high activity (both $O_2^{\cdot -}$ dismutase and peroxidase), and remains active under a broad range of pHs. The enzyme is stable to proteolytic digestion and 50% of its original activity is retained even after storing for 180 days at room temperature (25°C). Owing to the properties mentioned above, the recombinant PschSOD may find its utility in the medical, cosmetics, and food industries. It may also be valuable for producing transgenic plants that need additional tolerance to biotic and abiotic stresses [22]. Thus, the present study provides an excellent strategy to produce recombinant CuZn-SOD in its highly active and stable form.

Results and discussion

PCR amplification and cloning of PschSOD

To obtain the sequence of PschSOD without 'Chloroplast Transit Peptide', internal primers were synthesized and specific product was PCR amplified using Pea cDNA library as template. Amplified product was cloned into primary vector pGEM-T easy and confirmed by colony PCR and restriction digestion (Additional file 1A). To check the authenticity of insert, PschSOD clones were sequenced and matched for any mutation. Later, for heterologous protein expression, PschSOD gene was cloned into pET-28a vector (Additional file 1A and B). After confirming the right clone, PschSOD -pET-28a was transformed into protein expression strain i.e. BL21 (DE3). Transformed colonies obtained were checked for protein induction by IPTG (1 mM).

Growth culture condition optimization for overexpression and purification of protein

The production of properly folded native protein in *Escherichia coli (E. coli)* is often challenging due to aggregation of the overexpressed protein into inclusion bodies. Nevertheless, formation of recombinant proteins in soluble fraction (periplasmic and cytoplasmic) is favored at lower temperature [23]. In this case, the formation of recombinant protein was relatively higher at 37°C (induction period-5 h) as compared to 18°C (induction period-20 h), but a large amount of protein goes to inclusion body at 37°C (Additional file 2A-C, Table 1A). However, at 18°C most fraction of the protein remains in soluble fraction. When induced at 37°C, the amount of native (cytoplasmic or soluble) PschSOD protein, obtained with the purification procedure was ~ 2.8 mg L^{-1} of culture volume with bacterial wet biomass was about ~ 5 g L^{-1}, while induction at 18°C resulted in ~ 4 mg L^{-1} of culture volume with bacterial wet biomass was about ~ 4.5 g L^{-1} (Table 1A). Therefore, 18°C was selected for further expression study and overexpression of the protein.

The prosthetic copper and zinc are essential component of CuZn-SODs. Therefore, the effect of copper (Cu) and zinc (Zn) on the expression of PschSOD was investigated by exogenous supplementation of Cu ($CuSO_4$) and Zn ($ZnSO_4$) in the culture medium following induction with IPTG. The added concentration (250 μM) of each of these compounds does not interfere with the viability of bacterial cells as reflected in OD600 and CFU counting (Additional file 3: Figure S3A and B). Although, supplementation of copper has no effect on the overexpression of PschSOD in soluble form but zinc results in the increase in protein expression (Additional file 4: Figure S4A and B, Table 1B). However, addition of copper causes a nearly 10-fold increase of SOD activity, as observed in both crude extract and Ni-NTA purified protein samples (Additional file 4: Figure S4C and D, Table 1B). Thus, the concentration of both copper and zinc used here is not toxic to the recombinant BL21 cells and can be suitably used to over express PschSOD. It has been previously suggested that, copper and zinc supplementation in growth media can cause transcriptional and translational activation of CuZn-SOD, when overexpressed in bacteria, yeast and plants [16,17,24].

The purified recombinant protein was obtained with a stepwise combination of Ni-NTA affinity chromatography, desalting and buffer exchange with Zeba desalting

Table 1 Expression and activity of PschSOD in recombinant BL21 cells with variable parameters

A: Temperature optimization for IPTG induced bacterial expression of PschSOD

While monitoring effect of temperature, Cu or Zn was not added in the growth media during induction period

Induction temperature (°C)	Induction period (Hour)	Final cell density (g L^{-1})	Quantity of protein# (mg L^{-1})	SOD activity mg^{-1} of crude protein	SOD activity mg^{-1} of purified proteinᴪ
37	5	5	2.8	110	5,000-8,000
37	20	8	3.0	109	5,000-8,000
18	5	3	1.4	117	5,000-8,000
18	20	4.5	3.2	118	5,000-8,000

B: Effect of Cu ($CuSO_4$) and Zn ($ZnSO_4$) supplementation on expression and activity of PschSOD at 18°C

Final concentration of $CuSO_4$ (μM)	Final concentration of $ZnSO_4$ (μM)	Final cell density (g L^{-1})	Quantity of protein# (mg L^{-1})	SOD activity mg^{-1} of crude protein	SOD activity mg^{-1} of purified proteinᴪ
0	0	4.5-5.0	3.2	118	5,000-8,000
0	250	4.5-5.0	3.6	120	5,000-8,000
250	0	4.5-5.1	3.5	790	46,000-56,000
250	250	4.5-5.1	4.0	820	49,000-58,000

Effect of temperature and Cu/Zn supplementation during induction period is summarized. In all the cases, 1 mM of IPTG was used for induction. The protein was purified only from the soluble fraction (cytoplasmic) while inclusion body fraction was not considered. The results presented here represent average of 3 independent protein preparations.

#The purified PschSOD protein obtained with a stepwise combination of Ni-NTA affinity chromatography, desalting and buffer exchange.

ᴪSOD activity of purified PschSOD as measured by pyrogallol assay.

columns [25]. The homogeneity of the purified proteins was examined on 15% SDS-PAGE. Copper and zinc quantity as determined from ICP-MS, Zincon and Bathocuproine were found to be 1.84 ± 0.15 and 1.6 ± 0.12 respectively per dimer of the recombinant PschSOD as compared to 1.92 ± 0.10 and 1.9 ± 0.09 for bovine CuZn-SOD. The results of the analysis of metal content by ICP-MS, Zincon and Bathocuproine protocol has been shown in Table 2. There is no significant Fe/Mn/ Ni present in the recombinant PschSOD. ICP-MS analysis shows that copper and zinc content was found to be 1.84 ± 0.15 and 1.6 ± 0.12 respectively per dimer of the recombinant PschSOD as compared to 1.92 ± 0.10 and 1.9 ± 0.09 for the commercially available bovine CuZn-SOD. Copper content of PschSOD determined by Zincon and Bathocuproeine was found to be 1.82 ± 0.01 and 1.80 ± 0.08 respectively as compared to 1.92 ± 0.05 and 1.92 ± 0.06 for bovine SOD.

Determination of native and subunit molecular weight

Gel filtration chromatography of PschSOD shows a distinct single retention profile, corresponding to 36–38 kDa molecular mass (Figure 1A). However, the 15% SDS-PAGE analysis suggest that the sub-unit size PschSOD is ~18 kDa (Figure 1B). Western blot analysis by using both monoclonal anti-His antibody and polyclonal anti-CuZn-SOD, confirms the presence of Histidine tag in recombinant protein corresponding to the ~18 kDa molecular marker (Figure 1C and D respectively). These results suggest that the recombinant his-tagged PschSOD maintains its dimeric form in native condition during its recombinant synthesis.

Spectroscopic characterization

The recombinant PschSOD, display multiple electronic transition maxima referring to phenylalanine transitions at 252, 258, 264 and 268 nm and a shoulder at 278–280 nm (Figure 1E). These characteristics absorptions may be indicative of reduced quantity of tyrosine and tryptophan residues, as reported for other plant CuZn-SODs [6,7,26]. Deduced amino acid sequence from the cDNA, also corroborate this argument, where a single tyrosine residue is present at 98th position.

The spectrum in the visible region showed an electronic transition at around 670 nm (Figure 1E inset). Although, the maxima of d-d transition for Cu^{2+} co-ordination of chloroplastic CuZn-SOD is generally peaks at 680 nm [6], the 10 nm blue shift of this maxima in recombinant PschSOD suggest that the ligand environment of the prosthetic copper ion is different, resembling the cytosolic type of CuZn-SOD in its ligand to protein interaction.

Circular Dichroism (CD) spectrum of PschSOD (Figure 1F) was analyzed in 190–240 nm regions. A prominent negative band at 208–210 nm regions was discernible indicating the presence of small amount of α- helix but with high percentage of β-sheet content similar to human and other eukaryotic SODs [6,27]. CD spectra of chloroplast and cytosol CuZn-SODs from spinach also show similar spectral resolution [28]. Thus, it appears that the PschSOD conserves the β -barrel structure of the CuZn-SODs in its recombinant form.

Measurement of PschSOD superoxide dismutase and peroxidase activity

Superoxide dismutase specific activity of PschSOD, assayed by pyrogallol method was calculated to be 45,000-56,000 unitsmg^{-1} of protein (Figure 2A); an activity not reported in literature hitherto. This span of variation was noted from four different individual preparation of the recombinant protein (45, 520; 47, 320; 49, 220; 56, 103 units mg^{-1} protein). On comparative basis, the observed activity was recorded to be nearly 6-fold higher than the activity reported for commercial protein (Sigma-S8160). In addition to the usual superoxide dismutation activity, CuZn-SOD has also a peroxidative function, where the enzyme utilizes its own dismutation product, the H_2O_2 in presence of bicarbonate [25]. Bicarbonate mediated peroxidase activity was measured as increase in 2′, 7′ dichlorfluorescein (DCF) fluorescence intensity with increasing concentration of protein and compared with the activity obtained using commercial CuZn-SOD (Figure 2B and C). The peroxidase activity of the enzyme was also found to be nearly 5-fold higher than the commercial bovine CuZn-SOD (Sigma-S8160).

Inhibitor studies on PschSOD activity

Results on the effect of various inhibitors on the PschSOD activity have been shown in Figure 2D-G. The recombinant enzyme was least affected by Sodium azide (NaN_3), as only 40% of native activity was inactivated by 50 mM Potassium cyanide (KCN). Addition of 5 mM of Hydrogen peroxide (H_2O_2) completely inhibited of the enzyme

Table 2 Measurement of Copper (Cu) and Zinc (Zn) content of PschSOD by ICP-MS, Zincon and Bathocuproeine protocol

Proteins	ICP-MS		Zincon		Bathocuproeine	
	Cu	Zn	Cu	Zn	Cu	Zn
Bovine CuZn-SOD	1.9 ± 0.046	1.89 ± 0.05	1.92 ± 0.05	----	1.92 ± 0.06	----
PschSOD	1.84 ± 0.15	1.6 ± 0.12	1.82 ± 0.01	----	1.80 ± 0.08	----

The values have been depicted as the quotient of concentration of metal ions to protein (dimer). The data presented here represents metal content from 3 independent protein preparations. There is no significant amount of Fe/Mn/ Ni or any other metal was found in the analysis.

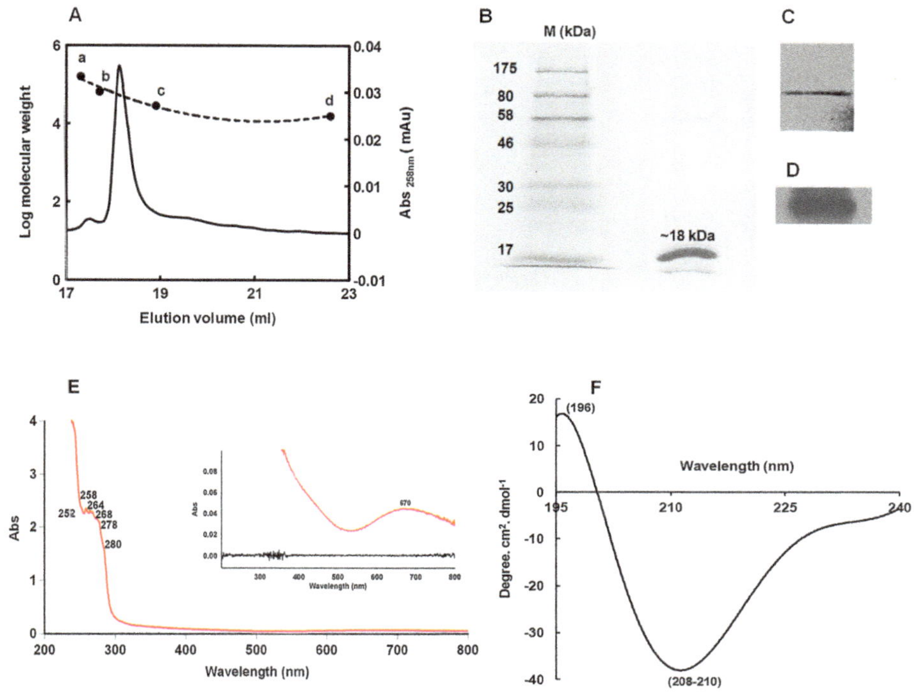

Figure 1 Molecular mass of purified PschSOD and spectral analysis. (A) Native molecular mass of recombinant PschSOD was measured with respect to standard protein marker as resolved in Biosil-250 analytical gel filtration column. Standard proteins used were: (a) 160 kDa-Glucose oxidase (b) 66.5 kDa-BSA fraction V (c) 29 kDa-Carbonic anhydrase and (d) 16 kDa-Bovine CuZn SOD (monomer) respectively **(B)**. The 15% SDS–PAGE showing PschSOD subunit. Subunit molecular mass as calculated using the molecular standard markers (M)suggest that the subunit mass of the protein is ~ 18 kDa. Western blot taking1μg of the protein and probed with poly-his antibody has been shown in **(C)** while probed using CuZn-SOD specific antibody has been depicted in **(D)**. **(E)** Room temperature (25°C) UV-Visible electronic transition spectra of 2 mg ml^{-1} PschSOD. The inset depicts the spectral resolution in an extended scale showing prominent broad peak at 670 nm, characteristics of d-d transition for copper in CuZn-SODs. CD spectral analysis of the protein at 25°C is represented in **(F)**. Negative band at 208 nm region and positive band at 196 nm is shown which suggest abundance of β-sheet in the protein.

activity. KCN was also able to inhibit 100% of the activity at very low concentration (160 μM), while Diethyldithiocarbamate (DDC) could completely abolish the enzyme activity at 1400 μM.

In general, the CuZn SODs purified from various sources exhibit a wide range of susceptibility to inhibitors like NaN_3, KCN and H_2O_2 [29]. DDC and KCN are specific inhibitors for CuZn-SODs and required relatively lesser concentrations to cease enzyme activity and the inhibition is also instant. However, NaN_3 and H_2O_2 mediated inhibition is slow and requires higher concentrations for complete inhibition.

Effect of pH and temperature on activity and stability of PschSOD

The effect of incubation temperature on PschSOD activity was studied and the results are shown in Figure 3A. The enzyme activity was found to be temperature resistant till 55°C, beyond which the activity declined sharply. The time kinetics of alternation in activity profile was also monitored at selected temperatures. The $t_{1/2}$ (reduction in activity to 50%) was calculated to be 62°C, which remains nearly comparable for 20, 40 and 60 min incubation. The enzyme activity showed a broad range of pH sustainability (Figure 3B). The optimal activity of the enzyme was found to be in the pH range of 6–8.0. About 70% of its optimal activity was retained in the acidic pH of 5.0 and in the alkaline pH of 10.0. While 90% of the maximal activity was lost at lower pH (pH-4.0), the remaining activity at pH 11.0, as compared to maximal was nearly 40%. Although, it is suggested that the CuZn-SOD is not pH dependent [30], but recent results from recombinant proteins came with contradicting result [31-33]. The decrease of the enzyme activity at acidic pH might be due to the dissociation of dimer into monomer, as the acidic pH favors the monomer formation, whereas the alkaline pH favors the dimer formation [34]. Charge interaction between the subunits has been also postulated to be a major determinant of CuZn-SOD activity at varying pHs [34].The lowering of activity at acidic pH in this group of SODs is also attributed to the release of Cu ion from the active site [25].

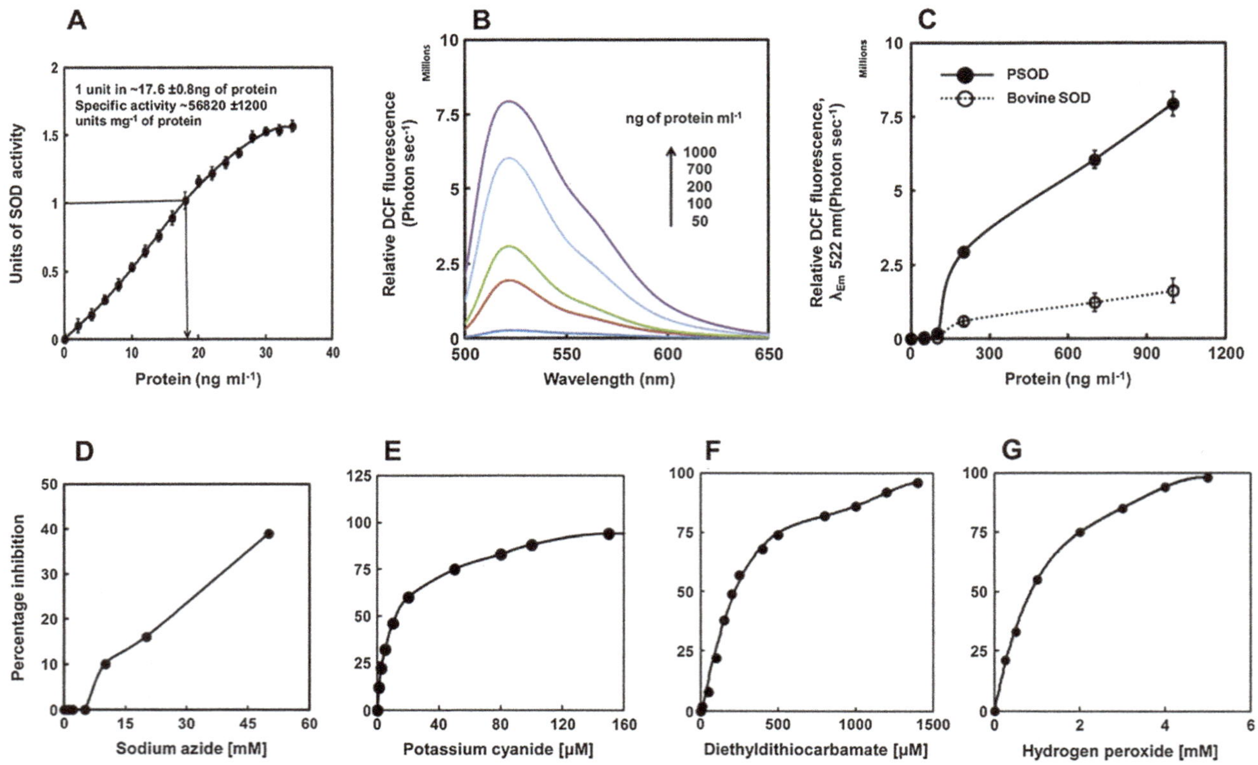

Figure 2 Measurement of superoxide dismutase (SOD) and peroxidase activity of PschSOD and effect of inhibitors on SOD activity. **(A)** Evaluation of SOD activity by pyrogallol assay with increasing protein concentrations of PschSOD. Specific activity (units/mg) of the protein was calculated in the linear range of superoxide scavenging activity and displayed in the figure inset. **(B)** Peroxidase activity of PschSOD measured as increase in relative 2′, 7′-dichlorfluorescein (DCF) fluorescence with increasing protein concentration (as mentioned in the figure). Comparative assessment of peroxidase activity with bovine CuZn-SOD is represented as the relative DCF fluorescence at emission maxima (λ_{max}-522 nm) under 480 nm excitation in **(C)**. Closed circle (•) represents the PschSOD while open circle (○) with solid line represents bovine SOD. Inhibitors tested were **(D)** Sodium azide **(E)** Potassium cyanide **(F)** Diethyldithiocarbamate and **(G)** Hydrogen peroxide. The residual SOD activity was determined taking 1 unit of SOD as 100%. Potassium cyanide and diethyldithiocarbamate acts immediately on CuZn-SOD while for sodium azide and hydrogen peroxide, PschSOD was incubated at desired concentration for 20 min at 37°C and assayed for residual activity.

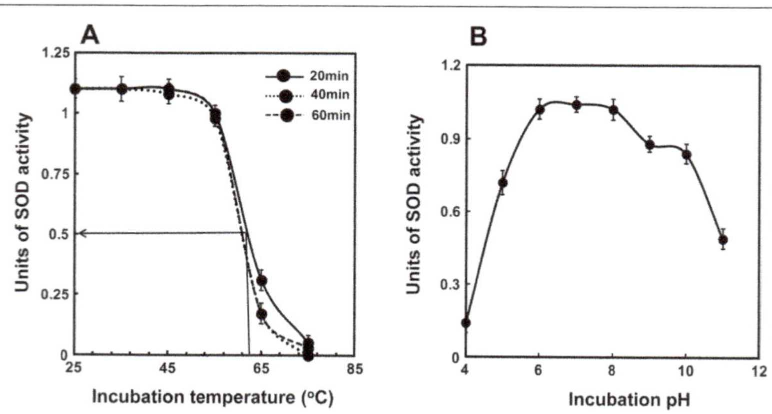

Figure 3 Effect of incubation temperatures and pHs on the PschSOD activity. Twenty units equivalent of protein was incubated at different temperatures (25°C-75°C) for different time interval (20, 40 and 60 min) as shown in **(A)** and the remaining activity was assayed. The $T_{1/2}$ of the activity for 60 min has been marked with the solid line. Effect of different pH on the activity of PschSOD superoxide dismutase activity has been shown in **(B)**. 20 units equivalent of protein were incubated in 50 mM of different buffer with respective pH at 37°C and the residual activity was assayed.

Effect of denaturating and proteolytic agents on PschSOD

PschSOD exhibited very high tolerance to denaturating and proteolytic agents. The activity was not significantly affected by 4 M of urea and imidazole (Figure 4A-B, Additional file 3A). The catalytic function of the enzyme remains unaltered by incubating the protein with trypsin and chymotrypsin for 3 hour (Figure 4C-D, Additional file 3B).

Proteolysis of proteins is often used as a selection tool to assess protein stability and to probe its energy states [35]. Extensive characterization of different prokaryotic and eukaryotic CuZn-SODs has established that monomeric and dimeric CuZn-SODs have very different activity, thermal stability, metal affinity, and resistance to proteases [36]. Various factors responsible for imparting high stability are relatively small solvent-exposed surface area, β-barrel fold, tight hydrophobic dimer interface and extensive ion-pair networks and better salt bridge formation [13]. Results of proteolytic susceptibility of PschSOD further suggested that the protein has limited access to partially and globally unfolded conformations under native conditions and the resistance might have been due to minimized occurrence of accessible conformations susceptible to proteolytic attack [37].

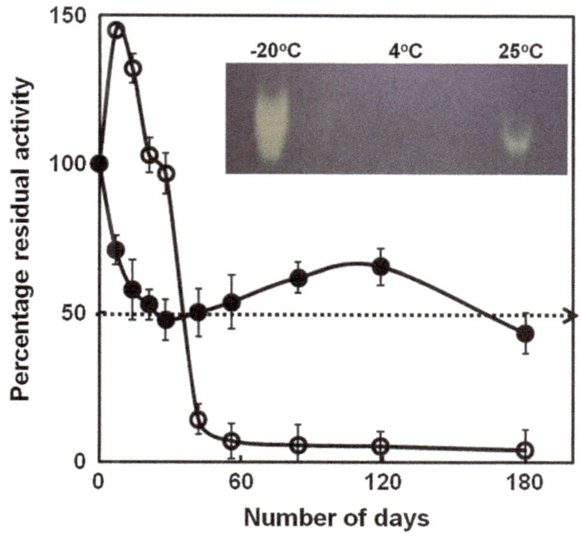

Figure 5 Longevity of PschSOD activity. Purified PschSOD was tested for its longevity both at room temperature and 4°C. Percentage residual activity was judged periodically with pyrogallol assay and further analyzed with native-PAGE and SOD specific staining after 180 days. The protein sample kept at −20°C was also judge for native activity and taken as control because its remains unaltered even after 180 days. Closed circle (•) represents the sample kept at room temperature while open circle (○) with solid line represents enzyme sample kept at 4°C.

Longevity of PschSOD stability and activity

Purified PschSOD was tested for its longevity through storage at room temperature (25°C) and 4°C and the protein samples kept at −20°C were considered as control (Figure 5). The enzyme did exhibit only 50% loss in activity even after 180 days of storage at room temperature. However, at 4°C the protein showed a transient elevation in activity, but undergoes a rapid inactivation in subsequent days. The final loss in activity is accounted for about 90% of activity as compared to activity of protein stored at −20°C. Higher reduction of PschSOD activity at 4°C suggests that the protein is not stable at this temperature and use of some cryo-protectants and/or suitable protease inhibitor may be able to maintain its original activity. The activity of the protein samples following 180 days of storage at different temperatures were also evaluated

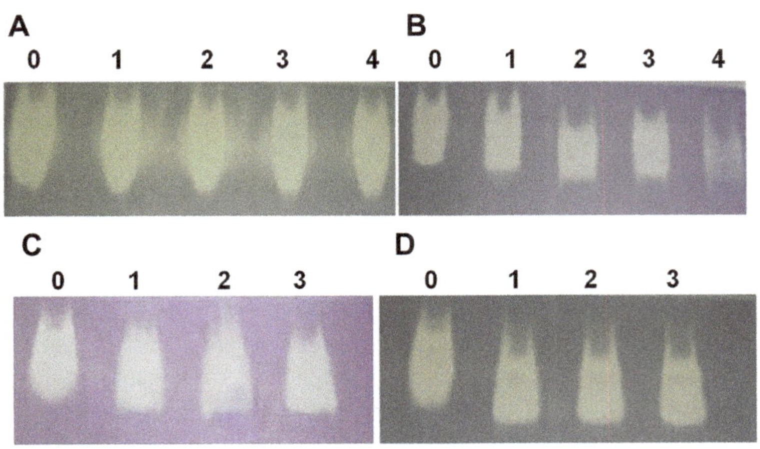

Figure 4 Effect of incubation of PschSOD with denaturating and proteolytic agents. PschSOD was incubated in the presence of **(A)** Urea **(B)** Imidazole at concentration 1–4 M as mentioned in the figure and with proteolytic enzyme **(C)** Trypsin (1/20 w/w) and **(D)** Chymotrypsin in 50 mM potassium phosphate buffer, pH 7.8, at 37°C with varying time intervals of 0–3 h. Residual activity was analyzed on 12% native-PAGE with NBT staining as described in method section. The densitometric analysis of the *in gel* SOD bands has been represented as Additional file 3.

in-gel activity assay and the results have been shown as inset to Figure 5 inset.

PschSOD sequence similarity and divergence

The percentage of identity and divergence in the amino acid sequence of selected SOD was identified using Clustal W program (Figure 6A). The results of this analysis showed that PschSOD was closely related to *Arabidopsis thaliana*-SOD sharing 77.23% similarity, while 71.29% similarity was evident for Rice-SOD, respectively. PschSOD also shared about 58% similarity with *S. cerevisiae* and *H. sapiens.* The Cu^{2+} site responsible for catalysis is also conserved as it is coordinated by four His residues (His-44, His-46, His-61, and His-118) and the Zn^{2+} site is bound to the protein through His-61, His-69, His-78, and Asp-81 (Figure 6A) which is also conserved. To understand the evolutionary relationship between SOD proteins from various phyla, an unrooted phylogram was constructed using Clustal X with 1000 bootstrap replicates. From the phylogenetic analysis, we found that plant SOD from various family made clades according to their family. Monocot and dicot fall in two different clades (Figure 6B). Bootstrap values of the nodes were found to be high in various families. A strong bootstrap value support for individual clades added confidence to the cladogram construction.

PschSOD: motif to structural analysis

In order to detect the conserved domains and motifs among various SODs, multiple sequence alignments were carried out. The N-terminal chloroplast localization signal was observed in all plant SOD, whereas none of the yeast, human and *E. coli* proteins possessed this signal (Figure 6A). Plant SODs were more similar to each other than to human and *E. coli.* The Histidine residue in alignment is conserved in all SODs across the phyla. On C-terminus, positively charged Arg-141 located in the active site channel was also conserved without any exception among all the members included in the presented figure (Figure 6A).

The search for structural homolog was carried out using PschSOD as candidate. PschSOD is a CuZn-SOD; it was modelled against four different CuZn-SODs which showed more than 90% sequence similarity (Expasy SWISS-Homology Modelling). Out of these models, two (3 km2 and 3s0p) were selected on the basis of presence of copper and zinc cofactor. In 3 km2 model, only zinc ions were present, so we also considered 3s0p model as it have copper ion in it. Both the models were superimposed to get both copper and zinc incorporated. Structure of PschSOD revealed 8 β sheets and 2 α helices mostly exposed. All the co-ordinate for Cu^{2+} and Zn^{2+} were found to be conserved. Histidine bound to copper and zinc are highlighted on multiple sequence alignment (Figure 6A). The authenticity of the model was confirmed from various scores like E value (7.6e-68). The color represented Estimated per-residue inaccuracy visualized using a color gradient from blue (more reliable regions) to red (potentially unreliable regions). Overall, the sequence identity is more than 90% and the entire characteristic residue forming the active site is quite conserved throughout the species. However, in the absence of molecular structure, it is hard to explain the structural basis for the enhanced kinetic behavior of PschSOD.

Conclusions

SODs are critical enzymes that manage oxidative stress in almost all organisms dealing with the inevitable metabolic product of aerobic respiration i.e.$O_2^{\cdot-}$ [8]. Plants contain multiple SOD isoenzymes, transcribe through distinct gene families. In plants, three forms of the enzyme exist, as classified by their active site metal ion: copper-zinc, manganese, and iron forms. The distribution of these enzymes has been studied both at the sub-cellular level and at the phylogenic level. It is only in plants that all three different types of SOD are known to coexist [3]. Their occurrence in the different sub-cellular compartments of plant cells allows us to study their molecular evolution and the possibility of understanding why three functionally equivalent but structurally different types of SOD have been maintained? Among all the isoforms of the enzyme, CuZn-SOD is the most abundant in plants and thoroughly dispersed throughout evolution. Phylogenetic analysis of plant sequences show that all chloroplast CuZn-SODs are more related to each other than to cytosolic CuZn-SODs, which also form their separate subgroup [38]. Chloroplast contains both Fe and CuZn-SOD and, distribution of these isoforms vary among different species [39]. Several functions have been postulated for SODs in chloroplasts like protection to oxidative stress (by controlling superoxide and hydrogen peroxide in the water–water cycle and control of free redox-active metals), photoprotection, regulation of electron transport and signaling. It is also remarkable to note that expression of Fe-SOD and CuZn-SOD is found to be reciprocal in response to Cu treatment, different stress treatments and development stages [8,39]. Thus, chloroplastic isoform of the enzyme remains an excellent choice to study kinetic behavior of this class of enzyme to gain deeper insight into its structural feature, which can help the enzyme to modulate its activity to perform these diverse functions. The rationale of the present study is to describe an efficient method for the recombinant expression of *Pisum sativum* chloroplastic CuZn-SOD (PschSOD) in the enzymatically active (soluble) form. The protein was evaluated for its biochemical characteristics and found to possess the highest activity, reported so far. Structural and signature motifs analysis through homology modeling further confirm that PschSOD retains the conserved architecture of this class

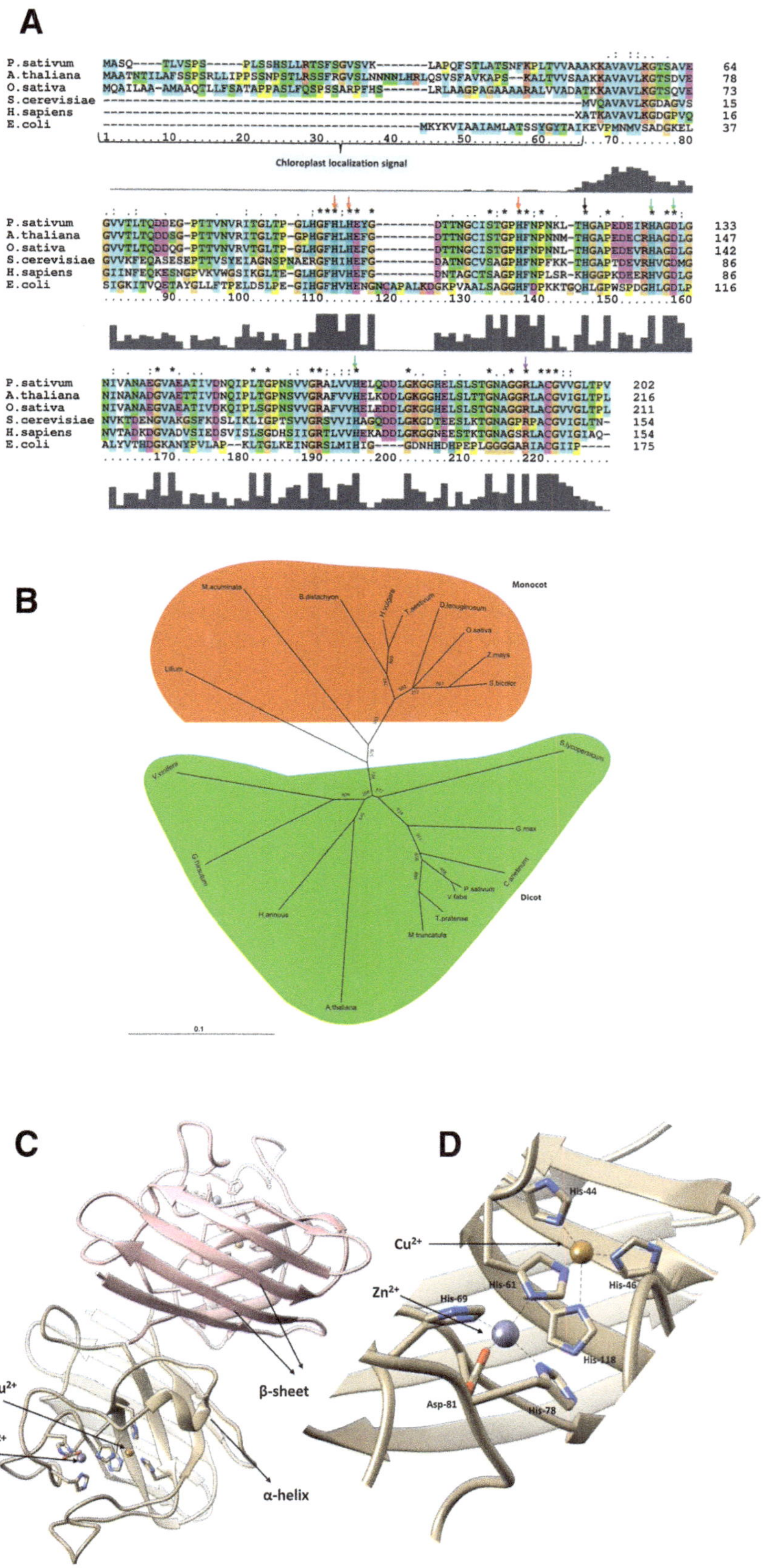

Figure 6 (See legend on next page.)

(See figure on previous page.)
Figure 6 Sequence analysis and Homology modelling of PschSOD (A) Multiple sequence alignment was performed using ClustalW software (http://www.ebi.ac.uk/Tools/msa/clustalw2/). Accession no. of corresponding gene sequence are: *Pisum sativum*: emb|CAA39819.1|, *Arabidopsis thaliana*: gb|AAD10208.1|, *Saccharomyces cerevisiae*: ref|NP_012638.1|, *Homo sapiens*: pdb|1L3N|A and *Escherichia coli*: ref|WP_000875886.1|). Various conserved residue (His-44, His-46, His-61, and His-118) for Cu^{2+} and (His-61, His-69, His-78, and Asp-81) for Zn^{2+} has been marked with red and green arrow. Common Histidine (His-61) is also marked by black arrow while positively charged Arg-141 is marked by purple arrow. Chloroplast localization signal is present in SOD from plants only. **(B)** Tree was constructed using neighbor-joining method with 1000 bootstrap replicates. Monocot and dicot families are separated in two different clades. **(C)** Homology modelling of PschSOD was prepared by using default parameter at SWISS model workspace (http://swissmodel.expasy.org). The figure shows the dimer of modelled PschSOD with both Cu^{2+} and Zn^{2+} in active site. Active site is exposed outside in both monomer. A helix and beta sheets are also labelled. **(D)** All the co-ordinates of Cu^{2+} (Brown) and Zn^{2+} (Black) with His-61 being common in both cofactors show.

of protein. But it is the non-conserved sequence compared to other which could be responsible for this high activity of this protein. Merely on the basis of homology modelling and multiple sequence alignment it is hard to elucidate the main character for the high activity of the protein. However, the higher kinetic properties in terms of its catalytic activity as observed in this study needs further understanding on the molecular aspects of substrata accessibly to the active site and its interaction kinetics. Thus, the purified PschSOD can be utilized for its structural study, has application in pharmaceutical industries and highly recommended for producing transgenic plants that need additional tolerance to oxidative stresses.

Methods

Plasmids, bacterial strain and reagents

Fresh seed of Pea (*Pisum sativum*) were sown on a 2:1:1 mixture of vermiculite, peat moss and perlite, and were maintained for 3 days in darkness before being transferred to normal growth conditions. Later the plants were grown at 26 ± 2°C temperature under 16 h light and 8 h darkness period. Leaves of 7-day old pea seedling were used to isolate RNA. *E. coli* DH5α was the host strain for plasmid construction and cloning. Plasmid vectors used for cloning were pGEM-T easy (Promega) and pET-28a (Novagen). Unless indicated, all analytical grade reagents were procured from Sigma. The expression vector pET-28a, *E. coli* BL21 (DE3) cells were obtained from Novagen. Ni-NTA columns and matrix were procured from Qiagen. Chromatography columns were procured either from Amersham Pharmacia or Pierce.

Construction of pea cDNA library and Cloning of *SOD* gene from pea

Pea cDNA library was constructed as described earlier [40]. In our recent study, PschSOD was found to be an interacting partner of PDH45 [41]. Complete sequence of this PschSOD matched with earlier reported SOD from pea (NCBI Accession no. CAA39819). Gene specific primers were designed using Primer 3 software. *NdeI* restriction site was inserted prior to PschSOD F (5′CAT ATGGCTGCCAAGAAAGCCGTCGCTG3′) and *BamHI* restriction site was inserted after PschSOD R (5′GGATC CTTATACTGGAGTCAAGCCAACC3′). Using pea cDNA library as template, PCR reactions were carried out in a final volume of 50 μl containing 5 μl 10X taq buffer (Agilent Technologies); 200 μM of dNTPs; 0.2 ng of each primer; 5 units of Taq DNA polymerase (Agilent Technologies). The PCR program used was as follows: 95°C for 5 min (1 cycle), 94°C for 30 s, 58°C for 30 s, 72°C for 1 min (32 cycles), and 72°C for 10 min (1 cycle). PCR product was separated on agarose gel. The expected band was excised from gel and purified using Midi gel extraction column (Advanced Microdevices Pvt. Ltd.). Using TA cloning vector, insert was cloned in pGEM-T easy vector. Confirmation of primary cloning of PschSOD -pGEM®-T by colony PCR and restriction digestion is shown in Additional file 1: Figure S1A.

Construction and transformation of the recombinant plasmid (pET-28a-PschSOD) for heterologous expression in *E. coli* system

PschSOD-pGEM-T clone and pET-28a vector were digested with both *NdeI* and *BamHI* restriction enzyme. Later, PschSOD -insert and pET-28a backbone were ligated and transformed in *E.coli* DH5α. Colony PCR and restriction digestion confirmed the final cloning of PschSOD in pET-28a expression vector. For heterologous protein expression, PschSOD -pET-28a clone was transformed in BL21 (DE3) cells. Transformed colonies obtained were checked for protein expression with the help of IPTG (1 mM).

Overexpression of PschSOD in BL21 bacterial cells

For overexpression of PschSOD, the overnight culture of recombinant BL21 (DE3) cells harboring pET-28a-PschSOD plasmid was diluted to 100 times in LB media, supplemented with 50 μg/ml kanamycin on an orbital shaker (240 rpm) at 37°C. When the OD_{600} of the culture reached a value of 0.4-0.6, IPTG was added at a final concentration of 1 mM and culture was grown for another 20-h at 18°C. Total bacterial proteins were fractionated into periplasmic, cytoplasmic and inclusion body fractions, following published protocol [42] and the presence of recombinant protein in soluble and/or insoluble fraction was checked on 15% SDS-PAGE taking 20 μg of protein and bands were visualized by staining with PhastGel Blue R(Amersham Biosciences, Sweden) and western blotting was done using Penta-His (Qiagen) and CuZn-SOD (Agrisera) antibody, by the process described earlier [43].

Optimization of the culture condition for PschSOD solubility, production and activity

The effect of temperature and copper-zinc supplementation on recombinant PschSOD overexpression and activity was examined. To check the solubility of PschSOD, growth temperature during induction period was maintained at 37°C and 18°C. Twenty hour post induction, cells were harvested by centrifugation (8000 × g for 20 min at 4°C) and fractionated as described earlier [42]. Both copper ($CuSO_4$) and zinc ($ZnSO_4$) were supplemented at final concentration of 250 μM in growth medium during induction period at 18°C. SDS-PAGE, western blot and activity measurement of the crude extract and purified protein (methods discussed in later sections) were performed. Other details are provided in the figure legends. Effect of copper-zinc supplementation on bacterial viability was also checked during induction period by taking OD at 600 and counting the colony forming units (CFU) [44]. Cultures with identical treatment, but without IPTG induction served as negative controls (non-induced).

Purification of the recombinant PschSOD

To purify the proteins in native confirmation, induced culture (supplemented with 250 μM of both $CuSO_4$ and $ZnSO_4$) was harvested by low speed centrifugation (8000 X g, 20 min), and the protein was affinity purified using Ni-NTA Fast Start columns (Qiagen) with slight modification from manufacturer's instructions. Briefly, the pellet was re-suspended in lysis buffer containing 50 mM NaH_2PO_4, 300 mM NaCl, 10 mM imidazole, 1 mg/ml lysozyme, 25 unit/ml of Benzonase Nuclease and protease inhibitor cocktail for histidine tagged proteins (Sigma). The suspension was further sonicated on ice to lyse the cells (six times for 10s each time with 5 s pause between). The clear lysate was loaded into the Ni-NTA column, and washed with optimized volume of wash buffer containing 50 mM NaH_2PO_4, 300 mM NaCl, 20 mM imidazole. The enzyme was eluted with elution buffer with 250 mM imidazole. The enzyme sample was desalted and buffer exchanged with low ionic strength phosphate buffer (5 mM K-PO_4, pH-7.2) using Zeba Desalt Spin Columns (Pierce) and further dialyzed for 48-h (12 hour periodic buffer change) with the same buffer at 4°C and concentrated using Amicon centriprep YM-10 centrifugal filter devices (Millipore).

Purity of the protein sample was analyzed on 15% SDS-PAGE taking 1 μg of protein and proteins bands were visualized by staining with PhastGel Blue R (Amersham Biosciences, Sweden) and polyhistidine tag was confirmed with western blotting using Penta-His antibody (Qiagen).

Protein concentration was determined according to Bradford method, by using bovine serum albumin (BSA) as the standard [45]. Metal content of the protein was measured by ICP-MS at SGS India Pvt. Ltd. Gurgaon, India and further analyzed following Zincon and Bathocuproine protocols [46-48]. In all the cases, metal content were analyzed from proteins that were desalted and buffer exchanged with 5 mM K-PO_4 buffer (pH-7.2). The evaluation was made using Cu and Zn content of bovine CuZn-SOD (Sigma-S8160) as reference standard and represented as the molar quotient of the metal ions and protein.

Determination of molecular weight (M_r and subunit size)

The native molecular weight of the purified PschSOD was determined by analytical gel filtration chromatography on a Bio-Sil 250 column (Bio-Rad). 5 mM K-PO_4 (pH-7.2) buffers containing 150 mM NaCl was used. The column was calibrated with the standard proteins: Glucose oxidase, BSA fraction V, Carbonic anhydrase and Bovine CuZn SOD (monomer). Subunit size was determined by 15% SDS-PAGE, using NEB's pre-stained protein ladder.

Spectroscopic measurement

UV–vis absorbance spectra were recorded from 800 to 200 nm with Varian Cary 300 Bio double - beam spectrophotometer using a cell with a 10 mm path length, at 25°C. The concentrations of SOD samples were 1 mg ml^{-1} in 5 mM K-PO_4, pH-7.2.

Spectropolarimetry

CD experiments were conducted in a J-815 Circular Dichroism Spectropolarimeter (JASCO International Co., Ltd. JAPAN) at 25°C using a quartz cuvette with a 0.1-cm path length. Spectra from three scans from 190 to 260 nm at 30 nm/min were averaged, and the buffer baselines were subtracted from their respective sample spectra. Measurements of the molar ellipticity were calculated following equation: $[\theta]_{MRW} = 100 \cdot \theta \cdot 10^{-3} / C_{mr} \cdot 0.1$. Where $[\theta]_{MRW}$ is the mean residue weight in degrees, C_{mr} represents the molar concentration multiplied by the number of amino acids, and 0.1 is the path length in cm. The final protein concentration of each sample used in the measurements was quantitated using a Bradford Assay kit and shown to be 0.2 mM.

SOD activity staining and assay

Native PAGE and specific staining for SOD were performed as described previously [49]. Briefly, each sample was separated by electrophoresis on 12% polyacrylamide gel and then the gel was first soaked in 1.23 mM NBT for 20 min, briefly washed, then soaked in the dark in

100 mM potassium phosphate buffer (pH 7.0) containing 28 mM N, N, N′, N′-tertamethylethylenediamine (TEMED) and 2.8×10^{-2} mM riboflavin for additional 20 min. The gel was briefly washed and illuminated on a light box for 15 min to initiate the photochemical reaction. All the procedures were carried out at room temperature and the SOD bands were visualize as clear white region in the dark formazon background.

SOD activity was assayed based on inhibitory effect of SOD on the spontaneous autoxidation of pyrogallol [50]. One unit of SOD was defined as the amount of protein required inhibiting the initial rate of pyrogallol autoxidation by 50% and was expressed as units/mg of protein. The generation of $O_2^{\cdot-}$ radical (Δ Absorbance min^{-1}- 0.20) was kept constant throughout the experiment. Concentration dependent inhibitory effect of various known inhibitors of CuZn-SODs like NaN_3, KCN, H_2O_2 and DDC were examined. KCN and DDC, the inhibitory effect of NaN_3 and H_2O_2 was observed only when the enzyme samples (equivalent to 5 units) were pre-incubated for 20 min at different concentrations of inhibitors and assayed with 1 unit equivalent of protein.

Measurement of bicarbonate mediated peroxidase activity

Bicarbonate mediated peroxidase activity of enzymes were measured according to published procedure [51]. Briefly, 2, 7-Dichlorfluorescein-diacetate (DCFH-DA) was first dissolved in methanol by vortexing and then hydrolyzed by 0.01 M NaOH in dark for 30 min at room temperature. Peroxidase activity was measured at increasing protein concentration in 100 mM K-PO_4 buffer, pH 7.4 containing 100 μM DTPA, 15 μM DCFH and 25 mM $NaHCO_3$ at 25°C. The reaction was initiated by adding 300 μM of H_2O_2. The reaction cocktail was incubated for 5 minutes, and the net peroxidative activity was measured as increase in 2, 7-Dichlorfluorescein (DCF) fluorescence emission intensity at 522 nm under 480 nm excitation at 25°C (excitation and emission slit width was maintained at 5 nm each).

Effect of temperature and pH on SOD activity

Thermostability of the protein was studied by incubating the enzyme samples at different temperature (25°C - 95°C) in 5 mM K-PO_4, pH-7.2. Aliquots required for the assay were removed at different time intervals and kept immediately on ice, for the determination of residual enzymatic activity.

The effect of pH on the stability and enzymatic activity were examined by incubating protein samples in 50 mM buffers at different pH values (4–12) (pH 4.0-6.0, citric-citrate; pH 7.0-8.0, potassium phosphate; pH 9.0-12.0 glycine-NaOH) for 30 min. The pH of the incubation mixtures was measured after the addition of the enzyme and also following different time intervals. The activity of the samples was assayed as described in earlier section.

Effect of denaturating and proteolytic agents on PschSOD activity

In order to study the effect of proteolytic enzymes, the enzyme was incubated with 1/20 w/w of trypsin and chymotrypsin in 50 mM K-PO_4 buffer, pH 7.8, at 37°C for defined time intervals. For chymotrypsin digestion, $CaCl_2$ was added to a final concentration of 20 mM. Similarly, defined concentration of denaturating agents; urea and imidazole (1-4 M) was added to the protein sample, incubated in 50 mM K-PO_4 buffer, pH 7.8, at 37°C for 1 hour and evaluated for its residual activity. Control samples were also incubated under identical conditions without the proteolytic enzyme/denaturating agents. Aliquots were removed at specified time intervals and analyzed for activity on 12% native-PAGE and stained for SOD activity. *In gel* SOD activity bands were also analyzed by quantity one software (Biorad) and represented in graphical format (Additional file 5A and B).

Effect of long-term storage on PschSOD stability and activity

PschSOD suspended in 5 mM K-PO_4 buffer, pH 7.2 was evaluated for its stability and activity under long-term storage at 25°C for 180 days. The measurement was made periodically with pyrogallol assay system. Control enzyme sample was incubated under identical conditions at 4°C. After 180 days both the enzyme sample were also evaluated for their *in gel* activity.

Multiple sequence alignment

Amino acid alignment of 5 CuZn-SODs from various organisms with several Histidine (His) signatures are shown. The conserved SOD motifs are indicated by black box on aligned sequences. The alignment was made using ClustalW software (http://www.ebi.ac.uk/Tools/msa/clustalw2/) [52] on entire protein sequence for each SOD type. CuZn-SODs are identified on left and numbers of amino acid residues are given on right. Asterisks indicate conserved amino acids while dots represent conservative amino acid changes. Gaps in the amino acid sequences are introduced to improve the alignment. An un-rooted NJ tree is made from the whole protein sequences of various plant CuZn-SODs. Tree was made using clustalX1.81 and viewed in Treeview 1.6.6 software. Analysis of the Phylogenetic tree revealed that various CuZn-SOD is divided into different classes i.e. monocot and dicot, each represented by a clade. CuZn-SOD from various genera belonging to same class falling in the same clades based on the bootstrap support value ≥50%. Scale bar represents 0.1 amino acid substitutions per site.

Protein analysis: Homology modelling of PschSOD

Homology modelling of PschSOD was prepared by using default parameter at SWISS model workspace (http://swiss-model.expasy.org/) [53,54]. We used automated mode for modelling as it selects desired template itself on the basis of highest similarity against the primary amino acid sequence. Templates used for modelling are pdb-id: *3 km2.2, 3hog.1, 3s0p.1, 1srd.1*. Each model generated was checked for various parameters which include Z score, GMQE (global quality estimation score) and QMEAN (model quality estimation) score to assess the accuracy of the model. Later, Molecular graphics visualization program PyMoL and Chimera were used for visualization and editing of PDB models.

Additional files

Additional file 1: Cloning and schematic representation of PschSOD cloning.

Additional file 2: Optimization of induction temperature for PschSOD recombinant expression.

Additional file 3: Effect of Copper ($CuSO_4$) and Zinc ($ZnSO_4$) supplementation on recombinant BL21 (DE3) viability.

Additional file 4: Effect of Copper ($CuSO_4$) and Zinc ($ZnSO_4$) supplementation on PschSOD expression and activity.

Additional file 5: Effect of denaturating and proteolytic agents on PschSOD O_2 dismutase activity (*in gel* densitometry analysis of SOD bands).

Abbreviations

CD: Circular Dichroism; DCFH-DA: 2′,7′-dichlorfluorescein-diacetate; DDC: Diethyldithiocarbamate; DTPA: Diethylenetriamine pentaacetic acid; *E.coli*: *Escherichia coli*; H_2O_2: Hydrogen peroxide; IPTG: Isopropyl β-D-1-thiogalactopyranoside; PschSOD: *Pisum sativum* SOD II; KCN: Potassium cyanide; NaN_3: Sodium azide; SOD: Superoxide dismutase.

Competing interests

The authors declare that they have no competing interests.

Authors' contributions

NT and SCS designed the experiments and interpreted the results. PM conducted the expression, purification of recombinant protein including biochemical and spectroscopic investigations of the purified enzyme.SY performed isolation, cloning, *in silico* analysis of protein. SY and MT analyzed the data and drafted the manuscript. SB worked on enzyme activity during long storage of the protein. Manuscript preparation was made with consultation and imputes from all the authors. All authors read and approved the final manuscript.

Acknowledgements

Work on *Pisum sativum* and SOD in NT's laboratory is partially supported by ICGEB, New Delhi, Department of Science and Technology (DST) and Department of Biotechnology (DBT), Government of India.PM, SB and SC are thankful to The Director, Institute of Life Sciences for providing financial support. The authors would like to acknowledge Adyasha Bharati and Rashmirekha Satapathy for their support in protein purification.

References

1. Fridovich I. Superoxide dismutases. Annu Rev Biochem. 1975;44:147–59.
2. Bannister JV, Bannister WH, Rotilio G. Aspects of the structure, function, and applications of superoxide dismutase. CRC Crit Rev Biochem. 1987;22(2):111–80.
3. Alscher RG, Erturk N, Heath LS. Role of superoxide dismutases (SODs) in controlling oxidative stress in plants. J Exp Bot. 2002;53(372):1331–41.
4. Gill SS, Tuteja N. Reactive oxygen species and antioxidant machinery in abiotic stress tolerance in crop plants. Plant Physiol Biochem. 2010;48(12):909–30.
5. Asada K. Production and scavenging of reactive oxygen species in chloroplasts and their functions. Plant Physiol. 2006;141(2):391–6.
6. Kanematsu S, Asada K. Characteristic amino acid sequences of chloroplast and cytosol isozymes of CuZn-superoxide dismutase in spinach, Rice and Horsetail. Plant Cell Physiol. 1990;31(1):99–112.
7. Kanematsu S, Asada K. Chloroplast and cytosol isozymes of Cu, Zn-superoxide dismutase: their characteristic amino acid sequences. Free Rad Res. 1991;12(1):383–90.
8. Bowler C, Van Camp W, Van Montagu M, Inzé D, Asada K. Superoxide dismutase in plants. Crit Rev Plant Sci. 1994;13(3):199–218.
9. Pelmenschikov V, Siegbahn PE. Copper-zinc superoxide dismutase: theoretical insights into the catalytic mechanism. Inorg Chem. 2005;44(9):3311–20.
10. Bourne Y, Redford SM, Steinman HM, Lepock JR, Tainer JA, Getzoff ED. Novel dimeric interface and electrostatic recognition in bacterial Cu, Zn superoxide dismutase. Proc Natl Acad Sci U S A. 1996;93(23):12774–9.
11. Li H-T, Jiao M, Chen J, Liang Y. Roles of zinc and copper in modulating the oxidative refolding of bovine copper, zinc superoxide dismutase. Acta Biochim Biophys Sin. 2010;42(3):183–94.
12. Carmela Bonaccorsi di Patti M, Carrı MT, Gabbianelli R, Da Gai R, Volpe C, Giartosio A, et al. A free cysteine residue at the dimer interface decreases conformational stability of *Xenopus laevis* copper,zinc superoxide dismutase. Arch Biochem Biophy. 2000;377(2):284–9.
13. Perry JJP, Shin DS, Getzoff ED, Tainer JA. The structural biochemistry of the superoxide dismutases. Biochim Biophys Acta. 2010;1804(2):245–62.
14. Li J, Fisher CL, Konecny R, Bashford D, Noodleman L. Density functional and electrostatic calculations of manganese superoxide dismutase active site complexes in protein environments. Inorg Chem. 1999;38(5):929–39.
15. Yogavel M, Mishra PC, Gill J, Bhardwaj PK, Dutt S, Kumar S, et al. Structure of a superoxide dismutase and implications for copper-ion chelation. Acta Crystallogr D Biol Crystallogr. 2008;64(8):892–901.
16. Krishnakumar R, Kim B, Mollo EA, Imlay JA, Slauch JM. Structural properties of periplasmic SodCI that correlate with virulence in *Salmonella enterica* serovar typhimurium. J Bacteriol. 2007;189(12):4343–52.
17. Stroppolo ME, Pesce A, D'Orazio M, O'Neill P, Bordo D, Rosano C, et al. Single mutations at the subunit interface modulate copper reactivity in Photobacterium leiognathi Cu, Zn superoxide dismutase. J Mol Biol. 2001;308(3):555–63.
18. Tanaka K, Takio S, Yamamoto I, Satoh T. Characterization of a cDNA encoding CuZn-superoxide dismutase from the liverwort Marchantia paleacea var. diptera. Plant Cell Physiol. 1998;39(2):235–40.
19. Salvemini D, Cuzzocrea S. Therapeutic potential of superoxide dismutase mimetics as therapeutic agents in critical care medicine. Crit Care Med. 2003;31(1):S29–38.
20. Angelova M, Dolashka-Angelova P, Ivanova E, Serkedjieva J, Slokoska L, Pashova S, et al. A novel glycosylated Cu/Zn-containing superoxide dismutase: production and potential therapeutic effect. Microbiology. 2001;147(6):1641–50.
21. Shuvaev VV, Han J, Tliba S, Arguiri E, Christofidou-Solomidou M, Ramirez SH, et al. Anti-inflammatory effect of targeted delivery of SOD to endothelium: mechanism, synergism with NO donors and protective effects *In vitro* and *In vivo*. PLoS One. 2013;8(10):e77002.
22. Gupta AS, Heinen JL, Holaday AS, Burke JJ, Allen RD. Increased resistance to oxidative stress in transgenic plants that overexpress chloroplastic Cu/Zn superoxide dismutase. Proc Natl Acad Sci U S A. 1993;90(4):1629–33.
23. Ghosh S, Rasheedi S, Rahim SS, Banerjee S, Choudhary RK, Chakhaiyar P, et al. Method for enhancing solubility of the expressed recombinant proteins in Escherichia coli. Biotechniques. 2004;37(3):418. 420, 422–413.
24. Tanaka K, Takio S, Yamamoto I, Satoh T. Purification of the cytosolic CuZn-superoxide dismutase (CuZn-SOD) of *Marachantia paleacea* var. diptera and its resemblance to CuZn-SOD from chloroplasts. Plant Cell Physiol. 1996;37(4):523–9.
25. Mishra P, Dixit A, Ray M, Sabat SC. Mechanistic study of CuZn-SOD from *Ipomoea carnea* mutated at dimer interface: Enhancement of peroxidase activity upon monomerization. Biochimie. 2014;97:181–93.
26. Kanematsu S, Asada K. CuZn-superoxide dismutases in rice: occurrence of an active, monomeric enzyme and two types of isozyme in leaf and non-photosynthetic tissues. Plant Cell Physiol. 1989;30(3):381–91.

27. Huang Y-H, Shih C-M, Huang C-J, Lin C-M, Chou C-M, Tsai M-L, et al. Effects of cadmium on structure and enzymatic activity of Cu, Zn-SOD and oxidative status in neural cells. J Cell Biochem. 2006;98(3):577–89.
28. Kanematsu S, Asada K. CuZn-superoxide dismutases from the fern *Equisetum arvense* and the green alga *Spirogyra* sp.: occurrence of chloroplast and cytosol types of enzyme. Plant Cell Physiol. 1989;30(5):717–27.
29. Misra HP, Fridovich I. Inhibition of superoxide dismutases by azide. Arch Biochem Biophys. 1978;189(2):317–22.
30. Ellerby LM, Cabelli DE, Graden JA, Valentine JS. Copper – Zinc Superoxide Dismutase: Why Not pH-Dependent? J Amer Chem Soc. 1996;118(28):6556–61.
31. Battistoni A, Folcarelli S, Cervoni L, Polizio F, Desideri A, Giartosio A, et al. Role of the dimeric structure in Cu,Zn superoxide dismutase. J Biol Chem. 1998;273(10):5655–61.
32. Holdom MD, Lechenne B, Hay RJ, Hamilton AJ, Monod M. Production and characterization of recombinant Aspergillus fumigatus Cu, Zn superoxide dismutase and its recognition by immune human sera. J Clin Microbiol. 2000;38(2):558–62.
33. Nedeva T, Dolashka-Angelova P, Moshtanska V, Voelter W, Petrova V, Kujumdzieva A. Purification and partial characterization of Cu/Zn superoxide dismutase from *Kluyveromyces marxianus* yeast. J Chromat B. 2009;877 (29):3529–36.
34. Lin C-T, Kuo T-J, Shaw J-F, Kao M-C. Characterization of the dimer – monomer equilibrium of the papaya copper/zinc superoxide dismutase and its equilibrium shift by a single amino acid mutation. J Agri Food Chem. 1999;47(7):2944–9.
35. Ahmad S, Kumar V, Ramanand KB, Rao NM. Probing protein stability and proteolytic resistance by loop scanning: a comprehensive mutational analysis. Protein Sci. 2012;21(3):433–46.
36. Parsell DA, Sauer RT. The structural stability of a protein is an important determinant of its proteolytic susceptibility in *Escherichia coli*. J Biol Chem. 1989;264(13):7590–5.
37. Manning M, Colón W. Structural basis of protein kinetic stability: resistance to sodium dodecyl sulfate suggests a central role for rigidity and a bias toward β-sheet structure†. Biochemistry. 2004;43(35):11248–54.
38. Fink RC, Scandalios JG. Molecular evolution and structure–function relationships of the superoxide dismutase gene families in angiosperms and their relationship to other eukaryotic and prokaryotic superoxide dismutases. Arch Biochem Biophys. 2002;399(1):19–36.
39. Pilon M, Ravet K, Tapken W. The biogenesis and physiological function of chloroplast superoxide dismutases. Biochim Biophys Acta. 2011;1807(8):989–98.
40. Pham XH, Reddy MK, Ehtesham NZ, Matta B, Tuteja N. A DNA helicase from *Pisum sativum* is homologous to translation initiation factor and stimulates topoisomerase I activity. Plant J. 2000;24(2):219–29.
41. Gill S, Tajrishi M, Madan M, Tuteja N. A DESD-box helicase functions in salinity stress tolerance by improving photosynthesis and antioxidant machinery in rice (*Oryza sativa* L. cv. PB1). Plant Mol Biol. 2013;82(1–2):1–22.
42. Mondal P, Ray M, Sahu S, Sabat SC. Co-expression of GroEL/ES enhances the expression of plant catalase in bacterial cytosol. Prot Pept Lett. 2008;15 (10):1075–8.
43. Ray M, Mishra P, Das P, Sabat SC. Expression and purification of soluble bio-active rice plant catalase-A from recombinant Escherichia coli. J Biotechnol. 2012;157(1):12–9.
44. Grey B, Steck TR. Concentrations of copper thought to be toxic to Escherichia coli can induce the viable but nonculturable condition. Appl Environ Microbiol. 2001;67(11):5325–7.
45. Bradford MM. A rapid and sensitive method for the quantitation of microgram quantities of protein utilizing the principle of protein-dye binding. Anal Biochem. 1976;72:248–54.
46. Nuevo Ordoñez Y, Montes-Bayón M, Blanco-González E, Sanz-Medel A. Quantitative analysis and simultaneous activity measurements of Cu, Zn-superoxide dismutase in red blood cells by HPLC – ICPMS. Anal Chem. 2010;82(6):2387–94.
47. Säbel CE, Neureuther JM, Siemann S. A spectrophotometric method for the determination of zinc, copper, and cobalt ions in metalloproteins using Zincon. Anal Biochem. 2010;397(2):218–26.
48. Rapisarda VA, Volentini SI, Farias RN, Massa EM. Quenching of bathocuproine disulfonate fluorescence by Cu(I) as a basis for copper quantification. Anal Biochem. 2002;307(1):105–9.
49. Beauchamp C, Fridovich I. Superoxide dismutase: improved assays and an assay applicable to acrylamide gels. Anal Biochem. 1971;44(1):276–87.
50. Marklund S, Marklund G. Involvement of the superoxide anion radical in the autoxidation of pyrogallol and a convenient assay for superoxide dismutase. Eur J Biochem. 1974;47(3):469–74.
51. Zhang H, Joseph J, Gurney M, Becker D, Kalyanaraman B. Bicarbonate enhances peroxidase activity of Cu,Zn-superoxide dismutase. Role of carbonate anion radical and scavenging of carbonate anion radical by metalloporphyrin antioxidant enzyme mimetics. J Biol Chem. 2002;277 (2):1013–20.
52. Larkin MA, Blackshields G, Brown NP, Chenna R, McGettigan PA, McWilliam H, et al. Clustal W and Clustal X version 2.0. Bioinformatics. 2007;23(21):2947–8.
53. Arnold K, Bordoli L, Kopp J, Schwede T. The SWISS-MODEL workspace: a web-based environment for protein structure homology modelling. Bioinformatics. 2006;22(2):195–201.
54. Kiefer F, Arnold K, Künzli M, Bordoli L, Schwede T. The SWISS-MODEL Repository and associated resources. Nucleic Acids Res. 2009;37 suppl 1: D387–92.

17

Characterization of competitive interactions in the coexistence of Bt-transgenic and conventional rice

Yongbo Liu[1*], Feng Ge[2], Yuyong Liang[3], Gang Wu[4] and Junsheng Li[1*]

Abstract

Background: Transgene flow through pollen and seeds leads to transgenic volunteers and feral populations in the nature, and consumer choice and economic incentives determine whether transgenic crops will be cultivated in the field. Transgenic and non-transgenic plants are likely to coexist in the field and natural habitats, but their competitive interactions are not well understood.

Methods: Field experiments were conducted in an agricultural ecosystem with insecticide spraying and a natural ecosystem, using *Bt*-transgenic rice (*Oryza sativa*) and its non-transgenic counterpart in pure and mixed stands with a replacement series.

Results: Insect damage and competition significantly decreased plant growth and reproduction under the coexistence of transgenic and conventional rice. Insect-resistant transgenic rice was not competitively superior to its counterpart under different densities in both agricultural and natural ecosystems, irrespective of insect infection. Fitness cost due to *Bt*-transgene expression occurred only in an agroecosystem, where the population yield decreased with increasing percentage of transgenic rice. The population yield fluctuated in a natural ecosystem, with slight differences among pure and mixed stands under plant competition or insect pressure. The presence of *Chilo suppressalis* infection increased the number of non-target insects.

Conclusions: Plant growth and reproduction patterns, relative competition ability and population yield indicate that *Bt*-transgenic and non-transgenic rice can coexist in agroecosystems, whereas in more natural habitats, transgenic rice is likely to outcompete non-transgenic rice.

Keywords: Coexistence, Transgenic rice, Competitive ability, Herbivory, Agricultural ecosystem, Natural ecosystem

Background

As genetically modified (GM) crops have been cultivated worldwide, concerns over ecological consequences emerge because transgenic crops can establish a feral populations in the natural habitat [1-5], which might lead to coexistence with or even extinction of local species. The adventitious presence of GM material in non-GM crops is also a concern due to its economic implications. Without regulations, farmers can grow either GM or GM-free plants in their fields. The coexistence of GM and conventional crops depends on choices by farmers, which is influenced by consumer choice and economic incentives [6,7]. Proper isolation distance or pollen barriers between GM and non-GM fields should be ensured to minimize cross-fertilization so that the adventitious presence of transgenic materials in conventional products will be kept below the legal tolerance threshold, e.g. 0.9% in European Union (Commission Recommendation, 2003). To prevent insects from developing resistance to transgenic crops, the refuge strategy, where transgenic and non-transgenic crops coexist with each other, has usually been adopted [8,9]. Thus, transgenes might exist in landraces and wild relatives through gene flow and introgression by pollen dispersal [1-5,10] or seed movement [11,12]. Therefore, transgenic plants might coexist with non-transgenic plants in and out of the fields [3,7,13-16].

Transgenic crops are usually resistant to pests, herbicides, or diseases, which might enhance their fitness and promote the spread of transgenes into natural populations [17]. In mixture stands, however, the persistence of transgenic

* Correspondence: liu.yongbo@yahoo.com; lijsh@craes.org.cn
[1]State Key Laboratory of Environmental Criteria and Risk Assessment, Chinese Research Academy of Environmental Sciences, Beijing 100012, China
Full list of author information is available at the end of the article

plants in a population depends on their relative fitness and competitive ability against non-transgenic neighbors. In general, insect-resistant plants have a competitive advantage and enhanced fecundity, particularly under high insect pressure [18-22]. Commercialization of *Bt*-transgenic crops, however, could reduce the abundance of target insects in agroecosystem [23,24]. This might influence the plant and insect relationship and the frequency of insect-resistant transgenic plants in mixed populations, which determines ecological risks and monitoring strategies [2].

Previous studies have been mainly focused on gene flow from transgenic crops to landraces and wild relatives and the coexistence of transgenic and non-transgenic progeny [1,3,20,22]. The long-term coexistence of transgenic and non-transgenic crops, however, has been largely ignored [15,25,26]. Moreover, ecological risk of crop-crop gene flow might be higher than that of crop-wild gene flow because no cross barrier exists between intraspecies crops. For example, pollen flow between transgenic and non-transgenic oilseed rapes (*Brassica napus*) results in the resistance of certain plants to three herbicides in the field [15].

GM rice (*Oryza sativa*) conferring insect-resistance, herbicide-tolerance, and high grain quality has been extensively developed in China [27], and two insect-resistant rice lines with *Bt* (*Bacillus thuringiensis*) transgenes were granted biosafety certificates in 2009. Thus, GM and non-GM rice are likely to coexist after the commercialization of GM crops, which might become a major concern over biosafety. This study aims to test (1) the relative competitive ability of insect-resistant *Bt*-transgenic to non-transgenic rice with different proportions of the former in mixed stands, (2) whether competition and herbivory pressures affect relative performance of *Bt*-transgenic to non-transgenic rice in pure and mixture stands, (3) whether increased frequency of transgenic plants in a population alters population productivity. The experiments were carried out in two settings: an agricultural ecosystem with insecticide spraying under local farming practice and a simulated natural ecosystem without insecticide use or other human disturbances.

Results

Plant growth and reproduction of transgenic and conventional rice

Overall, rice plants performed better in the agriculturally managed site, Wuhan, than that in the natural site, Nanchang. Plants in Wuhan site grew higher and produced 20% more valid tillers, 60% more biomass, 50% more viable seeds and 80% more seed weight per plant on average and had a lower rate of invalid tillers and hollow seeds than that in Nanchang (Figure 1). Thus, subsequent analysis was conducted for Wuhan and Nanchang sites respectively.

Herbivory damage and plant competition significantly decreased plant growth and reproduction. In Wuhan, infection decreased plant height and seed weight and increased the rate of invalid tillers (RIT), but it had little effect on other characteristics (Table 1, Figure 1). Plant height, tiller number, biomass and ear length were not significantly different among five percentages of *Bt*-transgenic rice (*Oryza sativa*) Huahui-1 (HH), but the number of hollow seeds increased and seed weight and reproductive allocation (RA) decreased with increasing HH percentage (Table 1, Figure 2). The fitness parameters in high density plots, except for RIT, were significantly lower than that of plants in low density plots (Table 1). There was a significant interaction between infection and density for RIT and seed weight (Table 1). Insect pressure and plant competition had effects on plant growth and reproduction (Figure 1). Compared to non-transgenic Minghui-63 (MH), transgenic rice HH had more valid tillers and hollow seeds, lower RIT, less viable seeds and smaller seed weight (Table 1). There was significant interaction between plant type and infection for predicting RA; between HH percentage and plant type for predicting biomass and seed weight; between plant type and density for predicting RIT, seed weight and RA, respectively (Table 1).

The increased frequency of transgenic rice had little effect on rice growth and reproduction. In the Nanchang site, as the percentage of transgenic plants increased, plants did not show significant difference in any variable (Table 2, Figure 2). Infection increased RIT and the number of hollow seeds and decreased seed weight and RA, but had little effect on other characteristics (Table 2, Figure 1). The plant fitness parameters in high density plots, except for RIT, were significantly lower than that of the plants in low density plots (Table 2). There was significant interaction between infection and density for RIT and seed weight. Insect pressure and plant competition had additive effects on plant growth and reproduction (Figure 1). There was little difference in the performance of HH and MH in the natural ecosystem (Table 2). Little interaction occurred among infection, percentage, density and plant type, except between percentage of transgenic plants and plant type for RIT and hollow seeds, and between density and plant type for biomass (Table 2).

Relative competitive ability of insect-resistant rice to conventional rice

There was little difference in competitive ability between *Bt*-transgenic and conventional rice. The relative difference (RD) between HH and MH did not significantly differ from 0 ($P > 0.05$, *t*-test) under three densities (0.30 m, 0.20 m and 0.10 m space) in the two insect experiments of Wuhan and Nanchang sites. The (relative

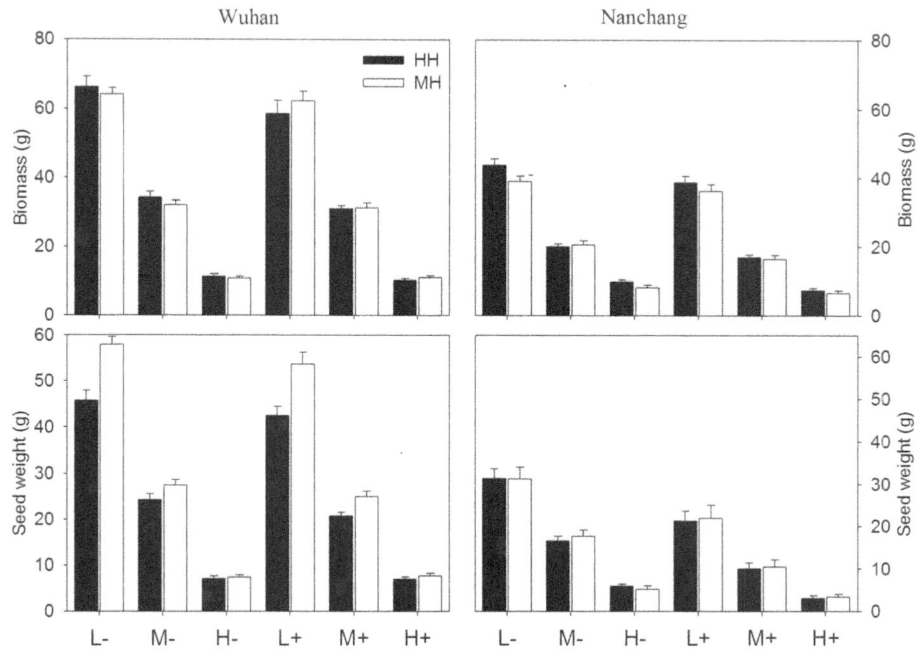

Figure 1 Biomass and seed weight of rice plants in Wuhan and Nanchang sites. Black bars, HH plants; white bars, MH plants. L, M, and H indicate low, medium and high densities respectively. "+" and "-" indicate the presence and absence of infection. Vertical bars denote SEM. See Table 2 for statistical significance.

competitive ability) RCC values did not significantly deviate from 1 ($P > 0.05$, *t*-test) under three densities for biomass and seed production, with or without insects (Table 3), suggesting that HH rice was not competitively superior to MH plants in the mixed stands.

Plot production under the coexistence of transgenic and conventional rice

Plot production for biomass and seed weight was lower in the Nanchang site than that in the Wuhan site. The variation of population yield among different percentages of

Table 1 Effect estimates and degrees of freedom of split-plot ANOVA for plant vegetation and reproductive growth in Wuhan site

	DF	RIT[a]	Biomass	Ear length	Viable seeds	Hollow seeds	Seed weight	RA[a]
Infection (I)	1	41.9^{*}	3.60	3.59	2.28	0.44	24.6^{*}	0.05
Percentage(P)	4	1.19	0.97	0.34	1.50	5.26^{**}	4.40^{*}	3.48*
I × P	4	0.73	0.65	0.10	0.92	0.85	1.14	0.21
Density (D)	2	51.9^{***}	1219^{***}	92.7^{***}	34.4^{***}	42.2^{***}	1710^{***}	22.5^{***}
I × D	2	4.44^{*}	2.18	1.49	0.75	0.24	3.90^{*}	2.72
P × D	8	0.58	1.05	1.28	0.54	1.83	2.23^{*}	1.80
I × P × D	8	0.46	0.56	0.52	0.68	0.37	2.03	1.36
Plant type (T)	1	24.6^{***}	0.00	0.30	19.9^{***}	64.4^{***}	29.8^{***}	107^{***}
I × T	1	0.04	3.10	0.01	0.01	0.16	0.27	5.88*
P × T	2	0.31	3.38^{*}	0.87	1.23	0.20	4.54^{*}	0.59
D × T	2	3.63^{*}	0.30	0.15	2.31	1.97	14.7^{***}	11.8^{***}
I × P × T	2	0.47	0.24	0.03	0.06	0.03	0.56	0.04
I × D × T	2	3.07	0.35	1.83	1.12	0.25	0.24	0.66
P × D × T	4	2.46	0.53	1.31	0.38	0.16	0.59	0.61
I × P × D × T	4	1.12	0.47	0.81	0.99	1.39	0.76	0.79

*, **and ***indicate significant difference at the $P < 0.05$, $P < 0.01$ and $P < 0.001$ levels.
[a]RIT, indicates the rate of invalid tillers; RA, reproductive allocation.

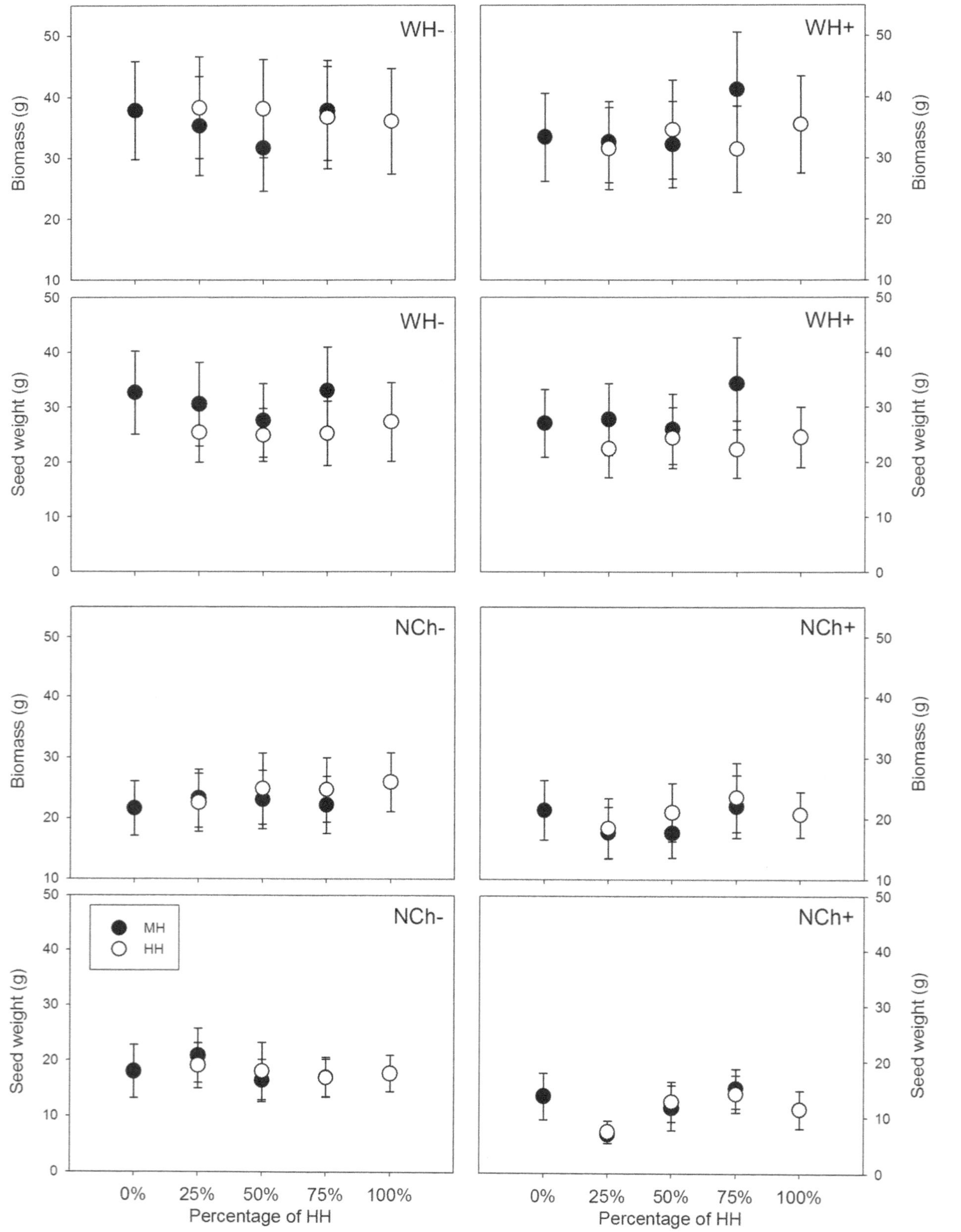

Figure 2 Biomass and seed weight in terms of the percentage of HH plants. Black points, MH plants; white points, HH plants. NCh, Nanchang; WH, Wuhan. "+" and "-" indicate the presence and absence of infection. Vertical bars denote SEM. See Table 1 and 2 for statistical significance.

Table 2 Effect estimates and degrees of freedom of split-plot ANOVA for plant vegetation and reproductive growth in Nanchang site

	DF	RIT[a]	Biomass	Ear length	Viable seeds	Hollow seeds	Seed weight	RA[a]
Infection (I)	1	30.1*	4.44	16.7	2.54	75.6*	26.0*	148**
Percentage(P)	4	1.35	1.31	2.17	0.61	1.86	0.33	0.62
I×P	4	1.63	1.25	1.28	1.43	0.96	1.64	1.60
Density (D)	2	317***	873***	55.1***	28.6***	17.4***	239***	23.8***
I×D	2	17.2***	0.91	1.70	0.31	0.01	6.78**	0.99
P×D	8	1.70	1.12	1.90	1.75	0.61	1.11	0.91
I×P×D	8	2.68*	1.44	2.23*	0.63	1.96	1.35	0.90
Plant type (T)	1	0.19	3.46	0.01	2.25	3.05	0.01	0.43
I×T	1	0.00	0.17	1.97	3.59	1.30	0.03	0.05
P×T	2	5.56**	0.89	2.72	2.88	4.38*	0.78	1.19
D×T	2	1.08	3.32*	0.08	1.07	1.34	0.08	1.30
I×P×T	2	0.33	0.27	2.27	0.36	4.21*	0.44	0.10
I×D×T	2	0.06	0.05	1.55	1.09	1.74	0.20	0.72
P×D×T	4	0.92	0.46	0.46	0.43	0.19	0.96	0.09
I×P×D×T	4	1.32	0.31	0.27	0.43	0.52	1.39	1.22

*, **and ***indicate significant difference at the $P < 0.05$, $P < 0.01$ and $P < 0.001$ levels.
[a]RIT, indicates the rate of invalid tillers; RA, reproductive allocation.

HH was higher in the Nanchang site than that in the Wuhan site (Figures 3, 4), suggesting that the population in the natural habitat was more dynamic than that in the habitat under human control, for both transgenic and non-transgenic rice.

Overall plot productivity biomass did not vary among different percentages of transgenic rice in the plots. The plot-level analysis showed little difference among pure plots and mixtures of the HH and MH plants for total biomass in Wuhan and Nanchang and for seed weight in Nanchang (Table 4; Figures 3, 4;). Total seed weight per plot decreased with increasing percentage of HH in Wuhan site (Table 4, Figure 3). Overall performance per plot including all the HH and MH plants was higher in low density than in high density (Table 4). Infection significantly decreased total seed weight of all the HH and MH plants per plot in the two sites, but had little effect on total biomass (Table 4). The insect damage and plant competition between transgenic and non-transgenic rice showed an additive effect on population production. A significant interaction existed between infection and density for the total seed weight in the two sites and the biomass in Wuhan site. Little interaction, however, was found among infection, HH percentage and density, except for seed weight in Wuhan (Table 4).

Insect survival under the coexistence of transgenic and conventional rice

The coexistence of transgenic and conventional rice had little effect on non-target insects in the simulated natural environment. There were more insects in infected plots than in the plots without infection in the Nanchang site, i.e., 29.5 *vs.* 10.6 planthoppers per two plants on average and 5.0 *vs.* 2.4 spiders per two plants on average (Figure 5). The number of planthoppers and spiders and the rate of infected plants by *Chilo suppressalis* (RPW) did not differ significantly among different percentages of HH. The RPW decreased with the increased percentage of HH plants in the plots without infection.

Table 3 Mean (±SD) of relative crowding coefficient (RCC) between Bt-transgenic and conventional rice

Infection	Density	Wuhan		Nanchang	
		Biomass	Seed weight	Biomass	Seed weight
No	Low	1.11 ± 0.05	1.33 ± 0.34	0.96 ± 0.24	1.10 ± 0.67
	Medium	1.02 ± 0.06	0.94 ± 0.23	0.94 ± 0.26	1.03 ± 0.50
	High	1.04 ± 0.09	1.08 ± 0.45	0.67 ± 0.24	1.14 ± 0.87
Yes	Low	1.02 ± 0.04	1.15 ± 0.49	1.39 ± 0.42	1.25 ± 0.40
	Medium	0.98 ± 0.03	0.89 ± 0.21	0.94 ± 0.32	1.43 ± 0.41
	High	1.02 ± 0.06	0.87 ± 0.28	0.94 ± 0.51	1.34 ± 0.66

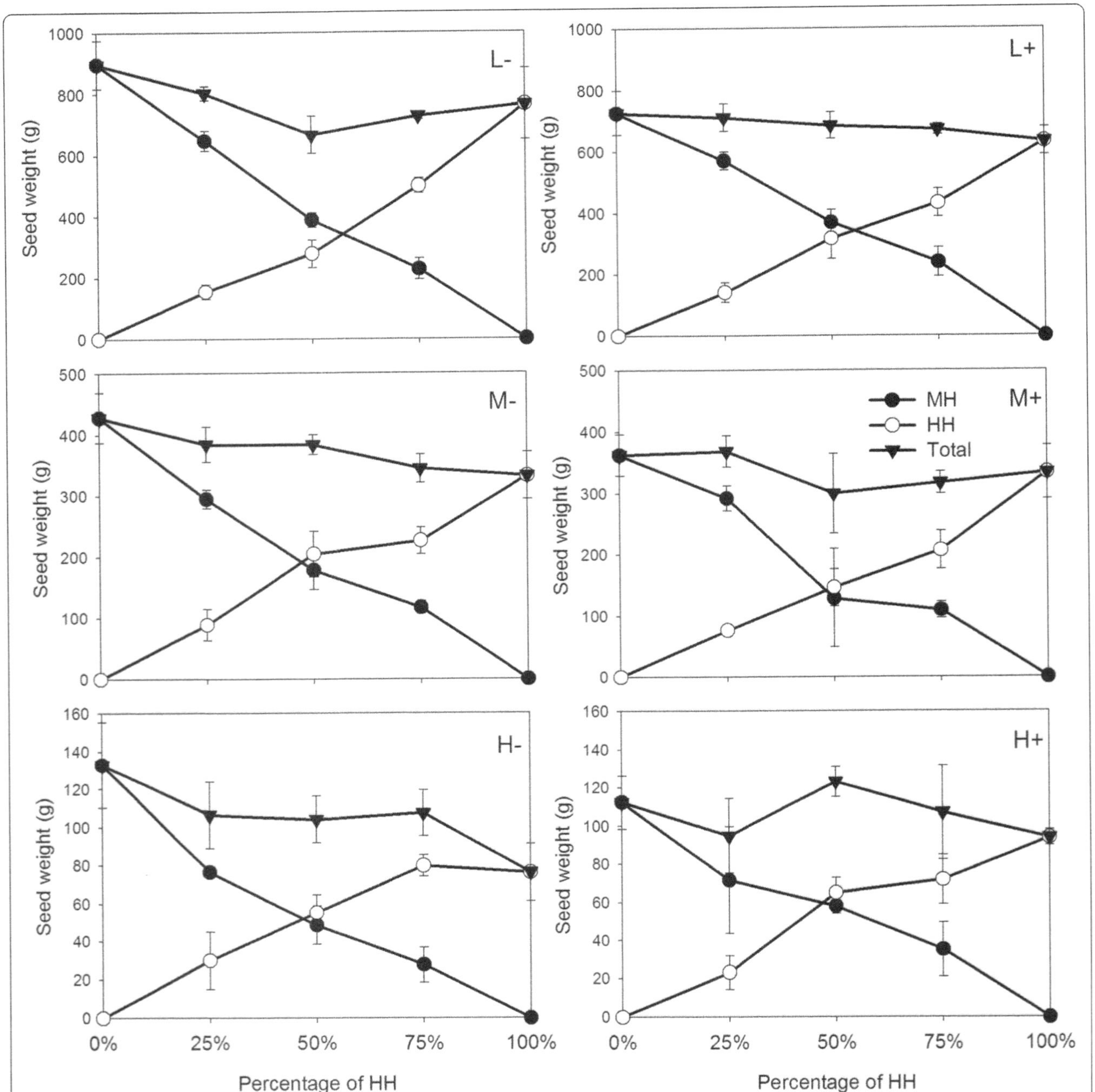

Figure 3 Total seed production per plot in terms of the percentage of HH in Wuhan site. Full circle symbols, total seed weight of all MH plants per plot; empty circle symbols, HH plants; full triangle symbols, all plants (including HH plus MH). L, M, and H indicate low, medium and high densities respectively. "+" and "-" indicate the presence and absence of infection. Vertical bars denote SEM.

Discussion

Effects of coexistence on plant growth and reproduction

The coexistence of transgenic and conventional rice has an effect on plant growth and reproduction, which depends on the local environment, herbivory festation and competition. In the agricultural ecosystem with insecticides to control non-targeted insects (Wuhan site), transgenic rice produced more aborted seeds and less viable seeds than non-transgenic plants. No difference of aborted and viable seeds was observed between transgenic and non-transgenic plants in the natural ecosystem, where non-targeted insects existed without insecticide application (Nanchang site). In another study, transgenic MH86 (a restorer line) with *Bt/CPTI* showed a yield loss in comparison with its non-*Bt* counterpart under low insect pressure [28,29], but little yield differences were observed between *cry1Ab/c Bt*-Shanyou63 (a hybrid variety) and its non-*Bt* counterpart [30,31]. Wang *et al.* [32] found *Bt*-transgenic

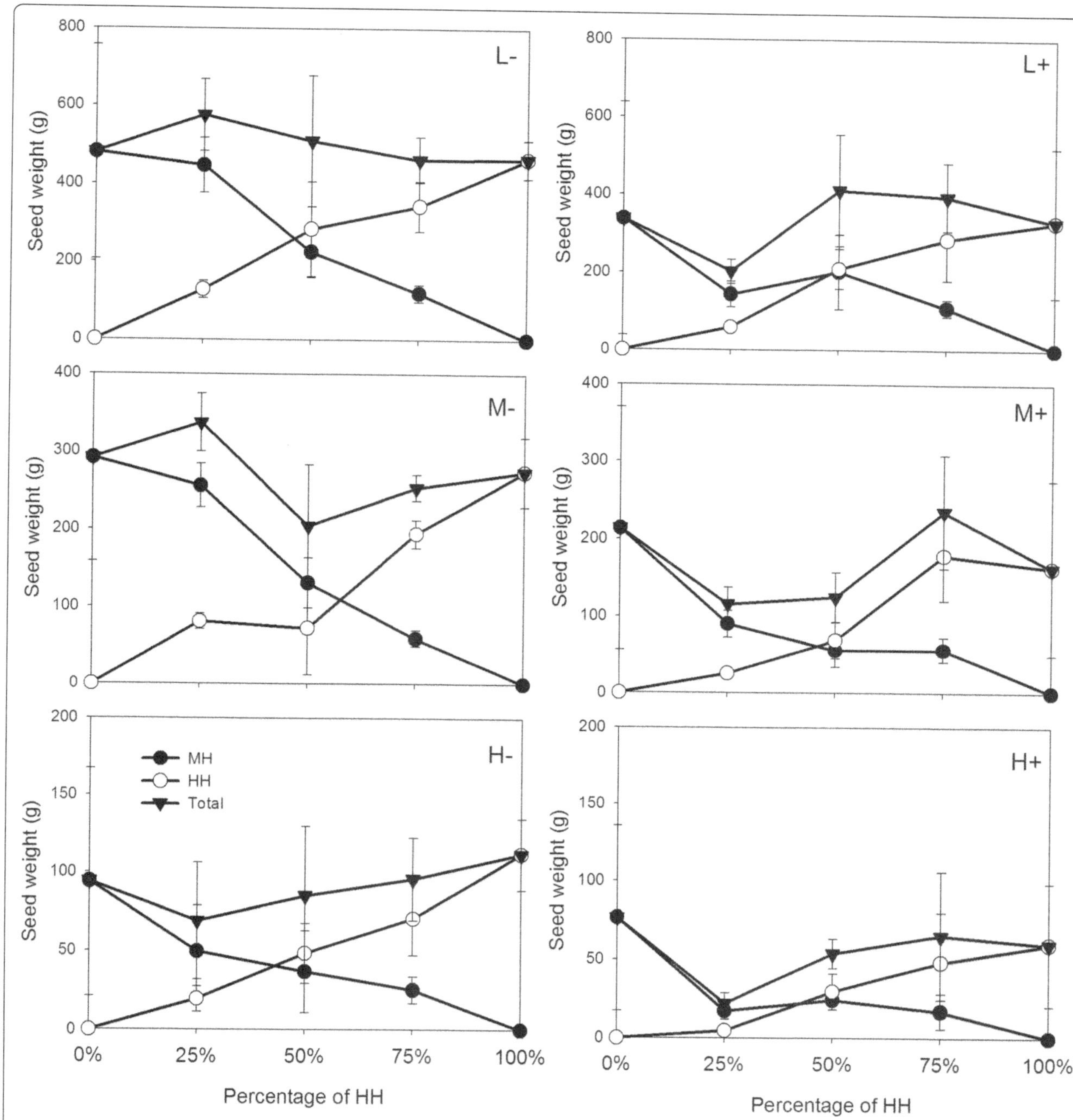

Figure 4 Total seed production per plot in terms of the percentage of HH in Nanchang site. Full circle symbols, total seed weight of all MH plants per plot; empty circle symbols, HH plants; full triangle symbols, all plants (including HH plus MH). L, M, and H indicate low, medium and high densities respectively. "+" and "-" indicate the presence and absence of infection. Vertical bars denote SEM.

Minghui63 with *cry2A* and *cry1C* had lower grain yield than Minghui63. They held that this yield loss was correlated with reduced grain filling percentage resulting from reduction of growth-promoting phytohormones. An alternative explanation is the fitness cost of expressing *Bt* transgene in the absence of insects, which has been shown in rice [29] and rice hybrids [20]. No fitness cost of expressing *Bt* transgene in the absence of selection pressure was detected in *Brassica* [18,33], in sunflower or in rice [21].

As shown in this study, herbivory has a negative effect on plant growth and reproduction [34,35]. Certain studies, however, show little effect of herbivory on plant performance because of compensatory growth [36,37]. Here, insect damage (*Chilo suppressalis*) decreased seed weight but

Table 4 Split-plot ANOVA for total biomass and seed weight per plot in Wuhan and Nanchang sites

		Wuhan		Nanchang	
	DF	Biomass	Seed weight	Biomass	Seed weight
Infection (I)	1	5.70	249**	4.84	20.8*
Percentage (P)	4	0.43	6.60**	1.45	0.17
I × P	4	0.38	1.61	1. 25	1.15
Density (D)	2	1379***	1943***	729***	147***
I × D	2	3.64*	7.66*	1.11	5.08*
P × D	8	1.07	2.11	1.31	0.90
Ix P × D	8	0.51	2.28*	1.31	0.86

*, ** and *** indicate significant difference at the P < 0.05, P<0.01 and P<0.001 levels.

had little effect on seed number. Vaughton and Ramsey also found that defoliation reduced seed mass but not seed number [38]. Insect infection increased the number of aborted seeds in the natural ecosystem, in accordance with previous results that foliar herbivory increased the rate of aborted seeds while floral herbivory decreased it [39].

Plant density, representing competition intensity, significantly reduced the production of both GM and non-GM rice. Competition and herbivory had an additive effect on plant performance, as shown in other reports [2,40]. The interaction could be antagonistic because the impacts of herbivory on neighbors could decrease competition from neighbors [41,42]. There might be no interaction if the disadvantage of the susceptible plants failed to correlate with the specific or combined pressure from competition and herbivory [43].

Relative competitive ability

Theoretically, plants harboring the *Bt* gene are expected to display insect-resistance and less herbivore damage, leading to greater competitive ability than insect-susceptible plants [18,19,21,44]. In this study, however, both RD and RCC values showed that *Bt*-transgenic rice failed to display competition advantage compared with non-transgenic rice. This is in agreement with other reports. For example, Chen *et al.* (2006) found no competition advantage of *Bt*-transgenic rice to control lines because of yield losses in transgenic rice [28].

There are four possible explanations for our failure to observe competitive advantage of *Bt*-transgenic rice. The first is the fitness cost of expressing *Bt* transgene discussed above. The second is that more non-target insects were observed in infected cages than in control cages. *Bt*-transgenic rice resisted to target insects but not to non-target insects (e.g. planthoppers), which affected the growth of *Bt*-transgenic rice. Thirdly, the presence of transgenic plants could protect non-transgenic plants, a "halo effect" proposed by Alstad & Andow [45] and observed in other reports [23,24]. Hutchison *et al.* [24] found that non-*Bt* corn was protected by the adjacent *Bt* corn from being damaged, resulting in billions of dollars of economic benefits for the United States over 14 years. *Bt* cotton has been cultivated over 10 years in Northern China, significantly decreasing the number of target insects (cotton bollworm, *Helicoverpa armigera*) on *Bt* cotton, non-*Bt* cotton and other host crops [23,46]. Fourthly, the relative competitiveness of insect-resistant plants to conventional plants depends on access to resources and allocation strategy. At Nanchang site, insect infection led to an increase in resources allocation to herbivory defence (including non-target insects), resulting in low production. Resistant

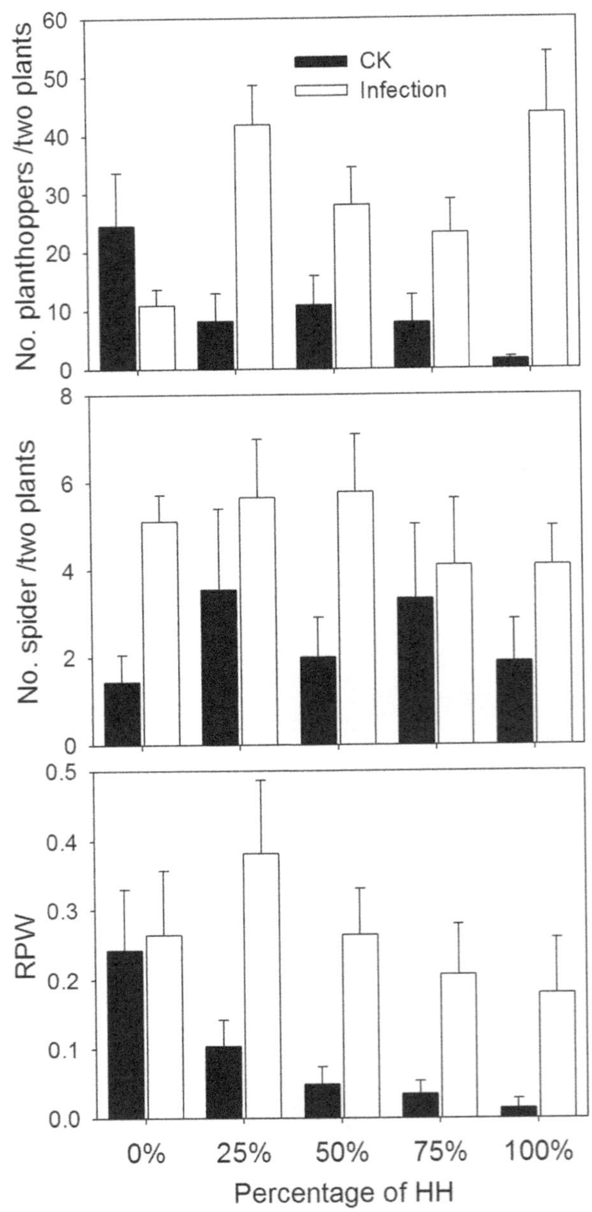

Figure 5 Insect number per two plants against the percentage of HH in Nanchang site. RPW, rate of plants having white straws, representing *C. suppressalis* damage. White bars, infection; black bars, control.

plants have better access to resources (including light, water, nutrient) than the susceptible counterpart because the former suffers less damage. However, there is a trade-off between resistance capacity and competitive ability for plants e.g. [40], since the allocation of resources to defence will decrease plant growth [47]. Kalinina *et al.* found that transgenic wheat resistant to powdery mildew was inferior to its non-transgenic counterpart in competition even with the presence of the pathogen, since transgenic wheat reduced yield and seed number under competition compared with its control [48].

Population yield in the coexistence of transgenic and conventional rice

Since volunteer plants could persist in or outside of cultivation through pollen-mediated gene flow or seed movement of transgenic plants [10-12], it seems inevitable that transgenic and non-transgenic plants would coexist in the field and natural habitats [3,7,13-16]. One concern of the coexistence is the shift from a susceptible population to a resistant one irrespective of insects e.g. [18].

Bt-transgenic rice failed to show competition advantage compared with non-transgenic rice in the present study. This would hamper the increased frequency of *Bt*-transgenic rice and encourage coexistence of transgenic and non-transgenic plants in mixed populations. In the Wuhan site, total seed weight per plot decreased with increased percentage of transgenic rice, due to the fitness cost of expressing *Bt* gene as discussed above. In an experiment to simulate insect-susceptible plants by mechanical wounding, the increase of healthy plants in a population of damaged plants did not increase the population productivity [2].

However, high fluctuation of population production in the natural ecosystem (Nanchang site) suggests its vulnerability, which might lead to a shift from non-transgenic to transgenic population and vice versa. Previous studies found that transgenic plants were more likely to invade a harsh environment [44,49]. The commercial planting of transgenic crops has increased non-target insects in the fields [23,46]. Such insect infestation is ubiquitous in natural habitats, which might facilitate the spread of insect-resistant transgenes. In this study, planthoppers increased when target insects (*C. suppressalis*) occurred, explaining high fluctuation of population production. In addition, rice volunteers frequently appear in the wastelands, orchards or other habitats near fields because of seed movement or seedlings dispersal during the farming. Thus, it is likely that cultivated transgenic rice establishes feral populations as a type of weedy rice if the transgene could persist in the nature over time.

Conclusions

Considering the results of plant growth and reproduction, relative competitive ability and population yield, it is likely that the *Bt*-transgenic and non-transgenic rice will coexist in agricultural ecosystems when commercial release of transgenic crops is permitted. A population in natural habitats is also possible to shift from non-transgenic to transgenic plants. The agricultural benefits and population productivity in diverse cultivation systems with coexisting GM and non-GM crops have been addressed [7,14]. Coexistence allows crop growers and consumers to produce or purchase conventional, organic or GM crops. In addition, as gene flow occurs naturally through pollen or seed flow, a transgenic crop may establish a feral population or volunteers that persist in natural ecosystems, such as oilseed rape, maize, flax and rice [4,10,11,15,25,44]. To mitigate the adventitious presence of GM plants caused by gene flow, strategies based on breeding, agronomic or molecular methods have been proposed e.g. [50]. The evaluation of the ecological consequences of *ex post* coexistence with GM plants, particularly in the natural ecosystem, is essential to construct sound regulations to reduce the adventitious presence of GM plants in agroecosystem and the invasion of transgenes in natural ecosystems.

Methods

Plant material

Bt-transgenic rice (*Oryza sativa*) Huahui-1 (HH) was granted a biosafety certificate in 2009. HH and its non-transgenic counterpart Minghui-63 (MH), donated by Prof. Y. Lin from Huazhong Agricultural University, were employed in this study. *Bt*-transgenic rice Huahui-1 conferred a fusion gene *Cry1Ab/Cry1Ac* driven by actin-I promoter [30]. The expression of *Bt* toxin protein in 120 HH seeds was confirmed with PCR and enzyme-linked immunosorbent assay (ELISA) before sowing.

Field trial

Bt-transgenic rice Huahui-1 (HH) and its counterpart Minghui-63 (MH) were sown in the fields. Rice seedlings were then transplanted into cages (3 m width × 3 m length × 2.5 m height, protected by a 2 mm-mesh nylon net), each cage containing three plots (6 × 6 = 36 plants per plot) with three densities: 0.30 m, 0.20 m and 0.10 m row spacing (Figure 6). There were thirty cages in total in five rows and six columns, and three-meter space was set between the cages to ensure the plants' access to sunlight. To simulate the colonization process of insect-resistant plants into a non-transgenic population, five percentages of HH were added to plots in a replacement series: 0 (T_0), 25% (T_{25}), 50% (T_{50}), 75% (T_{75}), and 100% of HH (T_{100}). The five percentages were placed randomly in a column (Figure 6). Plant types (HH or MH) were placed randomly in each plot, ensuring the same ratio of MH to HH for the 16 plants in the center and the 20 plants at the border. Thirty cages were separated into three blocks, with one

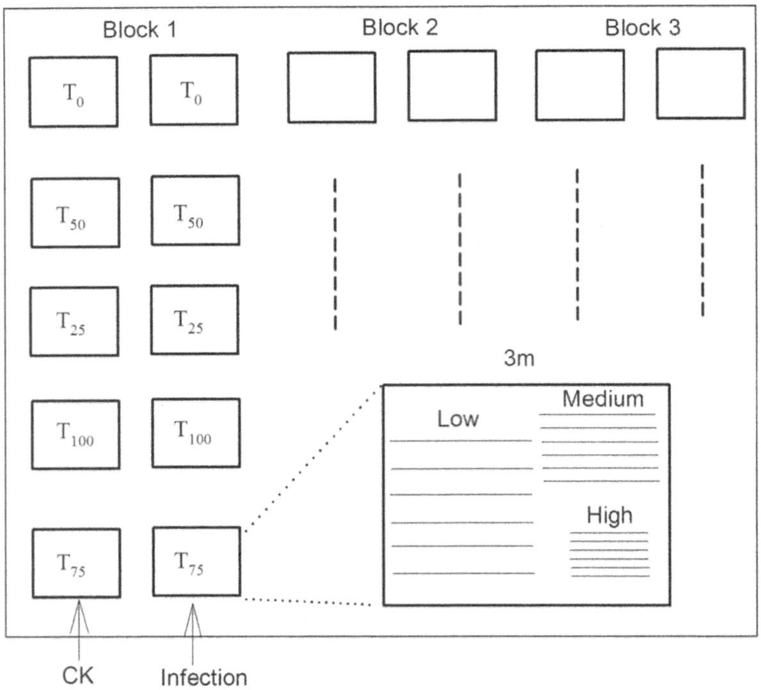

Figure 6 The map of the infection experiments in Wuhan and Nanchang sites. Infection, insect release of *C. suppressalis*; CK, without insect release. Low, medium and high densities with 0.3, 0.2 and 0.1 m space between plants. T_0, T_{25}, T_{50}, T_{75} and T_{100} indicate the percentages of transgenic rice.

column of cages infected by insects per block (Figure 6). Two second-instar larvae of *Chilo suppressalis* [Walker] were introduced to each plant.

The experiments were carried out at two sites, a transgenic rice research base of Huazhong Agricultural University, Wuhan, Hubei province (N30°27′58.8′′, E114°21′45.8′′), and a transgenic rice research base of Jiangxi Academy of Agricultural Sciences, Nanchang, Jiangxi province (N28°21′91.4′′, E115°55′49.7′′). The annual mean temperature and precipitation were 18.0°C and 2059 mm in Nanchang and 15.5°C and 1416 mm in Wuhan city in 2012 (China Statistical Yearbook). Rice seeds were sown in the fields on June 12, 2012 at Wuhan site and on June 5, 2012 at Nanchang site. The seedlings were transplanted in the fields after 28–30 days. After the seedlings survived, second-instar larvae of *C. suppressalis* were introduced to the plants. The site at Wuhan was considered as an agricultural ecosystem, with weeds removal by hands and insecticide spraying. Insecticide (22% a.i. buprofezin in 1000-fold dilution, 450 g ha^{-1}) was applied twice (at the third and sixth week after infection) to control rice planthoppers in infected cages during the growing season. Simultaneously, avermectin (1.8 g L^{-1} a.i. in 1000-fold dilution, 2.5 L ha^{-1}) and triazophos (20% a.i. in 1000-fold dilution, 1500 g ha^{-1}) for *C. suppressalis* were applied twice in the cages without infection. The site at Nanchang simulated a natural ecosystem, without insecticide spraying or weed removal in the cages during the growing season. Fertilization and irrigation was implemented whenever need according to local agricultural farming.

Measurements

Plant height, number of valid tillers, the rate of invalid tillers (dead tillers plus tillers without ears) to all tillers per plant (RIT), biomass (straw weight after being insolated ten days) and seed weight per plant were measured for all 16 plants (HH and/or MH) in the center of each plot. Reproductive allocation (RA) per plant was defined by dividing seed weight with the biomass. Six ears were chosen randomly for each plant type (HH *vs.* MH) in the plot center to measure ear length, number of viable seeds and hollow seeds per ear, and mean values were regarded as variables per plant.

Field insects (rice planthopper and spider) on two randomly selected plants per plot and the rate of plants with white straws (RPW, representing *C. suppressalis* damage) to 16 center plants (HH and/or MH) were surveyed before harvesting at the Nanchang site. In Wuhan, insects were not surveyed due to the application of insecticides.

Statistical analysis

The mean values of all HH or MH plants per plot were used for statistical analysis. The data were *log*-transformed to ensure a normal distribution of residuals. Split-plot ANOVA ($Y \sim Site + I \times P \times D \times T +$ Error (block/I/P/D)) was employed to test the effects of site (Wuhan *vs.* Nanchang), insect infection (I), percentage of HH plants (P), plant density (D) and plant types (T, HH vs. MH). Due to significant site effects, split-plot ANOVA ($Y \sim I \times P \times D \times T+$ Error (block/I/P/D)) was used in Wuhan and Nanchang, respectively. Tukey's HSD test was used for multiple comparisons among percentages of HH plants.

To further investigate the differences between HH and MH individual plant biomass, relative to the MH performance in pure stand T_0, the relative differences RD $= (R_{Ti} - S_{Ti})/S_{T0}$ were calculated (value at T_{100} was calculated as $RD = (R_{T100} - S_{T0})/S_{T0}$) [2]. RD values were subject to a *t*-test to determine their deviation from 0.

For replacement series analysis, the relative crowding coefficient (RCC) was calculated according to the equation in Ramachandran *et al.* [18]:

$$RCC = \left(\sum (R_{Ti}/S_{Ti})/3\right)/(R_{T100}/S_{T0})$$

where R_{Ti} and S_{Ti} are the biomass values for HH and MH respectively at percentage T_i (T_{25}, T_{50} and T_{75}) per unit area in the plot center. An RCC value of 1 indicates equal competitiveness between R and S plants. RCC >1 indicates that R plants are more competitive than S plants and vise versa. A *t*-test was performed to determine whether the RCC values were significantly deviated from 1.

As for biomass and full seed weight, a per-plot ANOVA ($Y \sim I \times D \times P+$ Error (block/I/D)), with block as a random factor and infection, density and percentage of HH plants as fixed factors, was carried out using the sum of the data of all MH and HH plants per plot, i.e. the whole population production. All statistic analyses were conducted in R software [51].

Competing interest
The authors declare that they have no competing interests.

Authors' contributions
YBL and JSL designed the study, carried out the experiment, analysed the data and wrote the manuscript; all authors contributed to the revisions. FG, YYL and GW provided field space for the experiment and involved themselves in the research. YBL and JSL provided funding, interpreted data and edited the manuscript. All authors read and approved the final manuscript.

Acknowledgements
This work was financially supported by the Special Program for New Transgenic Variety Breeding of the Ministry of Science and Technology, China (grant 2012ZX08011002), the National Environmental Protection Public Welfare Science and Technology Research Program of China (grant 201309038) and one Project of Chinese Research Academy of Environmental Sciences (2011YSKY-08).

Author details
[1]State Key Laboratory of Environmental Criteria and Risk Assessment, Chinese Research Academy of Environmental Sciences, Beijing 100012, China. [2]State Key Laboratory of Integrated Management of Pest and Rodents, Institute of Zoology, the Chinese Academy of Sciences, 8 Dayangfang, Beijing 100101, China. [3]Institute of Plant Protection, Jiangxi Academy of Agricultural Sciences, Nanchang 330200, China. [4]Hubei Insect Resources Utilization and Sustainable Pest Management Key Laboratory, College of Plant Science and Technology, Huazhong Agricultural University, Wuhan 430070, China.

References

1. Stewart JCN, Halfhill MD, Warwick SI. Transgene introgression from genetically modified crops to their wild relatives. Nat Rev Genet. 2003;4(10):806–17.
2. Liu Y, Wei W, Ma K, Darmency H. Spread of introgressed insect-resistance genes in wild populations of *Brassica juncea*: a simulated in-vivo approach. Transgenic Res. 2013;22:747–56.
3. Warwick SI, Légère A, Simard MJ, James T. Do escaped transgenes persist in nature? The case of an herbicide resistance transgene in a weedy *Brassica rapa* population. Mol Ecol. 2008;17(5):1387–95.
4. Ellstrand N, Meirmans P, Rong J, Bartsch D, Ghosh A, de Jong T, et al. Introgression of crop alleles into wild or weedy populations. Annu Rev Ecol Evol Syst. 2013;44:325–45.
5. Chen LJ, Lee DS, Song ZP, Suh HS, Lu BR. Gene flow from cultivated rice (*Oryza sativa*) to its weedy and wild relatives. Ann Bot. 2004;93:67–73.
6. Jank B, Rath J, Gaugitsch H. Co-existence of agricultural production systems. Trends Biotechno. 2006;24:198–200.
7. Demont M, Devos Y. Regulating coexistence of GM and non-GM crops without jeopardizing economic incentives. Trends Biotechnol. 2008;26(7):353–8.
8. Jongsma MA, Gould F, Legros M, Yang L, Loon JJA, Dicke M. Insect oviposition behavior affects the evolution of adaptation to Bt crops: consequences for refuge policies. Evol Ecol. 2010;24(5):1017–30.
9. Møldrup ME, Geu-Flores F, de Vos M, Olsen CE, Sun J, Jander G, et al. Engineering of benzylglucosinolate in tobacco provides proof-of-concept for dead-end trap crops genetically modified to attract Plutella xylostella (diamondback moth). Plant Biotechnol J. 2012;10(4):435–42.
10. Lu BR, Yang C. Gene flow from genetically modified rice to its wild relatives: Assessing potential ecological consequences. Biotechnol Adv. 2009;27(6):1083–91.
11. D'Hertefeldt T, Jorgensen RB, Pettersson LB. Long-term persistence of GM oilseed rape in the seedbank. Biol Lett. 2008;4:314–7.
12. van Heerwaarden J, Ortega Del Vecchyo D, Alvarez-Buylla ER, Bellon MR. New genes in traditional seed systems: Diffusion, detectability and persistence of transgenes in a maize metapopulation. PLoS One. 2012;7:e46123.
13. Hooftman DAP, Jong MJD, Oostermeijer JGB, Den Nijs HCM. Modelling the long-term consequences of crop-wild relative hybridization: a case study using four generations of hybrids. J Appl Ecol. 2007;44(5):1035–45.
14. Devos Y, Demont M, Dillen K, Reheul D, Kaiser M, Sanvido O. Coexistence of genetically modified (GM) and non-GM crops in the European Union. A review Agron Sustain Dev. 2009;29(1):11–30.
15. Hall L, Topinka K, Huffman J, Davis L, Good A. Pollen flow between herbicide resistant *Brassica napus* is the cause of multiple-resistant *B. napus* volunteers. Weed Sci. 2000;48:688–94.
16. Rong J, Song Z, Su J, Xia H, Lu B-R, Wang F. Low frequency of transgene flow from Bt/CpTI rice to its nontransgenic counterparts planted at close spacing. New Phytol. 2005;168(3):559–66.
17. Ellstrand NC. When transgenes wander, should we worry? Plant Physiol. 2001;125:1543–5.
18. Ramachandran S, Buntin GD, All JN, Raymer PL, Stewart Jr CN. Intraspecific competition of an insect-resistant transgenic canola in seed mixtures. Agron J. 2000;92:368–74.
19. Moon HS, Halfhill MD, Good LL, Raymer PL, Stewart Jr CN. Characterization of directly transformed weedy Brassica rapa and introgressed *B. rapa* with Bt cry1Ac and gfp genes. Plant Cell Rep. 2007;26(7):1001–10.
20. Yang X, Xia H, Wang W, Wang F, Su J, Snow AA, et al. Transgenes for insect resistance reduce herbivory and enhance fecundity in advanced generations of crop-weed hybrids of rice. Evol Appl. 2011;4(5):672–84.
21. Xia H, Lu B-R, Xu K, Wang W, Yang X, Yang C, et al. Enhanced yield performance of Bt rice under target-insect attacks: implications for field insect management. Transgenic Res. 2011;20(3):655–64.
22. Yang X, Wang F, Su J, Lu B-R. Limited fitness advantages of crop-weed hybrid progeny containing insect-resistant transgenes (Bt/CpTI) in transgenic rice field. PLoS One. 2012;7(7):e41220.
23. Wu KM, Lu YH, Feng HQ, Jiang YY, Zhao JZ. Suppression of Cotton Bollworm in Multiple Crops in China in Areas with Bt Toxin-Containing Cotton. Science. 2008;321(5896):1676–8.

24. Hutchison WD, Burkness EC, Mitchell PD, Moon RD, Leslie TW, Fleischer SJ, et al. Areawide Suppression of European Corn Borer with Bt Maize Reaps Savings to Non-Bt Maize Growers. Science. 2010;330(6001):222–5.
25. Jhala AJ, Bhatt H, Topinka K, Hall LM. Pollen-mediated gene flow in flax (*Linum usitatissmum* L.): can genetically engineered and organic flax coexist? Heredity. 2011;106:557–66.
26. Langhof M, Hommel B, Hüsken A, Schiemann J, Wehling P, Wilhelm R, et al. Coexistence in maize: Do nonmaize buffer zones reduce gene flow between maize fields? Crop Sci. 2008;48(1):305.
27. Huang JK, Hu RF, Rozelle S, Pray C. Insect-resistant GM rice in farmers' fields: Assessing productivity and health effects in China. Science. 2005;308:688–90.
28. Chen LY, Snow AA, Wang F, Lu BR. Effects of insectresistance transgenes on fecundity in rice (*Oryza sativa*, Poaceae): a test for underlying costs. Am J Bot. 2006;93:94–101.
29. Xia H, Chen L, Wang F, Lu B-R. Yield benefit and underlying cost of insect-resistance transgenic rice: Implication in breeding and deploying transgenic crops. Field Crops Res. 2010;118(3):215–20.
30. Tu J, Zhang G, Datta K, Xu C, He Y, Zhang Q, et al. Field performance of transgenic elite commercial hybrid rice expressing *Bacillus thuringiensis* δ-endotoxin. Nat Biotechnol. 2000;18:1101–4.
31. Wang Y, Zhang G, Du J, Liu B, Wang M. Influence of transgenic hybrid rice expressing a fused gene derived from *cry1Ab* and *cry1Ac* on primary insect pests and rice yield. Crop Protect. 2010;29(2):128–33.
32. Wang F, Ye C, Zhu L, Nie L, Cui K, Peng S, et al. Yield differences between Bt transgenic rice lines and their non-Bt counterparts, and its possible mechanism. Field Crop Res. 2012;126:8–15.
33. Di K, Stewart Jr CN, Wei W, Shen B-c, Tang Z-X, Ma K-P. Fitness and maternal effects in hybrids formed between transgenic oilseed rape (*Brassica napus* L.) and wild brown mustard [*B. juncea* (L.) Czern et Coss.] in the field. Pest Manage Sci. 2009;65(7):753–60.
34. Escarré J, Lepart J, Sans X, Sentuc JJ, Gorse V. Effects of herbivory on the growth and reproduction of *Picris hieracioides* in the Mediterranean region. J Veg Sci. 1999;10:101–10.
35. Rogers W, Siemann E. Effects of simulated herbivory and resource availability on native and invasive exotic tree seedlings. Basic Appl Ecol. 2002;3(4):297–307.
36. Hawkes CV, Sullivan JJ. The impact of herbivory on plants in different resource conditions a meta-analysis. Ecology. 2001;82:2045–58.
37. Boalt E, Lehtilä K. Tolerance to apical and foliar damage: costs and mechanisms in *Raphanus raphanistrum*. Oikos. 2007;116(12):2071–81.
38. Vaughton G, Ramsey M. Seed mass variation in the shrub *Banksia spinulosa* (Proteaceae): resource constraints and pollen source effects. Int J Plant Sci. 1997;158:424–31.
39. Wise MJ, Cummins JJ. Strategies of *Solanum carolinense* for regulating maternal investment in response to foliar and floral herbivory. J Ecol. 2006;94:629–36.
40. Weis AE, Hochberg ME. The diverse effects of intraspecific competition on the selective advantage to resistance: a model and its predictions. Am Nat. 2000;156:276–92.
41. Haag JJ, Coupe MD, Chahill JF. Antagonistic interactions between competition and insect herbivory on plant growth. J Ecol. 2004;92:156–67.
42. Schardler M, Brandl R, Haase J. Antagonistic interactions between plant competition and insect herbivory. Ecology. 2007;88:1490–8.
43. Steets JA, Salla R, Ashman TL. Herbivory and competition interact to affect reproductive traits and mating system expression in impatiens capensis. Am Nat. 2006;167:591–600.
44. Vacher C, Weis AE, Hermann D, Kossler T, Young C, Hochberg ME. Impact of ecological factors on the initial invasion of Bt transgenes into wild populations of birdseed rape (*Brassica rapa*). Theor Appl Genet. 2004;109(4):806–14.
45. Alstad DN, Andow DA. Implementing management of insect resistance to transgenic crops. AgBiotech News Information. 1996;8:177–81.
46. Lu Y, Wu K, Jiang Y, Xia B, Li P, Feng H, et al. Mirid Bug Outbreaks in Multiple Crops Correlated with Wide-Scale Adoption of Bt Cotton in China. Science. 2010;328(5982):1151–4.
47. Bergelson J, Purrington CB. Surveying patterns in the cost of resistance in plants. Am Nat. 1996;148:536–58.
48. Kalinina O, Zeller SL, Schmid B. Competitive performance of transgenic wheat resistant to powdery mildew. PLoS One. 2011;6:e28091.
49. Damgaard C, Kjaer C. Competitive interactions and the effect of herbivory on Bt-Brassica napus, Brassica rapa and Lolium perenne. J Appl Ecol. 2009;46(5):1073–9.
50. Liu Y, Wei W, Ma K, Li JS, Liang Y, Darmency H. Consequences of gene flow between oilseed rape (*Brassica napus*) and its relatives. Plant Sci. 2013;211:42–51.
51. R Development Core team. R: A language and environment for statistical computing. V. R foundation for statistical computing, Austria. 2008; http://www.r-project.org/ (Ed.).

A novel quadruplex real-time PCR method for simultaneous detection of *Cry2Ae* and two genetically modified cotton events (GHB119 and T304-40)

Xiang Li[1*†], Xiuxiu Wang[2†], Jielin Yang[1], Yueming Liu[1], Yuping He[1*] and Liangwen Pan[1]

Abstract

Background: To date, over 150 genetically modified (GM) crops are widely cultivated. To comply with regulations developed for genetically modified organisms (GMOs), including labeling policies, many detection methods for GMO identification and quantification have been developed.

Results: To detect the entrance and exit of unauthorized GM crop events in China, we developed a novel quadruplex real-time PCR method for simultaneous detection and quantification of GM cotton events GHB119 and T304-40 in cotton-derived products (based on the 5′-flanking sequence) and the insect-resistance gene *Cry2Ae*. The limit of detection was 10 copies for GHB119 and *Cry2Ae* and 25 copies for T304-40. The limit of quantification was 25 copies for GHB119 and *Cry2Ae* and 50 copies for T304-40. Moreover, low bias and acceptable standard deviation and relative standard deviation values were obtained in quantification analysis of six blind samples containing different GHB119 and T304-40 ingredients.

Conclusions: The developed quadruplex quantitative method could be used for quantitative detection of two GM cotton events (GHB119 and T304-40) and *Cry2Ae* gene ingredient in cotton derived products.

Keywords: *Cry2Ae*, Event-specific, GHB119, Quadruplex real-time PCR, T304-40

Background

Genetically modified (GM) crops have been commercially cultivated for over 17 years and the global planting area has increased 100-fold [1]. Along with the rapid development of GM organisms (GMO), food safety and environmental safety concerns about GMO increasingly attracted public attention worldwide. To respond to these concerns, a series of laws and regulations have been issued by many countries throughout the world, highlighting labeling policies for GMO-derived products to protect individual consumer's "right to know". To ensure the implementation of GMO labeling regulations, it is necessary to develop appropriate, rapid qualitative and quantitative methods for GMO detection.

Nucleic acid-based and protein-based techniques are two widely used methods for GMO detection. Due to their high specificity, sensitivity, operability and repeatability, nucleic acid-based methods (such as PCR, real-time PCR, isothermal amplification techniques and others) are widely employed for the detection of GMO events [2,3]. As the quantity of exogenous genes and GM crop events increase, several multiplex methods for simultaneous detection of more than one genetic element have been developed, such as multiplex PCR, microarray analysis, suspension array analysis, microdroplet-based PCR and others [4-6]. These techniques are especially useful for detecting GM events in complex or mixed samples that contain multiple exogenous insertions. Due to the limited stability and repeatability of microarray and suspension array-based PCR methods, such methods are unsuitable for routine analysis of GM events in imported and exported goods. Therefore, most methods based on multiplex techniques involve

* Correspondence: idealne@163.com; heyuping@shciq.gov.cn
†Equal contributors
[1]GMO Detection Laboratory, Shanghai Entry-Exit Inspection and Quarantine Bureau, 1208 Minsheng Road, Shanghai 200135, PR China
Full list of author information is available at the end of the article

conventional multiplex PCR, such as a six-fold screening method for detecting 35S, pNOS, tNOS, NPTII, CP4-EPSPS and PAT [7], methods for simultaneously detecting Cry1Ac, Cry2Ab, pCaMV35S, tNOS, NPTII, aadA, uidA, and MON15985 event [8] and multiplex PCR analysis for detection of four GM maize, five GM soybean and six GM canola elements, among others [9]. Compared with conventional PCR, real-time PCR has a higher sensitivity and a shorter detection time and is easier to perform. However, developing a multiplex real-time PCR system is proved to be difficult due to serious mutual interference between pairs of primers and probes that greatly reduces the detection sensitivity. Currently, only a few real-time multiplex PCR methods are reported and maximum four targets are simultaneously detected [10,11].

Cotton is an economically important crop with a global planting area of approximately 36 million hectares in 2012, 81% of which was planted with more than 30 GM cottons [1]. GM cotton is planted in 13 countries and six GM cottons, MON531, MON1445, MON15985, LLcotton25, MON88913 and GHB614, have been certificated for food and feed production in China since 2004 [12]. GHB119 and T304-40 are developed by Bayer Crop Science (Monheim, Germany) with both insect- resistance and herbicide-tolerance characteristics. GHB119 contains two exogenous genes (*Cry2Ae* and *bar*) while T304-40 contains two exogenous genes (*Cry1Ab* and *bar*). To date, GHB119 and T304-40 have been approved for cultivation in Japan and the United States and for food or feed processing in Canada, the United States, Australia and New Zealand [13]. However, to date, no detection methods for these two GM cotton events have been reported, which is also true for the *Cry2Ae* gene.

In this study, we analyzed the *Cry2Ae* gene sequence and the 5′-flanking sequence of GM cotton events GHB119 and T304-40. We then developed three detection systems based on a specific fragment of *Cry2Ae* and 5′ event-specific sequences of GHB119 and T304-40. A quadruplex quantitative real-time PCR method together with the cotton endogenous gene *ACP1* detection system was established for simultaneous detection and quantification of GHB119, T304-40 and *Cry2Ae* in cotton-derived products. Due to the high homology between *Cry2Ae* and other *Cry* genes, to improve the specificity and sensitivity of this system for *Cry2Ae* detection, locked nucleic acids (LNA) were included in the *Cry2Ae* probe.

Methods

Sample collection and preparation

Certified reference materials (CRMs) of GM cotton (*Gossypium*) GHB119 (ERM- BF428c), T304-40 (ERM-BF429c) and 281-24-236 × 3006-210-23 (ERM-BF422d), GM maize (*Zea mays* L.) Bt11 (ERM-BF412f), MON98140 (ERM-BF427d), NK603 (ERM-BF415f) and MON810 (ERM-BF413k) and GM soybean (*Glycine max*) GTS40-3-2 (ERM-BF410k), DP305423 (ERM-BF426d) and DP356043 (ERM- BF425d) were purchased from the Institute for Reference Materials and Measurements (IRMM), Joint Research Center (JRC) of European Commission (Geel, Belgium). Some GM materials, such as GM cotton MON88913 (0906-D), MON531 (0804-C), MON15985 (0804-D) and MON1445 (0804-B) and GM soybean MON89788 (0906-B) were purchased from the American Oil Chemists' Society (AOCS) (Urbana, IL, USA). Dry seed powders of GM rice (*Oryza sativa*) TT51-1, GM canola (*Brassica campestris L.*) T45, GM maize CBH351, GM cotton GK19 and SGK321 and a conventional cotton variety were provided by corresponding developers.

To verify the sensitivity of the established quadruplex real-time PCR method, six different percentages of mixed-cotton samples were produced using a Freezer Mill (Spex SamplePrep, Metuchen, NJ, USA). First, 10.0% CRM of GM cotton GHB119 powder (0.60 g) was mixed with non-GM cotton powder (0.60 g), resulting in a 5.0% (w/w) sample (labeled A1). Next, 0.60 g of A1 was mixed with 0.90 g of non-GM cotton powder, resulting in a 2.0% GHB119 sample (labeled A2). Then, a 1.0% GHB119 sample (labeled A3) was obtained by mixing 0.60 g of A2 with 0.60 g of non-GM dry cotton powder. Similarly, samples containing 5.0% (B1), 2.0% (B2) and 1.0% (B3; w/w) T304-40 powder were also prepared using the same mixing pattern. Then, six blind samples (S1-S6) containing both GHB119 and T304-40 powder were prepared. The mixing method and GM content of six samples were listed in Table 1.

DNA extraction

Genomic DNA samples were isolated and purified from approximately 200 mg dry cotton seed powder using a Plant Genomic DNA Extraction Kit (Tiangen Biotech Co., Ltd., Beijing, China) according to the manufacturer's instructions. The concentrations and quality of the extracted DNA samples were measured with a Nanovue Plus Spectrophotometer (GE Healthcare Bio-sciences, PA, USA) at 260, 280, 230 and 320 nm and evaluated by examining the A_{260}/A_{280} and A_{260}/A_{230} ratios, followed by electrophoresis on 0.8% agarose gels. All DNA solutions were stored at -20°C. In addition, the genomic DNA copy numbers were calculated by referring to the haploid genome size of cotton.

Design of primers and probes

After analyzing the sequences of *Cry2Ae* (Accession number: AX513526.1) and other Bt genes, such as *Cry2Ab*, *Cry1Ab*, *Cry1Ac*, *Cry3A* and *Cry9C* (Accession numbers are AB702969.1, KF303141.1, KF630361.1, X70979.1 and AY346129.1, respectively) by alignment using DNAMAN 6.0.40 software (Lynnon Biosoft Corp., Quebec, Canada) [14], a fragment of the *Cry2Ae* gene with minimum

Table 1 Mixing method and GM content of six prepared blind samples

Sample name	GM content (%) (w/w)		Mixing method
	GHB119	T304-40	
S1	0.5	5.0	0.2 g A3* + 0.2 g of 10.0% CRM of T304-40
S2	5.0	1.0	0.2 g of 10.0% CRM of GHB119 + 0.2 g of B2*
S3	2.5	0.5	0.2 g of A1* + 0.2 g of B3*
S4	1.0	2.5	0.2 g of A2* + 0.2 g of B1*
S5	0	0.5	0.2 g of non-GM cotton + 0.2 g of B3*
S6	0.5	0	0.2 g of A3* + 0.2 g of non-GM cotton

*A1: 5.0% (w/w) GM cotton GHB119 sample; A2: 2.0% (w/w) GM cotton GHB119 sample; A3: 1.0% (w/w) GM cotton GHB119 sample; B1: 5.0% (w/w) GM cotton T304-40 sample; B2: 2.0% (w/w) GM cotton T304-40 sample; B3: 1.0% (w/w) GM cotton T304-40 sample.

sequence similarity with the other *Cry* gene was chosen for the design of primers and probes. To improve the specificity of the probe for amplifying *Cry2Ae*, two nucleic acids on the probe were replaced by LNA (Table 2). The 5′-specific flanking sequences of the GM cotton events GHB119 and T304-40 (http://www.ncbi.nlm.nih.gov/nucleotide/) were analyzed, and primer pairs GHB119-F1/R2 and T304-F1/R2 and probes GHB119-P and T304-P were designed using Primer Express 3.0 software (Applied Biosystems, Foster City, CA, USA). The primer pair and probe 18SrRNA-F/R/P [15] that is specific to the *18SrRNA* gene of eukaryotes were used as an internal control. Primer pair ACP-FP/RP and probe ACP-P were designed to detect the endogenous *ACP1* gene (Accession number: U48777.1) of cotton. Four fluorescent channels were employed, using the fluorescent reporter dyes FAM, HEX, ROX and CY5 to label the 5′-ends of the probes GHB119-P, T304-P, ACP-P and CRY2E-P, respectively. Correspondingly, the quencher dyes BHQ1 (for FAM and HEX), BHQ2 (for ROX) and BHQ3 (for CY5) were used to label the 3′-ends of the probes, respectively. FAM and BHQ1 dyes were used to label 18SrRNA-P. The sequences of the primers and probes used in this study are listed in Table 2. The primers and fluorescence-labeled probes were synthesized and purified by Shanghai Huirui Biotechnology Co., Ltd. (Shanghai, China).

Real-time PCR conditions

Quadruplex real-time PCR was performed in an Applied Biosystems ViiA 7 Real-Time PCR System (Applied Biosystems, Foster City, CA, USA) in a 25 μl reaction volume. Each reaction included 12.5 μl of 2× HR qPCR Master Mix I (Shanghai HuiRui Biotechnology Co., Ltd, China), 5 μl of plant DNA templates and 200 nM of primer ACP-FP/RP with 100 nM of ACP-P probe and 240 nM each of primer GHB119-F1/R2, T304-F1/R2 or CRY2E-F1/R2 with 120 nM each of GHB119-P, T304-P or CRY2E-P probe. Real-time PCR amplification was carried out with

Table 2 Primers and probes used for quadruplex real-time PCR assays in this study

Target	Purpose	Primer/probe name	Sequence (5′-3′)	Amplicon length (bp)	Reference
18S rRNA	Eukaryotes gene	18SrRNA-F	CCTGAGAAACGGCTACCAT	137	19
		18SrRNA-R	CGTGTCAGGATTGGGTAAT		
		18SrRNA-P	FAM-TGCGCGCCTGCTGCCTTCCT –BHQ1		
ACP1	Cotton endogenous gene	ACP-FP	ATGAACCAGGGAAGAAGCACC	97	This study
		ACP-RP	CCTTATCCACGGTCTCTTGTTTG		
		ACP-P	ROX-CATTTACGATGCGTCCAATGCCTG-BHQ2		
GHB119	Event-specific detection	GHB119-F1	AAAACTTTGTGCAGCCTTCG	130	This study
		GHB119-R2	CGCAAACTAGGATAAATTATCGC		
		GHB119-P	FAM-TCCCCCTATCTTGCTAAATGGCTCC-BHQ1		
T304-40	Event-specific detection	T304-F1	GTCATTGTAGGGAGTTTGTCCAA	118	This study
		T304-R2	CTGTAGCCACAACACCACTTTG		
		T304-P	HEX-TTAATCCCAGTACTCGGCCGTC-BHQ1		
Cry2Ae	Gene-specific detection	CRY2E-F1	CTTGCTCTACTTTCCTTCCTCC	114	This study
		CRY2E-R2	CGAAAGACTCAGTTTGCCAGT		
		CRY2E-P	CY5-CAAGCCAAGA + C + CTAACGAAAGGAG-BHQ3*		

*The base before symbol "+" was decorated with LNA.

the following program: 95°C for 10 min followed by 45 cycles of 95°C for 15 s and 60°C for 1 min. The fluorescence was monitored during each PCR cycle at the annealing and extension step (60°C).

Specificity test

To determine the specificity of the three established real-time PCR assays for GM cotton GHB119, T304-40 and *Cry2Ae*, three primer pairs with the corresponding probe were tested by amplifying the genomic DNA samples of relevant GM plant events, including 10 GM cotton events (GHB119, T304-40, 281-24-236, 3006-210-23, MON88913, MON531, MON15985, MON1445, GK19 and SGK321), five GM maize events (Bt11, MON98140, NK603, CBH351 and MON810), four GM soybeans (GTS40-3-2, 305423, 356043 and MON89788), one GM rice (TT51-1), one GM canola (T45) and a conventional cotton variety (ZM-1).

Detection limit assay

To test the limit of detection (LOD) and quantification (LOQ) of the established quadruplex real-time PCR method, four gradient dilutions of DNA template were prepared. To obtain the DNA solution containing both GHB119 and T304-40 events, the DNA samples of 10% GHB119 (W/W) and 10% T304-40 (W/W) were mixed firstly. The same concentration (100 ng/μl) and equal volumes GM cotton GHB119 and T304-40 DNA solutions were mixed thoroughly, yielding the "first mixture" of 100 ng/μl DNA sample with 5 ng/μl GHB119 and 5 ng/μl T304-40 content. According to the mean molecular weight and size of the cotton genome, 1 ng cotton DNA equals approximately 464 copies of DNA. In the "first mixture", the concentration of the cotton haploid genomic DNA was 4,640,000 copies/μl, and those of GHB119 and T304-40 were 2320 copies/μl, respectively. The "first mixture" was serially diluted in 0.1 × TE buffer (1 mM Tris–HCl, 0.1 mM EDTA · Na, PH 8.0) to final concentrations of 200, 100, 40 and 20 copies/μl of total cotton haploid genomic DNA, which contained 10, 5, 2 and 1 copies/μl for both GHB119 and T304-40 content, respectively, because that the contents of GM cotton GHB119 and T304-40 were both 5% (W/W) in the "first mixture". Each sample was amplified in five parallel reactions, and the entire experiment was repeated four times. The absolute LOD and LOQ values represent the lowest amount or concentration of initial template DNA that could be reliably detected and quantified with a ≥95% confidence interval [16]. The LOD value was determined and validated by calculating the number of positive results in 20 reactions.

Construction of standard curves

To quantify GHB119 and T304-40 contents in the blind samples, four standard curves were first constructed through four fluorescent channels in one round of amplification. DNA solutions were prepared by serially diluting the prepared "first mixture" using 0.1 × TE, which contained 5 ng/μl GHB119 and 5 ng/μl T304-40. Six concentrations of cotton DNA, including 100, 20, 4, 2, 0.8 and 0.4 ng/μl, which equal approximately 46,400, 9,280, 1,856, 928, 371 and 185 copies/μl of haploid cotton genomic DNA, were employed for amplification using the quadruplex and simplex PCR assays. According to the content of GHB119 or T304-40, which was 5.0% in the above six concentrations, the number of copies was approximately 2,320, 464, 92.8, 46.4, 18.6 and 9.3 per μl for GHB119 or T304-40 content. In this assay, four parallel reactions were performed for each concentration, and the entire experiment was repeated three times. The Ct values that deviated considerably from the average value were rejected. Eight standard curves (four by quadruplex and four by simplex) were plotted using Ct values from six concentrations of genomic DNA against the logarithm of the number of copies of DNA. Furthermore, the constructed standard curves were evaluated by amplification efficiency ($E = [10^{(-1/\text{slope})} - 1] \times 100\%$) and square regression coefficient (R^2), respectively. The repeatability of four parallel reactions in one round and three experiments was validated by examining the standard deviation (SD) and relative standard deviation (RSD).

Analyses of blind samples using the quadruplex real-time PCR method

To evaluate the quadruplex PCR method for GM cotton GHB119, T304-40 and *Cry2Ae* quantification, six blind samples were prepared, including S1 (containing 0.5% GHB119 and 5.0% T304-40), S2 (5.0% GHB119 and 1.0% T304-40), S3 (2.5% GHB119 and 0.5% T304-40), S4 (1.0% GHB119 and 2.5% T304-40), S5 (0.5% GHB119 and 0% T304-40) and S6 (0% GHB119 and 0.5% T304-40). Four replicates were carried out per sample. The precision and accuracy of the quadruplex quantitative real-time PCR method were verified by examining the bias, SD and RSD values.

Results and discussion

Development of the quadruplex real-time PCR method for GM cotton GHB119, T304-40 and Cry2Ae detection

High-throughput, rapid detection methods have attracted increasing attention due to their considerable advantages. In this study, to rapidly detect two unapproved GM cotton events in China and their inserted gene, we developed a novel quadruplex quantitative real-time PCR detection method, combined with the LNA technique, for simultaneous detection and quantification of GM cotton events GHB119 and T304-40 as well as *Cry2Ae*. Primer pairs and probes for GHB119 and T304-40 detection were designed based on both 5′-flanking sequences of the two events

(Figure 1). Nearly 10 Cry proteins have been used to provide insect-resistance characteristics in GM crops (such as *cry1Ab*, *cry2Ab*, *cry9C* and so on), and the homology among these *Cry* genes is as much high as 95%. The alignments of five *Cry* genes in GM crops and the specific fragment used to design the primers and probe are shown in Figure 1. To improve the specificity of the detection method, two bases on the probe used for *Cry2Ae* gene detection were modified by decoration with LNA (Table 2).

Since quadruplex real-time PCR employs eight primers and four probes in a single amplification tube, mutual interference and inhibition may occur. Therefore, to obtain an amplification curve with the strongest fluorescent signals and the lowest Ct values, multiple pairs of primers and probes were designed and tested. We also tested four quantitative PCR (qPCR) master-mix solutions and used concentrations of primers ranging from 150 to 300 nM and probes ranging from 75 to 150 nM. We determined that the optimal amplification system comprises 1 × HR qPCR Master Mix I, 240 nM of each primer GHB119-F1/R2, T304-F1/R2 and CRY2E-F1/R2 and 200 nM of the primer pair ACP-FP/RP, 120 nM of each probe GHB119-P, T304-P and CRY2E-P and 100 nM of the probe ACP-P.

Specificity test of the real-time PCR method

To evaluate the specificity of the designed primer pair/probe for GHB119, T304-40 or *Cry2Ae*, 22 DNA samples from various plants including cottons were amplified. To determine if the extracted DNA samples were suitable for amplification, an *18SrRNA* gene fragment was first amplified, which showed significant fluorescent amplifications in all 22 DNA samples (data not shown). Furthermore, amplification curves of the cotton endogenous gene *ACP1* were observed only from cotton-derived samples (data not shown). The successful amplification of both *18SrRNA* and *ACP1* genes indicated that the extracted 22 DNA samples were suitable for real-time PCR analysis.

The specificity test for above mentioned three targets was carried out using simplex real-time PCR. In the test for primer pair/probe GHB119-F1/R2/P for GM cotton GHB119 detection, the amplification curve (with a Ct value of nearly 26) was present only in GHB119 genomic DNA sample (Figure 2A). Similarly, for primer pair/probe T304-F1/R2/P, the amplification curve was only observed in T304-40 genomic DNA sample (Figure 2B). In addition, the expected amplification curve for primer pair/probe CRY2E-F1/R2/P test (with a Ct value of approximately 23) was appeared only in GHB119 genomic DNA sample (Figure 2C). These results indicated that all three designed primer pairs/probes were highly specific and could be used for the quadruplex real-time PCR assay.

Sensitivity tests

LOD and LOQ are two of the most important parameters used to evaluate a newly established qPCR method. Furthermore, it is also important that the fluorescent signals and Ct values are similar in multiplex and simplex PCR when using the same DNA template.

In this study, to determine the LOD and LOQ of our quadruplex real-time PCR method, four concentrations of serially diluted DNA samples, including 10, 5, 2 and 1 copies/µl of GHB119 or T304-40 (the sample with a concentration of 10 copies/µl contained both 10 copies/µl

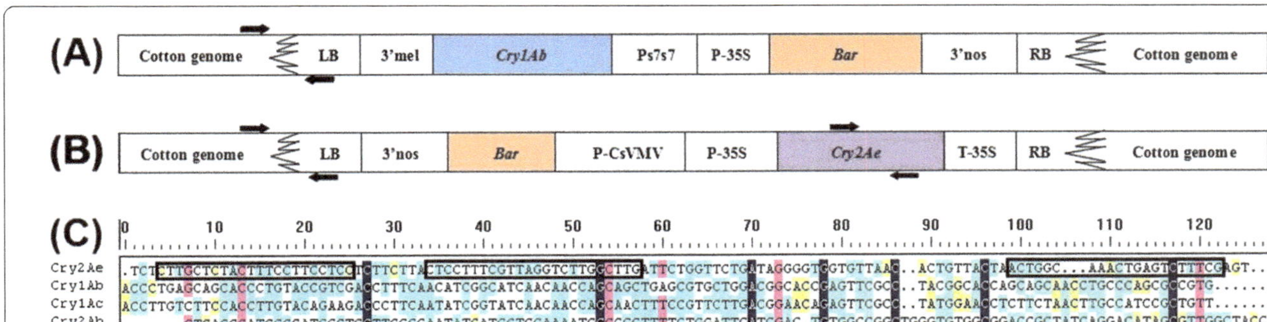

Figure 1 Schematic diagram of exogenous insertion of GM cotton events T304-40 and GHB119 and alignment of six Cry protein gene sequences. (A–B) Schematic diagram of exogenous insertion of GM cotton events T304-40 **(A)** and GHB119 **(B)**. LB: left border sequence of T-DNA; 3′mel: terminator from *Flaveria bidentis*; Cry1Ab: gene encoding Bt endotoxin of *Bacillus thuringiensis*; Ps7s7: duplicated promoter from subterranean clover stunt virus; P-35S: 35S promoter from cauliflower mosaic virus; Bar: sequence encoding the PAT (phosphinothricin acetyl-transferase) enzyme of *Streptomyces hygroscopicus*; 3′nos: 3′-terminator of nopaline synthase gene from *Agrobacterium tumefaciens*; Cry2Ae: gene encoding a Bt endotoxin of *Bacillus thuringiensis*; T-35S: 35S terminator from cauliflower mosaic virus; P-CsVMV: promoter derived from Cassava vein mosaic virus; RB: right border of transposon from *Agrobacterium tumefaciens*. Arrows show the primers location developed in this study. **(C)** DNA fragment sequences of six Cry protein genes showing the locations of the designed primers and probes.

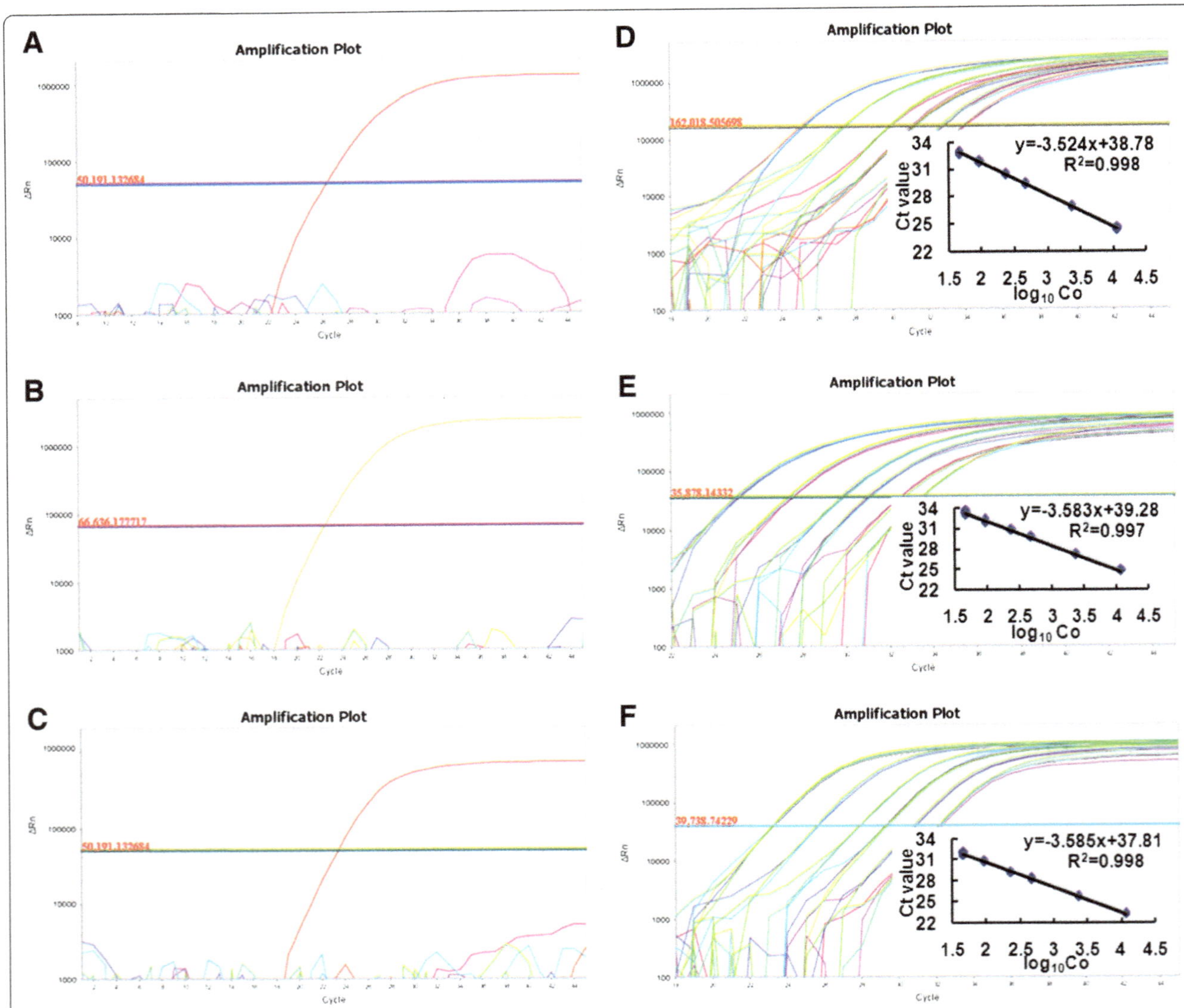

Figure 2 Specificity tests and standard curves of the real-time PCR method designed to detect GM cotton GHB119, T304-40 and *Cry2Ae*. (A–C) Specificity tests using 22 plant samples as templates for GHB119 **(A)**, T304-40 **(B)** and *Cry2Ae* **(C)** detection method. **(D–F)** Amplification plots and standard curves from six concentrations (2,320, 464, 92.8, 46.4, 18.6 and 9.3 copies/μl) of GHB119 **(D)**, T304-40 **(E)** or *Cry2Ae* **(F)** ingredient.

GHB119 and 10 copies/μl T304-40, and so on) were employed. When 10 and 5 copies/μl (with 5 μl templates) of DNA were used as template amplification curves were obtained in all 20 reactions for GHB119, T304-40 and *Cry2Ae* in either quadruplex or simplex PCR system (Figure 3A). When 2 copies/μl (with 5 μl templates) of DNA were used as template, amplification curves were obtained in all 20 reactions for GHB119 and *Cry2Ae* in either quadruplex or simplex PCR. However, when 10 copies of DNA was used as template for the T304-40 test (18 out of 20 reactions were positive in the quadruplex PCR assay while all 20 simplex PCR reactions were positive (Figure 3A). Differences were also observed in amplifications using five copies (5 μl templates) of DNA as template. Positive signals appeared 19 out of 20 reactions for all three targets using simplex PCR and in 17, 13 and 18 out of 20 reactions for GHB119, T304-40 and *Cry2Ae* detection by quadruplex PCR, respectively. In other words, the lowest levels of DNA that could be reliably detected using the quadruplex PCR method were 10, 25 and 10 copies (2, 5 and 2 copies/μl with 5 μl templates, respectively) for GHB119, T304-40 and *Cry2Ae*, respectively. These LOD values were similar to those obtained using simplex PCR. Note that 10 copies of haploid genomic DNA are equal to approximately 0.02 ng of cotton DNA.

We also calculated the mean Ct values using the above four concentrations of genomic DNA as template in both quadruplex and simplex PCR systems. No significant differences were observed between the quadruplex and simplex

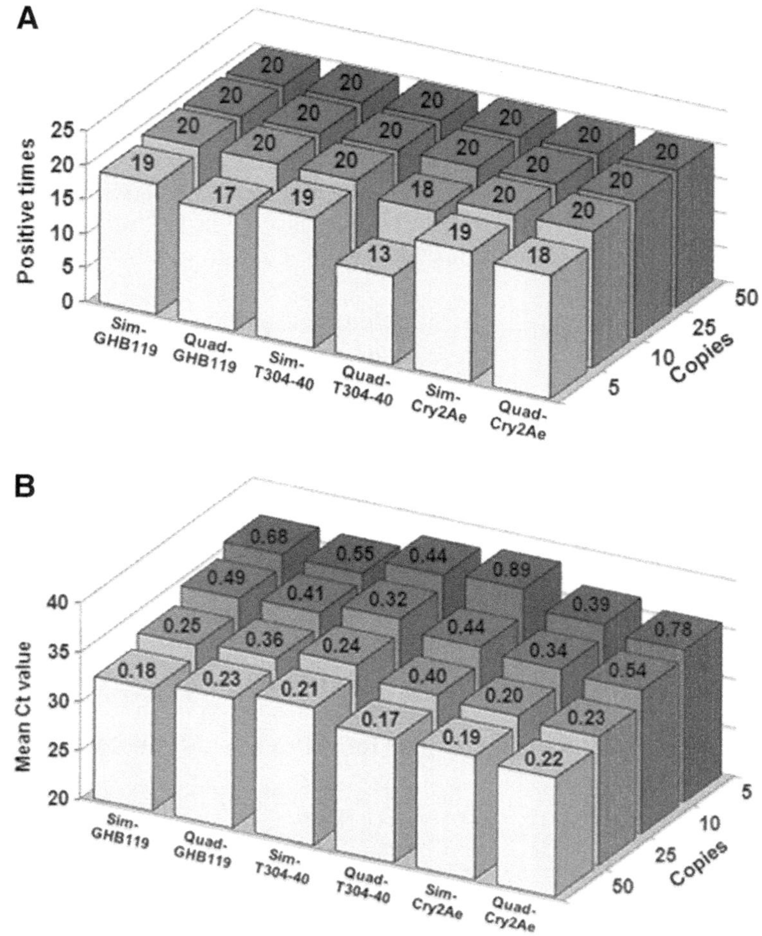

Figure 3 Results of sensitivity tests for the established real-time PCR method. (A) Number of positive amplifications out of 20 using four concentrations (10, 5, 2 and 1 copies/μl) of GHB119 or T304-40 ingredient by simplex and quadruplex PCR assays. **(B)** Mean Ct values and SD for GHB119, T304-40 and *Cry2Ae* detection via simplex and quadruplex PCR assays using four concentrations (10, 5, 2 and 1 copies/μl) of GHB119 or T304-40 DNA as template. The number above the columns is standard deviations.

PCR results for all three targets (Figure 3B). It is gratifying that the Ct values obtained by quadruplex PCR targeting GM cotton T304-40 and *Cry2Ae* from all four concentrations of cotton DNA tested were less than the results obtained using the simplex PCR method. In the amplifications targeting GHB119 event, the Ct values were similar in both quadruplex and simplex PCR systems. These results indicated that little or no interference occurred among the four groups of primers and probes when amplified in a single reaction tube.

To obtain the LOQ of the quadruplex PCR method, we calculated the SD of Ct values from 20 replicates using four concentrations of genomic DNA. The SD values were all <0.25 for GHB119, T304-40 and *Cry2Ae* detection when 50 copies of genomic DNA were used as templates (Figure 3B). For GHB119 and T304-40 quantification, reliable Ct values were observed when ≥25 copies of DNA were used as template in simplex PCR and ≥50 copies of DNA were used in quadruplex PCR. For *Cry2Ae*, the lowest concentration of template that could reliably be used for detection was 25 copies for both simplex and quadruplex PCR. Therefore, the LOQs of the quadruplex real-time PCR method were 50 copies of genomic DNA for GHB119 and T304-40 events and 25 copies for *Cry2Ae*. Thus, 0.44% to 100% GHB119 and T304-40 ingredient in 100 ng mixed samples could be accurately quantified and that value for *Cry2Ae* gene was a 0.22%-100% in 100 ng mixed samples.

Construction of standard curves

To quantify the GM content in blind samples, after optimizing the reaction conditions, we constructed standard curves using our quadruplex real-time PCR method. Six concentrations of mixed genomic DNA were used, including approximately 2,320, 464, 92.8, 46.4, 18.6 and 9.3 copies/μl for GHB119 or T304-40. Simplex real-time PCR was also carried out under the same conditions. Two parameters, that is, E and R^2, were employed to

Table 3 Parameters of standard curves constructed using the quadruplex and simplex PCR assays for GM cotton GHB119, T304-40 and *Cry2Ae* detection

Target		Linear equation	R^2	PCR efficiency (%)
GHB119	Simplex	y = −3.575x + 43.541	0.997	90.4
	Quadruplex	y = −3.524x + 38.780	0.998	92.2
T304-40	Simplex	y = −3.549x + 42.421	0.993	91.3
	Quadruplex	y = −3.583x + 39.275	0.997	90.1
Cry2Ae	Simplex	y = −3.516x + 42.034	0.998	92.5
	Quadruplex	y = −3.585x + 37.809	0.998	90.1

evaluate the standard curves. The amplification plots and standard curves were shown in Figure 2. The R^2 values were all >0.99 (0.993–0.998 range) for both the quadruplex and simplex PCR assays for GHB119, T304-40 and *Cry2Ae*, indicating that the six standard curves had good linearity and relativity (Table 3). The PCR efficiencies were high, ranging from 90 to 95% in both the quadruplex and simplex PCR assays. Specifically, for GM cotton GHB119 detection, PCR efficiencies were 90.4% and 92.2% for simplex and quadruplex PCR, respectively. For GM T304-40, PCR efficiencies were 91.3% and 90.1% for simplex and quadruplex PCR, respectively, and these values for *Cry2Ae* detection, were 92.5% and 90.1% for simplex and quadruplex PCR, respectively (Table 3). The good linearity observed between DNA quantity and fluorescence values (Ct) and the high PCR efficiencies indicated that the quadruplex real-time PCR method would be suitable for practical application in the quantitative detection GM cotton GHB119, T304-40 and *Cry2Ae* in blind samples.

Repeatability of the quadruplex real-time PCR method

The repeatability of the quadruplex real-time PCR assay for GHB119, T304-40 or *Cry2Ae* was evaluated by examining the SD and RSD values of four parallel replicates (SD^r and RSD^r) and three experiments (SD^R and RSD^R). The results revealed that the SD and RSD values were all in the acceptable range. Specifically, the SD^r values were less than 0.19 using six concentrations of DNA as template for detection of GHB119, T304-40 and *Cry2Ae*, while the RSD^r value ranged from 0.22% to 0.57% for the GHB119 assay, from 0.07% to 0.50% for the T304-40 assay and from 0.15% to 0.36% for the *Cry2Ae* assay (Table 4).

Table 4 Repeatability of the quadruplex real-time PCR method for GM cotton GHB119, T304-40 and *Cry2Ae* detection

Amount of DNA (copies/reaction)	Ct value					SD^{r*}	RSD^r (%)	SD^{R*}	RSD^R (%)
	REP1	REP2	REP3	REP4	Mean				
				GHB119					
11600	24.39	24.51	24.57	24.40	24.47	0.09	0.36	0.05	0.19
2320	26.80	26.83	26.96	26.78	26.84	0.08	0.30	0.03	0.12
464	29.23	29.28	29.46	29.49	29.37	0.13	0.43	0.03	0.11
232	30.26	30.41	30.50	30.35	30.38	0.10	0.34	0.06	0.19
92.8	31.69	31.72	31.83	31.69	31.73	0.07	0.22	0.08	0.25
46.4	32.53	32.89	32.93	32.69	32.76	0.19	0.57	0.11	0.33
				T304-40					
11600	24.68	24.70	24.72	24.69	24.70	0.02	0.07	0.05	0.20
2320	27.04	27.07	27.10	27.23	27.11	0.08	0.30	0.02	0.08
464	29.67	29.68	29.70	29.84	29.72	0.08	0.26	0.01	0.02
232	30.72	30.77	30.89	30.96	30.83	0.11	0.36	0.05	0.17
92.8	31.96	32.05	32.16	31.78	31.99	0.16	0.50	0.10	0.30
46.4	33.08	33.25	33.36	33.43	33.28	0.15	0.46	0.15	0.46
				Cry2Ae					
11600	23.29	23.32	23.34	23.21	23.29	0.06	0.24	0.04	0.17
2320	25.64	25.68	25.71	25.73	25.69	0.04	0.15	0.04	0.17
464	28.03	28.15	28.17	28.22	28.14	0.08	0.28	0.08	0.27
232	29.02	29.19	29.20	29.24	29.16	0.10	0.33	0.05	0.17
92.8	30.63	30.68	30.83	30.85	30.75	0.11	0.36	0.07	0.23
46.4	31.99	32.02	32.05	32.14	32.05	0.06	0.20	0.10	0.31

*SD^r: standard deviation of four parallel reactions; SD^R: standard deviation of three experiments; RSD^r: relative standard deviation of four parallel reactions; RSD^R: relative standard deviation of three experiment.

Table 5 Quantitative analysis of six blind samples

Sample code	Event or gene	GM content (%)	REP1 (%)	REP2 (%)	REP3 (%)	REP4 (%)	Mean value (%)	SD	RSD (%)	Bias (absolute value)	Bias (%)
	GHB119	0.5	0.47	0.48	0.51	0.45	0.48	0.02	4.17	−0.02	−3.96
S1	T304-40	5.0	4.86	4.87	5.02	5.03	4.95	0.09	1.82	−0.05	−1.08
	Cry2Ae	0.5	0.49	0.50	0.50	0.53	0.51	0.02	3.79	0.01	1.22
	GHB119	5.0	4.94	4.99	4.94	5.19	5.02	0.12	2.39	0.02	0.31
S2	T304-40	1.0	1.02	1.07	0.84	0.80	0.93	0.13	13.95	−0.07	−6.84
	Cry2Ae	5.0	4.67	5.14	4.85	4.82	4.87	0.20	4.07	−0.13	−2.61
	GHB119	2.5	2.46	2.41	2.51	2.50	2.47	0.05	2.02	−0.03	−1.14
S3	T304-40	0.5	0.46	0.49	0.48	0.45	0.47	0.02	4.25	−0.03	−5.86
	Cry2Ae	2.5	2.72	2.66	2.64	2.67	2.67	0.03	1.14	0.17	6.93
	GHB119	1.0	0.95	1.03	1.02	0.95	0.99	0.04	4.06	−0.01	−1.36
S4	T304-40	2.5	2.46	2.54	2.48	2.56	2.51	0.05	1.99	0.01	0.54
	Cry2Ae	1.0	1.06	1.08	1.09	1.09	1.08	0.01	1.22	0.08	8.23
	GHB119	0.0	0.00	0.00	0.00	0.00	0.00	0.00	0.00	0.00	0.00
S5	T304-40	0.5	0.44	0.44	0.46	0.53	0.47	0.04	8.52	−0.03	−6.10
	Cry2Ae	0.0	0.00	0.00	0.00	0.00	0.00	0.00	0.00	0.00	0.00
	GHB119	0.5	0.47	0.46	0.47	0.48	0.47	0.01	2.13	−0.03	−6.03
S6	T304-40	0.0	0.00	0.00	0.00	0.00	0.00	0.00	0.00	0.00	0.00
	Cry2Ae	0.5	0.52	0.51	0.52	0.56	0.53	0.02	4.31	0.03	5.80

The SD^R and RSD^R values for GHB119, T304-40 and *Cry2Ae* detection ranged from 0.01% to 0.15% and 0.02 to 0.46%, respectively (Table 4). The results of the repeatability test indicated that the quadruplex real-time PCR method is stable and repeatable for simultaneous quantitative detection of GHB119, T304-40 and *Cry2Ae.*

Blind sample analyses using the established quadruplex real-time PCR method

To verify the applicability of this quadruplex PCR method for quantification, six premixed samples (S1–S6) containing different percentages of GHB119 and T304-40 were prepared and analyzed to evaluate the accuracy and precision. The results showed that for quantification of GHB119, the mean values in the six blind samples were 0.48, 5.02, 2.47, 0.99, 0.00 and 0.47%, respectively (Table 5). The bias (between the experimental values and the given values) ranged from −6.03% to 0.31%. The SD value ranged from 0.01 to 0.12 and the RSD value ranged from 2.02 to 4.17%. Similarly, for T304-40 quantification, the mean values of six samples were 4.95, 0.93, 0.47, 2.51, 0.47 and 0.00%, respectively (Table 5). The bias ranged from −6.84% to 0.54%, the SD ranged from 0.02 to 0.13 and the RSD ranged from 1.82 to 13.95%. For *Cry2Ae* gene analysis, the mean GM contents of S1–S6 were found to be 0.51, 4.87, 2.67, 1.08, 0.00 and 0.53%, respectively. The bias values ranged from −0.03% to 0.17%, SD values ranged from 0.01 to 0.20 and RSD values from 1.14% to 4.31%. These values were all in the acceptable ranges [16].

Conclusions

To improve the implementation of the labeling policy, the supervision of unauthorized GM crop events in China was very important. In this study, a novel quadruplex real-time PCR method for simultaneous detection and quantification of GM cotton events GHB119 and T304-40 and Cry2Ae gene was developed. The low LODs, LOQs and the high accuracy and precision in quantification indicated that the developed quadruplex real-time PCR were suitable for the qualitative and quantitative detections of GHB119, T304-40 and Cry2Ae in cotton or cotton-derived products.

The accuracy and precision of the quantification for blind sample using our quadruplex quantitative PCR method are similar to those of many other quantitative analysis methods for GM soybean and maize [3,17]. Therefore, the quadruplex real-time PCR method developed in this study is reliable and suitable for the quantitative detection of GHB119, T304-40 and *Cry2Ae* in cotton or cotton-derived products.

Abbreviations

BHQ: The Black Hole Quencher; Ct: Cycle threshold; Cy5: Cyanine dye 5; FAM: 6-carboyfluorescein; HEX: 6-carboxy-2′, 4, 4′, 5′, 7, 7′-hexachlorofluorescein; LNA: Locked nucleic acid; LOD: Limit of detection; LOQ: Limit of quantification; PCR: Polymerase chain reaction; qPCR: Quantitative polymerase chain reaction; ROX: 5-carboxy-x-rhodamine; RSD: Relative standard deviation; SD: Standard deviation; SNP: Single nucleotide polymorphisms; Tm: Melting temperature.

Competing interests

The authors declare that they have no competing interests.

Authors' contributions

LX conceived of the study, participated in its design and coordination and performed the development of the quadruplex real-time PCR detection method. WX participated in the specificity and sensitivity tests of the quadruplex real-time PCR method. YJ participated in the blind samples analysis. LY participated in the screening of the primers and probes. HY participated in the design of the study and in drafting the manuscript. PL carried out the sequence alignments of *Cry* genes and the developed primers and probes. All authors read and approved the final manuscript.

Acknowledgements

We thank Dr. Litao Yang and Jianxin Shi of Shanghai Jiao Tong University for their kind help in revising of the vocabulary and grammar of the manuscript. This work was supported by grants from the Research Fund of General Administration of Quality Supervision, Inspection and Quarantine of China (2012IK185), the Shanghai Public Service Platform Special Fund (13DZ2290700) and the Science and Technology Program of Shanghai Entry-Exit Inspection and Quarantine Bureau (HK004-2012).

Author details

[1]GMO Detection Laboratory, Shanghai Entry-Exit Inspection and Quarantine Bureau, 1208 Minsheng Road, Shanghai 200135, PR China. [2]Institute of Biotechnology, East China University of Science and Technology, 130 Meilong Road, Shanghai 200237, PR China.

References

1. James C: **Global status of commercialized Biotech/GM crops.** *ISAAA Briefs* 2013, **2012**:33.
2. Lee D, Mura ML, Allnutt TR, Powell W: **Detection of genetically modified organisms (GMOs) using isothermal amplification of target DNA sequences.** *BMC Biotech* 2009, **9**(7):1–7.
3. Li X, Pan L, Li J, Zhang Q, Zhang S, Lv R, Yang L: **Establishment and application of event-specific polymerse chain reaction methods for two genetically modified soybean events, A2704-12 and A5547-127.** *J Agric Food Chem* 2011, **59**(24):13188–13194.
4. Guo J, Yang L, Chen L, Morisset D, Li X, Pan L, Zhang D: **MPIC: A high-throughput analytical method for multiple DNA targets.** *Anal Chem* 2011, **83**(5):1579–1586.
5. Xu J, Miao H, Wu H, Huang W, Tang R, Qiu M, Wen J, Zhu S, Li Y: **Screening genetically modified organisms using multiplex-PCR coupled with oligonucleotide microarray.** *Biosens Bioelectron* 2006, **22**(1):71–77.
6. Choi SH: **Hexaplex PCR assay and liquid bead array for detection of stacked genetically modified cotton event 281-24-236 × 3006-210-23.** *Anal Bioanal Chem* 2011, **401**(2):647–655.
7. Lu IJ, Lin CH, Pan TM: **Establishment of a system based on universal multiplex-PCR for screening genetically modified crops.** *Anal Bioanal Chem* 2010, **396**(6):2055–2064.
8. Randhawa GJ, Chhabra R, Singh M: **Multiplex PCR-based simultaneous amplification of selectable marker and reported genes for the screening of genetically modified crops.** *J Agric Food Chem* 2009, **57**(12):5167–5172.
9. James D, Schmidt A, Wall E: **Reliable detection and identification of genetically modified maize, soybean and canola by multiplex PCR analysis.** *J Agric Food Chem* 2003, **51**(20):5829–5834.
10. Pansiot J, Chaouachi M, Cavellini L, Romaniuk M, Ayadi M, Bertheau Y, Laval V: **Development of two screening duplex PCR assays for genetically modified organism quantification using multiplex real-time PCR master mixes.** *Eur Food Res Technol* 2011, **232**(2):327–334.
11. Samson MC, Gullí M, Marmiroli N: **Multiplex real-time PCR assays for simultaneous detection of maize MON810 and GA21 in food samples.** *Food Control* 2013, **30**(2):518–525.
12. *Imported agricultural genetically modified organisms approved by the Ministry of Agriculture as raw materials.* http://www.moa.gov.cn/ztzl/zjyqwgz/.
13. *Biosafety scanner system.* http://en. biosafetyscanner.org.
14. *Lynnon Corporation DNAMAN –Bioinformatics Solutions.* http://www.lynnon.com.
15. SN/T 1204–2003: *Protocol of the Real-Time PCR for Detecting Genetically Modified Plants and Their Derived Products.* Beijing, China: Standardization Administration of the People's Republic of China; 2003.
16. European Network of GMO Laboratories: *Definition of minimum performance requirements for analytical methods of GMO testing: Community Reference Laboratory for GM Food and Feed (CRL-GMFF).* 2008. http://gmo-crl.jrc.ec.europa.eu/doc/.Min_Perf_Requirements_Analytical_methods.pdf.
17. Li X, Yang L, Zhang J, Wang S, Shen K, Pan L, Zhang D: **Simplex and duplex PCR analysis of Herculex RW (59122) maize based on one reference molecule including separated fragments of 5′ integration site and endogenous gene.** *J AOAC Int* 2009, **92**(5):1472–1483.

19

Physiological effects of over-expressing compartment-specific components of the protein folding machinery in xylose-fermenting *Saccharomyces cerevisiae*

Basti Bergdahl*, Marie F Gorwa-Grauslund and Ed WJ van Niel

Abstract

Background: Efficient utilization of both glucose and xylose is necessary for a competitive ethanol production from lignocellulosic materials. Although many advances have been made in the development of xylose-fermenting strains of *Saccharomyces cerevisiae*, the productivity remains much lower compared to glucose. Previous transcriptional analyses of recombinant xylose-fermenting strains have mainly focused on central carbon metabolism. Very little attention has been given to other fundamental cellular processes such as the folding of proteins. Analysis of previously measured transcript levels in a recombinant XR/XDH-strain showed a wide down-regulation of genes targeted by the unfolded protein response during xylose fermentation. Under anaerobic conditions the folding of proteins is directly connected with fumarate metabolism and requires two essential enzymes: $FADH_2$-dependent fumarate reductase (FR) and Ero1p. In this study we tested whether these enzymes impair the protein folding process causing the very slow growth of recombinant yeast strains on xylose under anaerobic conditions.

Results: Four strains over-expressing the cytosolic (*FRD1*) or mitochondrial (*OSM1*) FR genes and *ERO1* in different combinations were constructed. The growth and fermentation performance was evaluated in defined medium as well as in a complex medium containing glucose and xylose. Over-expression of *FRD1*, alone or in combination with *ERO1*, did not have any significant effect on xylose fermentation in any medium used. Over-expression of *OSM1*, on the other hand, led to a diversion of carbon from glycerol to acetate and a decrease in growth rate by 39% in defined medium and by 25% in complex medium. Combined over-expression of *OSM1* and *ERO1* led to the same diversion of carbon from glycerol to acetate and had a stronger detrimental effect on the growth in complex medium.

Conclusions: Increasing the activities of the FR enzymes and Ero1p is not sufficient to increase the anaerobic growth on xylose. So additional components of the protein folding mechanism that were identified in transcription analysis of UPR related genes may also be limiting. This includes i) the transcription factor encoded by *HAC1* ii) the activity of Pdi1p and iii) the requirement of free FAD during anaerobic growth.

Keywords: *Saccharomyces cerevisiae*, Xylose fermentation, Protein folding, Transcription analysis, Fumarate reductase, Endoplasmic reticulum, Disulphide bond formation, FRD1, OSM1, ERO1

* Correspondence: Basti.Bergdahl@tmb.lth.se
Division of Applied Microbiology, Department of Chemistry, Lund University, P.O. Box 124, Lund SE-22100, Sweden

Background

The yeast *Saccharomyces cerevisiae* is one of few yeast species capable of growing under strict anaerobic conditions [1]. This trait together with a high tolerance toward inhibitory compounds, e.g. ethanol, weak acids, phenolics and furaldehydes, has made *S. cerevisiae* the organism of choice for bioethanol production [2]. The development of sustainable processes for biofuel production is an important step in the efforts to reduce greenhouse gas emissions and becoming independent of fossil fuels [3]. The utilization of lignocellulosic raw material for the production of fuel-grade ethanol is one such process currently under development [4,5]. These raw materials are generated as waste in e.g. agricultural and forestry industries and contain a large fraction of fermentable sugars. The composition of the sugar fraction varies between different materials, but the largest part often consists of glucose and xylose [4]. Efficient utilization of both these sugars is necessary for the ethanol production process to be economically feasible [6,7]. *S. cerevisiae* is well-known for its capability to ferment hexoses, especially glucose. However, *S. cerevisiae* cannot naturally utilize xylose and thus has to be genetically modified. The simultaneous expression of the *XYL1* and *XYL2* genes from *Scheffersomyces stipitis* encoding xylose reductase (XR) and xylitol dehydrogenase (XDH), respectively, is one pathway that enables xylose utilization by *S. cerevisiae* [8]. Although many advances have been made in the development of xylose-fermenting *S. cerevisiae* strains, the efficiency is still far from matching that of glucose [9,10].

The ability of *S. cerevisiae* to grow under strict anaerobic conditions is dependent on the activity of fumarate reductase (FR) enzymes [11,12]. *S. cerevisiae* has two genes, *FRD1* and *OSM1*, encoding the cytosolic and mitochondrial FR enzymes, respectively [13,14]. These enzymes catalyse the reduction of fumarate to succinate while regenerating FAD from $FADH_2$ [15]. This is the last step of the reductive branch of the TCA-cycle where mitochondrial NAD is regenerated under anaerobic conditions (Figure 1) [16]. Single deletion of *FRD1* or *OSM1* has no effect on the ability to grow under anaerobiosis, but a double deletion mutant cannot grow under such conditions unless an external electron acceptor is supplied (e.g. oxygen, menadione or phenazine methosulfate) [11,12,17]. It has thus been proposed that FR enzymes provide the only way for yeast to regenerate the FAD/FMN prosthetic group of flavin enzymes that are required for growth under anoxia [11]. Ero1p is one such essential flavin-containing oxidase, which normally uses oxygen as a final electron acceptor for FAD regeneration. Ero1p is a thiol oxidase [18,19] that operates together with Pdi1p [20] in the endoplasmic reticulum (ER) where they take part in the maturation of secretory proteins [21]. Both proteins are essential for the formation of disulphide bonds and together they form a classical proteinaceous electron relay system in which electrons are passed from the substrate polypeptide, via Pdi1p and the membrane-bound Ero1p, to eventually react with oxygen (Figure 1) [22-24].

Approximately one third of the yeast proteome is processed in the secretory pathway which makes this mechanism indispensable for cell viability, growth and function [21]. Many of the maturation processes taking place in the ER are regulated by a mechanism referred to as the unfolded protein response (UPR) under conditions of stress or increased folding requirements [25]. The ER contains a transmembrane protein encoded by *IRE1* which has a cytosolic ribonuclease domain and a luminal sensor domain. The sensor detects rising levels of unfolded proteins inside the ER and activates the cytosolic domain. This activity removes an intron from *HAC1* pre-mRNA and the fragments formed are ligated by Trl1p ligase to form mature *HAC1* mRNA. Hac1p is a potent transcriptional activator of UPR target genes which include both *PDI1* and *ERO1*, and dozens of genes related to translocation, glycosylation, folding and degradation [26]. Thus, the UPR balances the biosynthetic rate of secretory proteins and fine tunes the ER capacity to meet cellular demands.

Previous transcription analyses of recombinant xylose-fermenting strains have mainly focused on the changes that occur in genes encoding enzymes involved in central carbon metabolism [27-30]. Very little attention has been given to other fundamental cellular processes such as the folding of proteins. Recently we conducted a comprehensive analysis of the dynamic changes in intracellular metabolites that occur during batch fermentation of a glucose/xylose mixture using a recombinant XR/XDH-strain [31]. In that study we observed a marked accumulation of intracellular malate and fumarate during xylose fermentation. The direct connection between fumarate metabolism and protein folding under anaerobic conditions prompted us to further investigate this system during xylose fermentation. Due to the essential role of FR enzymes in anaerobic growth we hypothesized that the activity of FR is reduced on xylose which in turn would affect the capacity to regenerate the FAD required for proper folding of proteins in the ER. Such an impairment of the protein folding process could be the underlying cause for the slow anaerobic growth on xylose. In this study we examined this hypothesis by overexpressing the two genes encoding FR, alone and in combination with *ERO1*, and evaluated the effect on fermentation performance in two different media containing glucose and xylose.

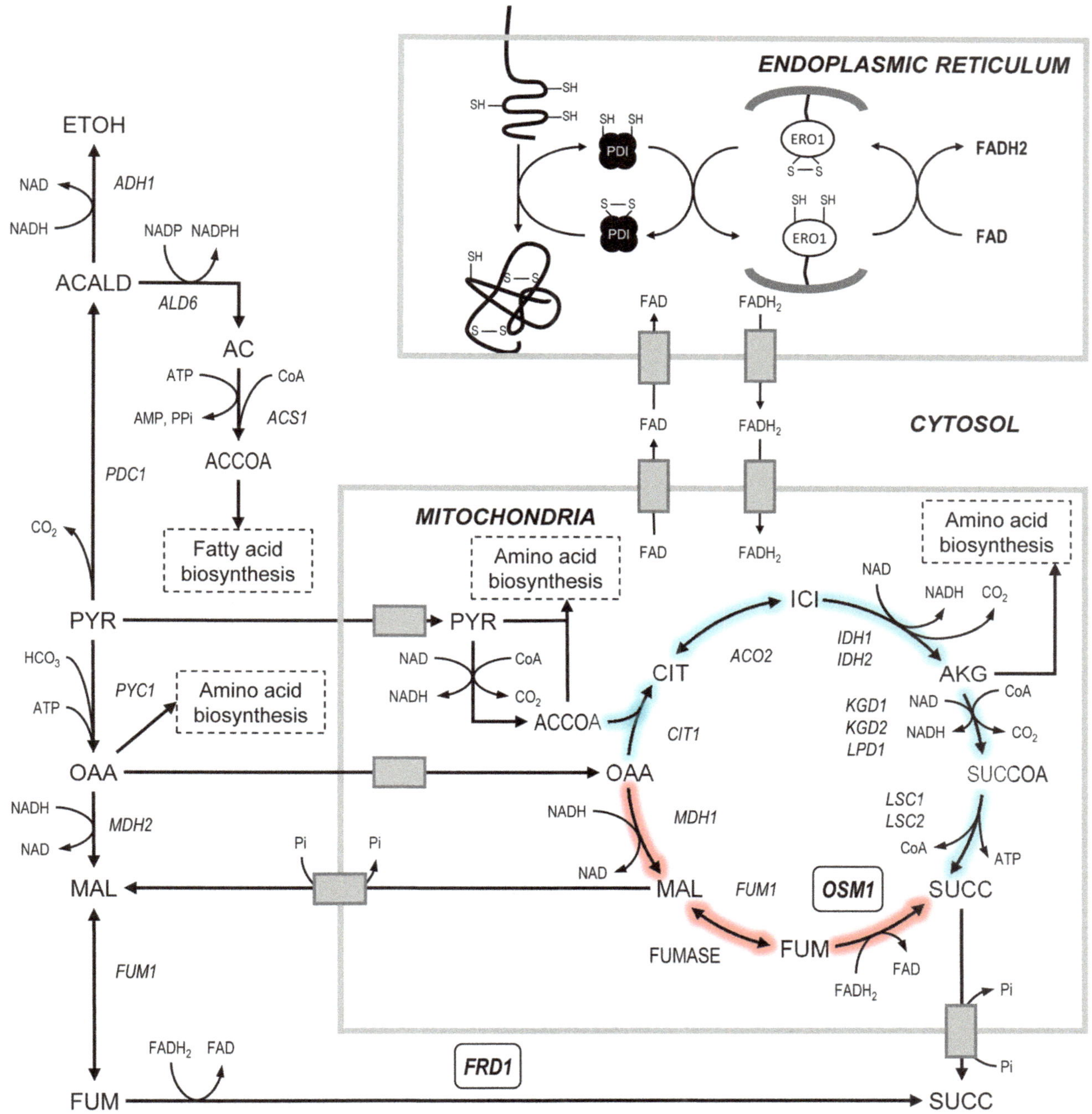

Figure 1 Metabolic pathways involving Frd1p, Osm1p, Pdi1p and Ero1p. Under anaerobic conditions the TCA cycle operates as two branches, a reductive path (red color) and an oxidative path (turqucise color). Osm1p catalyses the last step of the reductive branch of the TCA-cycle where fumarate is converted to succinate and FAD is regenerated from $FADH_2$. Frd1p catalyses the same reaction in the cytosol. The regenerated FAD is used inside the ER during the maturation of secretory proteins. The Pdi1p and Ero1p enzymes mediate the oxidative folding of proteins in the ER. When a nascent polypeptide is translocated into the ER the reduced cysteines are oxidised by Pdi1p giving rise to a disulphide bridge in the protein. The reduced Pdi1p is subsequently oxidised by the membrane bound Ero1p which transfers the electrons to oxygen via covalently bound FAD. *In vitro* experiments have shown that free FAD can act as the electron acceptor in the absence of oxygen.

Methods

Strains and culture conditions

Plasmids and yeast strains used in the study are listed in Table 1. Yeast strains were recovered from 20% glycerol stocks stored at –80°C on solid YPD plates (10 g/L yeast extract, 20 g/L peptone, 20 g/L glucose, 15 g/L agar) for two days at 30°C. Yeast cultures were grown in liquid YPD medium for 14–16 h, or less when required, at 30°C

Table 1 Strains and plasmids used in this study

Name	Relevant genotype	Reference
Plasmids		
p426TEF	*URA3*	[32]
YIplac128	*LEU2*	[33]
YIplac211	*URA3*	[33]
YIpDR1	YIplac128:*TDH3*p-*GXF1*-*CYC1*t	[34]
YIpDR7	YIplac211:*TDH3*p-*XYL1*(N272D)-*ADH1*t; *PGK1*p-*XYL2*-*PGK1*t	[35]
YIpBB1	YIplac128:*TDH3*p-*FRD1*-*CYC1*t	This study
YIpBB2	YIplac128:*TDH3*p-*OSM1*-*CYC1*t	This study
YIpBB3	YIplac128:*TDH3*p-*FRD1*-*CYC1*t; *TEF1*p-*ERO1*-*ERO1*t	This study
YIpBB4	YIplac128:*TDH3*p-*OSM1*-*CYC1*t; *TEF1*p-*ERO1*-*ERO1*t	This study
Yeast strains		
TMB3043	CEN.PK2-1C; *gre3*-Δ; *his3*::*PGK1*p-*XKS1*-*PGK1*t, *HIS3*; *tal1*::*PGK1*p-*TAL1*-*PGK1*t; *tkl1*::*PGK1*p-*TKL1*-*PGK1*t; *rki1*::*PGK1*p-*RKI1*-*PGK1*t; *rpe1*::*PGK1*p-*RPE1*-*PGK1*t; *ura3*, *leu2*	[36]
TMB3452	TMB3043; *ura3*::YIpDR7; *leu2*	This study
TMB3455	TMB3452; *leu2*::YIplac128	This study
TMB3456	TMB3452; *leu2*::YIpBB1	This study
TMB3457	TMB3452; *leu2*::YIpBB2	This study
TMB3458	TMB3452; *leu2*::YIpBB3	This study
TMB3459	TMB3452; *leu2*::YIpBB4	This study

and 180 rpm in an orbital shaker. Transformants were selected on solid Yeast Nitrogen Base (YNB) medium (6.7 g/L YNB with ammonium sulphate and without amino acids) with 20 g/L glucose as carbon source. Leucine was supplemented when required at 220 mg/L. Cultivations in shake flasks had a culture volume of 10% or less of the flask volume. The defined medium used in fermentation experiments contained 2X YNB, 20 g/L glucose, 50 g/L xylose and 50 mM potassium hydrogen phthalate buffer at pH 5.5. The complex medium used in fermentation experiments contained 10 g/L yeast extract, 20 g/L glucose, 50 g/L xylose, 50 mM potassium hydrogen phthalate buffer at pH 5.5, 10 g/L $(NH_4)_2SO_4$, 1 g/L $MgSO_4$ and 6 g/L KH_2PO_4. All media used in anaerobic cultivations were supplemented with Tween 80 (400 mg/L) and ergosterol (10 mg/L).

E. coli strain NEB5α (New England Biolabs, USA) was used for sub-cloning of plasmid DNA. Transformants were selected on solid LB plates (5 g/L yeast extract, 10 g/L tryptone, 10 g/L NaCl, 15 g/L agar, pH 7.0), supplemented with 100 mg/L of ampicillin, for 16 h at 37°C. Cultures of transformed *E. coli* were recovered from 25% glycerol stocks stored at −80°C and grown in liquid LB medium, supplemented with ampicillin, for 14–16 h at 37°C and 180 rpm in an orbital shaker.

Cloning procedures

Standard molecular biology techniques were used for all cloning procedures [37]. Plasmid purification was performed using GeneJet Plasmid Mini-prep Kit (Thermo Scientific, USA). DNA fragments digested with restriction enzymes were excised from agarose gel and purified using QIAquick Gel Extraction Kit (Qiagen, Germany) when needed. DNA fragments amplified by PCR were purified using GeneJet PCR Purification Kit (Thermo Scientific, USA). All enzymes for PCR amplifications, ligations and restriction digestions were obtained from Thermo Scientific unless stated otherwise. PCR products were amplified using the High Fidelity Phusion Hotstart II Polymerase or High Fidelity Taq Polymerase Mix (Thermo Scientific, USA). Competent yeast cells were prepared and transformed according to the PEG/LiAc method [38] and heat shock competent *E. coli* cells were prepared according to the Inoue method [37] and transformed according to the supplier's instructions.

The gene *FRD1* was amplified from yeast genomic DNA by PCR using primers FRD1-F1-SpeI (5′-GAC TAG TAA ATG TCT CTC TCT CCC GTT G-3′) and FRD1-R1-SalI (5′-GTT GTC GAC TTA CTT GCG GTC ATT GGC-3′). The gene *OSM1* was amplified from yeast genomic DNA by PCR using primers OSM1-F1-SpeI (5′-GAC TAG TAA AAT GAT TAG ATC TGT GAG-3′) and OSM1-R1-SalI (5′-GTT GTC GAC TCA GTA CAA TTT TGC TAT G-3′). The amplified *FRD1* and *OSM1* genes were digested with *Spe*I and *Sal*I and inserted between the *TDH3* promoter and *CYC1* terminator of the previously constructed vector YIpDR1, yielding the two plasmids YIpBB1 and YIpBB2 (Table 1). The cloned genes were verified by DNA sequencing.

The *ERO1* gene, including a 300 bp downstream sequence, was amplified from yeast genomic DNA by PCR using primers ERO1-F1-BamHI (5′-GAG GAT CCA AAA TGA GAT TAA GAA CCG CCA TTG-3′) and ERO1-R1-XhoI (5′-GTC TCG AGC CCT TGA AGA TGG TAC C-3′). The amplified *ERO1* ORF and terminator was digested with *Bam*HI and *Xho*I and inserted between the *TEF1* promoter and *CYC1* terminator of the vector p426TEF (Table 1). The *TEF1*p-*ERO1*-*ERO1*t cassette was amplified from plasmid p426TEF-ERO1 using primers p426TEF-F1-SacI (5′-GGA GCT CAT AGC TTC AAA ATG-3′) and p426ERO1t-R1-NdeI (5′-CAT ATG CCC TTG AAG ATG GTA C-3′). The amplified cassette was digested with *Bam*HI and *Xho*I and ligated into plasmids YIpBB1 and YIpBB2, digested with the same enzymes, yielding vectors YIpBB3 and YIpBB4, respectively (Table 1). The cloned cassette was verified by DNA sequencing.

Strain construction

S. cerevisiae strain TMB3452 was constructed by transforming strain TMB3043 with YIpDR7, linearized with

*Eco*RV. The integrative plasmids YIpBB1-4 were linearized with *Eco*RV, *Afl*II, *Eco91*I and *Afl*II, respectively, and used to transform strain TMB 3452, yielding strains TMB3456 (*FRD1*), TMB3457 (*OSM1*), TMB3458 (*FRD1 ERO1*) and TMB3459 (*OSM1 ERO1*) (Table 1). The control strain TMB3455 was constructed by transforming TMB3452 with YIplac128, linearized with *Eco*RV. Correct integration of the vectors was verified by PCR using primer pairs that annealed on the plasmid and on the chromosome outside the homologous regions of *URA3* or *LEU2* loci. The functionality of the over-expressed *FRD1* and *OSM1* genes was verified by measuring the specific fumarate reductase activity in cell-free protein extracts from strains TMB3455-59 grown anaerobically on xylose. Strains TMB3456-59 had between 9- and 30-fold higher activities than the control showing that the over-expressed genes generated increased levels of functional enzymes (Additional file 1: Figure S1). The functionality of the over-expressed *ERO1* gene was verified by determining the sensitivity toward DTT. Single colonies of strains TMB3455-59 were inoculated in 5 mL YNB medium with 2% glucose and grown over-night. 200 μL of this culture was used to inoculate fresh medium and cultivated until the cells reached $OD_{620\ nm} \approx 1$. A volume containing 2×10^6 cells (150–200 μL) was spread on solid YPD medium after which a filter disc was placed in the centre of the plate and soaked with 20 μL or 30 μL of 1 M DTT solution. The plates were incubated 48 h at 30°C before measuring the diameter of the halo formed around the disc. The halos formed with strains TMB3458 and TMB3459 were 20-30% smaller compared with the other strains showing that the over-expressed *ERO1* gene generated increased level of a functional enzyme (Additional file 1: Figure S2).

Anaerobic batch fermentations

Yeast strains were pre-grown in YNB medium with 2% glucose until mid-exponential phase and inoculated into 50 mL medium with 2% glucose and 5% xylose in 250 mL crimp-sealed serum flasks at a concentration of 0.04 g CDW/L. Prior to inoculation the headspace was flushed with N_2-gas containing less than 5 ppm oxygen (AGA Gas AB, Sweden) for at least 30 min to ensure anaerobic conditions. Cultivations were carried out in a 30°C water bath with continuous stirring at 180 rpm using magnetic stirrers. To prevent diffusion of oxygen into the culture during sampling, the excess gas was released through a 0.2 μm filter connected to Norprene tubing submerged under water. Cell dry weight was calculated from OD measurements using a pre-determined relationship between OD and cell dry weight. Fermentation experiments were performed in biological duplicates.

Preparation of cell-free protein extracts

Yeast strains were cultivated anaerobically in crimp-sealed serum flasks as described above using 2X YNB medium with 2% glucose and 5% xylose. After 36 h, when xylose was the only carbon source, the cells were harvested by centrifugation at 4000 rpm and 4°C for 10 min. The resulting cell pellets were resuspended in 1X Lysis buffer (20 mM K,Na-PO_4, 20 mM EDTA, 1 mM DTT, pH 7.0) to obtain a final suspension of 1 g wet cells/mL. Protein extracts were prepared using a Precellys®24 homogenizer with cooling provided by the Cryolys® unit (Bertin Technologies, France) and Precellys Lysing Kit VK05 containing 0.5 mm glass beads. The unit was operated for 3 × 45 s at 5000 rpm with 30 s pauses between each cycle and at max 8°C. Cell debris was pelleted by centrifugation at 14,000 × *g* and 4°C for 15 min and the supernatant was transferred to a small glass vial. Before sealing the vial the headspace was replaced with N_2-gas in an anaerobic chamber. The protein extracts were stored at 4°C until the following day. Protein concentrations were determined with the Coomassie Protein Assay Reagent (Thermo Scientific, USA) using BSA as standard (Thermo Scientific, USA).

Fumarate reductase activity assay

Fumarate reductase activity was determined in a 1.2 mL reaction mixture containing: 50 mM K,Na-PO_4 buffer (pH 7.3), 10 mM fumaric acid (pH 7.3), 0.4 mM FMN-Na, 2% (w/v) sodium hydrosulfite. All reagents, except protein extract, were added to a rubber-sealed cuvette inside an anaerobic chamber. The assay buffer was prepared by dissolving 2.762 g of KH_2PO_4 and 2.837 g of $Na_2HPO_4 \cdot 2H_2O$ in 50 mL anaerobic water (separately) inside the anaerobic chamber, creating one 10X solution and one 2X solution, respectively. A 200 mM stock solution at the correct pH was made by mixing 5 mL 10X with 25 mL 2X and completing the volume to 50 mL with anaerobic water. A 40 mM stock solution of fumaric acid was prepared by dissolving 0.464 g fumaric acid in 50 mL of 5 mM K,Na-PO_4 buffer at pH 7.3. The pH was adjusted to 7.3 using 5 M NaOH and the volume was completed to 100 mL with water. Solutions of the cofactor and reducing agent were prepared fresh before each assay. Both compounds were weighed and brought into the anaerobic chamber. The FMN cofactor was dissolved in anaerobic water and sodium hydrosulfite was dissolved in 100 mM K,Na-PO_4 buffer, but only after all other reagents had been added to the cuvette to ensure that FMN becomes completely reduced to $FMNH_2$. The reaction was started by addition of protein extract using a gas-tight syringe. The formation of FMN was measured at 445 nm in cuvettes with light path length of 1 cm using an Ultrospec 2100 Pro spectrophotometer (Amersham

Biosciences Corp., USA) controlled by computer software SWIFT Reaction Kinetics v. 2.05 (Biochrom Ltd., Cambridge, UK). The absorbance was converted to concentration units using the molar extinction coefficient of FMN ($\varepsilon_{445\ nm}$ = 12,500 L/mol/cm). The slope was determined within an interval of at least 1 min, using at least three different dilutions of the protein extract. Only the slopes within the range 0.045-0.40 A/min were considered acceptable. One unit is defined as the conversion of 1 μmol of fumarate per minute.

Biomass determination and analysis of metabolites

Optical density was measured at 620 nm using an Ultrospec 2100 Pro spectrophotometer (Amersham Biosciences Corp., USA). Cell dry weight was measured in triplicate by filtering a known volume of the culture through a pre-weighed nitrocellulose filter with 0.45 μm pore size. The filters were washed with three volumes of water, dried in a microwave oven and weighed after equilibrating to room temperature in a desiccator.

Concentrations of glucose, xylose, xylitol, glycerol, succinate, acetate and ethanol were analysed by HPLC (Waters, USA). Aminex HPX-87H ion exchange column (Bio-Rad, USA) was used at 45°C with a mobile phase of 5 mM H_2SO_4 at a flow rate of 0.6 mL/min. All compounds were detected with a RID-10A refractive index detector (Shimadzu, Japan).

Calculation of metabolic rates

In defined medium (YNB) the volumetric production and consumption rates (q_i and q_s), the maximum specific consumption rates of xylose (r_{xyl}) and glucose (r_{glc}), as well as the biomass yield per consumed substrate ($Y_{x/s}$) were calculated through linear regression according to Equations 1–4. The product yields per consumed substrate ($Y_{i/s}$) were calculated using the determined volumetric rates according to Equation 5. The product yields per biomass ($Y_{i/x}$) were calculated as the ratio between $Y_{i/s}$ and $Y_{x/s}$ when xylose was the carbon source and according to Equation 6 when glucose was the carbon source. Due to the non-exponential growth on xylose, the maximum specific growth rate ($\mu_{\max}$) was calculated as the product between r_{xyl} and $Y_{x/s}$ according to Equation 7. For comparative reasons $\mu_{\max}$ on glucose was calculated in the same way. These values corresponded well with those calculated through regression of ln(OD) vs. time. The specific production rates were calculated as the product between $\mu_{\max}$ and $Y_{i/x}$ according to Equation 8.

$$q_i = \frac{dC_i}{dt}, q_s = \frac{dC_s}{dt} \quad (1)$$

$$r_{xyl} = \frac{d(C_{xyl}/C_x)}{dt} \quad (2)$$

$$r_{glc} = \frac{d(dC_{glc}/dt)}{dC_x} \quad (3)$$

$$Y_{x/s} = \frac{dC_x}{dC_{s,tot}} \quad (4)$$

$$Y_{i/s} = \frac{q_i}{q_s} \quad (5)$$

$$Y_{i/x} = \frac{dC_i}{dC_x} \quad (6)$$

$$\mu_{\max} = r_s \cdot Y_{x/s} \quad (7)$$

$$r_i = \mu_{\max} \cdot Y_{i/x} \quad (8)$$

In complex medium all calculations were made as described above except for the maximum specific consumption rate of glucose which could not be determined accurately enough. Hence, the $\mu_{\max}$ under these conditions was determined as the slope in a linear regression of ln(OD) vs. time and r_{glc} was calculated using Equation 7. The significance of a difference between two mean values was calculated using a two-tailed t-test assuming equal variance of the two samples with two degrees of freedom ($n_1 = n_2 = 2$) unless stated otherwise. Balances of carbon and degree of reduction, determined for the glucose and xylose phases using calculated yield coefficients, are given as supplementary information in Additional file 2: Table S1.

Transcriptional analysis of UPR related genes

The list of genes reported as targets of the UPR were taken from [26]. Transcript data from cultivations of a XR/XDH-strain using glucose or xylose in aerobic and anaerobic conditions was taken from [27]. UPR target genes that had a significant change in expression ($P \leq 0.1$) during the transition from aerobic to anaerobic conditions with xylose as carbon source were included in the analysis. Two genes commonly used as targets of the UPR (*KAR2* and *INO1*) [27] did not pass the statistical requirement and were thus not included in the analysis (Additional file 2: Table S5). Hierarchical clustering was performed using the *clustergram* function in MATLAB R2010b with $\log_2$ values of transcriptional changes. Euclidean distance was used to calculate the distance between the data points and average linkage was used to generate the dendrogram. Data used for the hierarchical clustering is given as supplementary information in Additional file 2: Table S2.

Results and discussion

Genes targeted by the unfolded protein response are differentially expressed on xylose compared to glucose

A previously performed transcripome study in a recombinant XR/XDH-strain [27] was used to investigate the expression level of genes associated with the UPR. It revealed that *IRE1* and *HAC1* levels were ~65% and ~50% higher ($P = 0.02$ and $P = 0.05$, respectively) in cells growing anaerobically on xylose compared to cells growing aerobically (Figure 2). However, the transcript level of *TRL1*, encoding the ligase, was ~50% lower ($P = 0.02$) (Figure 2), suggesting that *HAC1* is mainly present as fragments rather than mature mRNA molecules. The transcript levels of these genes did not change significantly during the same transition (i.e. from aerobic to anaerobic conditions) on glucose (Figure 2) indicating that the response was associated with the type of carbon source available. The cluster analysis illustrated in Figure 3 indeed shows that a majority (61%) of the included UPR target genes were down-regulated on xylose when the conditions changed from aerobic to anaerobic (Group A). The genes belonging to this group represent functions taking place in all steps of the maturation process (translocation, glycosylation, disulphide bond formation, protein folding, vesicle budding and vesicle transport to the cell membrane) (Additional file 2: Table S2) and do not change during the same transition on glucose. The smallest group (Group B) contained those genes which are regulated equally on glucose and xylose, with only small differences in abundance between the two carbon sources. A small number of genes were up-regulated on xylose while remaining largely unchanged on glucose (Group C). These genes are related to peroxisome biogenesis (*PEX4*), ubiquitin recycling (*DOA4*) and the cell integrity signalling pathway (*TUS1*). The three genes *ECM3*, *ECM8* and *YET2* encode proteins with unknown functions. *RIB1* encodes GTP cyclohydrolase II which catalyses the first step of the riboflavin (FAD and FMN) biosynthesis pathway. The fourth group contains genes which are strongly up-regulated in response to anaerobiosis in both conditions (Group D). However, in contrast to the genes in Group B, the transcript levels of these genes were in general lower on xylose compared to glucose. This analysis shows that the UPR responded differently to xylose compared to glucose. Furthermore, the repression of a majority of the target genes suggests that the capacity of the secretory pathway could be limited during xylose fermentation.

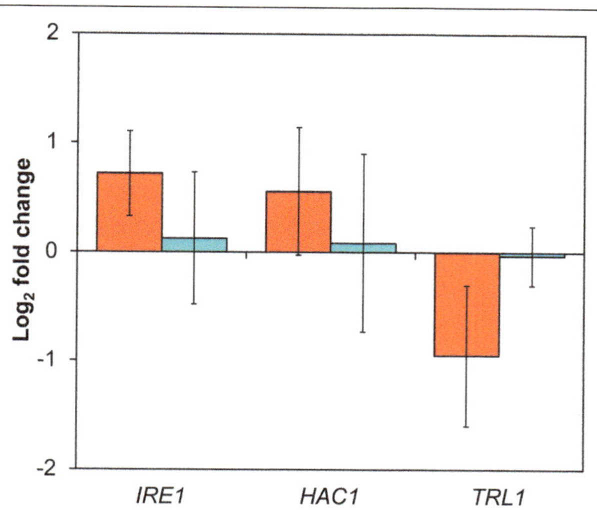

Figure 2 Transcriptional response of genes involved in the regulation of the UPR. Bars show the $\log_2$ fold change of the indicated genes during a transition from aerobic to anaerobic conditions in a XR/XDH-strain utilizing xylose (red bars) or glucose (turquoise bars) as carbon source [27]. Values are given as mean fold change of two independent measurements in each conditions and errors represent the 95% confidence interval of the mean ($df = 2$).

Consistent with previously reported expression patterns on non-fermentable carbon sources and entry into stationary phase [39,40], the expression of *ERO1* was reduced by ~30% ($P < 0.02$) on xylose compared to glucose, both under anaerobic and aerobic conditions (Additional file 2: Table S5) [27]. Hence, a reduced capacity to form disulphide bonds in nascent polypeptides translocated to the ER could be one underlying cause for the low anaerobic growth rate on xylose. The previously reported accumulation of intracellular fumarate during xylose fermentation [31] suggested that recombinant XR/XDH strains have a limited FR activity under such conditions, which would lead to insufficient recycling of the FAD co-factor required by Ero1p. This hypothesis was investigated by evaluating the physiological effect of over-expressing the cytosolic and mitochondrial FR enzymes, alone and in combination with *ERO1*, during fermentation of a mixture of glucose and xylose.

Over-expression of cytosolic FR has limited impact on xylose fermentation

Due to the higher contribution of the cytosolic FR to the total cellular reductase activity the over-expression of *FRD1*, alone and in combination with *ERO1*, was evaluated first. When cultivated in defined medium (2X YNB) with 20 g/L glucose and 50 g/L xylose, TMB3456 (*FRD1*) and TMB3458 (*FRD1 ERO1*) consumed the glucose within 20 h and ca. 70% of the xylose within 90 h (Figure 4B and D). The two strains behaved very similar to the control strain (Figure 4A) and no significant physiological differences from this strain were observed (Additional file 2: Table S3). These results suggest that the implemented genetic modifications had no positive effect on the anaerobic growth on xylose in defined medium. However, the modifications only ensured that the genes required for regeneration of FAD co-factors were present (Figure 1). If substrates are not available in sufficient amounts (in this case unfolded polypeptides) these modifications are not effective in regenerating the co-factor. The poor repressive capability of xylose leads to an up-regulation of *MDH2* [27] (Additional file 2: Table S5),

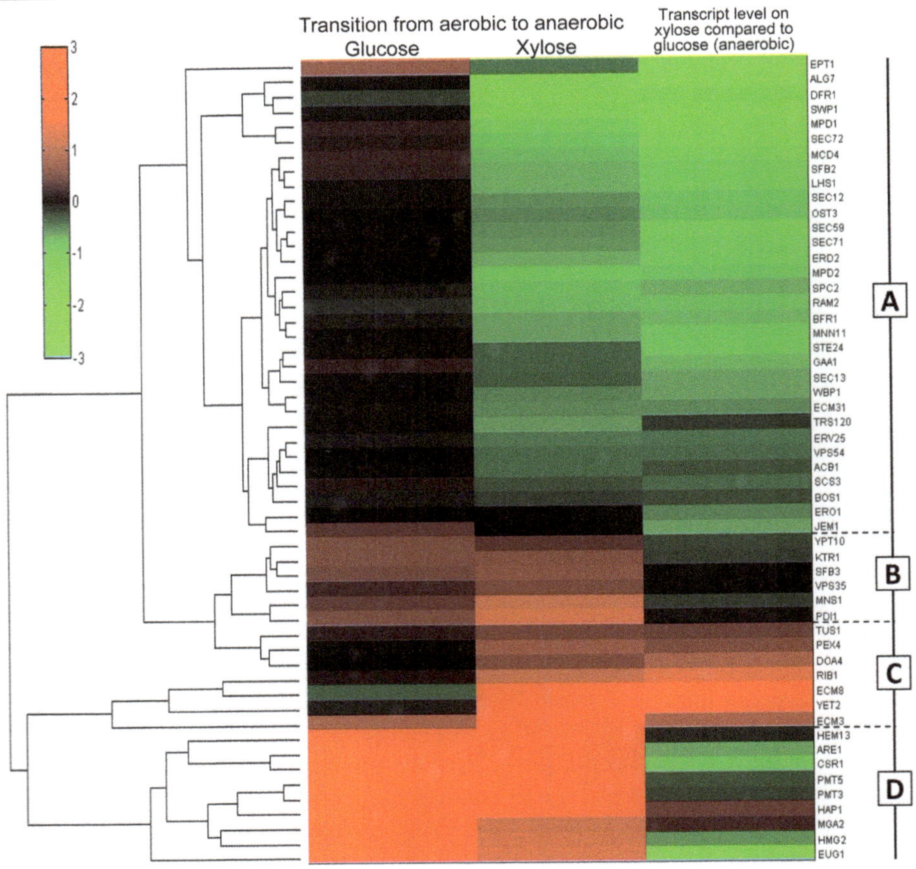

Figure 3 Hierarchical clustering of genes induced by the UPR. Genes that are induced by the UPR [26] and changed significantly ($P \leq 0.1$) during a transition from aerobic to anaerobic conditions in a XR/XDH-strain utilizing xylose [27] were analysed by hierarchical clustering. The heat map was generated using the $\log_2$ fold change between the different conditions. The analysis shows that a majority (61%) of the included UPR target genes are down-regulated on xylose when the conditions change from aerobic to anaerobic (Group A). These genes do not change during the same transition on glucose indicating that the UPR responds differently to xylose compared to glucose.

encoding cytosolic malate dehydrogenase, which has previously been suggested as an unbeneficial reaction that directs carbon away from amino acid synthesis by converting oxaloacetate into malate [31]. A reduced synthesis of amino acids could thus hamper the formation of polypeptides and mask the effect of the genetic modifications. To investigate whether the strains were limited by amino acid synthesis they were cultivated in a complex medium containing yeast extract. Changing from a defined medium to a complex medium led to several significant changes in the physiology of all strains, e.g. increased maximum growth rate, increased biomass yield and reduced by-product yields (Additional file 2: Table S3 and Table S4). Hence, the glucose was consumed within 15 h and ca. 80% of the xylose was consumed within 65 h (Additional file 1: Figure S3). However, despite the increased nutrient availability no significant differences were observed between the control strain and the *FRD1*-overexpressing strains TMB3456 and TMB3458 (Additional file 2: Table S4).

Over-expression of mitochondrial FR has a detrimental effect on xylose fermentation

Although the change from a defined medium to a complex medium led to a general increase in the growth rate on xylose, the over-expression of *FRD1*, alone or in combination with *ERO1*, had little effect on the cellular physiology. One explanation could be that the previously observed accumulation of fumarate [31] does not occur in the cytosol, but is restricted to the mitochondria. To investigate this possibility we over-expressed the *OSM1* gene, alone and in combination with *ERO1*, and evaluated the fermentation performance in both defined and complex media with 20 g/L glucose and 50 g/L xylose.

In defined medium, TMB3457 (*OSM1*) and TMB3459 (*OSM1 ERO1*) also consumed the glucose within 20 h and ca. 60% of the xylose within 90 h (Figure 4C and E). During glucose fermentation these strains behaved similar to the other strains and gave nearly the same product yields as the control strain (Table 2). When xylose became the sole carbon source the physiology changed

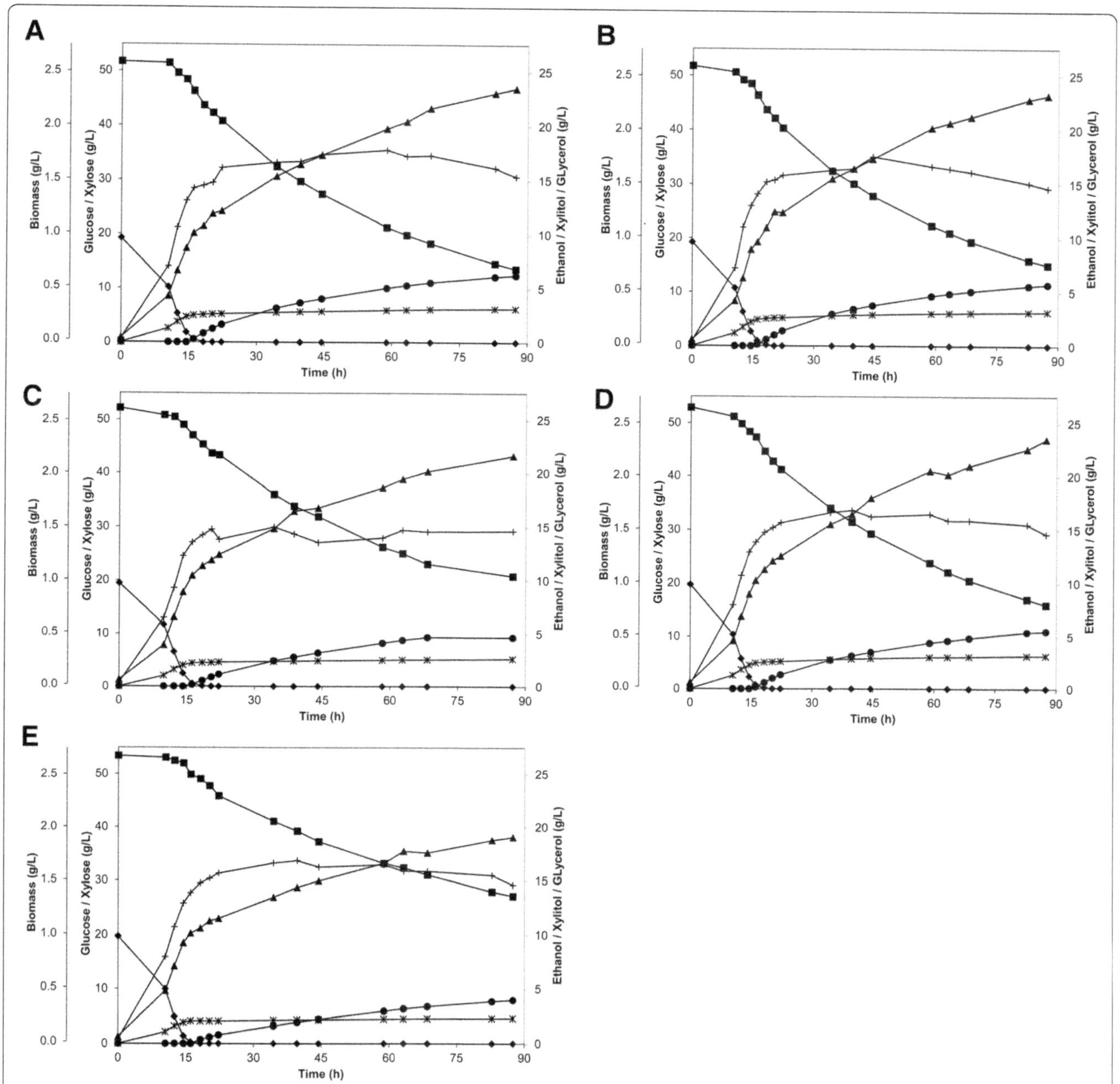

Figure 4 Fermentation of a glucose/xylose mix in defined medium. A mix of 20 g/L glucose and 50 g/L xylose was fermented in defined medium (2X YNB) by strains **A)** TMB3455 (control), **B)** TMB3456 (*FRD1*), **C)** TMB3457 (*OSM1*), **D)** TMB3458 (*FRD1 ERO1*) and **E)** TMB3459 (*OSM1 ERO1*). Symbols: diamonds, glucose; squares, xylose; plus, biomass; triangles, ethanol; circles, xylitol; stars, glycerol. Figures illustrate one representative experiment out of two biological replicates.

drastically and the two strains behaved differently. The maximum growth rate of strain TMB3457 (*OSM1*) was 47% lower ($P = 0.03$) compared with the control strain (Table 2) and the biomass-specific yields of acetate and ethanol increased by 130% ($P = 0.04$) and 69% ($P = 0.05$), respectively (Table 2). As a result a smaller fraction of the carbon was distributed to biomass and glycerol in favour of acetate and ethanol (Table 2). Strain TMB3459 (*OSM1 ERO1*) did not display any difference in growth rate on xylose compared with the control strain (Table 2). The acetate yield per biomass also increased by 44% ($P = 0.008$) whereas the glycerol and xylitol yields decreased by 39% ($P = 0.08$) and 20% ($P = 0.002$) compared with the control strain (Table 2). Hence, this strain diverted the carbon away from glycerol, xylitol and ethanol for a higher yield of acetate and maintained biomass yield (Table 2).

In complex medium strains TMB3457 (*OSM1*) and TMB3459 (*OSM1 ERO1*) consumed glucose within 15 h but only 50-60% of xylose was consumed within 65 h (Additional file 1: Figure S3). The performance of the

Table 2 Maximum specific growth rates and product yields in anaerobic batch fermentation of 20 g/L glucose and 50 g/L xylose in defined medium (2X YNB) using strains TMB3455, TMB3457 and TMB3459

	Glucose phase			Xylose phase		
	TMB3455 (control)	**TMB3457 (*OSM1*)**	**TMB3459 (*OSM1 ERO1*)**	**TMB3455 (control)**	**TMB3457 (*OSM1*)**	**TMB3459 (*OSM1 ERO1*)**
μ_{max} (1/h)	0.29 ± 0.012	0.30 ± 0.019	0.32 ± 0.044	0.029 ± 0.002	0.018 ± 0.001	0.028 ± 0.003
Yields per substrate (mol/mol sugar)						
$Y_{xylt/s}$	–	–	–	0.24 ± 0.03	0.25 ± 0.01	0.20 ± 0.02
$Y_{glyc/s}$	0.22 ± 0.02	0.20 ± 0.02	0.20 ± 0.05	0.020 ± 0.000	0.016 ± 0.002	0.012 ± 0.001
$Y_{ac/s}$	0.020 ± 0.001	0.012 ± 0.000	0.017 ± 0.005	0.019 ± 0.001	0.031 ± 0.003	0.029 ± 0.003
$Y_{etoh/s}$	1.87 ± 0.14	1.96 ± 0.21	1.96 ± 0.37	1.03 ± 0.00	1.24 ± 0.15	0.87 ± 0.00
$Y_{succ/s}$	0.003 ± 0.000	0.001 ± 0.000	0.002 ± 0.001	0.009 ± 0.000	0.007 ± 0.002	0.006 ± 0.000
$Y_{x/s}$ (g/mol)	10.4 ± 0.5	10.5 ± 0.2	11.9 ± 0.9	3.8 ± 0.4	2.7 ± 0.6	3.9 ± 0.4
Yields per biomass (mmol/g CDW)						
$Y_{xylt/x}$	–	–	–	63.1 ± 0.3	92.9 ± 25.8	50.3 ± 0.8
$Y_{glyc/x}$	20.2 ± 0.8	18.3 ± 0.5	16.6 ± 0.2	5.2 ± 0.6	5.8 ± 0.7	3.2 ± 0.6
$Y_{ac/x}$	2.1 ± 0.1	1.8 ± 0.1	1.8 ± 0.0	5.1 ± 0.3	11.7 ± 1.8	7.3 ± 0.0
$Y_{etoh/x}$	153 ± 8	153 ± 11	147 ± 7	271 ± 30	458 ± 53	223 ± 23
$Y_{succ/x}$	0.19 ± 0.04	0.28 ± 0.18	0.22 ± 0.04	2.5 ± 0.4	2.4 ± 0.3	1.6 ± 0.22

Values are given as mean ± standard deviation of two independent experiments. Data for strains TMB3456 and TMB3458 is reported in Additional file 2: Table S3.

strains did not differ significantly from the control strain during the glucose phase (Table 3). With xylose as the carbon source, on the other hand, both strains grew markedly slower than the control strain. The growth rates of TMB3457 (*OSM1*) and TMB3459 (*OSM1 ERO1*) on xylose were reduced by 25% ($P = 0.09$) and 51% ($P = 0.12$), respectively (Table 3). Both strains also increased the biomass-specific yield of acetate by 73% ($P = 0.003$) and 157% ($P = 0.03$), respectively (Table 3). Although both strains diverted more carbon toward acetate, different by-product yields were reduced by each strain. TMB3457 (*OSM1*) gave a reduced substrate-specific

Table 3 Maximum specific growth rates and product yields in anaerobic batch fermentation of 20 g/L glucose and 50 g/L xylose in complex medium (10 g/L yeast extract) using strains TMB3455, TMB3457 and TMB3459

	Glucose phase			Xylose phase		
	TMB3455 (control)	**TMB3457 (*OSM1*)**	**TMB3459 (*OSM1 ERO1*)**	**TMB3455 (control)**	**TMB3457 (*OSM1*)**	**TMB3459 (*OSM1 ERO1*)**
μ_{max} (1/h)	0.37 ± 0.003	0.38 ± 0.001	0.38 ± 0.003	0.033 ± 0.004	0.024 ± 0.000	0.016 ± 0.008
Yields per substrate (mol/mol sugar)						
$Y_{xylt/s}$	–	–	–	0.14 ± 0.01	0.16 ± 0.01	0.13 ± 0.01
$Y_{glyc/s}$	0.13 ± 0.01	0.11 ± 0.01	0.11 ± 0.01	0.013 ± 0.001	0.009 ± 0.001	0.009 ± 0.002
$Y_{ac/s}$	0.030 ± 0.006	0.022 ± 0.000	0.025 ± 0.003	0.022 ± 0.003	0.037 ± 0.005	0.037 ± 0.001
$Y_{etoh/s}$	2.7 ± 0.6	2.6 ± 0.2	2.6 ± 0.2	1.4 ± 0.1	1.4 ± 0.0	1.1 ± 0.0
$Y_{succ/s}$	0.005 ± 0.004	0.003 ± 0.001	0.003 ± 0.001	0.003 ± 0.001	0.006 ± 0.000	0.005 ± 0.001
$Y_{x/s}$ (g/mol)	17.5 ± 1.4	18.4 ± 0.9	19.5 ± 0.6	6.4 ± 0.8	6.3 ± 0.7	4.2 ± 0.8
Yields per biomass (mmol/g CDW)						
$Y_{xylt/x}$	–	–	–	22.3 ± 0.8	25.5 ± 1.2	31.0 ± 8.7
$Y_{glyc/x}$	7.3 ± 0.5	6.4 ± 0.2	6.4 ± 0.0	2.1 ± 0.0	1.5 ± 0.4	2.3 ± 0.9
$Y_{ac/x}$	1.3 ± 0.4	1.3 ± 0.1	1.5 ± 0.1	3.4 ± 0.0	5.9 ± 0.2	8.8 ± 1.4
$Y_{etoh/x}$	94.2 ± 5.0	91.3 ± 5.9	93.0 ± 3.5	222 ± 11	228 ± 21	272 ± 62
$Y_{succcxxx}$	0.12 ± 0.07	0.17 ± 0.08	0.21 ± 0.02	0.54 ± 0.03	0.91 ± 0.03	1.2 ± 0.5

Values are given as mean ± standard deviation of two independent experiments. Data for strains TMB3456 and TMB3458 is reported in Additional file 2: Table S4.

yield of glycerol by 31% ($P = 0.09$) whereas TMB3459 (*OSM1 ERO1*) gave reduced yields of ethanol and biomass by 21% ($P = 0.06$) and 34% ($P = 0.11$), respectively (Table 3).

These results indicate that the reactions taking place in the mitochondria have a significant role in xylose metabolism and that they are more sensitive to changes in metabolite levels (leading to altered regulatory signals, degree of allosteric inhibition of enzymes etc.) on xylose compared to glucose. Hence, the over-expression of *OSM1* caused physiological responses that were not observed in strains over-expressing *FRD1*. In strain TMB3457 (*OSM1*) these responses were: i) a 20-40% reduction in glycerol yield, ii) a 63-68% increase in acetate yield and iii) a 27-38% reduction in growth rate during xylose fermentation (Table 2 and Table 3). A possible explanation for this phenotype could be that Osm1p interacts with Gut2p and Ach1p which are both located specifically inside the mitochondria (Additional file 1: Figure S4). *GUT2* encodes a FAD-dependent glycerol 3-phosphate dehydrogenase located in the mitochondrial inner membrane [41] and expression of the gene is repressed in the presence of glucose and induced during growth on non-fermentable carbon sources [42]. An induction of the *GUT2* gene has also been shown to occur during xylose fermentation with transcript levels more than 7-fold higher compared to glucose [27] (Additional file 2: Table S5). Recently, the protein encoded by *ACH1* was shown to function as a Coenzyme A (CoA) transferase rather than an acetyl-CoA hydrolase [43]. Hence, this enzyme catalyses the following reaction: acetyl-CoA + succinate → succinyl-CoA + acetate ($\Delta_r G'^{o} = -8.1$ kJ/mol) and is also induced nearly 5-fold during xylose fermentation compared to glucose [27] (Additional file 2: Table S5). If both Gut2p and Ach1p are active on xylose, the pathway formed together with Osm1p could explain the diversion of carbon from glycerol to acetate as well as the reduced growth rate due to a reduced acetyl-CoA availability (Additional file 1: Figure S4). However, this hypothesis needs to be validated through further investigations.

Frd1p and Osm1p do not influence regeneration of cytosolic NAD

An early study on the role of FR showed that the addition of oxidised methylene blue or phenazine methosulfate, which chemically oxidises NADH to NAD, to cultures of a *frd1*Δ *osm1*Δ double mutant rescued the non-growing phenotype under anaerobic conditions [17]. It was thus proposed that the two FR enzymes are involved in the oxidation of excess NADH formed in the cytosol and the mitochondria under anaerobic conditions. The formation of xylitol during xylose fermentation has been ascribed to an inability of the cells to regenerate the NAD needed for the conversion of xylitol to xylulose by XDH [44]. The addition of acetoin, a substrate which can be reduced by yeast to 2,3-butanediol using NADH, to cultures fermenting xylose reduces the xylitol yield significantly [45,46]. If the FR enzymes participate in the conversion of excess NADH a reduction of the xylitol yield during xylose fermentation would be expected when the FR activity is increased. However, the over-expression of *FRD1* or *OSM1* did not affect the amount of xylitol produced from xylose (Table 2, Table 3, Additional file 2: Table S3 and Table S4). This is in agreement with other studies showing that addition of acetoin to a culture of a *frd1*Δ *osm1*Δ double mutant does not rescue the non-growing phenotype under anaerobic conditions [11]. In addition, the proposed mechanism of NADH oxidation requires the presence of an enzyme capable of transferring electrons from NADH to FAD to generate the reduced cofactor used as substrate by fumarate reductase [11,17]. As of yet, no such enzyme has been identified in *S. cerevisiae* that is expressed under anaerobic conditions. There are, however, FR enzymes that use NADH as co-factor instead of $FADH_2$. The NADH-dependent FR enzyme from *Trypanosoma brucei* has recently been expressed in a xylose-utilizing *S. cerevisiae* strain which resulted in increased ethanol yield and decreased xylitol yield when fermenting xylose [47]. Based on these results we conclude that the FR enzymes from *S. cerevisiae* do not influence the regeneration of cytosolic NAD.

Other potential limitations in the secretory pathway in *S. cerevisiae* during xylose fermentation

The results obtained in this study indicate that the activities of FR and Ero1p enzymes are not limiting during xylose fermentation. Measurements of the FR activity in strain TMB3455 when growing on glucose and when metabolizing xylose indeed showed equal activities around 3 mU/mg protein in both conditions (Additional file 1: Figure S5). This is contradicting previous studies showing that the activity and expression of *FRD1* and *OSM1* decrease upon entry into stationary phase [11,39] and during carbon starvation [48]. It is, however, possible that the FR activity is maintained high on xylose in response to cellular stress. The up-stream regions of the *FRD1* and *OSM1* genes contain binding sites for stress-related transcription factors Yap1p and, in the case of *FRD1*, Msn2p/Msn4p [49]. Due to the involvement of Yap1p, an oxidative stress would be most likely in this case [50]. This possibility is supported by much higher intracellular concentrations of both reduced and oxidized glutathione [31] and higher transcript levels of *MSN4* and *YAP1* [27] in recombinant *S. cerevisiae* strains during xylose fermentation compared to glucose fermentation (Additional file 2: Table S5).

The up-regulation of *RIB1* (Figure 3) could be an indication that the need for flavin-containing complexes and/or the free co-factor is increased on xylose. Ero1p has been shown to use free FAD as electron acceptor *in vitro* under anaerobic conditions [51] which is in line with the ability of the FR enzymes to use free $FADH_2$ as cofactor and their essential nature under such conditions. As *S. cerevisiae* possesses one gene encoding a mitochondrial FAD transporter (*FLX1*) [52] and three genes encoding putative transporters associated with the ER (*FLC1-3*) [53], FAD-exchange between compartments should indeed be possible. This highlights free FAD as a potentially crucial metabolite for sustaining anaerobic growth.

Pdi1p is, together with Ero1p, essential for disulphide bond formation in the ER [54]. Although transcriptional analysis did not show any difference in expression of *PDI1* between glucose and xylose fermentation [27] (Additional file 2: Table S5), proteome analysis of a mutant strain with improved xylose-fermenting capability identified Pdi1p as significantly more abundant in the improved strain compared with the wild-type [55]. This strongly indicates that there indeed is a need for a higher capacity in the protein folding mechanism for efficient xylose fermentation. Whether this capacity is for regulation of disulphide bond formation [56], reduction of non-native disulphide bonds [57] or targeting unfolded glycoproteins for degradation [58] remains to be investigated.

Conclusions

Over-expression of *FRD1*, encoding the cytosolic FR, did not improve the growth rate on xylose, even when the strain was provided with all amino acids and an additional copy of the *ERO1* gene controlled by a strong promoter. Over-expression of the *OSM1* gene, encoding the mitochondrial FR, even had a detrimental effect and caused a reduced growth rate. Hence, we conclude that increasing the activities of the FR enzymes and Ero1p is not sufficient to increase the anaerobic growth on xylose. However, our transcription analysis of UPR related genes showed a wide down-regulation of the secretory pathway during xylose fermentation indicating that additional components of the protein folding mechanism may be limited. Two immediate factors should notably be investigated further: i) the actual form of *HAC1* inside the cells during xylose fermentation to determine the regulatory component and ii) the activity of Pdi1p which is acting together with Ero1p in the formation of disulphide bridges. Furthermore, understanding the requirement of free FAD for anaerobic growth could prove to be essential for the development of efficient xylose-utilizing *S. cerevisiae* strains.

Additional files

Additional file 1: Supplementary Figures S1-S5. Description: FR activity measurements, DTT sensitivity results, fermentation profiles.

Additional file 2: Supplementary Tables S1-S5. Description: Genes included in the cluster analysis, maximum specific growth rates and product yields using strains TMB3455, TMB3456 and TMB3458.

Abbreviations

AC: Acetate; ACALD: Acetaldehyde; ACCOA: Acetyl-CoA; AKG: Alpha-ketoglutarate; CoA: Coenzyme A; CIT: Citrate; ER: Endoplasmic reticulum; ETOH: Ethanol; FR: Fumarate reductase; FUM: Fumarate; ICIT: Isocitrate; MAL: L-malate; OAA: Oxaloacetate; PYR: Pyruvate; SUCC: Succinate; SUCCOA: Succinyl-CoA; UPR: Unfolded protein response; XDH: Xylitol dehydrogenase; XR: Xylose reductase.

Competing interests

The authors declare that they have no competing interests.

Authors' contribution

The study was conceived by BB who designed the experiments and supervised the master student who performed most of the molecular biology work. The enzymatic assay was developed by BB who also performed the fermentation experiments, analysed the data and wrote the manuscript. BB, MGG and EVN have discussed the results and implications throughout the work. MGG and EVN have commented on the manuscript at all stages. All authors read and approved the final manuscript.

Acknowledgements

We wish to acknowledge Mégane Cléret for her work in the initial characterization of strain TMB3456 and Tarinee Boonyawan for the amplification and cloning of the *OSM1* and *ERO1* genes. The Swedish National Energy Agency (Energimyndigheten, Project No. P35350-1) is gratefully acknowledged for financial support.

References

1. Visser W, Scheffers WA, der Vegte WH B-v, van Dijken JP: **Oxygen requirements of yeasts.** *Appl Environ Microbiol* 1990, **56**(12):3785–3792.
2. Jönsson L, Alriksson B, Nilvebrant N-O: **Bioconversion of lignocellulose: inhibitors and detoxification.** *Biotechnol Biofuels* 2013, **6**(1):16.
3. Festel GW: **Biofuels - economic aspects.** *Chem Eng Technol* 2008, **31**(5):715–720.
4. Girio FM, Fonseca C, Carvalheiro F, Duarte LC, Marques S, Bogel-Lukasik R: **Hemicelluloses for fuel ethanol: a review.** *Bioresour Technol* 2010, **101**(13):4775–4800.
5. Soccol CR, Faraco V, Karp S, Vandenberghe LPS, Thomaz-Soccol V, Woiciechowski A, Pandey A: **lignocellulosic bioethanol: current status and future perspectives.** In *Biofuels: alternative feedstocks and conversion processes.* Edited by Pandey A, Larroche C, Ricke SC, Dussap CG, Gnansounou E. Oxford, UK: Elsevier Ltd; 2011:101–122.
6. Sassner P, Galbe M, Zacchi G: **Techno-economic evaluation of bioethanol production from three different lignocellulosic materials.** *Biomass Bioenergy* 2008, **32**(5):422–430.
7. Jeffries TW: **Emerging technology for fermenting D-xylose.** *Trends Biotechnol* 1985, **3**(8):208–212.
8. Kötter P, Amore R, Hollenberg CP, Ciriacy M: **Isolation and characterization of the *Pichia stipitis* xylitol dehydrogenase gene, *XYL2*, and construction of a xylose-utilizing *Saccharomyces cerevisiae* transformant.** *Curr Genet* 1990, **18**(6):493–500.
9. Van Vleet JH, Jeffries TW: **Yeast metabolic engineering for hemicellulosic ethanol production.** *Curr Opin Biotechnol* 2009, **20**(3):300–306.
10. Hahn-Hägerdal B, Karhumaa K, Fonseca C, Spencer-Martins I, Gorwa-Grauslund MF: **Towards industrial pentose-fermenting yeast strains.** *Appl Microbiol Biotechnol* 2007, **74**(5):937–953.
11. Camarasa C, Faucet V, Dequin S: **Role in anaerobiosis of the isoenzymes for *Saccharomyces cerevisiae* fumarate reductase encoded by *OSM1* and *FRDS1*.** *Yeast* 2007, **24**(5):391–401.
12. Arikawa Y, Enomoto K, Muratsubaki H, Okazaki M: **Soluble fumarate reductase isoenzymes from *Saccharomyces cerevisiae* are required for anaerobic growth.** *FEMS Microbiol Lett* 1998, **165**(1):111–116.

13. Muratsubaki H, Enomoto K: **One of the fumarate reductase isoenzymes from *Saccharomyces cerevisiae* is encoded by the *OSM1* gene.** *Arch Biochem Biophys* 1998, **352**(2):175–181.
14. Enomoto K, Ohki R, Muratsubaki H: **Cloning and sequencing of the gene encoding the soluble fumarate reductase from *Saccharomyces cerevisiae*.** *DNA Res* 1996, **3**(4):263–267.
15. Rossi C, Hauber J, Singer TP: **Mitochondrial and cytoplasmic enzymes for the reduction of fumarate to succinate in yeast.** *Nature* 1964, **204**(4954):167–170.
16. Camarasa C, Grivet JP, Dequin S: **Investigation by ^{13}C-NMR and tricarboxylic acid (TCA) deletion mutant analysis of pathways for succinate formation in *Saccharomyces cerevisiae* during anaerobic fermentation.** *Microbiology* 2003, **149:**2669–2678.
17. Enomoto K, Arikawa Y, Muratsubaki H: **Physiological role of soluble fumarate reductase in redox balancing during anaerobiosis in *Saccharomyces cerevisiae*.** *FEMS Microbiol Lett* 2002, **215**(1):103–108.
18. Frand AR, Kaiser CA: **The *ERO1* gene of yeast is required for oxidation of protein dithiols in the endoplasmic reticulum.** *Mol Cell* 1998, **1**(2):161–170.
19. Pollard MG, Travers KJ, Weissman JS: **Ero1p: A novel and ubiquitous protein with an essential role in oxidative protein folding in the endoplasmic reticulum.** *Mol Cell* 1998, **1**(2):171–182.
20. Freedman RB: **Protein disulfide isomerase - multiple roles in the modification of nascent secretory proteins.** *Cell* 1989, **57**(7):1069–1072.
21. Barlowe CK, Miller EA: **Secretory protein biogenesis and traffic in the early secretory pathway.** *Genetics* 2013, **193**(2):383–410.
22. Frand AR, Kaiser CA: **Ero1p oxidizes protein disulfide isomerase in a pathway for disulfide bond formation in the endoplasmic reticulum.** *Mol Cell* 1999, **4**(4):469–477.
23. Tu BP, Ho-Schleyer SC, Travers KJ, Weissman JS: **Biochemical basis of oxidative protein folding in the endoplasmic reticulum.** *Science* 2000, **290**(5496):1571–1574.
24. Tu BP, Weissman JS: **The FAD- and O_2-dependent reaction cycle of Ero1-mediated oxidative protein folding in the endoplasmic reticulum.** *Mol Cell* 2002, **10**(5):983–994.
25. Bernales S, Papa FR, Walter P: **Intracellular signaling by the unfolded protein response.** *Annu Rev Cell Dev Biol* 2006, **22:**487–508.
26. Travers KJ, Patil CK, Wodicka L, Lockhart DJ, Weissman JS, Walter P: **Functional and genomic analyses reveal an essential coordination between the unfolded protein response and ER-associated degradation.** *Cell* 2000, **101**(3):249–258.
27. Runquist D, Hahn-Hägerdal B, Bettiga M: **Increased expression of the oxidative pentose phosphate pathway and gluconeogenesis in anaerobically growing xylose-utilizing *Saccharomyces cerevisiae*.** *Microb Cell Fact* 2009, **8:**49.
28. Salusjärvi L, Pitkänen JP, Aristidou A, Ruohonen L, Penttilä M: **Transcription analysis of recombinant *Saccharomyces cerevisiae* reveals novel responses to xylose.** *Appl Biochem Biotechnol* 2006, **128**(3):237–261.
29. Jin YS, Laplaza JM, Jeffries TW: ***Saccharomyces cerevisiae* engineered for xylose metabolism exhibits a respiratory response.** *Appl Environ Microbiol* 2004, **70**(11):6816–6825.
30. Wahlbom CF, Otero RRC, van Zyl WH, Hahn-Hägerdal B, Jönsson LJ: **Molecular analysis of a *Saccharomyces cerevisiae* mutant with improved ability to utilize xylose shows enhanced expression of proteins involved in transport, initial xylose metabolism, and the pentose phosphate pathway.** *Appl Environ Microbiol* 2003, **69**(2):740–746.
31. Bergdahl B, Heer D, Sauer U, Hahn-Hägerdal B, van Niel EW: **Dynamic metabolomics differentiates between carbon and energy starvation in recombinant *Saccharomyces cerevisiae* fermenting xylose.** *Biotechnol Biofuels* 2012, **5**(1):34.
32. Mumberg D, Muller R, Funk M: **Yeast vectors for the controlled expression of heterologous proteins in different genetic backgrounds.** *Gene* 1995, **156**(1):119–122.
33. Gietz RD, Sugino A: **New yeast-*Escherichia coli* shuttle vectors constructed with *in vitro* mutagenized yeast genes lacking six-base pair restriction sites.** *Gene* 1988, **74**(2):527–534.
34. Runquist D, Fonseca C, Rådstrom P, Spencer-Martins I, Hahn-Hägerdal B: **Expression of the Gxf1 transporter from *Candida intermedia* improves fermentation performance in recombinant xylose-utilizing *Saccharomyces cerevisiae*.** *Appl Microbiol Biotechnol* 2009, **82**(1):123–130.
35. Runquist D, Hahn-Hägerdal B, Bettiga M: **Increased ethanol productivity in xylose-utilizing *Saccharomyces cerevisiae* via a randomly mutagenized xylose reductase.** *Appl Environ Microbiol* 2010, **76**(23):7796–7802.
36. Karhumaa K, Hahn-Hägerdal B, Gorwa-Grauslund MF: **Investigation of limiting metabolic steps in the utilization of xylose by recombinant *Saccharomyces cerevisiae* using metabolic engineering.** *Yeast* 2005, **22:**359–368.
37. Sambrook J, Fritsch EF, Maniatis T: *Molecular cloning: A laboratory manual.* Cold Spring Harbor Laboratory Press: Cold Spring Harbor; 1989.
38. Gietz RD, Schiestl RH: **High-efficiency yeast transformation using the LiAc/SS carrier DNA/PEG method.** *Nat Protoc* 2007, **2**(1):31–34.
39. Gasch AP, Spellman PT, Kao CM, Carmel-Harel O, Eisen MB, Storz G, Botstein D, Brown PO: **Genomic expression programs in the response of yeast cells to environmental changes.** *Mol Biol Cell* 2000, **11**(12):4241–4257.
40. Roberts GG, Hudson AP: **Transcriptome profiling of *Saccharomyces cerevisiae* during a transition from fermentative to glycerol-based respiratory growth reveals extensive metabolic and structural remodeling.** *Mol Genet Genomics* 2006, **276**(2):170–186.
41. Rønnow B, Kielland-Brandt MC: ***GUT2*, a gene for mitochondrial glycerol 3-phosphate dehydrogenase of *Saccharomyces cerevisiae*.** *Yeast* 1993, **9**(10):1121–1130.
42. Grauslund M, Rønnow B: **Carbon source-dependent transcriptional regulation of the mitochondrial glycerol-3-phosphate dehydrogenase gene, *GUT2*, from *Saccharomyces cerevisiae*.** *Can J Microbiol* 2000, **46**(12):1096–1100.
43. Fleck CB, Brock M: **Re-characterisation of *Saccharomyces cerevisiae* Ach1p: fungal CoA-transferases are involved in acetic acid detoxification.** *Fungal Genet Biol* 2009, **46**(6–7):473–485.
44. Hahn-Hägerdal B, Hallborn J, Jeppsson H, Meinander N, Walfridsson M, Ojamo H, Penttilä M, Zimmermann FK: **Redox balances in recombinant *Saccharomyces cerevisiae*.** In *Ann N Y Acad Sci, vol. 782.* Edited by Asenjo JA, Andrews BA, Asenjo JA, Andrews BA. New York: New York Acad Sciences; 1996:286–296.
45. Wahlbom CF, Hahn-Hägerdal B: **Furfural, 5-hydroxymethyl furfural, and acetoin act as external electron acceptors during anaerobic fermentation of xylose in recombinant *Saccharomyces cerevisiae*.** *Biotechnol Bioeng* 2002, **78**(2):172–178.
46. Sonderegger M, Jeppsson M, Hahn-Hägerdal B, Sauer U: **Molecular basis for anaerobic growth of *Saccharomyces cerevisiae* on xylose, investigated by global gene expression and Metabolic Flux Analysis.** *Appl Environ Microbiol* 2004, **70**(4):2307–2317.
47. Salusjärvi L, Kaunisto S, Holmström S, Vehkomäki M-L, Koivuranta K, Pitkänen J-P, Ruohonen L: **Overexpression of NADH-dependent fumarate reductase improves D-xylose fermentation in recombinant *Saccharomyces cerevisiae*.** *J Ind Microbiol Biotechnol* 2013, **40:**1383–1392.
48. Bradley PH, Brauer MJ, Rabinowitz JD, Troyanskaya OG: **Coordinated concentration changes of transcripts and metabolites in *Saccharomyces cerevisiae*.** *PLoS Comput Biol* 2009, **5**(1):e1000270.
49. Abdulrehman D, Monteiro PT, Teixeira MC, Mira NP, Lourenço AB, dos Santos SC, Cabrito TR, Francisco AP, Madeira SC, Aires RS, Oliveira AL, Sá-Correia I, Freitas AT: **YEASTRACT: providing a programmatic access to curated transcriptional regulatory associations in *Saccharomyces cerevisiae* through a web services interface.** *Nucleic Acids Res* 2011, **39**(suppl 1):D136–D140.
50. Herrero E, Ros J, Bellí G, Cabiscol E: **Redox control and oxidative stress in yeast cells.** *BBA General Subjects* 2008, **1780**(11):1217–1235.
51. Gross E, Sevier CS, Heldman N, Vitu E, Bentzur M, Kaiser CA, Thorpe C, Fass D: **Generating disulfides enzymatically: Reaction products and electron acceptors of the endoplasmic reticulum thiol oxidase Ero1p.** *Proc Natl Acad Sci U S A* 2006, **103**(2):299–304.
52. Tzagoloff A, Jang J, Glerum DM, Wu M: ***FLX1* codes for a carrier protein involved in maintaining a proper balance of flavin nucleotides in yeast mitochondria.** *J Biol Chem* 1996, **271**(13):7392–7397.
53. Protchenko O, Rodriguez-Suarez R, Androphy R, Bussey H, Philpott CC: **A screen for genes of heme uptake identifies the *FLC* family required for import of FAD into the endoplasmic reticulum.** *J Biol Chem* 2006, **281**(30):21445–21457.
54. Farquhar R, Honey N, Murant SJ, Bossier P, Schultz L, Montgomery D, Ellis RW, Freedman RB, Tuite MF: **Protein disulfide isomerase is essential for viability in *Saccharomyces cerevisiae*.** *Gene* 1991, **108**(1):81–89.
55. Karhumaa K, Påhlman A-K, Hahn-Hägerdal B, Levander F, Gorwa-Grauslund M-F: **Proteome analysis of the xylose-fermenting mutant yeast strain TMB3400.** *Yeast* 2009, **26**(7):371–382.
56. Kim S, Sideris DP, Sevier CS, Kaiser CA: **Balanced Ero1 activation and inactivation establishes ER redox homeostasis.** *J Cell Biol* 2012, **196**(6):713–725.

57. Laboissiere MCA, Sturley SL, Raines RT: **The essential function of protein-disulfide isomerase is to unscramble nonnative disulfide bonds.** *J Biol Chem* 1995, **270**(47):28006–28009.
58. Gauss R, Kanehara K, Carvalho P, Ng DTW, Aebi M: **A complex of Pdi1p and the mannosidase Htm1p initiates clearance of unfolded glycoproteins from the endoplasmic reticulum.** *Mol Cell* 2011, **42**(6):782–793.

Production and characterization of *in planta* transiently produced polygalacturanase from *Aspergillus niger* and its fusions with hydrophobin or ELP tags

Eridan Orlando Pereira[1,2,4], Igor Kolotilin[1], Andrew Jonathan Conley[3] and Rima Menassa[1,2*]

Abstract

Background: Pectinases play an important role in plant cell wall deconstruction and have potential in diverse industries such as food, wine, animal feed, textile, paper, fuel, and others. The demand for such enzymes is increasing exponentially, as are the efforts to improve their production and to implement their use in several industrial processes. The goal of this study was to examine the potential of producing polygalacturonase I from *Aspergillus niger* in plants and to investigate the effects of subcellular compartmentalization and protein fusions on its accumulation and activity.

Results: Polygalacturonase I from *Aspergillus niger* (*An*PGI) was transiently produced in *Nicotiana benthamiana* by targeting it to five different cellular compartments: apoplast, endoplasmic reticulum (ER), vacuole, chloroplast and cytosol. Accumulation levels of 2.5%, 3.0%, and 1.9% of total soluble protein (TSP) were observed in the apoplast, ER, and vacuole, respectively, and specific activity was significantly higher in vacuole-targeted *An*PGI compared to the same enzyme targeted to the ER or apoplast. No accumulation was found for *An*PGI when targeted to the chloroplast or cytosol. Analysis of *An*PGI fused with elastin-like polypeptide (ELP) revealed a significant increase in the protein accumulation level, especially when targeted to the vacuole where the protein doubles its accumulation to 3.6% of TSP, while the hydrophobin (HFBI) fusion impaired *An*PGI accumulation and both tags impaired activity, albeit to different extents. The recombinant protein showed activity against polygalacturonic acid with optimum conditions at pH 5.0 and temperature from 30 to 50°C, depending on its fusion. *In vivo* analysis of reducing sugar content revealed a higher release of reducing sugars in plant tissue expressing recombinant *An*PGI compared to wild type *N. benthamiana* leaves.

Conclusion: Our results demonstrate that subcellular compartmentalization of enzymes has an impact on both the target protein accumulation and its activity, especially in the case of proteins that undergo post-translational modifications, and should be taken into consideration when protein production strategies are designed. Using plants to produce heterologous enzymes for the degradation of a key component of the plant cell wall could reduce the cost of biomass pretreatment for the production of cellulosic biofuels.

Keywords: Polygalacturonase, *Aspergillus niger*, *Nicotiana benthamiana*, Elastin-like polypeptide, Hydrophobin I, Cellulosic biofuels

* Correspondence: Rima.Menassa@agr.gc.ca
[1]Agriculture and Agri-Food Canada, 1391 Sandford Street, London, ON N5V 4T3, Canada
[2]Department of Biology, Western University, London, ON N6A 5B7, Canada
Full list of author information is available at the end of the article

Background

The rising demand for sustainable energy in the world and the availability of abundant biomass have attracted an increasing interest in bioconversion of plant cell walls into ethanol and other biofuels [1]. There are many challenges to developing the capability for producing ethanol in an effective and economical fashion. After years of research on the production of ethanol from lignocellulosic biomass, one of the key impediments in this field is the recalcitrance of the plant cell wall to breakdown. The digestibility of cellulose present in lignocellulosic biomass is hindered by many physicochemical, structural, and compositional factors [2] and current pretreatments rely on thermochemical technologies using high temperatures, toxic acids, peroxides, and ammonia. The goal of pretreatment is to break down the physical structure of the cell wall and make the crystalline structure of cellulose more accessible to cellulases. This pretreatment process may account for up to 30% of the cost of biofuel production [1]. Enzymatic pretreatment represents a more promising and environmentally friendly technology.

The plant cell wall is mainly composed of cellulose, hemicellulose, lignin, pectin and smaller amounts of several other inorganic materials [3]. A cell wall component that, particularly in dicots, is critical for tissue integrity and accessibility to cell wall-degrading enzymes (CWDEs) is the cohesive pectin matrix embedding the cellulose-hemicellulose network that confers rigidity to the cell wall [4]. Pectin is the most complex class of plant cell wall polysaccharides comprised mainly of three pectic polysaccharides named homogalacturonan (HG), rhamnogalacturonan I and rhamnogalacturonan II [5]. Complete pectin degradation requires a battery of pectinases, including pectate lyases, polygalacturonases (PGs) and rhamogalacturonases, pectin methylesterases and pectin acetylesterases. However polygalacturonases, the enzymes responsible for catalyzing hydrolysis of α-1,4-glycosidic linkage in the α-(1,4)-linked D-galactopyranosyluronic acid (Gal*p* A) residues of HG, are considered a key factor in plant tissue maceration, especially during phytopathogen infection [6]. It has been observed that the addition of a pectinase cocktail, which has PG as a component, to cellulase cocktails increases the yields of glucose by 7.5% [7]. However, due to the different degree of acetylation of the Gal*p* A units in the HG region it is essential that the candidate enzyme presents a certain degree of flexibility for efficient hydrolysis. *In silico* models in agreement with mass spectrometry studies using 3 different PGs demonstrated that PG I from *Aspergillus niger* (*An*PGI) has a higher tolerance towards acetylated pectin [8].

The cost of biomass degrading enzymes is widely considered an important factor in the commercialization of lignocellulosic biomass-to-ethanol processes and the use of microorganisms in the production of enzymes is the main platform currently available. One of the reasons for high enzyme cost is the need for high volume bioreactors which substantially increase the cost of production. The use of plants represents an alternative for enzyme production [9]. Transient expression of recombinant proteins in *Nicotiana benthamiana* constitutes an ideal system for screening and analysis of recombinant enzymes due to its low cost and short time of production, where from cloning to expression data can be achieved in only two weeks. The most promising enzymes and constructs are then usually stably transformed into plants, which offer an almost unlimited scale-up potential. Nonetheless, depending on the purpose, large scale protein production using transient expression systems has been reported [10], where 450–750 kg of *N. benthamiana* biomass can be infiltrated in 8 hours, producing 1–10 g of recombinant protein/kg of fresh weight within 7–14 days [11].

Although the expression of CWDEs in plants has been demonstrated, accumulation levels in nuclear transformed plants were low [12] and due to the complexity of the plant cell wall, other enzymes remain to be explored using this system. Several strategies have been proposed to improve protein production in plants. Among them, protein fusions can address issues of stability and aid in purification [13]. Targeting heterologous proteins to the appropriate subcellular compartment can be critical for obtaining high levels of accumulation, since the structure and stability of the recombinant protein is affected by its route and final destination in the cell. In plant leaves, heterologous proteins have been typically targeted to the apoplast, ER, vacuole, chloroplasts and cytosol [14] by including a combination of targeting and retention signal sequences in the expression construct.

In the case of protein fusions, elastin-like polypeptide (ELP), a pentapeptide repeat polymer (Val-Pro-Gly-Xaa-Gly) that forms an aggregate above its transition temperature and hydrophobin (HFBI), a small and surface-active protein derived from filamentous fungi have proven valuable for improving recombinant protein accumulation in plants. Despite the fact that these peptides were originally designed for purification, they have also been shown to increase recombinant protein accumulation in plants [15-17].

Using a series of constructs targeting the polygalacturonase I from *Aspergillus niger* (*An*PGI) to different subcellular compartments, we report here the effects of subcellular targeting on the accumulation and activity of *An*PGI in leaves of *N. benthamiana*. We also analyze the effects of ELP and HBFI on protein accumulation in the ER and vacuole. Lastly we show through self-hydrolysis analysis that leaves producing *An*PGI have a higher content of reducing sugars (up to 20 fold more) in comparison with wild type leaves. These results provide an important step towards inexpensive production of cell wall degrading enzymes for the bioconversion of biomass into fermentable sugars.

Results

Analysis of heterologous polygalacturonase I transiently expressed in *N. benthamiana*

Several factors need to be taken into consideration for achieving high levels of recombinant protein accumulation in plants, including high transcript levels, correct post translational modifications, and protein turnover. To maintain high levels of transcripts, the double enhanced cauliflower mosaic virus (CaMV) 35S constitutive promoter was used in a series of *AnpgI* gene expression vectors that were designed to target the recombinant protein to five different subcellular compartments: apoplast, ER, vacuole, chloroplasts and cytosol (Figure 1A). Agrobacterium cultures containing each of these vectors were transiently co-infiltrated in *N. benthamiana* with an Agrobacterium culture containing a p19-encoding construct. P19 is a suppressor of post-transcriptional gene silencing from *Cymbidium* ringspot virus [18]. Leaves infiltrated only with p19 were used as the negative control.

Overexpression of recombinant proteins tends to result sometimes in chlorosis or even necrosis of the infiltrated leaf sectors [15]. Although pectic enzymes targeted to the apoplast in transgenic plants have been shown to impair plant growth and fitness [19], leaves infiltrated with the *An*PGI constructs did not show any chlorotic or necrotic phenotype by harvest time.

Analysis of transiently expressed protein in *N. benthamiana* was performed by western blot analysis of leaf total soluble protein extracts, using a monoclonal antibody against the c-Myc tag to determine *An*PGI accumulation.

Densitometry analysis showed high accumulation of *An*PGI at 2.5%, 3.0% and 1.9% of total soluble protein (TSP) in the apoplast, ER and vacuole respectively (Figure 2A). These analyses also demonstrated that the *An*PGI recombinant protein does not accumulate to detectable levels when targeted to the chloroplasts or cytosol of *N. benthamiana* leaves. Accumulation of *An*PGI only occurred in the secretory pathway, and the fact that *An*PGI has two *N*-glycosylation sites led us to ask if glycosylation may be related to stability of the protein.

Recombinant *An*PGI glycosylation analysis

To determine the presence of glycans, a deglycosylation experiment on purified proteins was performed by digesting *An*PGI with Endoglycosidase H (Endo H) followed by SDS-PAGE and western blot analysis. After deglycosylation with Endo H, the downshift in the band size indicates that *N*-glycans were incorporated in the protein (Figure 3). Since *N*-glycosylation does not occur in the cytosol or chloroplast, this data suggests that glycosylation could be essential for the *in vivo* stability of *An*PGI. A similar effect was observed for the expression of *Acidothermus cellulolyticus* E1 endo-β-1,4-glucanase in tobacco leaves [20]. Although in that case, the protein was observed in the cytosol and chloroplasts but with average levels of 0.004 and 0.0003% of TSP respectively, which was more than 100 fold lower than its accumulation in the secretory pathway.

Results of western blot analysis showed that plant-produced *An*PGI has a molecular mass of about 60 kDa

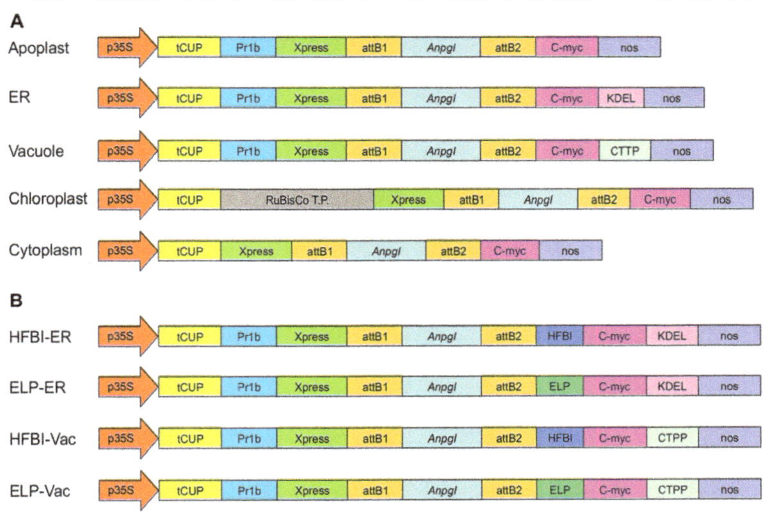

Figure 1 Schematic representation of the diverse *An*PGI heterologous expression constructs evaluated in this study: A) constructs used for protein targeting experiments and B) constructs used to analyze the effect of fusion tags on protein expression. p35S, double enhanced 35S promoter from Cauliflower Mosaic Virus 35S gene; tCUP, translation enhancer from the tobacco cryptic upstream promoter; nos, nopaline synthase transcription terminator; Pr1b, tobacco pathogenesis related 1b protein secretory signal peptide; C-myc, detection/purification tag; KDEL, endoplasmic reticulum retrieval tetrapeptide; CTPP, vacuole sorting peptide; RuBisCo T.P., rubisco small subunit transit peptide; HFBI, hydrophobin I; ELP, elastin-like polypeptide. Schematic not drawn to scale.

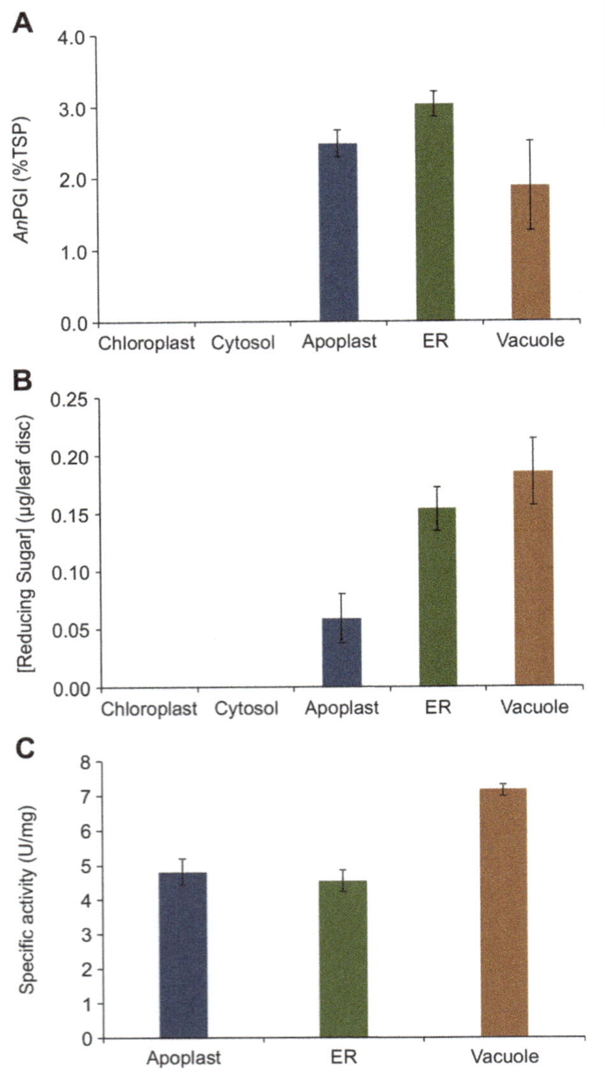

Figure 2 Accumulation and activity of *An*PGI in different subcellular compartments. A) Accumulation of *An*PGI in different subcellular compartments. **B)** A *N. benthamiana* agroinfiltrated leaf disc was incubated in 50 mM sodium acetate at 50°C for 24 h and used to analyze self hydrolysis and release of reducing sugars using the dinitro salicylic acid (DNS) method. **C)** Purified *An*PGI was used to confirm the difference in polygalacturonase activities when targeted to different subcellular compartments using polygalacturonic acid as the substrate. Accumulation results represent the average of *An*PGI levels in five different plants, and the release of reducing sugars and protein activity were determined in triplicate. Reducing sugar concentration was normalized using the assay results for wild-type plants. Error bars represent ± standard error.

(Figure 3). This molecular mass is substantially larger than the predicted molecular weight of *An*PGI constructs of 41.9 kDa (36.4 kDa of *An*PGI plus 5.5 kDa of the added tags). This result is in accordance with previous results reported in the literature, where a 55 kDa molecular mass was determined for *An*PGI by SDS-PAGE [21].

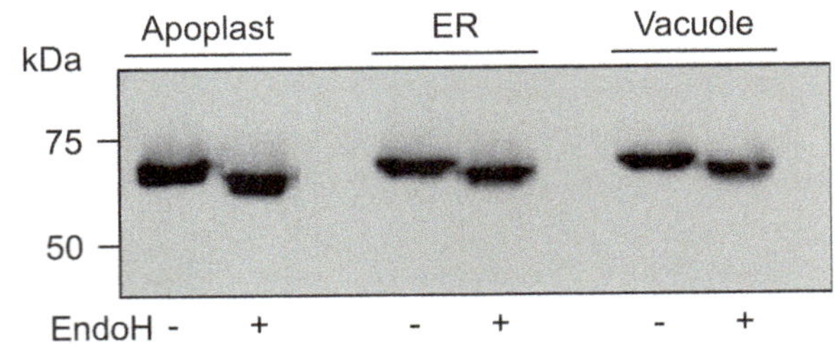

Figure 3 Deglycosylation analysis of plant-expressed *An*PGI targeted to different compartments. Purified *An*PGI was incubated for 3 h in denaturing conditions in the absence (–) or presence (+) of EndoH, analysed by SDS-PAGE under reducing conditions, and subjected to Western blot analysis with anti-C-myc antibody.

This difference may reflect the acidic nature of *An*PGI protein, which has an estimated pI of 4.0.

The influence of subcellular targeting on *An*PGI activity

N. benthamiana leaves producing *An*PGI were analyzed for self-hydrolysis by incubating leaf discs at 50°C for 24 h and using the dinitrosalicylic acid (DNS) assay. As expected, no significant difference in the release of reducing sugars was observed in leaves infiltrated with constructs targeting the chloroplasts or cytosol, compared with the negative control infiltrated with p19 alone. However, reducing sugars were released from tissues accumulating *An*PGI in the apoplast, ER and vacuole. Also, a higher amount of reducing sugars was released when *An*PGI was targeted to the vacuole, despite the lower amount of accumulated protein compared with the apoplast and ER (Figure 2A and B). This result could mean that post-translational modifications specific to the vacuole might affect the activity of the enzyme, or that the acidic environment in the vacuole is more conducive to proper folding of *An*PGI, and led to a more active enzyme. This led us to determine the specific activity of purified *An*PGI targeted to the three compartments.

Transiently produced *An*PGI was purified by affinity chromatography using the c-Myc tag fused at the C-terminus of the protein, and purified protein was used for a comparative analysis of *An*PGI catalytic activity when localized to the apoplast, ER or vacuole. The results showed that indeed the specific activity of *An*PGI against polygalacturonic acid is 30% higher when the protein is targeted to the vacuole (Figure 2C) than when it is targeted to the apoplast or ER.

The effect of ELP and HFBI fusions on *An*PGI accumulation

Elastin-like polypeptides (ELPs) are pentapeptide repeat polymers of Val-Pro-Gly-Xaa-Gly, where the guest residue Xaa can be any amino acid except proline [22], that aggregate above a transition temperature, T_t. Due to this property, ELPs have been explored as fusion partners for an inexpensive non-chromatographic method for protein purification [16]. Besides their utility for purification, ELP fusions have been shown to increase accumulation levels of several heterologous proteins by 2- to 100-fold when expressed in plants [16,17].

Hydrophobins are small fungal proteins with a size of approximately 10 kDa, which contain a large proportion of hydrophobic residues and eight cysteines connected by disulfide bonds [23]. Due to their propensity to self-assemble into an amphipathic protein membrane at hydrophilic-hydrophobic interfaces, and their ability to alter the hydrophobicity of their fusion partners, hydrophobins have been explored for purification purposes [23]. Hydrophobins were also demonstrated to be useful in improving the accumulation of fusion partners [24]. For example, a hydrophobin I (HFBI) gene from *Trichoderma reesei* was used as a green fluorescent protein (GFP) fusion which led to an increase of GFP accumulation in *N. benthamiana* leaves from 18% to 38% of TSP [15].

To analyze the effects of ELP and HFBI fusions on *An*PGI accumulation, fusion constructs (Figure 1B) were transiently co-expressed in *N. benthamiana* with p19. All proteins were produced, and although some degradation was observed with the ELP-ER and HFBI-ER constructs, most of the protein was in full size form (Figure 4A). Plants analyzed 4 days post infiltration showed a higher accumulation of *An*PGI fusion with ELP in both the ER and vacuole when compared with unfused *An*PGI (Figure 4B). Although the ELP fusion enhanced *An*PGI accumulation in both compartments, the effect of the fusion was more significant in the vacuole, where the ELP enhanced the accumulation of *An*PGI by almost 2-fold from 1.9 to 3.6% of TSP (Figure 4B). The presence of HFBI on the other hand was deleterious to the accumulation of *An*PGI in both compartments, with very low accumulation in the vacuole.

To further characterize the effects of the protein fusions, we assessed *in planta* self-hydrolysis and enzymatic activity for *An*PGI::ELP and *An*PGI::HFBI. We found that the presence of protein fusions impaired the activity of *An*PGI resulting in a reduction of the amount of reducing sugars released *in planta* (Figure 4C). The same reduction was confirmed with the purified enzyme against its substrate, polygalacturonic acid (Figure 4D).

Biochemical characterization of recombinant *An*PGI and its fusions

Temperature and pH optima of purified plant-produced *An*PGI were determined and were compared with previously described results of *Aspergillus niger*-produced enzyme [25]. Recombinant *An*PGI, independent of its subcellular localization, showed similar biochemical properties as the native *Aspergillus* enzyme with respect to its pH optimum (pH 5.0) (Figure 5). However, the temperature optimum was found to be at least 10°C lower than results previously reported for native *An*PGI [25], presenting 80-100% activity at a broad temperature range, from 20-40°C (Figure 6A).

Temperature and pH optima of the ELP and HFBI fusion proteins were compared with those of unfused *An*PGI. No differences of activity at various pH values were observed, with all proteins showing a pH optimum at pH 5.0 (Figure 5). However, the fused *An*PGI recovered its native temperature optimum of 50°C and displayed a narrower temperature range than the unfused enzyme (Figure 6B).

Discussion

Because of the increased interest in the conversion of biomass to biofuels, several enzymes useful in the bioconversion of plant tissues to biofuels have already been expressed using plants as bioreactors. However, the main emphasis until now has been on the expression of cellulases and xylanases targeting cellulose and hemicellulose [26]. Pectins, especially in dicots, compose 35% of primary plant cell walls and up to 5% of walls in wood tissue [27] and are critical for tissue integrity and accessibility to cell wall-degrading enzymes (CWDEs). Previous studies have focused on the heterologous expression of polygalacturonase as a bioregulator for pathogen defense or as a modifier of plant cell wall pectin structure [4,28]. To test the potential of plants as bioreactors for the production of polygalacturonase, we produced *Aspergillus niger* polygalacturonase I (*An*PGI) in *Nicotiana benthamiana* and characterized the produced enzyme.

The subcellular accumulation of *An*PGI was higher in the ER, followed by the apoplast and vacuole while no detectable *An*PGI was found when targeted to the chloroplasts or cytosol. These results confirm the impact of different subcellular compartments on the accumulation of cell wall degrading enzymes, especially in the case of a protein that requires post-translational modifications. Similar results have been previously described for another *Aspergillus niger* cell wall degrading enzyme, β-glucosidase I, produced in stably transformed tobacco plants [29]. β-glucosidase I accumulated to very low levels when targeted to the chloroplasts or cytosol, and was not functional possibly due to lack of glycosylation [29]. The glycosylation process is initiated during protein translocation into the ER and the addition of *N*-glycans to the protein has been linked to diverse roles, including stability against denaturation and proteolysis, fine-tuning the charge and isoelectric point, regulating protein turnover [30], and has also been linked to protein activity [31]. Although some glycoproteins may accumulate well in the cytosol [32] and chloroplasts [33], and perform the required biological functions, others such as *An*PGI

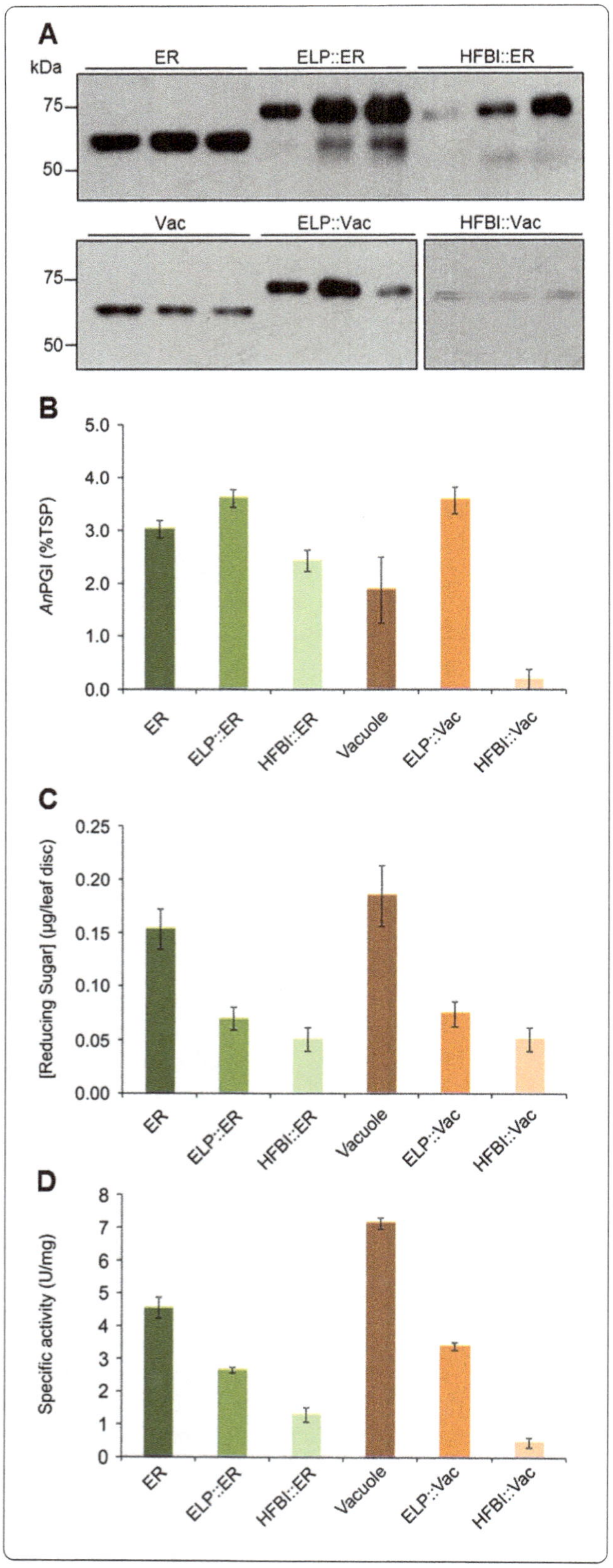

Figure 4 Accumulation and activity of *An*PGI and its fusions with ELP and HFBI targeted to the ER and vacuole. A) Western blot analysis of AnPGI and its fusions targeted to the ER or the vacuole. Each lane was loaded with 5 µg TSP from three replicates, and detected with anti c-myc antibody. **B)** The effect of fusion partner on the accumulation of *An*PGI. **C)** A *N. benthamiana* leaf disc agroinfiltrated with *An*PGI and its fusions was incubated in 50 mM of sodium acetate at 50°C for 24 h and used to analyze self hydrolysis and release of reducing sugars using the dinitro salicylic acid (DNS) method. **D)** The specific activity against polygalacturonic acid was used to analyse the effect of ELP and HFBI fusions on *An*PGI activity. Accumulation results represent the average of *An*PGI levels in five different plants, and the release of reducing sugars and protein activity were determined in triplicate. Reducing sugar concentration was normalized using the assay results for wild-type plants. Error bars represent ± standard error.

may require glycosylation for stability and will not accumulate in those compartments.

The subcellular location of *An*PGI within the secretory pathway had a distinct effect on its specific activity. *An*PGI targeted to the vacuole showed 30% higher specific activity than *An*PGI targeted to the apoplast or ER. This result is consistent with at least two other CWDEs produced in sugar cane, where activities of cellobiohydrolase I (CBH I) and CBH II were, respectively, five and two times more active when targeted to the vacuole [34] when compared with ER-targeted proteins. As well, the effect of subcellular targeting on protein activity has been observed with human gastric lipase produced in tobacco; however, in this case, the vacuolar-targeted enzyme exhibited two-fold lower specific activity compared with the same protein targeted to the apoplast or ER, probably due to proteolytic-induced instability of this specific protein when targeted to the vacuole [35]. In the case of glycosyl hydrolases, higher specific activity was expected when targeted to the vacuole, as proteomic analysis of the vegetative vacuole of *Arabidopsis thaliana* uncovered several glycosyl hydrolases, including a polygalacturonase [36], which indicates that the vacuole is a natural compartment for these enzymes and may provide a chemical environment that favors appropriate protein folding and activity. Therefore, and although most of the attention when analyzing protein expression in plants has been directed towards protein accumulation, protein activity may play a bigger role in the final analysis, especially when compartmentalization results in protein with low or no activity, which in many cases is due to a lack of post-translational modifications or protein truncation [29]. When compared to the previously reported specific activity of native *An*PGI [21], plant-produced *An*PGI showed 75 times lower specific activity. However, direct comparison of specific activities of enzymes characterized in different studies is problematic due to experimental differences as was pointed out by Kester et al. [21] and a side-by-side comparison would need to be conducted.

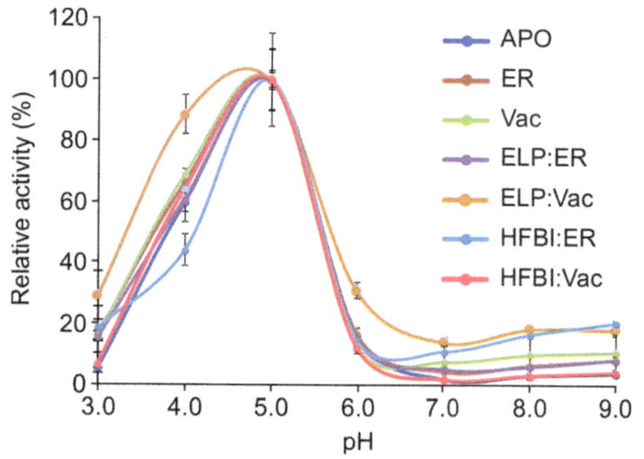

Figure 5 The pH profile of *An*PGI transiently expressed in *N. benthamiana* leaves. The purified polygalacturonase, *An*PGI and its fusions with ELP or HFBI were assayed at pH ranging from 3.0 to 6.0, in 50 mM sodium acetate buffer and at pH 6.0 to 9.0, in 50 mM sodium-phosphate buffers at 40°C. Each point was determined in triplicate and shown as the average ± standard error.

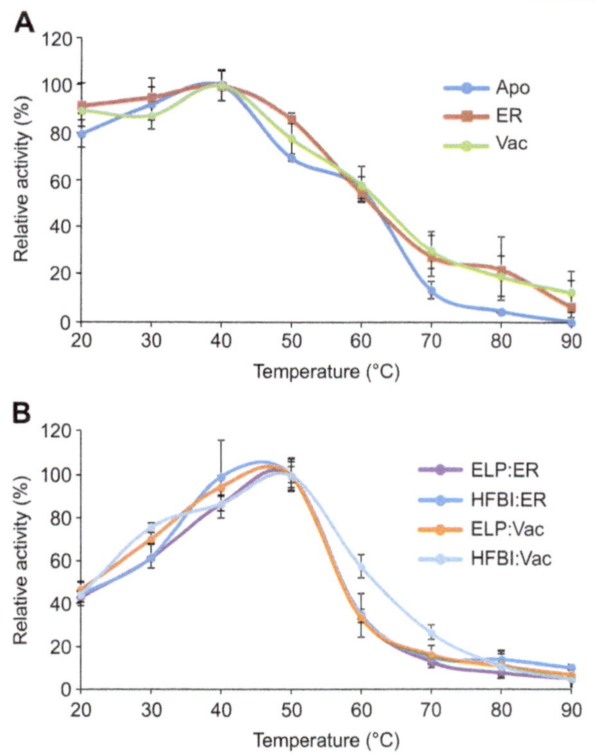

Figure 6 Temperature profile of *An*PGI transiently expressed in *N. benthamiana* leaves. A) Purified polygalacturonase I targeted to different subcellular compartments and **B)** *An*PGI fusions with ELP or HFBI targeted to the ER and vacuole. The enzymes were assayed at pH 5.0 and determined as the release of reducing ends of the 0.5% polygalacturonan solution in 50 mM sodium acetate buffer. The highest polygalacturonase activity is set to 100%. Each point was determined in triplicate and shown as the average ± standard error.

Recently, diverse protein fusions have been explored to increase the level of heterologous protein production in plants using technologies such as Zera [37], ELPs [16,38] and HBFI [15]. The advantage of using these protein fusions is that besides enhancing protein accumulation in plants, they also provide a means for their purification. In the study presented here, ELP and HFBI tags were fused to *An*PGI and targeted to the ER and vacuole. *An*PGI::ELP transient expression in *Nicotiana benthamiana* leaves led to an increase in accumulation in both compartments, while *An*PGI::HFBI transient expression led to a small reduction in the ER and to a drastic reduction in the vacuole. This is the first report of ELP and HFBI fusions targeted to the vacuole and the observation that ELP increases accumulation while HFBI reduces accumulation is interesting and should be further investigated. One of the hypotheses for the lower accumulation of *An*PGI::HFBI is that HFBI, which is a globular protein with four disulfide bonds [39], might be affecting *An*PGI folding, leading the misfolded protein to ER-associated protein degradation (ERAD). Specifically for this study, a flexible linker consisting of $(GGGS)_3$ was used to separate HFBI from *An*PGI and was probably not effective enough in separating the two protein domains. One of the possible solutions for this problem is the optimization of the linker used to separate the two proteins. Optimization of linkers has been shown to improve both protein expression yield and biological activity [40]. A recent study comparing several linkers demonstrated the need for empirical evaluation of different linkers, and the possible beneficial effect of twistable linkers on activity of the fusion partners, possibly by allowing a reorientation of the functional domains [32]. Moreover, the addition of ELP and HFBI fusions had adverse effects on enzyme activity. Although these results were not predicted, they were not completely unexpected. Previous studies using ELP tags have shown different outcomes on protein activity [38,41] and impact of fusion tags on recombinant protein activity needs to be assessed on a case-by-case basis.

More specifically in the case of *An*PGI::ELP targeted to the vacuole, because the loss of activity was directly proportional to the increase in accumulation, the ELP fusion may still offer a viable option in the production of this enzyme for industrial purposes. ELPs are thermally responsive synthetic biopolymers [22] that are valuable for simple nonchromatographic bioseparation of recombinant proteins. Because of this property, and taking into consideration the amount of enzyme that will be necessary for the production of soluble sugars, the use of this technology could represent a potential solution to lower the enzyme cost in the process by recycling the enzyme of interest and should be further investigated.

As the HFBI fusion can also be used for bioseparation [24], the same mechanism could be used for recycling the enzyme; however in this case, as HFBI showed a detrimental effect on both protein accumulation and activity of *An*PGI, improvements to the construct design for increasing accumulation of the protein fusion should be assessed prior to further developing this technology.

Although in the long term, *An*PGI should be produced directly into energy crops to reduce the costs of production of second generation biofuels, the results achieved here represent a step forward towards the development of this technology. The selection of the best compartment to store and maintain bioactive enzymes was demonstrated, no necrosis was observed in plants producing *An*PGI, indicating that this enzyme was not toxic at the accumulation levels obtained and can be potentially produced in stable transgenic plants. However, as the expression of pectic enzymes in the apoplast has been shown to affect plant growth and fitness [19,42] , genetic engineering strategies like the use of inducible promoters should be explored in order to avoid negative effects on plant growth. Further, storing this enzyme in the ER or vacuole of stably transformed plants could prevent negative effects on growth and should be explored. Moreover, the combination of plant-produced cell wall degrading enzymes with an ELP fusion can be promising for reducing the costs associated with enzymatic deconstruction of plant cell walls.

Conclusions

In conclusion, this study demonstrated the feasibility of using plants as bioreactors for the production of active *An*PGI at high levels. Both the accumulation and bioactivity of plant-produced *An*PGI were affected by different sub-cellular targeting of the recombinant protein, as well as by fusion with ELP and HFBI. Such analyses allow us to select the best combination of a sub-cellular compartment and a fusion partner for production of enzymes in plants, thus maximizing their catalytic potential in downstream applications.

Methods

Anpgl plant expression vectors

The polygalacturanase I gene from *A. niger* (Accession # XP_001389562) was chemically synthesized (GenScript, Piscateway, NJ, USA) (*Anpgl*) and inserted into a series of pCaMGate plant binary expression vectors (Conley et al., in preparation) using the Gateway technology® to generate 9 expression constructs targeting 5 different compartments (apoplast, ER, vacuole, chloroplast and cytosol) and ELP and HFBI fusions targeting the ER and vacuole compartments. To produce these vectors, first *Anpgl* was cloned into the Gateway donor vector pDONR/Zeo™ (Invitrogen, Carlsbad, USA) and the integrity of the construct was validated by sequence analysis. Using the Gateway cloning system, *Anpgl* was subsequently subcloned into the pCaMGate plant binary expression vectors (Conley et al., in preparation), derived from the pCaMterX [43] binary vector which placed the gene of interest under control of the double enhanced cauliflower mosaic virus 35S promoter [44] and the nopaline synthase (nos) terminator [45].

For constructs destined to the secretory pathway, downstream of the promoter, pCaMGate vectors harbor the tCUP translation enhancer [46], the Pr1b secretory signal peptide from tobacco [47] and the Xpress tag (for detection) followed by the *AnPgl* gene, a C-Myc detection/purification tag and a KDEL signal peptide, used for ER-retrieval. For the vacuole targeting construct, the C-terminal propeptide (CTPP) of tobacco chitinase vacuolar sorting signal [48] replaces the C-terminal KDEL; and no additonal sequence was used for secretion of *An*PGI to the apoplast. For the chloroplast-targeted construct, the transit peptide from the tobacco RuBisCo small subunit replaces the secretory signal peptide and for the cytosol no extra sequence was present downstream of the translation enhancer (Figure 1A). For the fusion constructs, ELP and HFBI are placed in-frame upstream of the c-Myc tag in the pCaMGate vectors, with a $(GSSS)_3$ linker separating *An*PGI from either ELP or HFBI (Figure 1B). The final expression clones were used to transform *Agrobacterium tumefaciens* strain EHA105 [49].

Transient expression in *N. benthamiana* and Western blot analysis

A suspension of *Agrobacterium tumefaciens* strain EHA105 carrying the expression construct was mixed with an equal amount of *Agrobacterium* culture containing the suppressor of post-transcriptional gene silencing p19 from *Cymbidium* ringspot virus [18] and co-infiltrated into leaves of 5–6 week old *N. benthamiana* plants through the stomata of abaxial leaf epidermis using a syringe [16]. Infiltrated plants were maintained in a controlled growth chamber for 2 to 9 days at 22°C, with a 16 h photoperiod. For each experiment, three leaf disks (7 mm diameter) from infiltrated tissue were collected from five different plants and ground in liquid nitrogen using 2.3 mm ceramic beads (BioSpec Products, 11079125z, Bartlesville, USA) in a TissueLyser (Qiagen®). Total soluble protein was extracted from the ground tissue in ice-cold phosphate- buffered saline (PBS), pH 7.4 supplemented with 0.1% Tween-20, 2% PVPP (polyvinyl polypyrrolidone), 1 mM EDTA (ethylenediaminetetraacetic acid), 100 mM sodium acorbate, 1 mM PMSF (phenylmethylsulfonyl fluoride) and 1 μg/ml leupeptin. Protein extraction from a control plant infiltrated with p19 as a negative control was also performed under similar conditions. Total soluble protein (TSP) concentration was spectrophotometrically determined using the Bradford assay [50] with bovine serum albumin as standard.

Equal volume of plant extract was separated by sodium dodecylsulphate-polyacrylamide gel electrophoresis (SDS-PAGE) (10%) and transferred to PVDF membrane. To detect the recombinant protein the membrane was incubated with primary mouse anti-C-myc monoclonal antibody (Genscript, A00864, Piscataway, USA). The primary antibody was detected with HRP-conjugated goat anti-mouse IgG antibody (Bio-Rad, 170–6516, Hercules, USA) and visualized using the ECL detection kit (GE healthcare, Mississauga, Canada) and autoradiography as described by the manufacturer. Western blots were analysed using image densitometry with TotalLab TL100 software (Nonlinear Dynamics, Durhan, USA). Intensities were determined by comparison with known amounts of a synthetic positive control protein containing a cellulose-binding domain and a C-myc tag (synthesized by Genscript, Piscataway, USA).

Protein purification

For the protein activity assays, total soluble protein was extracted from plants producing *An*PGI 4 days post infiltration (dpi) by grinding the agroinfiltrated samples in liquid nitrogen using ceramic beads in the TissueLyser (Qiagen®) as previously described. Samples were mixed with six volumes (w/v) of cold PBS buffer, and the homogenate was clarified twice by centrifugation (20,000 g, 10 min at 4°C) to obtain the total soluble protein (TSP). The c-Myc purification was performed by affinity chromatography using the c-Myc tagged Protein MILD PURIFICATION KIT (MBL, 3305, Woburn, USA) according to the manufacturer's instructions.

Deglycosylation analysis

Enzymatic deglycosylation of transiently produced *An*PGI targeted to the apoplast, ER, and vacuole was carried out on purified protein using Endoglycosidase H (Sigma-Aldrich, A 0810, St. Louis, USA) accordingly to the manufacturer's instructions. The Endo H specificity includes all high-mannose and hybrid type of glycans from *N*-linked glycoproteins. The digestion was carried out at 37°C for 3 h followed by SDS-PAGE under reducing conditions and western-blot analysis with anti-C-myc antibody.

Enzyme assays

In planta polygalacturonase activity was assayed by estimating the amount of reducing sugar released from a 7 mm diameter leaf disk incubated in 100 µl of 50 mM of sodium acetate solution pH 5.0 at 50°C for 24 h. The amount of reducing sugar released was quantified by the DNS method [51] and the reducing sugar concentration was normalized using the assay results for control plants. Specific activity was quantitatively estimated by measuring the hydrolysis of polygalacturonic acid using an equimolar quantity of *An*PGI. The reaction was carried out in 96 well flat bottom microplates using 0.5% polygalacturonic acid substrate in 50 mM sodium acetate buffer pH 5.0 incubated at 50°C for 30 min. Galacturonic acid was used as the standard and the amount of galacturonic acid released was quantified by using the DNS method [51]. One unit of PGase activity was defined as the amount of enzyme required to release one micromole of galacturonic acid per minute per mg of protein under standard assay conditions.

Effects of pH and temperature

The optimum enzyme pH was measured using polygalacturonic acid (0.5%) as the substrate in 50 mM acetate buffer, pH 3.0-6.0 and 50 mM sodium-phosphate buffer, pH 6.0-9.0 incubated at 40°C for 30 min. The optimum temperature was determined in the range of 20°C to 90°C in 50 mM sodium acetate buffer pH 5.0 incubated for 30 min.

Plant cell wall saccharification

Four millimeter diameter leaf discs were collected at 4 dpi and sterilized in a 1% sodium hypochlorite solution for 5 min and washed twice with sterilized water. The plant material was incubated at 50°C for 24 hours in a filter-sterilized solution containing 50 mM sodium acetate buffer, pH 5.0. Enzymatic saccharification efficiency was determined as the amount of reducing sugars released and compared with untreated plant material using the DNS assay.

Abbreviations

CWDE: Cell wall degrading enzyme; HG: Homogalacturonan; PG: Polygalacturonase; CBH: Cellobiohydrolase; *Anpgl*: Gene encoding polygalacturonase I protein from *Aspergillus niger*; HFBI: Hydrophobin I; ELP: Elastin-like polypeptide; TSP: Total soluble protein; SDS-PAGE: Sodium dodecylsulphate-polyacrylamide gel electrophoresis; ER: Endoplasmic reticulum; PVPP: Polyvinyl polypyrrolidone; PBS: Phosphate-buffered saline; EDTA: Ethylenediaminetetraacetic acid; PMSF: Phenylmethylsulfonyl fluoride; HRP: Horse radish peroxidase; IgG: Immunoglobulin G; DNS: Dinitro salicylic acid; tCUP: Tobacco cryptic upstream promoter; Pr1b: Pathogenesis-related protein 1b of tobacco; CTPP: C-terminal propeptide; nos: Nopaline synthase; Endo H: Endoglycosidase H; dpi: Day post infiltration.

Competing interests

The authors declare that they have no competing interests.

Authors' contributions

EOP designed the research, performed the experiments and drafted the manuscript, AJC designed the pCaMGate vectors, IK provided ideas and feedback and RM conceived the study and participated in its design. EOP, IK, AJC and RM edited the manuscript. All authors read and approved the final manuscript.

Acknowledgements

We thank Angelo Kaldis and Hong Zhu for technical support. We are grateful to Alex Molnar for help with preparing the figures. This study was supported by the Agriculture and Agri-Food Canada's A-base funding Program and the Agricultural Bioproducts Innovation Program through the Cellulosic Biofuels Network. The authors declare no conflict of interest.

Author details

[1]Agriculture and Agri-Food Canada, 1391 Sandford Street, London, ON N5V 4T3, Canada. [2]Department of Biology, Western University, London, ON N6A 5B7, Canada. [3]VTT Technical Research Centre of Finland, Tietotie 2, Espoo 02044, Finland. [4]Current address: Universidade Estadual do Ceará, Av. Dr. Silas Munguba, 1700 - Campus do Itaperi, Fortaleza 60714-903, Brazil.

References

1. Himmel ME, Ding SY, Johnson DK, Adney WS, Nimlos MR, Brady JW, Foust TD: **Biomass recalcitrance: engineering plants and enzymes for biofuels production.** *Science* 2007, **315:**804–807.
2. Kumar P, Barrett DM, Delwiche MJ, Stroeve P: **Methods for pretreatment of lignocellulosic biomass for efficient hydrolysis and biofuel production.** *Ind Eng Chem Res* 2009, **48:**3713–3729.
3. Jørgensen H, Kristensen JB, Felby C: **Enzymatic conversion of lignocellulose into fermentable sugars: challenges and opportunities.** *Biofuel Bioprod Bior* 2007, **1:**119–134.
4. Lionetti V, Francocci F, Ferrari S, Volpi C, Bellincampi D, Galletti R, D'Ovidio R, De Lorenzo G, Cervone F: **Engineering the cell wall by reducing de-methyl-esterified homogalacturonan improves saccharification of plant tissues for bioconversion.** *Proc Natl Acad Sci U S A* 2010, **107:**616–621.
5. Ridley BL, O'Neill MA, Mohnen D: **Pectins: structure, biosynthesis, and oligogalacturonide-related signaling.** *Phytochemistry* 2001, **57:**929–967.
6. Abbott DW, Boraston AB: **Structural biology of pectin degradation by Enterobacteriaceae.** *Microbiol Mol Biol Rev* 2008, **72:**301–316.
7. Zhang M, Su R, Qi W, He Z: **Enhanced enzymatic hydrolysis of lignocellulose by optimizing enzyme complexes.** *Appl Biochem Biotechnol* 2010, **160:**1407–1414.
8. Andre-Leroux G, Tessier D, Bonnin E: **Endopolygalacturonases reveal molecular features for processivity pattern and tolerance towards acetylated pectin.** *Biochim Biophys Acta* 2009, **1794:**5–13.
9. Carroll A, Somerville C: **Cellulosic biofuels.** *Annu Rev Plant Biol* 2009, **60:**165–182.
10. Lai H, Chen Q: **Bioprocessing of plant-derived virus-like particles of Norwalk virus capsid protein under current good manufacture practice regulations.** *Plant Cell Rep* 2012, **31:**573–584.
11. Pogue GP, Vojdani F, Palmer KE, Hiatt E, Hume S, Phelps J, Long L, Bohorova N, Kim D, Pauly M, Velasco J, Whaley K, Zeitlin L, Garger SJ, White E, Bai Y, Haydon H, Bratcher B: **Production of pharmaceutical-grade recombinant aprotinin and a monoclonal antibody product using plant-based transient expression systems.** *Plant Biotechnol J* 2010, **8:**638–654.
12. Yu LX, Gray BN, Rutzke CJ, Walker LP, Wilson DB, Hanson MR: **Expression of thermostable microbial cellulases in the chloroplasts of nicotine-free tobacco.** *J Biotechnol* 2007, **131:**362–369.
13. Conley AJ, Joensuu JJ, Richman A, Menassa R: **Protein body-inducing fusions for high-level production and purification of recombinant proteins in plants.** *Plant Biotechnol J* 2011, **9:**419–433.
14. Ahmad A, Pereira EO, Conley AJ, Richman AS, Menassa R: **Green biofactories: recombinant protein production in plants.** *Recent Pat Biotechnol* 2010, **4:**242–259.
15. Joensuu JJ, Conley AJ, Lienemann M, Brandle JE, Linder MB, Menassa R: **Hydrophobin fusions for high-level transient protein expression and purification in *Nicotiana benthamiana*.** *Plant Physiol* 2010, **152:**622–633.
16. Conley AJ, Joensuu JJ, Jevnikar AM, Menassa R, Brandle JE: **Optimization of elastin-like polypeptide fusions for expression and purification of recombinant proteins in plants.** *Biotechnol Bioeng* 2009, **103:**562–573.
17. Patel J, Zhu H, Menassa R, Gyenis L, Richman A, Brandle J: **Elastin-like polypeptide fusions enhance the accumulation of recombinant proteins in tobacco leaves.** *Transgenic Res* 2007, **16:**239–249.
18. Silhavy D, Molnar A, Lucioli A, Szittya G, Hornyik C, Tavazza M, Burgyan J: **A viral protein suppresses RNA silencing and binds silencing-generated, 21- to 25-nucleotide double-stranded RNAs.** *EMBO J* 2002, **21:**3070–3080.
19. Capodicasa C, Vairo D, Zabotina O, McCartney L, Caprari C, Mattei B, Manfredini C, Aracri B, Benen J, Knox JP, De Lorenzo G, Cervone F: **Targeted modification of homogalacturonan by transgenic expression of a fungal polygalacturonase alters plant growth.** *Plant Physiol* 2004, **135:**1294–1304.
20. Ziegelhoffer T, Raasch J, Austin-Phillips S: **Dramatic effects of truncation and sub-cellular targeting on the accumulation of recombinant microbial cellulase in tobacco.** *Mol Breed* 2001, **8:**147–158.
21. Kester HCM, Visser J: **Purification and characterization of polygalacturonases produced by the hyphal fungus aspergillus-niger.** *Biotechnol Appl Biochem* 1990, **12:**150–160.
22. Urry DW: **Entropic elastic processes in protein mechanisms. I. Elastic structure due to an inverse temperature transition and elasticity due to internal chain dynamics.** *J Protein Chem* 1988, **7:**1–34.
23. Linder MB, Szilvay GR, Nakari-Setala T, Penttila ME: **Hydrophobins: the protein-amphiphiles of filamentous fungi.** *FEMS Microbiol Rev* 2005, **29:**877–896.
24. Joensuu J, Conley A, Linder M, Menassa R: **Bioseparation of recombinant proteins from plant extract with hydrophobin fusion technology.** *Methods Mol Biol* 2012, **824:**527–534.
25. Lang C, Looman AC: **Efficient expression and secretion of *Aspergillus niger* RH5344 polygalacturonase in *Saccharomyces cerevisiae*.** *Appl Microbiol Biotechnol* 1995, **44:**147–156.
26. McDonald KA: ***In planta* production of cell wall degrading enzymes.** In *Handbook of Bioenergy Crop Plants.* Edited by Shonnard DR. Boca Raton, FL: CRC Press; 2012:55–73.
27. Mohnen D: **Pectin structure and biosynthesis.** *Curr Opin Plant Biol* 2008, **11:**266–277.
28. Ferrari S, Galletti R, Pontiggia D, Manfredini C, Lionetti V, Bellincampi D, Cervone F, De Lorenzo G: **Transgenic expression of a fungal endo-polygalacturonase increases plant resistance to pathogens and reduces auxin sensitivity.** *Plant Physiol* 2008, **146:**669–681.
29. Wei S, Marton I, Dekel M, Shalitin D, Lewinsohn E, Bravdo B-A, Shoseyov O: **Manipulating volatile emission in tobacco leaves by expressing *Aspergillus niger* β-glucosidase in different subcellular compartments.** *Plant Biotechnol J* 2004, **2:**341–350.
30. Helenius A, Aebi M: **Roles of N-linked glycans in the endoplasmic reticulum.** *Annu Rev Biochem* 2004, **73:**1019–1049.
31. Jeoh T, Michener W, Himmel M, Decker S, Adney W: **Implications of cellobiohydrolase glycosylation for use in biomass conversion.** *Biotechnol Biofuels* 2008, **1:**1–12.
32. Sainsbury F, Varennes-Jutras P, Goulet MC, D'Aoust MA, Michaud D: **Tomato cystatin SlCYS8 as a stabilizing fusion partner for human serpin expression in plants.** *Plant Biotechnol J* 2013, **11:**1058–1068.
33. Tran M, Zhou B, Pettersson PL, Gonzalez MJ, Mayfield SP: **Synthesis and assembly of a full-length human monoclonal antibody in algal chloroplasts.** *Biotechnol Bioeng* 2009, **104:**663–673.
34. Harrison MD, Geijskes J, Coleman HD, Shand K, Kinkema M, Palupe A, Hassall R, Sainz M, Lloyd R, Miles S, Dale JL: **Accumulation of recombinant cellobiohydrolase and endoglucanase in the leaves of mature transgenic sugar cane.** *Plant Biotechnol J* 2011, **9:**884–896.
35. Mokrzycki-Issartel N, Bouchon B, Farrer S, Berland P, Laparra H, Madelmont J-C, Theisen M: **A transient tobacco expression system coupled to MALDI-TOF-MS allows validation of the impact of differential targeting on structure and activity of a recombinant therapeutic glycoprotein produced in plants.** *FEBS Lett* 2003, **552:**170–176.
36. Carter C, Pan S, Zouhar J, Avila EL, Girke T, Raikhel NV: **The vegetative vacuole proteome of *Arabidopsis thaliana* reveals predicted and unexpected proteins.** *Plant Cell* 2004, **16:**3285–3303.
37. Llop-Tous I, Ortiz M, Torrent M, Ludevid MD: **The expression of a xylanase targeted to ER-protein bodies provides a simple strategy to produce active insoluble enzyme polymers in tobacco plants.** *PLoS One* 2011, **6:**e19474.
38. Floss DM, Sack M, Stadlmann J, Rademacher T, Scheller J, Stöger E, Fischer R, Conrad U: **Biochemical and functional characterization of anti-HIV antibody–ELP fusion proteins from transgenic plants.** *Plant Biotechnol J* 2008, **6:**379–391.
39. Hakanpaa J, Paananen A, Askolin S, Nakari-Setala T, Parkkinen T, Penttila M, Linder MB, Rouvinen J: **Atomic resolution structure of the HFBII hydrophobin, a self-assembling amphiphile.** *J Biol Chem* 2004, **279:**534–539.
40. Chen X, Zaro JL, Shen W-C: **Fusion protein linkers: Property, design and functionality.** *Adv Drug Deliver Rev* 2012, **65:**1357–1369.
41. Kaldis A, Ahmad A, Reid A, McGarvey B, Brandle J, Ma S, Jevnikar A, Kohalmi SE, Menassa R: **High-level production of human interleukin-10 fusions in tobacco cell suspension cultures.** *Plant Biotechnol J* 2013, **11:**535–545.

42. Biswal AK, Soeno K, Gandla ML, Immerzeel P, Pattathil S, Lucenius J, Serimaa R, Hahn MG, Moritz T, Jonsson LJ, Israelsson-Nordstrom M, Mellerowicz EJ: **Aspen pectate lyase PtxtPL1-27 mobilizes matrix polysaccharides from woody tissues and improves saccharification yield.** *Biotechnol Biofuels* 2014, **7**:11.
43. Harris LJ, Gleddie SC: **A modified *Rpl3* gene from rice confers tolerance of the *Fusarium graminearum* mycotoxin deoxynivalenol to transgenic tobacco.** *Physiol Mol Plant Pathol* 2001, **58**:173–181.
44. Kay R, Chan A, Daly M, McPherson J: **Duplication of CaMV 35S promoter sequences creates a strong enhancer for plant genes.** *Science* 1987, **236**:1299–1302.
45. Bevan M, Barnes WM, Chilton MD: **Structure and transcription of the nopaline synthase gene region of T-DNA.** *Nucleic Acids Res* 1983, **11**:369–385.
46. Wu K, Malik K, Tian L, Hu M, Martin T, Foster E, Brown D, Miki B: **Enhancers and core promoter elements are essential for the activity of a cryptic gene activation sequence from tobacco, tCUP.** *Mol Genet Genomics* 2001, **265**:763–770.
47. Cutt JR, Dixon DC, Carr JP, Klessig DF: **Isolation and nucleotide sequence of cDNA clones for the pathogenesis-related proteins PR1a, PR1b and PR1c of *Nicotiana tabacum* cv. Xanthi nc induced by TMV infection.** *Nucleic Acids Res* 1988, **16**:9861.
48. Vitale A, Raikhel NV: **What do proteins need to reach different vacuoles?** *Trends Plant Sci* 1999, **4**:149–155.
49. Hood EE, Gelvin SB, Melchers LS, Hoekema A: **New *Agrobacterium* helper plasmids for gene transfer to plants.** *Transgenic Res* 1993, **2**:208–218.
50. Bradford MM: **A rapid and sensitive method for the quantitation of microgram quantities of protein utilizing the principle of protein-dye binding.** *Anal Biochem* 1976, **72**:248–254.
51. Miller GL, Blum R, Glennon WE, Burton AL: **Measurement of carboxymethylcellulase activity.** *Anal Biochem* 1960, **1**:127–132.

21

GTP-binding protein Era: a novel gene target for biofuel production

Gerben P Voshol[1,2], Vera Meyer[1,3] and Cees A M J J van den Hondel[1*]

Abstract

Background: Biodiesel production using cyanobacteria is a promising alternative to fossil fuels. In this study we created a transposon library of *Synechococcus elongatus* PCC 7942 in order to identify novel gene targets for enhanced fatty acid and hydrocarbon production. The transposon library was subsequently screened for desirable traits using macro- and microscopic observations as well as staining with the lipophilic dye Nile Red.

Results: Based on the screening results, we selected a single mutant, which has an insertion in the gene encoding for the GTP-binding protein Era. We subsequently verified the phenotype-genotype relation by overexpression, reintroducing and complementing the mutation. Overexpression of *era* caused a reduction in the cell size in the late exponential phase of growth and an increase in the total amount of intracellular fatty acids. Reintroduction of the inactivated transposon caused a significant increase in the cellular length as well as changes in the amounts of individual hydrocarbons and fatty acids. Ectopic complementation of this mutation fully restored the hydrocarbon production profile to that of wild-type and partially restored the fatty acid production. Moreover, the cellular size was significantly smaller than that of the inactivated transposon mutant.

Conclusions: The GTP-binding protein Era has never been studied in cyanobacteria and proved to be an essential gene for *S. elongatus* PCC 7942. We also found that this protein is important for hydrocarbon and fatty acid metabolism as well as determination of the cell size in PCC 7942. Our results suggest that the GTP-binding protein Era can be used as a novel target for further improvement of biofuel precursors production.

Keywords: Biofuels, Cyanobacteria, GTP-binding protein Era, Fatty acids, Hydrocarbons, *Synechococcus elongatus* PCC 7942

Background

Biodiesel produced by photosynthetic microorganisms, such as eukaryotic algae and cyanobacteria, provides a promising alternative to reduce our reliance on fossil fuels. Biodiesel can be produced either directly by these organisms or from their biomass. This fuel is carbon neutral, renewable and its use requires minimal changes in the current structure of fuel delivery and consumption. However, the relatively low productivity and the high cost of harvesting the biomass present mayor limitations for commercialization of cyanobacteria-derived biodiesels [1,2].

These limitations have been mainly addressed by using a rational approach [3-5]. The rational approach involves designing strains with enhanced biodiesel precursor production (e.g. fatty acids, hydrocarbons) by modifying and/or introducing known metabolic pathways. The most common modifications include the introduction of a heterologous thioesterase and removal of the endogenous acyl-ACP synthetase. The introduced thioesterase is capable of hydrolysing the acyl-ACP molecule and thus releases the fatty acid, the major precursor of biodiesel [6]. The acyl-ACP synthetase can reactivate free fatty acids by attaching them to the ACP-molecule [7]. By introducing a thioesterase in combination with disrupting the endogenous acyl-ACP synthetase, one can significantly enhance fatty acid production and secretion in cyanobacteria [1,4,5].

However, despite some success, the reported fatty acid yields are still not sufficient for large-scale production. This is partly due to the fact that the bulk of the fatty acids remain within the cell. Changing the cell structure in such a way that the cells can auto-flocculate or elongate might help to harvest the remaining biomass more

* Correspondence: c.a.m.van.den.hondel@biology.leidenuniv.nl
[1]Molecular Microbiology and Biotechnology, Institute of Biology Leiden, Leiden University, Leiden, The Netherlands
Full list of author information is available at the end of the article

efficiently [8]. Since the rational approach is guided by existing knowledge about the cellular processes involved in fatty acid biosynthesis and/or transport, it can limit strain improvement due to missing information on key intermediates, regulators, competing pathways etc. Therefore, combining the rational approach with random approaches (e.g. transposon mutagenesis) may result in the identification of novel genes involved in efficient biodiesel precursor production.

Random mutagenesis has been successfully used in cyanobacteria to isolate filamentous [9] and grazing resistant mutants [10] as well as to identify environmentally responsive genes [11] and genes involved in polyhydroxybutyrate synthesis [12]. Furthermore, this approach was previously applied to isolate genes involved in fatty acid production in *Escherichia coli* [13]. However, genes identified by Hoover and colleagues (2012) in *E. coli* lack obvious orthologous genes in cyanobacteria. Moreover, to our knowledge, random mutagenesis has never been applied to isolate lipid-overproducing mutants in cyanobacteria.

In this study we used *Synechococcus elongatus* PCC 7942 which is amenable to genetic modification and has a fully sequenced genome. This strain produces the major biodiesel precursors and lacks the ability to synthesize PHB (a competing pathway for biofuel production) [12]. These properties make *S. elongatus* PCC 7942 a good model strain to isolate mutants with an enhanced fatty acid production. The construction and analysis of a transposon generated mutant library, resulted in the identification of a gene that directly or indirectly affects fatty acid and hydrocarbon production. Moreover, this gene, named *era*, is also involved in determining the cellular morphology of PCC 7942. To our knowledge this gene has never been analysed in cyanobacteria.

Results

Isolation of a putative fatty acid overproducing mutant by random mutagenesis

To obtain a mutant library, we haven chosen to construct a transposon library using plasmid pRL1063a, which contains a Tn5 transposon with a kanamycin resistance marker, promoterless *luxAB* reporter genes and an origin of replication which functions in *E. coli*, but not in PCC 7942. This library, contained approximately 600 individual mutants, was subsequently screened for mutants with beneficial phenotypes such as filamentous and autoflocculating growth, higher lipid content and phenotypic stability. Out of the initial 600 mutants a single strain was selected for further study based on its (i) elongated cell morphology, (ii) enhanced fluorescence after staining with the lipophilic dye Nile Red and (iii) phenotypic stability.

The selected mutant, 2A01, was both phenotypically and genotypically stable for a period of at least 3 months (verified by liquid to liquid sub-cultivation and PCR). This mutant showed a different colony morphology compared to PCC 7942 (Figure 1A and B). 2A01 has a round, slightly elevated colony morphology with irregular shaped edges on plate (Figure 1B) while PCC 7942 is round, raised with an entire smooth edge (Figure 1A). Phase contrast microscopy revealed that cultures of mutant 2A01 contained a heterogeneous mix of short cells similar in size to PCC 7942 and long elongated cells some more than 30 times the length of PCC 7942 (Figure 1C and D).

We also used flow cytometry to determine cellular morphology (size and complexity) of PCC 7942 and 2A01 stained with Nile Red (Figure 2B and D) or unstained (Figure 2A and A). Stained cells are very similar to unstained cells in respects to their cellular size (forward scatter), however the complexity (side scatter) after staining with Nile Red is slightly, but significantly higher. Comparison of cells of PCC 7942 (Figure 2A) and 2A01 (Figure 2C), furthermore indicates that cells of 2A01 are both larger in overall size and complexity than PCC 7942.

The largest differences between PCC 7942 and 2A01 where observed when comparing the yellow fluorescence (indication of lipid content) of Nile Red stained cells. In general, unstained cells showed a low amount of yellow fluorescence compared to both red and green fluorescence (Additional file 1: Figure S1 and S2). Cells stained with Nile Red showed an increase of yellow fluorescence (Figure 3A and C versus B and D) and 2A01 showed a higher increase in yellow fluorescence even at comparable cellular size (forward scatter) than PCC 7942.

Transposon insertion site

To identify the insertion site of the transposon, genomic DNA was isolated, digested, re-ligated and transformed into *E. coli* DH5α. Plasmid DNA was extracted from colonies that acquired antibiotic resistance and restriction patterns were compared to identify additional transposition events. Based on their distinct restriction patterns, DNA of two isolated plasmids was sequenced. Both sequences showed the insertion of the transposon into the 3 prime region of gene Synpcc7942_0160 annotated as a GTP-binding protein Era. The insertion caused the formation of an early stop codon leading to a truncated protein lacking the last 20 amino acids. The GTP-binding protein era of PCC 7942 contains both a GTP binding and ATP hydrolysing domain at its N terminus and a type 2 K homology (KH) domain at its C-terminus (Additional file 1: Figure S3). In PCC 7942, this gene is the first gene in an operon containing a total of 8 genes [14].

Verification genotype-phenotype relationship

To identify the genotype, phenotype relationship of mutant 2A01 and to elucidate the function of the GTP-binding protein Era in PCC 7942, several rational approaches were used.

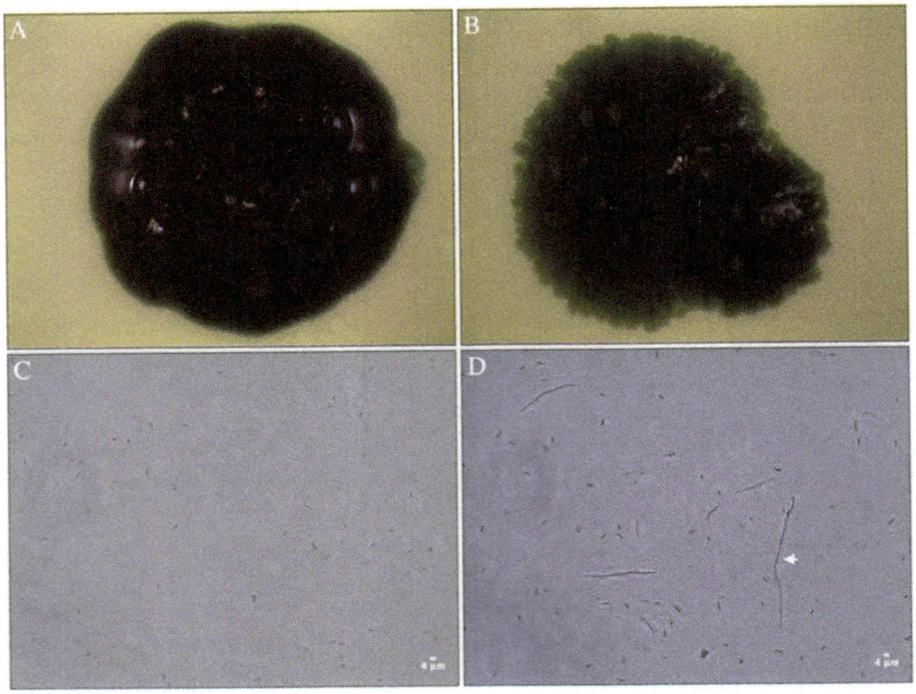

Figure 1 Macro (A and B) and microscopic morphology (C and D) of wild-type PCC 7942 (A and C) and mutant 2A01 (B and D). Macroscopic morphology was visualized after growth on plate using a stereomicroscope. Microscopic morphology was visualized by phase-contrast microscopy. Scale bars represent 4 μm. Example of an elongated cell is indicated with an arrow (97 μm long).

To test the effect of overexpression of the GTP-binding protein Era in PCC7942, a construct containing an IPTG inducible promoter followed by a ribosomal binding site and the full-length wild type era gene was made (pNS3: era). Insertion of this construct into the neutral site 3 from PCC 7942 [15] was obtained after transformation, resulting in strain Se:era. Full segregation was verified by PCR amplification. As a negative control, a similar construct lacking the ribosomal binding site and era gene (pNS3) was inserted into the neutral site 3 from PCC 7942 (PCC 7942).

To reconstruct the original phenotype, the plasmid containing the transposon flanked by genomic DNA isolated from 2A01 was reinserted into the genome of PCC 7942 after inactivating the transposase by introducing a frame shift mutation. This mutant showed the correct insertion and was fully segregated (as shown by PCR amplification). This mutant was complemented by introducing the overexpression construct pNS3:era, resulting in strain se: Δera + era and as a negative control the construct lacking the ribosomal binding site and era gene was inserted into neutral site 3, mutant se:Δera.

In addition, a knockout construct was made in which the coding region of the era gene was replaced by a Km resistance cassette. Mutants obtained revealed an elongated phenotype (Additional file 1: Figure S4). However after subcultivation, these initial mutants either lost viability, or the wild type phenotype was restored.

Growth

All strains show similar specific growth rates as measured using OD750 (Figure 4). The growth rates of mutant strains were 0.041 ± 0.002, 0.042 ± 0.001 and 0.039 ± 0.001 h^{-1} for Se:era, Se:Δera and Se:Δera + era respectively. These specific growth rates were not significantly different from that of PCC 7942 (0.041 ± 0.002 h^{-1}). The final biomass of mutant strains at time of harvesting was also not significantly different compared to PCC 7942.

Cellular morphology: length distribution during growth

To monitor changes in cell size, aliquots from cultures were examined using a phase contrast microscopy (Figure 5).

Strain se:era, containing an additional copy of era, is significantly smaller at the start of the experiment (at 0 hours) compared to WT with an median of 3.3 and 3.8 μm, respectively. At time points 24, 48 (exponential phase) and 72 hours (early stationary phase) both PCC 7942 and se:era have a similar median cell length of 3.1, 3.3 and 3.4-3.5 μm (se:era-PCC 7942), respectively. At 96 hours, the median cell length of PCC 7942 is significantly larger compared to se:era (3.5 μm versus 3.3 μm). At the time of harvesting (120 hours) the cell length of PCC 7942

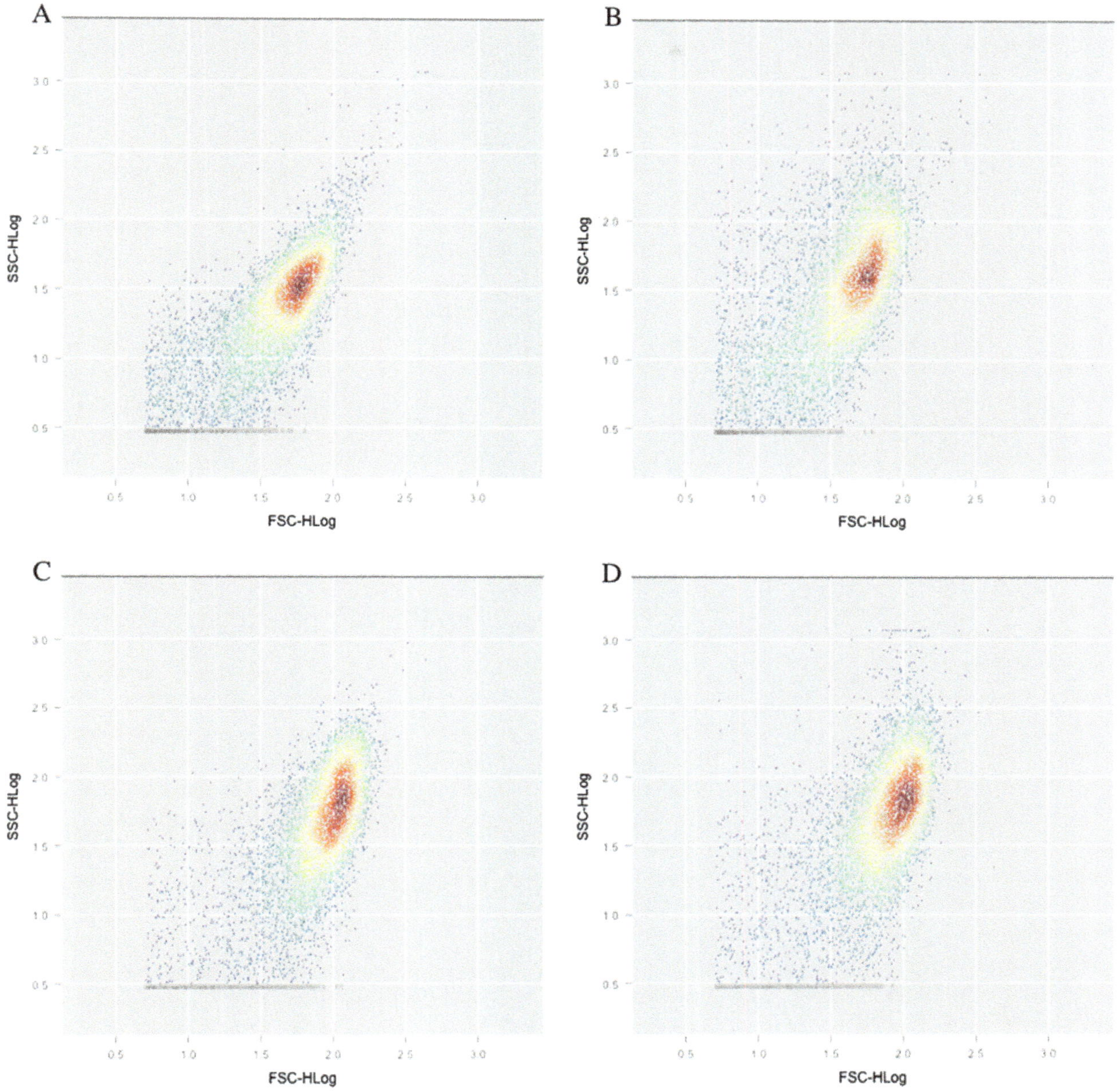

Figure 2 Side scatter (cellular complexity) versus front scatter (cell length) as determined using flow cytometry. PCC 7942 **(A and B)** and mutant 2A01 **(C and D)** grown in liquid media for 48 hours either unstained **(A and C)** or stained **(B and D)** with the lipophilic dye Nile red.

elongated further to 3.8 μm while cells from se:era had a median cell length of 3.4 μm.

Strain se:Δera has a significantly larger median cell size compared to PCC 7942 at all time points. Se:Δera had a median cell length of 6.4 μm at the time of inoculation (0 hours). This cell length decreased in the first 24 hours of growth to 5.2 μm. After this time point the cell length increase to 5.3, 5.6, 6.0 and 7.0 μm at 48, 72, 96 and 120 hours, respectively.

Strain se:Δera + era has a significantly reduced cell length compared to se:Δera, but is larger than PCC 7942 at all time points (Figure 5).

Cellular morphology: microscopy

Differential interference microscopic analysis of the mutant and wild-type strains revealed that strain PCC 7942, se:era and se:Δera:era appear very similar to each other in cell size at 48 hours (Figure 6A, B and F). This result is similar to the results obtained by measuring the cell size (Figure 5C). Furthermore, septum formation occurs in a symmetric manner for all these mutants.

Mutant se:Δera however has both elongated cells and cell similar in size to PCC 7942. Constrictions can occur similar to PCC 7942, at the middle of the cell (Figure 6D). However, constrictions are often observed more towards

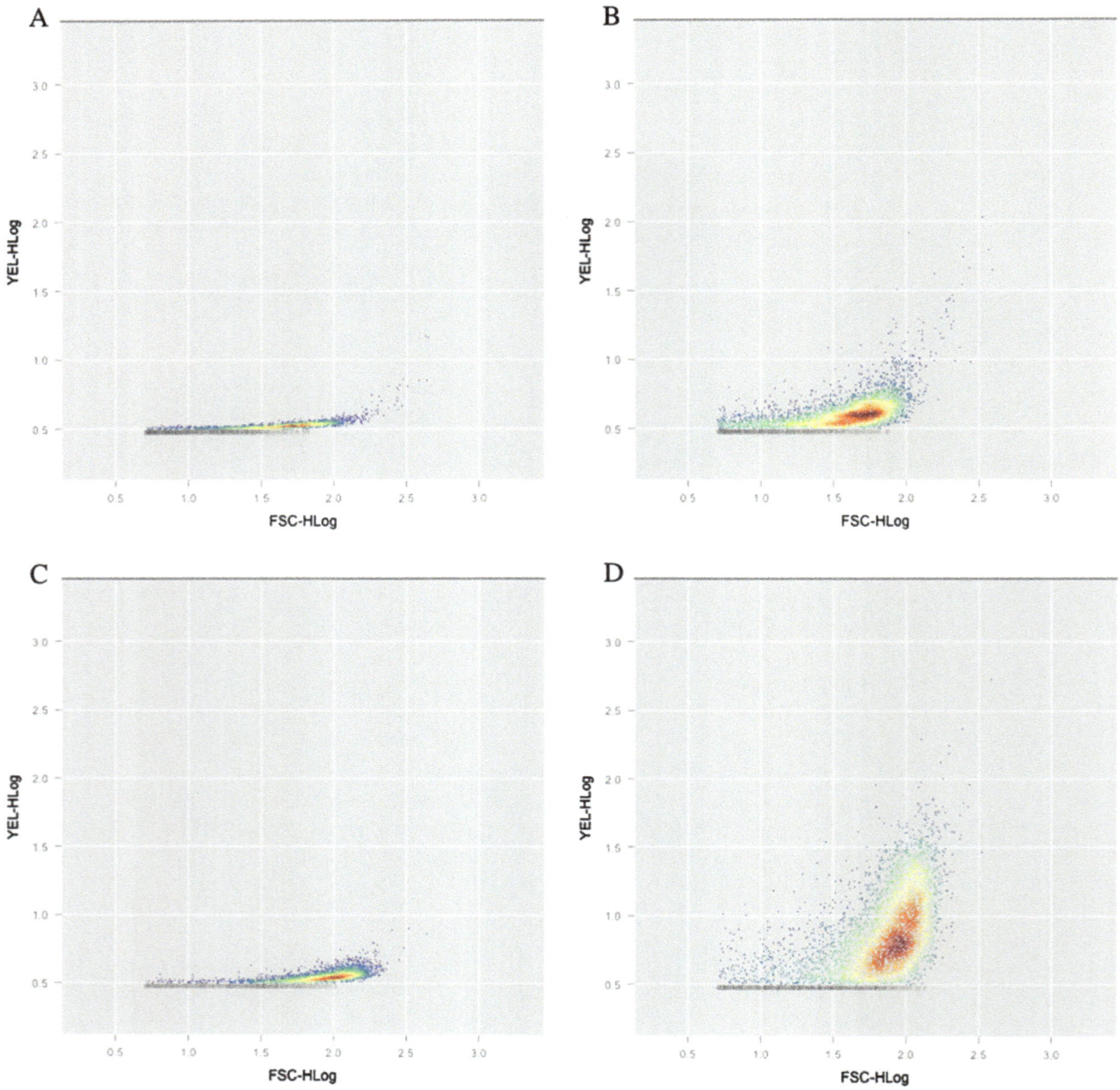

Figure 3 Yellow fluorescence versus front scatter (cell length) as determined using flow cytometry. PCC 7942 **(A and B)** and mutant 2A01 **(C and D)** grown in liquid media for 48 hours either unstained **(A and C)** or stained **(B and D)** with the lipophilic dye Nile red.

the cellular pole (Figure 6E) or sometimes multiple times within a single cell (Figure 6C). Furthermore, full separation of the daughter cells does also not always occur leading to the formation of cells arrested at the pre-divisional two-cell stage.

Fatty acid and hydrocarbon profile PCC 7942

PCC 7942 contains a mixture of saturated and unsaturated fatty acids ranging from C14 to C18. Palmitic (C16:0) and palmitoleic acid (C16:1) represent the main intracellular fatty acids of PCC 7942 (Figure 7 and Additional file 2: Table S1). Together these fatty acids represent approximately 89% (35.7 ± 1.2 μg C16:0/mg DW and 30.1 ± 1.2 μg C16:1/mg DW) of the total fatty acids (73.6 ± 2.7 μg/mg DW). The remainder mainly consists of saturated and unsaturated C14 and C18 fatty acids followed by trace amounts of odd chain C15:0, C17:0 and C17:1.

PCC 7942 is able to synthesize both alkanes and alkenes (Figure 8 and Additional file 2: Table S1). The alkanes synthesized by PCC 7942 are mostly heptadecane (C17:0),

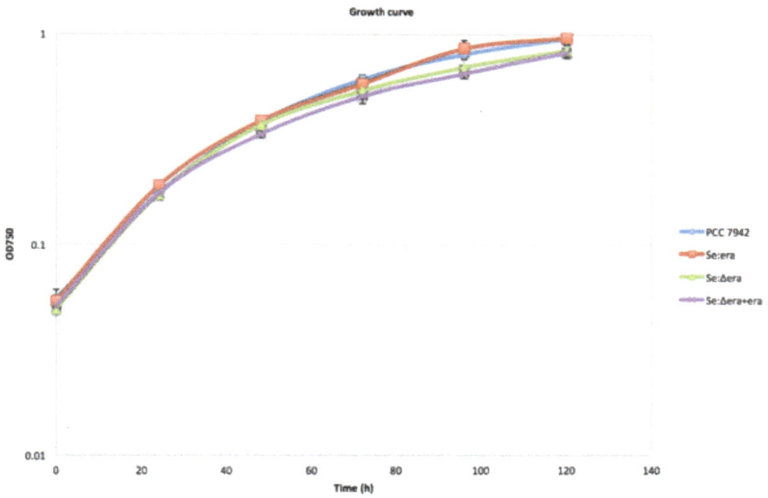

Figure 4 Cellular growth of wild type PCC 7942, strain Se:era, Se:Δera and Se:Δera + era. Growth was measured at 24 hour intervals using the optical density at 750 nm. Error bars represent the standard deviation of four replicates.

followed by pentadecane (C15:0) and hexadecane (C16:0). The only alkene that we detected was 8-heptadecene at levels comparable to those of hexadecane.

Fatty acid and hydrocarbon profile Se:era

Strain Se:era, which contains an additional era gene inserted at neutral site 3, has a significant increase in the total amount of fatty acids (Figure 7). The amount of fatty acids increased to 79. 6 ± 1.9 μg/mg in Se:era versus 73.6 ± 2.7 μg/mg DW in PCC7942 (an 8% increase). This increase in fatty acids is mostly due to a significant rise in the amount of C16:0 (7%), C18:1 (34%) and to a far lesser extend C17:1 (26%). Together they account for 68% of the increase in total fatty acids, 42%, 26% and 1% for C16:0, C18:1 and C17:1 respectively.

No differences were observed in the amount of hydrocarbons compared to PCC 7942 (Figure 8).

Fatty acid and hydrocarbon profile Se: Δera

Strain Se:Δera, containing the inactivated transposon, contains a similar amount (75.2 ± 1.0 μg/mg) of total fatty acids as PCC 7942 (73.6 ± 2.7 μg/mg DW) (Figure 7). Nevertheless, there are some significant changes in the amount of individual fatty acids. This strain has an 162% increase in the amount of C14:0 and an 93% increase in C14:1 compared to PCC7942. Furthermore, it has a 39% reduction in the amount of C18:0 and a 21% reduction in the amount of C18:1 versus PCC 7942.

This strain is the only strain with significant changes in the hydrocarbon profile (Figure 8). It contains a higher amount of pentadecane and 8-heptadecene compared to PCC 7942, 25% and 76% respectively. Furthermore, it has a substantial reduction (50%) in the amount of heptadecane.

Fatty acid and hydrocarbon profile Se: Δera + era

Strain Se:Δera + era, containing the inactivated transposon complemented ectopically by era, has an almost identical amount of total fatty acids (73.8 ± 1.2 μg/mg) compared to PCC 7942 (73.6 ± 2.7 μg/mg DW) (Figure 7). The only fatty acid, which showed a significant difference, is myristic acid (C14:0). This fatty acid was increased by 31% compared to PCC 7942.

The hydrocarbons were similar compared to PCC 7942 (Figure 8).

Discussion

The *era* gene is essential for *S. elongatus* PCC 7942

In this study we generated a transposon library of *S. elongatus* PCC 7942 in order to isolate mutants with beneficial biodiesel production traits, including enhanced fatty acid and hydrocarbon content. The transposon mutant 2A01 was selected based on (i) increased staining with the lipophilic dye Nile red (indicating a higher lipid content), (ii) elongated cell morphology (might facilitate harvesting) and (iii) phenotypic stability. The transposon of mutant 2A01 was inserted into the 3 prime region of the Synpcc7942_0160 (GTP-binding protein Era) gene. Due to this insertion, a premature stop codon was introduced into the KH-domain of this protein. The KH domain is involved in RNA binding in *E. coli* and the GTP binding domain regulates its activity [16]. Moreover, all GTP-binding protein Era homologs that have been studied to date for membrane binding activity (*E.coli, S. pneumonia and S. mutans*) are associated with the cytoplasmic membrane [17]. This activity is most likely also linked with the KH domain which is well conserved in all GTP-binding protein Era homologs including that of *S. elongatus* PCC 7942 [18].

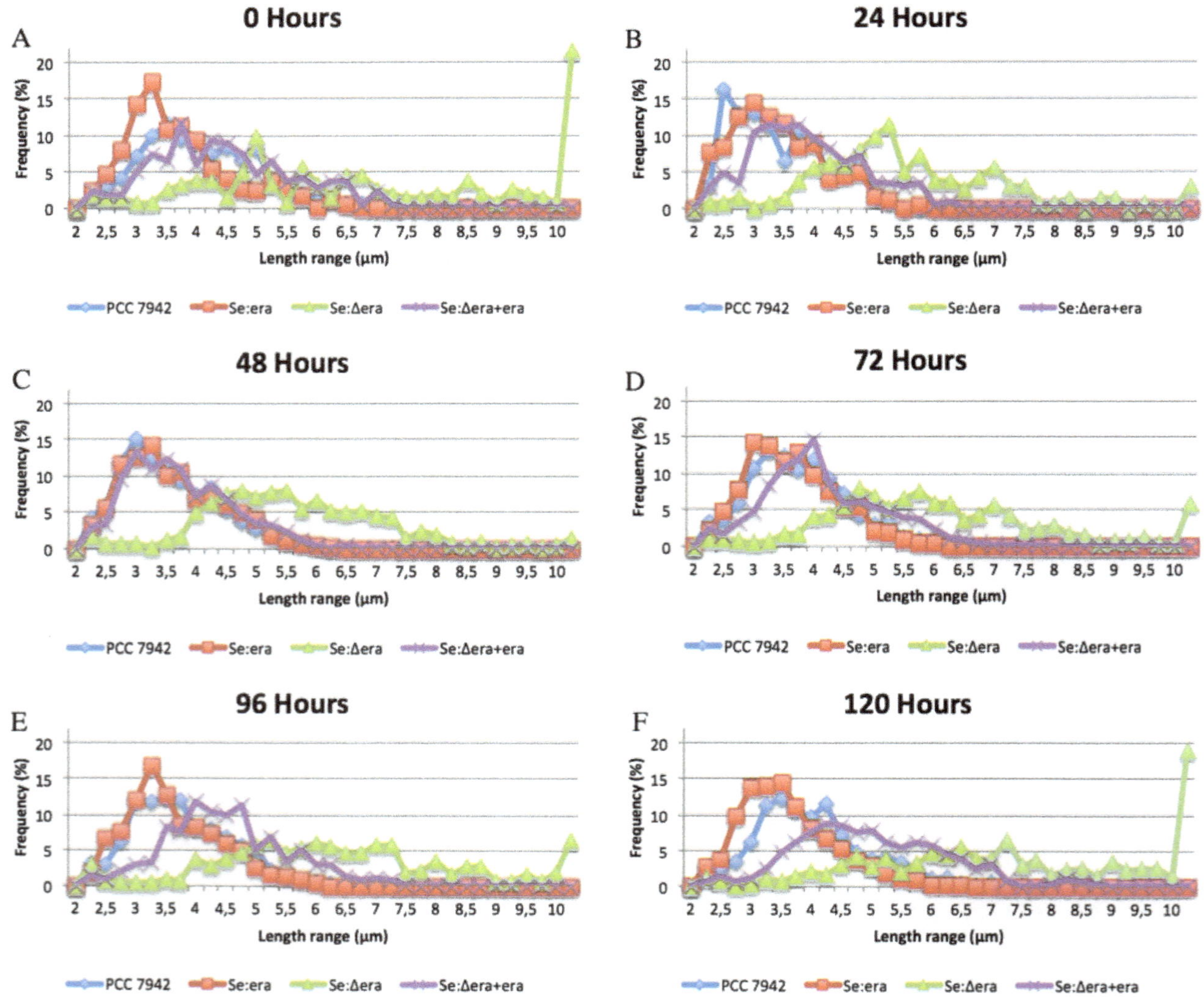

Figure 5 Distribution of cell lengths in liquid cultures of PCC 7942 and mutant Se:era, Se:Δera and Se:Δera + era after 0 (A), 24 (B), 48 (C), 72 (D), 96 (E) and 120 (F) hours. Lengths of cells of each strain was determined at 24 hour intervals by taking pictures using a phase contrast microscope and using imageJ to determine their cell size.

The *era* gene is the first gene of an operon containing a total of 8 genes. These include 5 genes encoding hypothetical proteins (most likely involved in nutrient stress, ribosome assembly and photosynthesis) and 3 that encode a tRNA-Met, an Iojap-like protein and the GTP-binding protein Era itself (Additional file 2: Table S3). Genes, which are grouped into operons together in multiple species, likely have similar functions even if transcribed individually in other species [19]. This suggests that *era* has a similar function in cyanobacteria as in *E. coli* in linking nutrient stress, to other cellular processes such as protein synthesis (ribosome assembly), photosynthesis and cell division [16].

To study whether the phenotype of the transposon insertion is (also) caused by a possible effect on the expression of downstream genes, we attempted to create a knockout of the *era* gene. We prevented transcription of all downstream genes by replacing the *era* gene with an antibiotic cassette flanked by two transcriptional terminators. Unfortunately, several transformation experiments with this cassette did not result in the isolation of the expected mutant, suggesting that one or several of the genes downstream of *era* are essential. However, subsequent introduction of a knockout cassette lacking the transcriptional terminators resulted in the isolation of a transient knockout mutant of *era*. After subcultivation of this knockout mutant, cells either failed to grow or the wild type phenotype was restored. Restoration of the wild type phenotype was associated with the removal of the resistance cassette from the *era* gene. These results indicates that *era* might also be an essential gene in PCC 7942 as was previously shown for *E. coli* [20]. Moreover, we found that the transient mutant

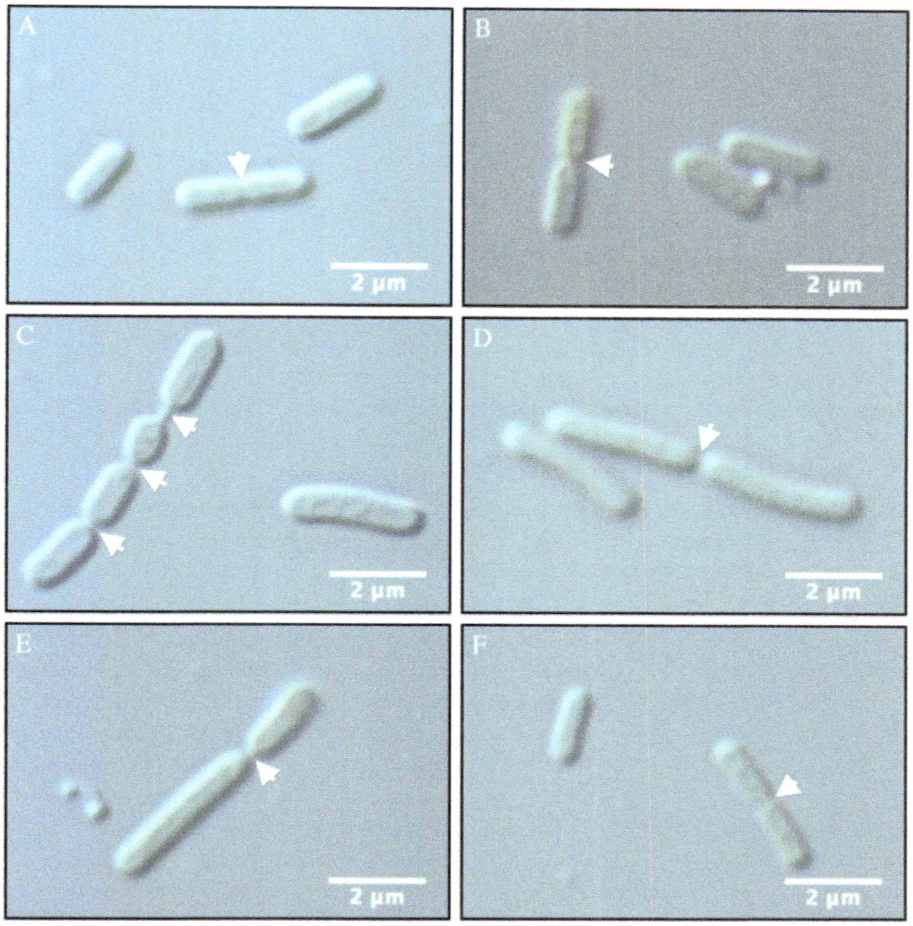

Figure 6 Morphology of PCC 7942 (A), mutant Se:era (B), Se:Δera (C, D and E) and Se:Δera + era (F) grown in liquid media for 48 hours. Scale bars represent 2 µm. The site of septa formation are indicated with arrows.

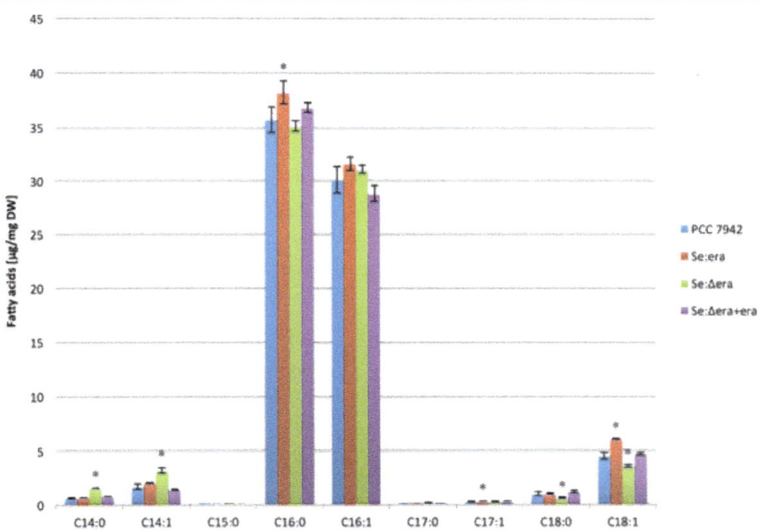

Figure 7 Quantitative measurement of intracellular fatty acids in PCC 7942, strain Se:era, Se:Δera and Se:Δera + era. Metabolites indicated with an asterisk * are significantly different ($p < 0.05$) from those of strain PCC 7942. Error bars represent the standard deviation of four replicates.

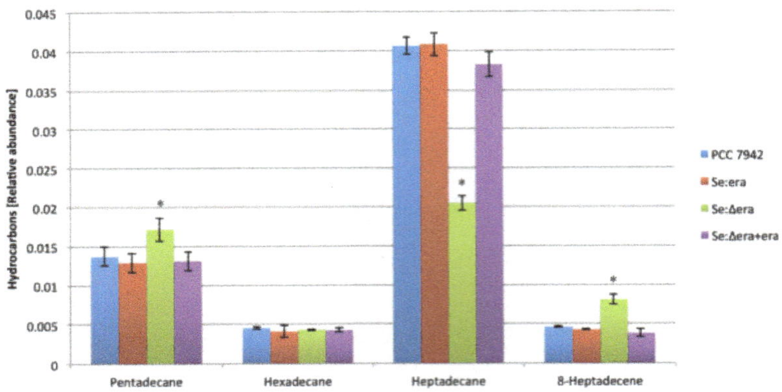

Figure 8 Quantitative measurement of intracellular hydrocarbons in PCC 7942, strain Se:era, Se:Δera and Se:Δera + era. Metabolites indicated with an asterisk * are significantly different ($p < 0.05$) from those of strain PCC 7942. Error bars represent the standard deviation of four replicates.

also showed elongated cells and problems in septa formation. In addition, the deletion mutation was fully complemented by the introduction of an ectopic *era* gene (Additional file 1: Figure S4). We further verified the phenotype-genotype relationship by a) overexpressing era b) recreating the original transposon mutant, and c) complementing the recreated transposon mutant with an ectopic copy of era.

GTP-binding protein Era is involved in lipid and hydrocarbon production of PCC 7942

Analysis of an *era* overexpression strain, Se:era, showed a significantly higher amount of fatty acids compared to PCC 7942 (Figure 7). These results demonstrate the importance of the GTP-binding protein Era for the production of fatty acids in *S. elongatus* PCC 7942.

To further verify the role of Era in biodiesel precursor production, we reintroduced the inactive transposon in PCC 7942 and complemented the resulted Se:Δera mutant with an ectopic copy of era. Strain Se:Δera does not show a significant difference in the total amount of fatty acids, compared to PCC7942 (Figure 7). However, because this strain has significantly longer cells (Figure 5), there are fewer cells per mg of dry weight compared to PCC 7942. These results indicate that there is a significantly higher amount of lipids per cell, supporting the original Nile red results. Moreover, several individual fatty acids are significantly different from PCC 7942 (Figure 7), suggesting that era plays a role in fatty acid synthesis in this strain. Complementation of Se:Δera with the ectopically integrated *era* gene (se:Δera + era) fully restored the fatty acid profile to PCC 7942 (except for C14:0), supporting the previous observation that Era is responsible for the changes in fatty acid profile.

In Se:era we did not observe any significant changes in hydrocarbon production. However, we found significant changes in hydrocarbon production in Se:Δera, which were restored when complemented with era (Figure 8), suggesting that Era is also involved in hydrocarbon metabolism.

Rational strain improvement for biodiesel precursor production of strain PCC 7942 has been attempted in the past by the introduction of thioesterase, acetyl-coA carboxylase and ribulose-1,5-bisphosphate carboxylase/oxygenase. However, these modifications failed to increase the net FA production [21]. Moreover, identified targets using RNA-seq to enhance the strain productivity of PCC 7942 led only to minor improvements in FA production [3]. However, our results show a substantial increase in fatty acid production without causing negative effects on biomass accumulation, suggesting that the GTP-binding protein Era is a new promising gene target for further strain improvement for the production of fatty acids (and possibly hydrocarbons).

GTP-binding protein Era is important for cell size of PCC 7942

Several studies in *E. coli* showed that overexpression of *era* up to 5% of total cellular protein does not alter cellular growth, cAMP levels or protein production [22,23]. This is possibly due to the fact that the activity of the GTP-binding protein Era is regulated by binding of GTP [24]. However, we observed a significantly effect on cellular size during late exponential growth of PCC 7942 containing an additional *era* gene. Although the effect on cell size is subtle and can easily be missed during microscopic examination, cells of Se:era are significantly smaller than those of PCC 7942 (>11%; Figure 5). To our knowledge, this is the first report demonstrating that overexpression of Era causes a change in cell size.

Depletion of Era in *E. coli* causes a decrease in growth, loss of viability, septum formation and elongated cells [17]. In the mutant se:Δera, containing the inactivated transposon, we did observe an effect on septum formation and cell size. However, this mutant did not show a difference in growth rate nor viability compared to PCC

7942. Analysis of the phenotype of se:Δera suggests that the expression of *era* might be reduced to a level which is slightly lower than that required for normal cell division [23].

Introduction of an ectopic era in strain se:Δera resulted in restoring the formation of septa in the middle of a cell, but not entirely the cell length. While more than 96% of all the cells were within the same length range as PCC 7942, some cells were up to 26% larger than WT. However, the median length of these cells was 31% smaller compared to strain Se:Δera. These results suggest the insertion in era itself and/or one or more downstream genes is responsible for the elongated cell morphology.

The fact that introduction of an additional *era* gene did not fully complement the Se:Δera mutant, might have several reasons. First of all, the expression of the era gene used for complementation is not regulated by its native promoter. For example, it has been shown that expression of *era* is coupled to cellular growth (fast growth > Era) in *E. coli* [23]. Another possibility is that downstream genes might be responsible for part of the phenotype and/or presence of a truncated era interferes with the complementation.

Conclusions

It was previously shown that Era plays a role in many cellular processes including ribosome assembly and cell cycle regulation [17]. This protein has been well studied in diverse bacterial species, but never in cyanobacteria. In this report we showed that *era* is an essential gene for *S. elongatus* PCC 7942. In this strain, the GTP-binding protein Era is involved in hydrocarbon and fatty acid metabolism as well as the determination of cell size. A possible explanation for the observed changes in fatty acid and hydrocarbons is the role that Era plays in coupling nutrient stress, cell division and photosynthesis. In a normal situation Era delays cell division and synthesis of fatty acids, causing cells to become longer and synthesize less proteins and fatty acids. In the Era overexpression strain (Se:era), this regulation is slightly disrupted causing almost no change in cellular size (3.3 μm versus 3.4 μm in the stationary phase) and a relative increased fatty acid synthesis.

We propose that the Era protein can be used as a novel target for improving biofuel production traits. It would be interesting to evaluate the effect of overexpression of *era* in combination with other genetic modifications, for example in a strain expressing a thioesterase and containing a knockout of the acyl-ACP synthetase. Furthermore, the question remains open whether overproduction of the GTP-binding protein Era also influences cellular size and fatty acid/hydrocarbon content in other bacteria.

Methods

Bacterial strains and plasmids

Wild-type and mutant *E. coli* and *S. elongatus* PCC 7942 strains are shown in Table 1. Table 1 further lists the plasmids that were used in this study. Primers used for vector construction and verification are listed in Table 2.

Culture conditions

PCC 7942 and its mutants were cultured in BG-11 medium at 30°C. Liquid cultures were incubated on a rotary shaker at 250 rpm under continuous light (60 μE/m^2/s). When needed antibiotics were added to a final concentration of 25 μg/ml kanamycin and/or 10 μg/ml chloramphenicol. Cellular growth was routinely determined by taking an aliquot of 1 ml every 24 hours for 5 days, diluted to an appropriate optical density and the absorption was determined at 750 nm.

Transposon library construction

To perform transposon mutagenesis, plasmid pRL1063a was used [11]. Plasmid pRL1063a contains a transposon based on Tn5, which bears several antibiotic resistance genes (kanamycin (Km), bleomycin (Ble) and streptomycin (Sm)), promoterless luciferase (*luxAB*) reporter genes, an oriV not recognized by PCC 7942 and a transposase gene. The oriT for conjugative transfer is present on the plasmid but is not part of the transposon. Conjugation was used to introduce pRL1063a into PCC 7942, using a method similar to that of Clerico et al. 2007 [26]. After 3 days of growth under low light conditions (5 μE/m^2/s), kanamycin was added underneath the plates to a final concentration of 25 μg/ml and plates were re-incubated at 30°C under normal light conditions (60 μE/m^2/s) until colonies formed.

Screening for increased lipids using flow cytometry and nile red

To get a qualitative indication of lipid content, mutant cells were stained using the lipophilic dye Nile Red [13]. Nile Red is a dye that becomes strongly fluorescent when present in a hydrophobic environment (e.g. membrane lipids). Depending on the hydrophobicity of the compounds, the dye emits either a yellow or red fluorescence. The intensity of the signal is an indication for the amount of lipids present. An aliquot of cell (1ml) from an exponentially growing culture was taken, transferred to an eppendorf tube and Nile red (1 mg/ml in DMSO) was added to a final concentration of 1 μg/ml. Cells were examined using a guave easyCyte flow cytometer (Merck), by excited with a 485nm laser and determining several parameters such as cellular size (forward scatter), complexity (side scatter) and yellow fluorescence (583/26 nm).

Table 1 Strains and plasmids used in this study

	Description	Reference
Strain		
E. coli DH5α	Used for molecular cloning	[25]
S. elongatus PCC 7942	Wild-type freshwater cyanobacterium	gifted by Susan S. Golden (UCSD, California)
2A01	*S. elongatus* PCC 7942 containing transposon pRL1063a	This study
PCC 7942	*S. elongatus* PCC 7942 containing pNS3	This study
Se:era	*S. elongatus* PCC 7942 expressing the *era* gene from neutral site 3	This study
Se:Δera	*S. elongatus* PCC 7942 with a insertion of the inactivate transposon (pSE3) and expressing the *era* gene from neutral site 3	This study
Se:Δera + era	*S. elongatus* PCC 7942 expressing a truncated thioesterase from *E. coli* DH5α ('*tesA*) and containing a disrupted acyl-ACP synthetase (SynPCC 7942_0918)	This study
Plasmid		
pJet1.2	Used for the cloning of blunt PCR products	Thermo-Fisher Scientific (Waltham, MA, USA)
pRL1063a	Contains a transposon based on Tn5, which bears several antibiotic resistance genes (kanamycin (Km), bleomycin (Ble) and streptomycin (Sm)), promoterless luciferase (*luxAB*) reporter genes, an oriV not recognized by PCC 7942 and a transposase gene. The oriT for conjugative transfer is present on the plasmid but is not part of the transposon.	[11]
pHN1-LacUV5	Targets Neutral site 3, confers resistance to chloramphenicol antibiotic (Cm) and contains a strong isopropyl-β-D-thiogalactopyranoside (IPTG)-regulated lacUV5 promoter followed by a ribosome binding site (RBS) and an unique HindIII restriction site	[15]
pNS3	Derived from pHN1-LacUV5 with the original RBS removed and lacking the ATG start codon near the multiple cloning site	This study
pNS3:ERA	Derived from pNS3 with *era* gene (Synpcc7942_0160) cloned into the unique HindIII restriction site behind the inducible lacUV5 promoter	This study
pSe:Δera	Inactivate transposon TN5-1063a originally isolated from strain 2A01, with the transposase mutate by digestion with NotI and blunting using T4 DNA polymerase	This study

Determination of transposon insertion site

To determine the site of transposition, cyanobacterial genomic DNA from mutant 2A01 was extracted using the method of Clerico *et al.* [26]. Subsequently, 10 μg of chromosomal DNA was digested overnight with EcoRI, re-ligated using T4 ligase and 5 μg DNA from this mixture was transformed into chemically competent *E. coli* DH5α cells. After plaiting the cells on LB medium containing 50 μg/ml kanamycin, colonies that formed after overnight cultivation at 37°C were inoculated into liquid LB medium and plasmid DNA was isolated and analysed by restriction analysis and sequencing.

Construction of Era overexpression mutant and control

To elucidate the function of Era for *S. elongatus* PCC 7942, an overexpression vector (pNS3:ERA) was constructed. This vector was constructed using vector pNS3, which contains amongst others a chloramphenicol resistance marker, IPTG inducible promoter and homologous regions for integration at neutral site III. The *era* gene was amplified from

Table 2 Primers used in this study

Primer	Sequence[a]	Remarks
5pERAfw	ctcgaggtaggggttgatctcgtgga	Used in combination with pRL1063a_rev and ERA3pRev2 for segregation check
ERA3pRev2	ggcaaacgctgaaagtcttc	Amplifies the 3 prime region outside the inactive transposon construct pSe:Δera. Used to check the segregation in combination with pRL1063a_fw and 5pERAfw
ERAfw	*aagctt***aaggaggaaaaa**atgtccgaccttttcaccac	Primer for amplifying the *era* gene, addition of HindIII restriction site and ribosome binding site
ERArv	*aagctt*ttactcactctcaggtcggtagc	Primer for amplifying the *era* gene, addition of HindIII restriction site
pRL1063a_fw	aggaggtcacatggaatatcagat	Used for sequencing flanking regions of TN5-1063a
pRL1063a_rev	tactagattcaatgctatcaatgag	Used for sequencing flanking regions of TN5-1063a

[a]Restriction sites are indicated in italics and the ribosomal binding site is indicated in bold.

PCC 7942 genomic DNA with primers ERAfw (containing a HindIII restriction site and ribosomal binding site) and ERArv (containing a HindIII restriction site) (Table 2). The resulting product was digested and ligated into a pNS3 vector, which resulted in the final plasmid pNS3: ERA. Transformation of this vector into PCC 7942 resulted in strain Se:era. At the same time, transformation of pNS3 into PCC 7942 was done to create a negative control.

Construction of Era disruptions and complementation

To reconstruct the original phenotype caused by the transposon found in 2A01, the re-isolated transposon of 2A01 was inactivated and reintroduced into PCC 7942 as follows. The plasmid was digested using EcoRV to reduce the size of the vector (from app. 20 kb to 11 kb by removing part of the genomic DNA from this plasmid) and thereby making subsequent cloning steps easier. This smaller plasmid was linearized using NotI, which cuts in the transposase gene blunted using T4 DNA Polymerase and re-ligated. This vector (pSe:Δera) was transformed into strain Se:era, creating strain Se:Δera + era and PCC 7942 containing pNS3 creating strain Se:Δera.

Analysis of cellular length

To see the effect of the introduced mutations on bacterial cell length, pictures were taken with a phase contrast microscope (Carl-Zeiss, Sliedrecht, The Netherlands) and analysed using ImageJ [27] with the Coli-Inspector plugin (Norbert Vischer, Bacterial Cell Biology, University of Amsterdam). This plugin is able to determine the cell length and diameter of individual cells in an automated manner. From this data a histogram was constructed and the median cell length was determined.

Extraction of hydrocarbons and fatty acids for GC-MS

Extraction of hydrocarbons and fatty acids was performed as previously described in Voshol et al. 2014. Briefly, cells from a 5 day old culture were diluted to a start OD750 of approximately 0.05 and grown in BG-11 for 5 days as described above (in quadruplo). Bacterial cells were subsequently harvested by centrifugation (20 min, 4500 rcf), the supernatant was discarded and the pellet was freeze-dried. To each 20 mg of lyophilized biomass, 50 µl of internal standard (C10:0, 5 mg/ml), 1 ml Hexane and 2 ml methanol containing 0.5 M sodium methoxide was added. The samples were vortexed for 30 seconds and sonicated for 5 minutes. Subsequently they were incubated at 50°C for 10 minutes after which 3 ml 5% HCL in methanol was added to stop the reaction. The samples were then vortexed (30 seconds), sonicated (5 minutes) and incubated at 70°C (20 minutes). Samples were allowed to cool to room temperature and were extracted twice with 4 ml of hexane containing 50 mg/L butylated hydroxytoluene. The samples were then vacuum evaporation, dissolved in 1 ml of hexane and transferred to GC-MS vials.

GC-MS analysis

These extracted hydrocarbons and fatty acids mixtures were analysed using an Agilent model 7890A gas chromatograph as previously described (Voshol et al. 2014). Briefly, samples were inserted into a model 7693 autosamples, separated using a DB-WAX column (10 m, 0.25 mm, 0.25 µm) and detected using an inert XL mass spectrometer (model 5975C) using helium as carrier gas. The initial oven temperature was 50°C for 1 minute, then the temperature was increased to 230°C (25°C/min until 200°C and 3°C/min until 230°C) after which this temperature was held for 8 minutes. Retention time and the mass spectrum of authentic samples were used for identification. Hydrocarbons were quantified relative to the internal standard while absolute quantities of fatty acids were determined based on calibration curves constructed using authentic standards.

Statistics

R *Statistical* Software (Foundation for *Statistical* Computing, Vienna, Austria) was used to carry out the data analysis. Data were first tested using Leven's test to indicate whether their variances were significantly different. Data that did not show a significant difference using Leven's test were subsequently compared using ANOVA and post-hoc Tukey Honest Significant Differences test. Data that were not normally distributed or showed unequal variances were analysed using the nonparametric Mann–Whitney-Wilcoxon test. Differences were considered significant if their p values were smaller than 0.05.

Additional files

Additional file 1: Additional figures. Red **(Figure S1)** and Green **(Figure S2)** fluorescence versus front scatter as determined using flow cytometry. Domain organization of the GTP-binding protein Era of *S. elongatus* PCC 7942 **(Figure S3)** and phase-contrast microscopic view of *S. elongatus* PCC 7942 and *era* knockout mutant where the complete coding region of the *era* gene was replaced by a Km resistance cassette.

Additional file 2: Additional tables. Quantity of fatty acids **(Table S1)** and hydrocarbons **(Table S2)** of PCC 7942 and its mutants. **Table S3** contains the original annotation of the eight genes of the *era* operon and their putative function.

Competing interests

The authors declare that they have no competing interests.

Authors' contributions

GPV conceived the study, designed and carried out the construction, cultivation, cellular length determination and the intracellular fatty acid extraction and analysis of *S. elongatus* PCC 7942 and its mutants. Furhtermore GPV drafted the manuscript and CAMJJvdH and VM critically revised it for intellectual content. All authors read and approved the final manuscript.

Acknowledgements
The Dutch Ministry of Economic Affairs (EOSLT 07039) financially supported the research described in this paper. The Government of The Russian Federation, Grant 074-U01, also financially supported this work. We like to thank Dr. Natalia V. Malfanova for critically reading the manuscript and Ir. Ellen L. Lagendijk for her help in performing some of the experiments. We also like to thank Prof. Dr. Susan S. Golden for kindly gifting S. elongatus PCC 7942 and Prof. Dr. Pamela A. Silver for providing the Neutral site targeted vectors. We thank Prof. Dr. Robert Verpoorte and Dr. Young H. Choi (Natural Products Lab) for their help determining the fatty acid and hydrocarbon profile of PCC 7942 and Dr. Paul A. Passarinho and Dr. Miranda D. van de Rhee (Genetwister Technologies B.V.) for their fruitful discussions.

Author details
[1]Molecular Microbiology and Biotechnology, Institute of Biology Leiden, Leiden University, Leiden, The Netherlands. [2]International Research Centre "Biotechnologies of the Third Millennium", ITMO University, Saint-Petersburg, Russia. [3]Applied and Molecular Microbiology, Institute of Biotechnology, Berlin University of Technology, Berlin, Germany.

References

1. Ruffing AM, Jones HDT. Physiological effects of free fatty acid production in genetically engineered Synechococcus elongatus PCC 7942. Biotechnol Bioeng. 2012;109:2190–9.
2. Slade R, Bauen A. Micro-algae cultivation for biofuels: Cost, energy balance, environmental impacts and future prospects. Biomass Bioenergy. 2013;53:29–38.
3. Ruffing AM. RNA-Seq analysis and targeted mutagenesis for improved free fatty acid production in an engineered cyanobacterium. Biotechnol Biofuels. 2013;6:113.
4. Liu X, Sheng J, Curtiss R. Fatty acid production in genetically modified cyanobacteria. Proc Natl Acad Sci U S A. 2011;108:6899–904.
5. Hu P, Borglin S, Kamennaya NA, Chen L, Park H, Mahoney L, et al. Metabolic phenotyping of the cyanobacterium Synechocystis 6803 engineered for production of alkanes and free fatty acids. Appl Energy. 2013;102(null):850–9.
6. Cho H, Cronan JE. Defective export of a periplasmic enzyme disrupts regulation of fatty acid synthesis. J Biol Chem. 1995;270:4216–9.
7. Kaczmarzyk D, Fulda M. Fatty acid activation in cyanobacteria mediated by acyl-acyl carrier protein synthetase enables fatty acid recycling. Plant Physiol. 2010;152:1598–610.
8. Markou G, Georgakakis D: Cultivation of filamentous cyanobacteria (blue-green algae) in agro-industrial wastes and wastewaters: A review. Appl Energy 2011;88:3389–3401
9. Campellone KG, Leong JM. Nck-independent actin assembly is mediated by two phosphorylated tyrosines within enteropathogenic Escherichia coli Tir. Mol Microbiol. 2005;56:416–32.
10. Simkovsky R, Daniels EF, Tang K, Huynh SC, Golden SS, Brahamsha B. Impairment of O-antigen production confers resistance to grazing in a model amoeba-cyanobacterium predator-prey system. Proc Natl Acad Sci U S A. 2012;109:16678–83.
11. Wolk CP. Use of a Transposon with Luciferase as a Reporter to Identify Environmentally Responsive Genes in a Cyanobacterium. Proc Natl Acad Sci. 1991;88:5355–9.
12. Miyake M, Takase K, Narato M, Khatipov E, Schnackenberg J, Shirai M, et al. Polyhydroxybutyrate production from carbon dioxide by cyanobacteria. Appl Biochem Biotechnol. 2000;84–86:991–1002.
13. Hoover SW, Tyler Youngquist J, Angart PA, Withers ST, Lennen RM, Pfleger BF. Isolation of improved free fatty acid overproducing strains of Escherichia coli via nile red based high-throughput screening. Environ Prog Sustain Energy. 2012;31:17–23.
14. Vijayan V, Jain IH, O'Shea EK. A high resolution map of a cyanobacterial transcriptome. Genome Biol. 2011;12:R47.
15. Niederholtmeyer H, Wolfstädter BT, Savage DF, Silver PA, Way JC. Engineering cyanobacteria to synthesize and export hydrophilic products. Appl Environ Microbiol. 2010;76:3462–6.
16. Britton RA. Role of GTPases in bacterial ribosome assembly. Annu Rev Microbiol. 2009;63:155–76.
17. Verstraeten N, Fauvart M, Versées W, Michiels J. The universally conserved prokaryotic GTPases. Microbiol Mol Biol Rev. 2011;75:507–42. second and third pages of table of content.
18. Hang JQ, Zhao G. Characterization of the 16S rRNA- and membrane-binding domains of Streptococcus pneumoniae Era GTPase. Eur J Biochem. 2003;270:4164–72.
19. Price MN, Arkin AP, Alm EJ. Correction: The Life-Cycle of Operons. PLoS Genet. 2006;2:e126.
20. March PE, Lerner CG, Ahnn J, Cui X, Inouye M. The Escherichia coli Ras-like protein (Era) has GTPase activity and is essential for cell growth. Oncogene. 1988;2:539–44.
21. Ruffing AM: Borrowing genes from Chlamydomonas reinhardtii for free fatty acid production in engineered cyanobacteria. J Appl Phycol 2013;25:1495-1507
22. Gollop N, March PE. A GTP-binding protein (Era) has an essential role in growth rate and cell cycle control in Escherichia coli. J Bacteriol. 1991;173:2265–70.
23. Britton RA, Powell BS, Dasgupta S, Sun Q, Margolin W, Lupski JR, et al. Cell cycle arrest in Era GTPase mutants: a potential growth rate-regulated checkpoint in Escherichia coli. Mol Microbiol. 1998;27:739–50.
24. Shimamoto T, Inouye M. Mutational analysis of Era, an essential GTP-binding protein of Escherichia coli. FEMS Microbiol Lett. 1996;136:57–62.
25. Hanahan D. Studies on transformation of Escherichia coli with plasmids. J Mol Biol. 1983;166:557–80.
26. Clerico E, Ditty J, Golden S: Specialized Techniques for Site-Directed Mutagenesis in Cyanobacteria. Methods Mol. Biol. 2007;362:155–171
27. Schneider CA, Rasband WS, Eliceiri KW. NIH Image to ImageJ: 25 years of image analysis. Nat Methods. 2012;9:671–5.

Morphology and ploidy level determination of *Pteris vittata* callus during induction and regeneration

Blake L Joyce[1], Shigetoshi Eda[2], John Dunlap[3] and C Neal Stewart Jr[1*]

Abstract

Background: Morphological and ploidy changes of the arsenic hyperaccumulator, Chinese brake fern (*Pteris vittata*) callus tissue are described here to provide insight into fern life cycle biology and for possible biotechnology applications. *Pteris vittata* callus was studied using transmission and scanning electron microscopy, and flow cytometry.

Results: Callus induction occurred both in light and dark culture conditions from prothallus tissues, whereas rhizoid formation occurred only in dark culture conditions. Callus tissues contained two types of cells: one actively dividing and the other containing a single large vacuole undergoing exocytosis. Sporophytes regenerated from callus asynchronously form clusters of cells in a manner apparently analogous to direct organogenesis. Extracellular matrices were observed in actively-growing callus and at the base of regenerating sporophytes. Callus tissue nuclei were found to be primarily diploid at induction and throughout maintenance of cultures indicating that callus cell fate is determined at induction, which closely follows apogamous sporophyte development. Presence of a dense extracellular matrix in conjunction with sporophyte development suggests a link between the suspensor-like activity of the embryonic foot during normal fern embryo development and the suspected functions of extracellular matrices in angiosperms.

Conclusions: Further investigation could lead to a better understanding of genes involved in *P. vittata* embryo development and apogamous sporophyte development. The methodology could be useful for in vitro propagation of rare and valuable fern germplasm.

Keywords: Extracellular matrix, Phytoremediation, Chinese brake fern, *Pteris vittata*, Scanning electron microscopy, Tissue culture and transformation, Transmission electron microscopy

Background

The Chinese brake fern (*Pteris vittata*) is an arsenic hyperaccumulator capable of storing arsenic in leaves up to 2.3% dry weight [1]. As such, this species has been a focus for use in phytoremediation. A fundamental problem in developing tissue culture and transformation systems *de novo* is ill defined tissue stages; i.e., distinguishing proliferating tissues as callus, somatic embryos or other differentiated tissue. Scanning electron microscopy (SEM), transmission electron microscopy (TEM), and histology have been used to characterize tissue culture systems [2-6] as well as the fern life cycle [7-12]. Even though *P. vittata* transformation has recently been accomplished using spores as transformation targets [13], there remains interest in developing tissue culture systems in this and other fern species. In this respect, fern tissue culture has lagged behind that of many other plants. There have been several reports of callus induction from *P. vittata* tissue [14-18]. Callus has been induced from fern explants ranging from prothalli, croziers, rhizomes, and even pinnae strips. These reports focus on culture conditions and nutritional requirements necessary for tissue culture, but do not fully characterize the events occurring during development, maintenance, and differentiation of callus.

Fern tissue culture and transformation technologies would yield possibilities for new areas of research investigating fern physiology and developmental biology through study of prothalli genes using overexpression and knockout

* Correspondence: nealstewart@utk.edu
[1]Department of Plant Sciences, The University of Tennessee, Knoxville, Tennessee 37996, USA
Full list of author information is available at the end of the article

analysis. Research investigating fern genomics and proteomics could yield new tools for molecular biology including promoters and genes responsible for arsenic accumulation pathways that can be transferred to other organisms to confer value-added traits for academic research and industrial applications.

The objective of our study was to characterize *P. vittata* callus induction, maintenance, and the subsequent sporophytes recovered from calli through use of transmission and scanning electron microscopy and flow cytometry.

Methods

Environmental conditions and tissue culture

Spores were harvested from mature P. vittata (courtesy of Edenspace Systems Corporation) fronds, using a straight-edged razor blade and collected into 1.7 ml centrifuge tubes (Denville, Metuchen, NJ, USA). A volume of 0.1 mL of spores was surface sterilized by sequential washes of 10% dilution of commercial chlorine bleach (5.25% sodium hypochlorite, Fisher Scientific, Pittsburgh, PA, USA) and 70% ethanol. They were then rinsed three times with sterile water and plated onto medium which contained half-strength Murashige-Skoog salts [19], B5 vitamins [20], 2.0% sucrose, and solidified with 0.2% Gelzan (Sigma) hereto referred to as 0.5MS20gS. Plates of sterilized spores were placed in lighted conditions to germinate. After about four weeks prothalli formed and were transferred onto callus induction medium which consisted of 0.5MS20gS supplemented with 0.5 mg L^{-1} gibberellic acid (GA_3) and 0.5 mg L-1 6-benzylaminopurine (BA): Yang medium [21].

Light and dark environmental conditions were used for tissue culture. The light culture environment consisted of 16 hours of light at 79 μmol m^{-2} s^{-1} and 8 hours of darkness. The dark environment consisted of 24 hours of darkness. Both environments had a constant temperature of 25℃.

Fern tissue culture plates were then placed in either light or dark culture conditions. Every two weeks the prothalli were subcultured until they formed calli. Calli were excised and subcultured and maintained on the same medium in the same conditions or on liquid callus induction medium lacking Gelrite gellan gum in dark conditions.

To induce differentiation, calli were moved to medium containing half-strength MS salts supplemented with 6.0% maltose (Fisher Scientific), 0.5% activated charcoal (Sigma-Aldrich, St. Louis, MO, USA), and solidified with 0.2% Gelrite gellan gum. The calli were placed in lighted conditions and subcultured every two weeks until sporophytes were regenerated.

Fixation of tissues

Fifteen samples from each stage of tissue culture were randomly selected and put into 3% glutaraldehyde in 0.1 M cacodylate buffer for 90 min at room temperature. They were then rinsed three times in the cacodylate buffer and transferred to 2% OsO_4 in 0.1 M buffer for 90 min [22]. The samples were then dehydrated in a graded acetone series (25%, 50%, 75%, 95%, 100%, and dry 100%) for 30 min at each step. The samples were then divided into two groups according to preparation needed for the scanning or transmission electron microscope.

Scanning electron microscopy

Samples were prepared using the critical-point drying method with carbon dioxide. Each tissue sample was affixed to two-sided carbon tape and then coated with 50 nm of gold. At least five samples were viewed from each stage of tissue culture and tissue culture condition.

Transmission electron microscopy

Samples were transferred into a 1:3 acetone/Spurr mixture overnight. The samples were then moved to a 3: 1 acetone/Spurr mixture for 4 h with the beaker lid on and 4 h with the beaker lid removed. The samples were then placed in 100% Spurr's resin and left overnight. Individual callus pieces were put into molds with fresh resin and placed in a 68°C oven for 24 h polymerize. At least three blocks for each condition were randomly selected and trimmed for thin-sectioning with glass knives on a microtome. Sections were collected on copper grids coated with glue. All grids were post-stained with uranyl acetate in 50% methanol for 30 min, and washed three times by 30 dips into distilled water. They were then stained in lead citrate for 5 min, washed again and then loaded for viewing using a Zeiss Libra 200MC transmission electron microscope.

Flow cytometry

P. vittata callus, frond, and prothallus tissues were collected and weighed for nuclei extraction. For each preparation, approximately 0.1 g of each sampled tissue, in triplicate, was chopped in Otto buffer I using straight-edge razor blades [23]. Prothallus and callus controls were chopped independently, whereas the prothallus and callus tissues for genome size estimation comparison were co-chopped. After chopping, the suspension of each replicate was filtered by pipetting into a 5 ml polystyrene round-bottom tube with a 35 μm cell strainer cap (BD Falcon). Propidium iodide (50 μg ml^{-1}) was added to the filtrate of each replicate.

The nuclei stained with propidium iodide were analyzed using a LSR II flow cytometer (BD Biosciences, San Jose, CA, USA) with the photomultiplier tube voltages set at 410 eV for forward scatter channel, 229 eV for side scatter channel, and 639 eV for the FL2 channel. The flow cytometer was equipped with a coherent sapphire 20 mW solid state 488 nm laser for the FL2 channel. Fluorescence of the stained nuclei was detected by using 555DCLP dichroic mirror and 575/26 nm bandpass filter. The threshold was set at 30,000 on the FL2 channel to exclude unstained particles in the suspension. The flow rate was set to 1.0 μl/s. Data, obtained by counting 5000 particles for each sample, were analyzed using DiVa software (BD Biosciences, San Jose, CA, USA). Signals from cellular debris were gated out on a 2-D dot plot of forward scatter (FSC) and side scatter (SSC) (Additional file 1).

Results

Callus induction

Prothalli began to swell after four to six weeks of culture on callus induction medium in both light and dark conditions (Figure 1A and 1B). Adventitious prothalli, antheridia, archegonia, and rhizoids were present on the surface of the prothallus after six weeks of culture on callus induction medium in light and dark culture conditions (Figure 1A – D). Antheridia were the primary sex organ present on any surface of the prothalli, however a few archegonia were observed. In dark conditions, the prothalli formed thick mats of rhizoids which had to be removed for visualization with scanning electron microscopy. Rhizoids were observed elongating in a radial pattern from the basal cells of antheridia-like structures (Figure 1C and D). Undifferentiated cells forming callus were present on the surface of prothalli at eight weeks (Figure 1D). An extracellular matrix as well as fibrillar and granular structures were present on clusters of developing calli in both lighted and dark conditions (Figure 1E and 1F). Swelling gametophytes were excised and subcultured until the tissue evident in cultures appeared to consist exclusively of callus tissue (Additional file 1).

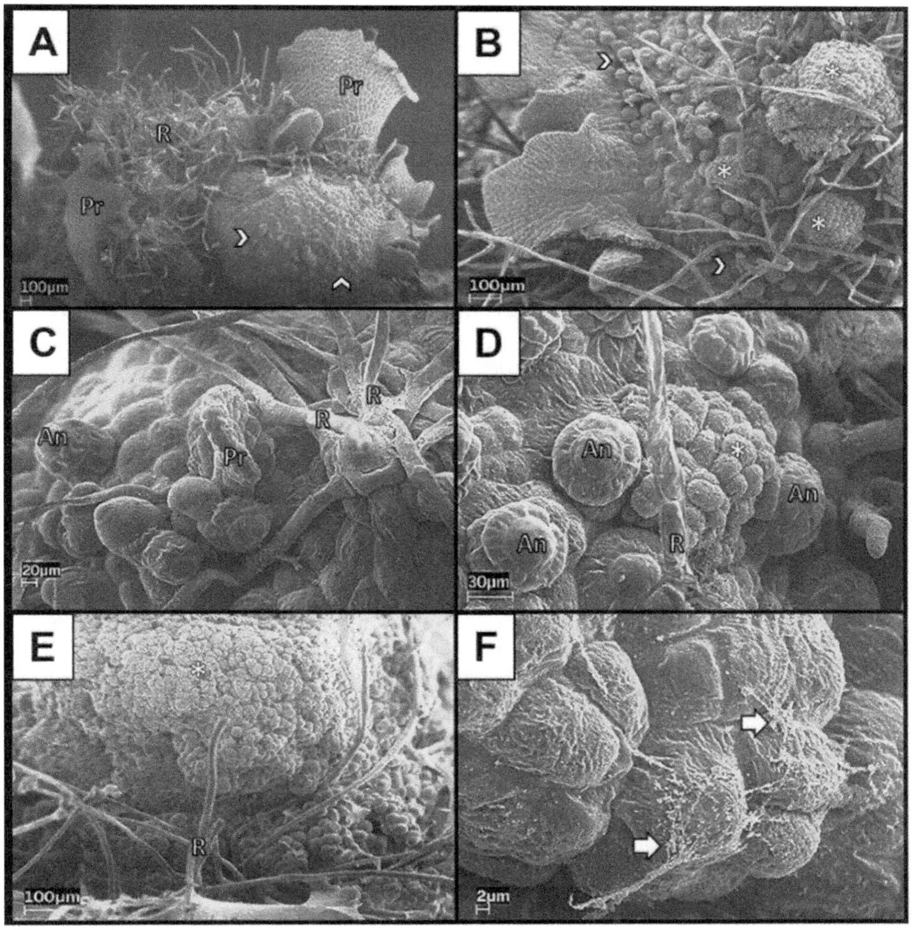

Figure 1 Scanning electron microscopy of representative *Pteris vittata* prothalli grown under light culture conditions for six weeks (A, C, E) and dark conditions for six weeks (B, D, F). A) Swollen prothallus with adventitious prothallus blades, rhizoids, and sex organs. Archegonia were observed, although the majority of sex organs were antheridia. **B)** Callus tissue forming on a swollen prothallus under dark conditions. Rhizoids were removed so that the prothallus surface could be viewed. **C)** Early formation of adventitious prothalli and rhizoids from antheridia-like structures. **D)** Callus tissue developed around antheridia cells. **E)** Formation of undifferentiated callus cells from the swollen prothallus. **F)** Fibrillar extracellular matrix covering dividing callus. Antheridia (An), callus (asterisk), extracellular matrix (arrow), groups of antheridia (chevron), rhizoids (R), adventitious prothallus blades (Pr).

Callus maintenance on semi-solid medium

Callus cultures grew rapidly on semi-solidified medium in both light and dark conditions. Callus differentiated to form sporophytes if the subculture interval exceeded two weeks (Additional file 1). On occasion, the explants died on the semi-solidified medium, but typically generated healthy callus tissue for subsequent subculturing.

Transmission electron microscopy of callus tissue showed the presence of two different cell types (Figure 2). Larger cells contained large vacuoles with electron dense regions, thin cytoplasm, and cell wall segments with few plasmodesmata (Figure 2A). Smaller cells each contained dense cytoplasm with large protein bodies and starch grains, small vacuoles, and a nucleus with two to several nucleoli in section and sparce heterochromatin (Figure 2B and 2D). Plasmodesmata are more abundant in the walls of the smaller cells than in the larger cells (Figure 2B). These cells also underwent cell wall biosynthesis as evidenced by nearby Golgi bodies (Figure 2F).

Maintenance and regeneration of callus tissue

During callus maintenance, large differentiated terete structures formed from the center of the callus tissue (Figure 3A). Extracellular material seemed to be found

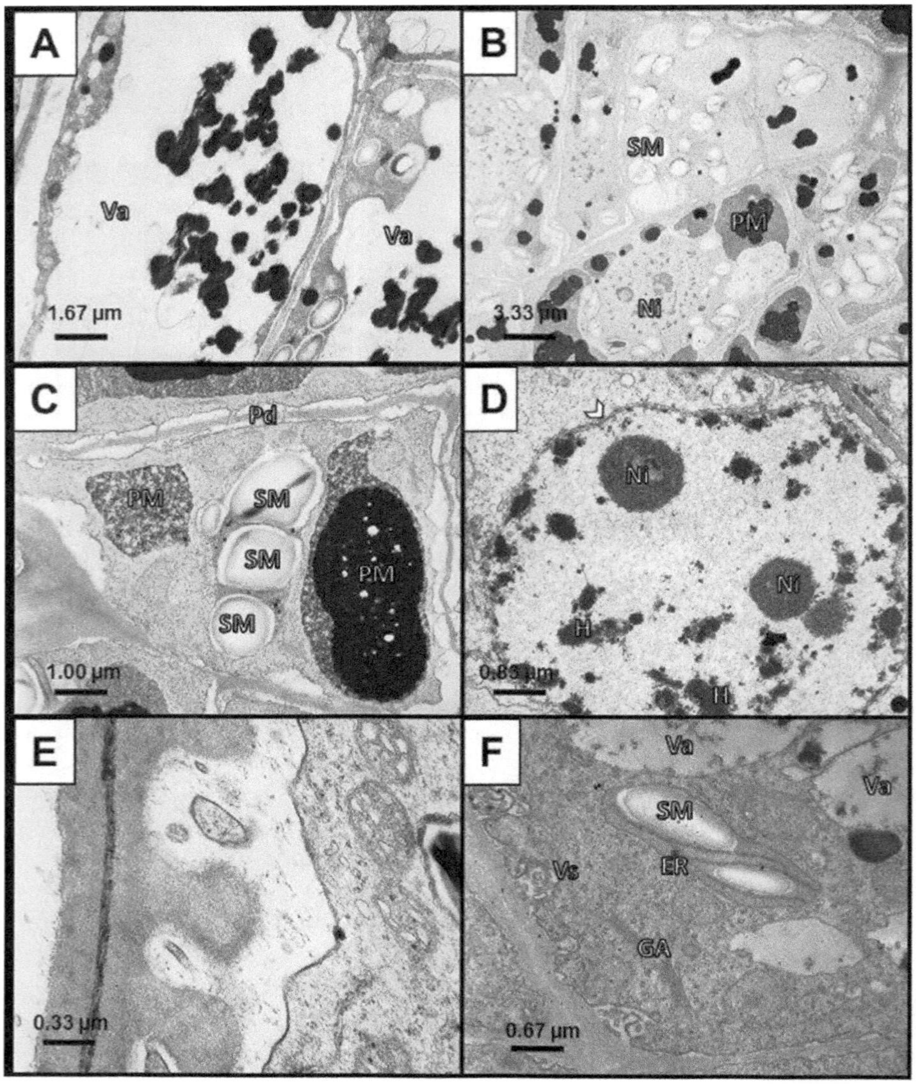

Figure 2 Transmission electron microscopy of two cell types observed in *Pteris vittata* callus maintained on semi-solid callus induction medium. A) The first cell type had a characteristically large vacuole with electron-dense vacuolar tannins and a thin layer of peripheral cytoplasm. **B)** The second cell type had a multitude of small vacuoles, nuclei with nucleoli, endoplasmic reticulum associated with starch granules, and Golgi apparati. **C)** Smaller cells of the second cell type contained more plasmodesmata, and dense cytoplasm with large amyloplasts and protein-filled vacuoles. **D)** Heterochromatin and two nucleoli were observed in all nuclei of the second cell type. **E)** Invaginations in the cell wall containing vesicles. **F)** Plastid surrounded by Golgi apparati and vesicles near the cell wall. Endoplasmic reticulum (ER), Golgi apparatus (GA), heterochromatin (H), nuclear envelope (NE), nucleoli (Ni), protein bodies (PM), plasmodesmata (Pd), starch granules (SM), vacuole (Va), vesicles (Vs).

Figure 3 Representative callus on Yang semi-solid medium for six weeks (A and B) and regenerating callus grown on differentiation medium under light culture conditions for four weeks (C) and eight weeks (D). A) Spontaneous sporophytic growth occurred on callus during maintenance on callus induction medium. **B)** Typical round callus cells present on the callus before sporophytic growth. An extracellular matrix extends from cells. **C)** Asynchronous differentiation of sporophytes begins with clusters of cells from the callus surface when induced on differentiation medium. Differentiation of sporophytes was asynchronous. **D)** A crozier of a maturing sporophyte.

on the surface of some callus (Figure 3B). Sporophyte differentiation occurred in clusters of cells that did not have the extracellular material (Figure 3C). Sporophyte regeneration was asynchronous and occurred across the surface of the callus tissue. After six weeks of regeneration, distinct mature sporophytes with croziers formed (Figure 3D). The cells at the base of sporophytes had thick, fibrillar extracellular matrix and were not spherical like the cells on the surface of the callus.

Flow cytometry of prothallus and callus nuclei

As seen in Figure 4, nuclei from callus tissue contained approximately twice the fluorescence intensity, and hence, approximately twice the DNA of the prothallus nuclei when quantified based on fluorescence pulse area (PI-A). Several small peaks were present at the cutoff limit of callus fluorescence (Figure 4A). These peaks were not continuous and did not have a mean at the expected gametophyte 1n level (50,000), which could represent numerous poorly-resolved haploid peaks (see also the Additional file 2). Alternatively, these peaks might represent cellular debris. Nonetheless, the data were reproducible (see also the Additional file 2). The ungated callus tissue flow diagrams did not have peaks, but rather a flat line near the 1n number (Additional file 2). The fluorescence pulse width (PI-W) formed a single peak at the nearly same value (approximately 75,000) in all the tested nuclei (Figure 4B,D, and F) demonstrating that the nuclei were approximately the same size.

Discussion

Antheridia and archegonia were present across the entire gametophytic surface in both light and dark culture conditions even though in nature antheridia and archegonia are spatially separated. GA_3 and ethylene interact to induce formation of antheridia in the fern *Anemia phyllitidis* [24], and ubiquitous production of sex organs has also been observed in other axenically cultured fern prothalli [9]. The presence of GA_3 in the medium and production of ethylene in sealed plates could account for the ubiquitous presence of antheridia and rhizoids. Typically, the cap cells of antheridia open to release spermatozoa for fertilization, but no opened antheridia were observed on any prothalli in our experiments, which indicated that the fusion of sex cells did not occur prior to or during callus development.

SEM also revealed the presence of a fibrillar extracellular matrix on the outside the callus during callus induction and at the base of sporophytes during differentiation and development. Similar extracellular matrices have been observed in embryogenic and organogenic tissue cultures of flowering plants such as *Coffea arabica* [25], *Papaver somniferum* [4], *Brassica napus* [5], and *Actinidia deliciosa* [26]. Although the biological significance of extracellular

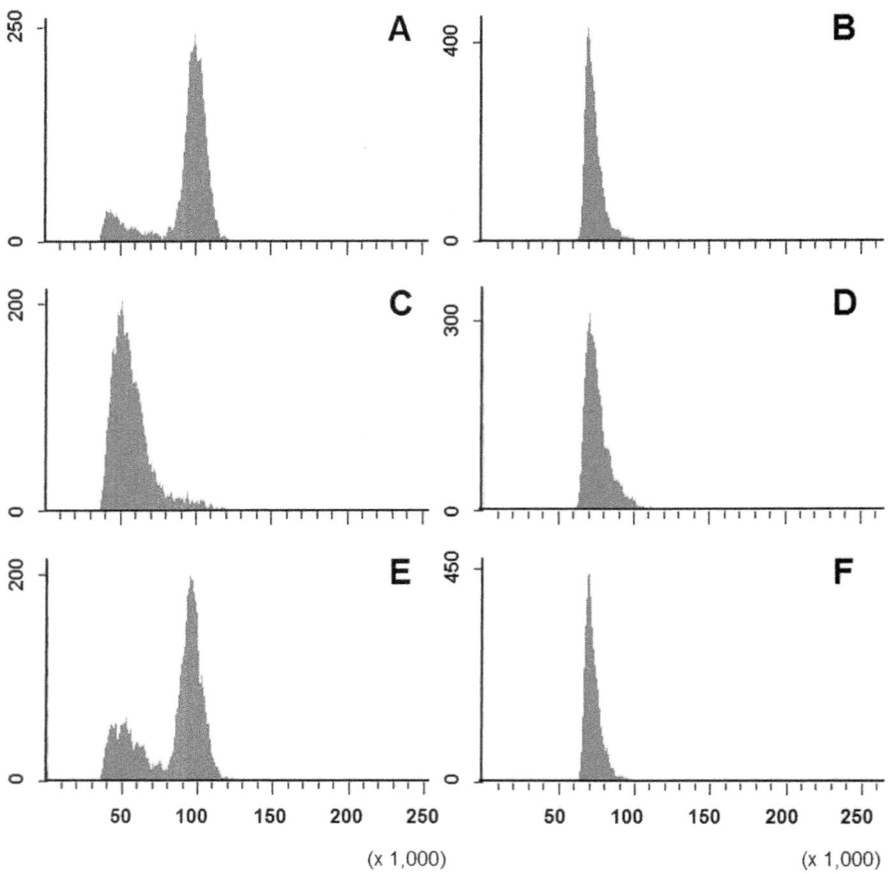

Figure 4 Flow cytometry histograms of suspended *Pteris vittata* callus and prothallus nuclei stained with propidium iodide. A) Isolation and staining of *P. vittata* callus control (2n) nuclei using PI-A calculation and **B)** PI-W. **C)** Isolation and staining of *P. vittata* prothallus control (1n) using PI-A calculation and **D)** PI-W. **E)** Dual isolation (co-chop) and staining of *P. vittata* prothallus (1n) nuclei exhibit one-half the fluorescence seen in callus nuclei (2n) using the PI-A. **F)** Co-chopping resulted prothallus and callus nuclei having the same peak in the PI-W histogram suggesting that the same particle size was counted for all events. Histograms are representative of three technical and biological replicates.

matrices in tissue cultures is unknown, they have been considered as a biomarker for pre-embryogenic tissue [5] and involved in cell-cell adhesion and signaling during embryo development in some species [4,27]. The extracellular matrix, including collapsed cells at the base of the developing prothalli, in our cultures (Figure 3D) might be orthologous to the 'foot' development at the base of sporophytes on developing prothalli. The foot has been implicated in nutrient movement from gametophyte to sporophyte cells, which as also been compared to the suspensor of higher plant embryos [27].

Three types of calli were derived from *P. vittata* prothalli on basal medium the dark [18]. The type 2 callus described in Kato [18] had morphological similarities to our callus grown on Yang medium; both regenerated sporophytes. Kato [14] showed that including hormones such as indole-3-acetic acid (IAA) and GA_3 promoted thickening of the prothallus pad and the formation of apogamous sporophytes. A mass of tissue derived from thickened prothalli of the fern *Pteridium* also resulted in apogamous sporophytes [12]. In this study, we observed the same thickening of prothalli exposed to BA and GA_3 (Figure 1) during the formation of callus. From these calli, we recovered diploid sporophytic tissue, even though the callus was derived from haploid prothalli. We do not know by what mechanism this transition occurred. One explanation was that spontaneous apogamy occurred, since no antheridia or archegonia were observed during the experiments. While tissue culture-derived apogamy has not been shown in ferns before, it has been documented in the moss *Physcomitrium coorgense* [28]. Similar to our experiments the moss callus that produced apogamous sporophytes also accumulated large amounts of starch and contained an abundant number of plasmodesmata [29].

Another arsenic-hyperaccumulating fern, *Pteris cretica* [30], reproduces by obligate apogamy. Our *P. vittata* callus TEM images were congruent with observations of developing *P. cretica* egg cells with regards to the large number of free ribosomes, lipid bodies, and the many small vacuoles observed in each [31].

The production of sporophytes from *P. vittata* callus is an opportunity to investigate the genetic and morphological characteristics of prothalli during life cycle changes in pteridophytes. Histology of *P. vittata* rhizome callus has previously revealed distinct nodules of meristematic cells that gave rise to shoots, roots, or both, depending on concentrations of sugar and 2,4-D present in the medium [15]. In our experiments, *P. vittata* callus tissue was also comprised of cell clusters that differentiated into sporophytes. This type of callus most closely resembles embyrogenic callus of monocots and dicots, however, no secondary embyrogenesis deriving from single cells was observed in this work. Apogamous sporophyte production from gametophytes has been suggested as a strategy to continue fern life cycle either during dry environmental conditions that would not allow sperm motility or as a way for polyploid species to forgo mispaired chromosomes during meiosis. As regenerated sporophytes in this work produced spores and had no abnormal physiology they likely resulted from the natural apogamy life cycle during tissue culture [32].

Additionally, it is important to note that induced apogamy *in vitro* usually yields sporophytes with abnormal morphology [33], but the sporophytes derived from callus tissue in our experiments were indistinguishable from sporophytes grown on soil and produced viable spores that could germinate.

Conclusion

Apogamous fern embryo development has been suggested as an early adaptation for colonizing dry land. Apogamous *P. vittata* callus could be further investigated to achieve a better understanding of genes and the physiology involved in *P. vittata* embryo development and apogamous sporophyte development. Additionally, the callus derived from *P. vittata* sporophytes was previously suggested as explants for transformation. However, callus does not seem to be suitable for use in the stable genetic transformation of *P. vittata*, and as such other targets for fern transformation are likely better.

Additional files

Additional file 1: Development of callus from *Pteris vittata* gametophytes and regeneration of sporophytes from callus.

Additional file 2: Flow cytometry dot blot and histograms showing gating on total events from extractions.

Competing interests

The authors declare that they have no competing interests.

Authors' contributions

BLJ: Performed experiments and drafted the paper. SE: Helped perform flow cytometry experiments. JD: Helped perform the microscopy studies. CNS: Along with BLJ, conceived the study and oversaw the research. All authors helped with data analysis, editing the manuscript, and approved of its content.

Acknowledgements

We would like to thank Keith Gwathney, Dianne Trent, and Dr. Stephen Kania for flow cytometry consultation. We would also like to thank Muthukumar Balasubramaniam for advise during manuscript preparation and revision. Funding for this study was provided by Edenspace Systems Corporation via an NSF STTR grant. We wish to thank Jennifer Hinds for help with formatting the manuscript. The manuscript was improved by the comments by Warren Hauk and anonymous reviewers.

Author details

[1]Department of Plant Sciences, The University of Tennessee, Knoxville, Tennessee 37996, USA. [2]Center for Wildlife Health, Department of Forestry, Wildlife and Fisheries, the University of Tennessee, Knoxville, Tennessee 37996, USA. [3]Advanced Microscopy and Imaging Center, The University of Tennessee, Knoxville, TN 37996, USA.

References

1. Ma LQ, Komar KM, Tu C, Zhang W, Cai Y, Kennelley ED: **A fern that hyperaccumulates arsenic.** *Nature* 2001, **409**(6820):579–579.
2. Basu S, Gangopadhyay G, Mukherjee B, Gupta S: **Plant regeneration of salt adapted callus of indica rice (var. Basmati 370) in saline conditions.** *Plant Cell Tissue Organ Cult* 1997, **50**(3):153–159.
3. Eudes F, Acharya S, Laroche A, Selinger LB, Cheng KJ: **A novel method to induce direct somatic embryogenesis, secondary embryogenesis and regeneration of fertile green cereal plants.** *Plant Cell Tissue Organ Cult* 2003, **73**(2):147–157.
4. Ovečka M, Bobák M: **Structural diversity of *Papaver somniferum* L. cell surfaces *in vitro* depending on particular steps of plant regeneration and morphogenetic program.** *Acta Physiologiae Plantarum* 1999, **21**(2):117–126.
5. Namasivayam P, Skepper J, Hanke D: **Identification of a potential structural marker for embryogenic competency in the *Brassica napus* spp. *oleifera* embryogenic tissue.** *Plant Cell Rep* 2006, **25**(9):887–895.
6. Šamaj J, Baluška F, Bobák M, Volkmann D: **Extracellular matrix surface network of embryogenic units of friable maize callus contains arabinogalactan-proteins recognized by monoclonal antibody JIM4.** *Plant Cell Rep* 1999, **18**(5):369–374.
7. Bondada B, Tu C, Ma L: **Surface structure and anatomical aspects of Chinese brake fern (Pteris vittata; Pteridaceae).** *Brittonia* 2006, **58**(3):217–228.
8. Duckett JG, Ligrone R: **The structure and development of haustorial placentas in Leptosporangiate ferns provide a clear-cut distinction between Euphyllophytes and Lycophytes.** *Ann Bot* 2003, **92**(4):513–521.
9. Elmore HW, Adams RJ: **Scanning electron-microspopic observations of gametophyte and sperm of bracken fern, *Pteridium aquilinum* (L) Kuhn.** *New Phytologist* 1976, **76**(3):519–522.
10. Gantt E, Arnott HJ: **Spore germination and development of the young gametophyte of the ostrich fern (*Matteuccia struthiopteris*).** *Am J Bot* 1965, **52**(1):82–94.
11. Sheffield E, Laird S, Bell P: **Ultrastructural aspects of sporogenesis in the apogamous fern *Dryopteris borreri*.** *J Cell Sci* 1983, **63**(1):125–134.
12. Whittier DP: **Rapid gametophyte maturation in *Ophioglossum crotalophoroides*.** *Am Fern J* 2009, **93**(3):137–145.
13. Muthukumar B, Joyce BL, Elless MP, Stewart CN Jr: **Stable transformation of ferns using spores as targets: *Pteris vittata* and *Ceratopteris thalictroides*.** *Plant Physiol* 2013, **163**:648–658.
14. Kato Y: **Physiological and morphogenetic studies of fern gametophytes and sporophytes in aseptic culture.12: sporophyte formation in dark-cultured gametophyte of *Pteris vittata* L.** *Botanical Gazette* 1970, **131**(3):205–210.
15. Kshirsagar MK, Mehta AR: ***In vitro* studies in ferns - growth and differentiation in rhizome callus of *Pteris vittata*.** *Phytomorphology* 1978, **28**(1):50–58.
16. Kwa SH, Wee YC, Loh CS: **Production of aposporus gametophytes and calli from *Pteris vittata* L pinnae strips cultured *in vitro*.** *Plant Cell Rep* 1991, **10**(8):392–393.
17. Trotta A, Mantovani M, Fusconi A, Gallo C: ***In vitro* culture of *Pteris vittata*, an arsenic hyperaccumulating fern, for screening and propagating strains useful for phytoremediation.** *Caryologia* 2007, **60**(1–2):160–164.
18. Kato Y: **Physiological and morphogenetic studies of fern gametophytes in aseptic culture. I. Callus tissues from dark-cultured *Pteris vittata*.** *Botanical Gazette* 1963, **124**:413–416.

19. Murashige T, Skoog F: **A revised medium for rapid growth and bio assays with tobacco tissue cultures.** *Physiologia Plantarum* 1962, **15**(3):473–497.
20. Gamborg OL, Miller RA, Ojima K: **Nutrient requirements of suspension cultures of soybean root cells.** *Exp Cell Res* 1968, **50**(1):151–158.
21. Yang X, Chen H, Xu W, He Z, Ma M: **Hyperaccumulation of arsenic by callus, sporophytes and gametophytes of *Pteris vittata* cultured *in vitro*.** *Plant Cell Rep* 2007, **26**(10):1889–1897.
22. Dawson NS, Dunlap JR, Walne PL: **Structure and elemental composition of pellicular warts of Euglena spirogyra (Euglenophyceae).** *Br Phycol J* 1988, **23**(1):61–69.
23. Dolezel J, Bartos J: **Plant DNA flow cytometry and estimation of nuclear genome size.** *Ann Bot* 2005, **95**(1):99–110.
24. Kazmierczak A: **Studies on morphology and metabolism of prothalli during GA(3)-induced formation of antheridia in anemia phyllitidis.** *Acta Physiologiae Plantarum* 1998, **20**(3):277–283.
25. Sondahl MR, Salisbury JL, Sharp WR: **SEM characterization of embryogenic tissue and globular embyros during high-frequency somatic embryogenesis in coffe callus cells.** *Zeitschrift Fur Pflanzenphysiologie* 1979, **94**(2):185–188.
26. Popielarska-Konieczna M, Kozieradzka-Kiszkurno M, Świerczyńska J, Góralski G, Ślesak H, Bohdanowicz J: **Ultrastructure and histochemical analysis of extracellular matrix surface network in kiwifruit endosperm-derived callus culture.** *Plant Cell Rep* 2008, **27**(7):1137–1145.
27. Verdeil JL, Hocher V, Huet C, Grosdemange F, Escoute J, Ferrière N, Nicole M: **Ultrastructural changes in coconut calli associated with the acquisition of embryogenic competence.** *Ann Bot* 2001, **88**(1):9–18.
28. Johnson G, Renzaglia K: **Evaluating the diversity of pteridophyte embryology in the light of recent phylogenetic analyses leads to new inferences on character evolution.** *Plant Systemat Evol* 2009, **283**(3):149–164.
29. Lal M, Narang A: **Ultrastructural and histochemical-studies of transfer cells in the callus and apogamous sporophytes of *Physcomitrium coorgense* broth.** *New Phytologist* 1985, **100**(2):225–231.
30. Baldwin PR, Butcher DJ: **Phytoremediation of arsenic by two hyperaccumulators in a hydroponic environment.** *Microchem J* 2007, **85**(2):297–300.
31. El Desouky FA, Laird SM, Sheffield E: **Oogenesis in the apogamous fern *Pteris cretica*.** *Ann Bot* 1990, **65**(3):297–303.
32. Chao Y, Liu H, Chiang Y, Chiou W: **Polyploidy and speciation in *Pteris* (Pteridaceae).** *J Bot* 2012.
33. Cordle AR, Bui LT, Irish EE, Cheng CL: **Laboratory-induced apogamy and apospory in *Ceratopteris richardii*.** In *Working with Ferns.* Edited by Kumar A, Fernandez H, Revilla MA. New York: Springer; 2010:25–36.

Saccharification and hydrolytic enzyme production of alkali pre-treated wheat bran by *Trichoderma virens* under solid state fermentation

Reda M. El-Shishtawy[1*], Saleh A. Mohamed[2*], Abdullah M. Asiri[1,3], Abu-bakr M. Gomaa[4], Ibrahim H. Ibrahim[2] and Hasan A. Al-Talhi[2]

Abstract

Background: In continuation of our previously interest in the saccharification of agriculture wastes by *Bacillus megatherium* in solid state fermentation (SSF), we wish to report an investigation and comparative evaluation among *Trichoderma* sp. for the saccharification of four alkali-pretreated agricultural residues and production of hydrolytic enzymes, carboxymethyl cellulase (CMCase), filter paperase (FPase), pectinase (PGase) and xylanase (Xylase) in SSF. The optimization of the physiological conditions of production of hydrolytic enzymes and saccharification content from *Trichoderma virens* using alkali-pretreated wheat bran was the last goal.

Methods: The physico-chemical parameters of SSF include incubation time, incubation temperature, moisture content of the substrate, incubation pH, supplementation with carbon and nitrogen sources were optimized.

Results: Saccharification of different solid state fermentation sources wheat bran, date's seeds, grass and palm leaves, were tested for the production of fermentable sugar by *Trichoderma* sp. The maximum production of hydrolytic enzymes CMCase, FPase, PGase and Xylase and saccharification content were obtained on wheat bran. Time course, moisture content, optimum temperature, optimum pH, supplementation with carbon and nitrogen sources were optimized to achieve the maximum production of the hydrolytic enzymes, protein and total carbohydrate of *T. virens* using alkali pre-treated wheat bran. The maximum production of CMCase, FPase, PGase, Xylase, protein and carbohydrate content was recorded at 72 h of incubation, 50-70 % moisture, temperature 25-35 °C and pH 5. The influence of supplementary carbon and nitrogen sources was studied. While lactose and sucrose enhanced the activity of PGase from 79.2 to 582.9 and 632.6 U/g, starch inhibited all other enzymes. This was confirmed by maximum saccharification content. Among the nitrogen sources, yeast extract and urea enhanced the saccharification content and CMCase, PGase and Xylase.

Conclusions: The results of this study indicated that alkali pre-treated wheat bran was a better substrate for saccharification and production of hydrolytic enzymes CMCase, FPase, PGase and xylase by *T. virens* compared to other alkali-pretreated agricultural residues tested.

Keywords: *Trichoderma* sp, Saccharification, Hydrolytic enzymes, Agriculture wastes

* Correspondence: elshishtawy@hotmail.com; saleh38@hotmail.com
[1]Chemistry Department, Faculty of Science, King Abdulaziz University, P.O. Box 80203, Jeddah 21589, Saudi Arabia
[2]Biochemistry Department, Faculty of Science, King Abdulaziz University, Jeddah, Kingdom of Saudi Arabia
Full list of author information is available at the end of the article

Background

Agricultural residues, forests and agro industrial practices generally accumulated in the environment and caused pollution problem. Active efforts were being made to convert these organic waste resources into either glucose or alcohol, and use this either as fuel or as a valuable starting material for chemical synthesis [1]. Saccharification of polysaccharides to glucose by microbial hydrolytic enzymes which had attracted the attention of the researchers, as this was the first step of bioconversion of organic material into valuable products such as sugar, fine chemicals and biofuels [2]. As the cost of cellulosic substrates play the central role in determining the economy of the saccharification process, lot of emphasis had been given to the usage of low price substrates and therefore screening of the agricultural wastes for release of sugars as organic wastes from renewable forest and agricultural residues [3]. The saccharification of different agro-wastes had been reported by other workers employing enzymes from different organisms [4–6].

Recently, a significant interest raised in using solid state fermentation (SSF) instead of submerged fermentation (SmF). The advantages of SSF in comparison to traditional SmF were better yields, easier recovery of products, the absence of foam formation and smaller reactor volumes. Moreover, contamination risks were significantly reduced due to the low water contents and, consequently, the volume of effluents decreases [7]. Another very important advantage was that, it permits the use of agricultural and agro-industrial residues as substrates which were converted into products with high commercial value like secondary metabolites, organic acids, pesticides, aromatic compounds, fuels and enzymes [8]. Furthermore, the utilization of these compounds helps in solving pollution, which otherwise caused their disposal [9]. For enzyme production, the costs of these techniques were lower and the production higher than submerged cultures [10,11].

Structural properties of cellulose such as the degree of crystallinity, the degree of polymerization and the surface area, limit accessibility of substrate to enzyme and had been demonstrated [12] to affect the rate of enzymatic hydrolysis of cellulose. Pretreatment methods, which disrupted the highly-ordered cellulose structure and the lignin-carbohydrate complex, remove lignin, and increase the surface area accessible to enzymes, promoted the hydrolysis, and increased the rate and extent of hydrolysis of cellulose in various lignocellulosic residues. The enzymatic hydrolysis of cellulosic materials correlated with the level of cellulose crystallinity [13] complete enzymatic hydrolysis of the polysaccharides of lignocelluloses required a concerted action of a complex array of hydrolases including cellulase, xylanase, pectinase, and other side-group cleavage enzymes [14].

Several cell-decomposing microorganisms produce cellulases which were the most economic and available sources for fermentable sugar production, because these microorganisms could grow on inexpensive media. The genus *Trichoderma,* filamentous ascomycetes were widely used in industrial applications because of high secretary capacity and inducible promoting characteristics [15]. The structural complexity were often easily degraded by xylanases, mannanases etc. which were present in some cellulase preparations, so that their presence may actually lead to increased production of reducing sugars and greater susceptibility of the residual cellulose [16–18].

In continuation of our interest in the saccharification of agriculture wastes by SSF [19], we wished to report an investigation and comparative evaluation among *Trichoderma* sp., *T. reesei, T. viride, T. harzianum* and *T. virens* for the saccharification of four alkali-pretreated agricultural residues, wheat bran, date's seeds, wild grass and palm's leaves under solid state fermentation for the production of hydrolytic enzymes, carboxymethyl cellulase (CMCase), filter paperase (FPase), pectinase (PGase) and xylanase (Xylase). The polysaccharide composition of these agricultural residues included different concentrations from cellulose, hemicelluloses and lignin [20–22]. The optimization of the physiological conditions of production of hydrolytic enzymes and saccharification content from *T. virens* using alkali-pretreated wheat bran was the last goal.

Methods

Microorganisms

T. reesei, T. viride, T. harzianum and *T. virens* were obtained from National Research Centre, Cairo, Egypt and maintained on potato dextrose agar. The slants were grown at 28 °C for seven days and stored at 4 °C.

Pretreatment of agricultural wastes

Wheat bran, date's seeds, grass and palm leaves were chosen as the sole nutrient source for solid-state fermentation (SSF). They dried in an oven at 80 °C for 24 h. The dried substrates were then milled in a commercial mill and sieved. The mean diameter of the dried substrates was 1.0 mm. The substrates were pretreated with 1.0 M NaOH at 121 °C and 15 psi pressure for 1 hr at the ratio of 1:10 (w/v) [23,24]. The pretreated materials were washed with tap water until the pH of the filtrate reached 7.0. The washed materials were dried at 60 °C overnight to constant weight and stored at room temperature for further use.

Inoculum medium

The medium used for inoculum of *Trichoderma* sp. preparation contained (g l^{-1}): KH_2PO_4, 28; $(NH_4)_2SO_4$, 19.6; Urea, 4.2; $MgSO_4$. $7H_20$, 4.2; $CoCl_2$, 4.2; $FeSO_4$. $7H_20$, 0.07; $MnSO_4$. $7H_20$, 0.021; ZnSO4 $7H_20$, 0.019; $CaC1_2$, 0.028;

yeast extract, 7; and glucose, 15; pH 5.0 ± 0.2. The media were sterilized by autoclaving at 121 °C pressure of 15 psi for 15 min. The culture was incubated and shaken at 30 °C for 48 hr in an orbital shaking incubator at 150 rpm before transferring to the production medium [25].

Solid state fermentation

SSF was performed to study the effect of various physicochemical parameters required for the optimum production of enzymes and saccharification content by *Trichoderm* sp. prior to inoculation, the agriculture waste was sterilized in an autoclave for 20 min at 121 °C and 1.2 atmospheres. To each 50 ml Erlenmeyer flask, 5 g of sterilized agriculture waste, 5×10^5 spores/g, and appropriate amount of water (10 % moisture) were added. The physico-chemical parameters included incubation time, incubation temperature (20, 30, 35, 40, 45 °C), moisture content of the substrate (10 %, 20 %, 40 %, 60 %, 100 %) and incubation pH (4 to 8) were optimized. The pH was adjusted using 0.1 M NaOH or HCl. Studies were also performed to evaluate the influence of different carbon sources (glucose, maltose, starch, sucrose, lactose at 1 % w/v) and nitrogen sources (yeast extract, urea, sodium nitrate, ammonium sulphate, ammonium chloride at 1 % w/v) when added to the fermentation medium contained agriculture waste. Each experiment is done in 3 sets.

Enzyme extraction

Crude enzyme was extracted by mixing a 5 g of fermented matter with 50 ml distilled water on a rotary shaker (180 rpm/min) overnight. The suspension was then centrifuged at 12000 rpm for 10 min and the supernatant was designated as a crude extract.

Enzyme assays

Carboxymethylcellulase (CMCase), filter paperase (FPase), pectinase (PGase) and xylanase (Xylase) activities were assayed by determining the liberated reducing end products using glucose, glucose, galacturonic acid and xylose as standards, respectively [26]. Substrates used were CM-cellulose, filter paper, polygalacturonic acid and birchwood xylan for CMCase, FPase, PGase and Xylase, respectively. The reaction mixture (0.5 ml) contained 1 % substrate, 0.05 M sodium acetate buffer pH 5.5 and 0.1 ml crude extract. Assays were carried out at 37 °C for 1 h. Then 0.5 ml dinitrosalicylic acid reagent was added to each tube. Then the reaction mixture was mixed well, and heated in a boiling water bath for 10 min. After cooling to room temperature, the absorbance was measured at 560 nm. One unit of enzyme activity is defined as the amount of enzyme which liberated one μmol of reducing sugar per min under standard assay conditions. All the experimental work was run in triplicates.

Protein determination

Protein concentration was determined according to the dye binding method of Bradford [27] with bovine serum albumin as standard.

Determination of total reducing sugars

Total reducing sugars were determined by the method of Miller [26]. The reaction mixture contained 0.5 ml of crude extract and 0.5 ml dinitrosalicylic acid reagent. The tubes were heated in a boiling water bath for 10 min. After cooling to room temperature, the absorbance was measured at 560 nm. Glucose served as the calibration standard for total reducing sugar determination.

Determination of total soluble carbohydrates

Total soluble carbohydrates were determined by the method of Dubois et al. [28]. The reaction mixture contained 25 μl of a 4:1 mixture of phenol and water, 0.8 ml of crude extract and 2 ml of concentrated sulfuric acid. Then mixed well, and heated in a boiling water bath for 30 min. The absorbance was determined at 480 nm. Glucose served as the calibration standard for total carbohydrate determination.

Statistical analysis

The obtained data were statistically analyzed as a randomized complete block design with three replicates by analysis of variance (ANOVA) using the statistical package software SAS (SAS Institute Inc., 2000, Cary, NC., USA). Comparisons between means were made by *F*-test and the least significant differences (LSD) at level $P = 0.05$. Correlations coefficient among the different parameters were also calculated by SAS.

Results and discussion

Production of CMCase, FPase, PGase and Xylase were tested using alkali pretreated wheat bran, date's seeds, wild grass and palm leaves as substrates by *T. reesei*, *T. viride*, *T. harzianum* and *T. virens* in solid state fermentation (SSF). Figure 1 showed that the maximum production of CMCase, FPase, Xylase were obtained by *T. virens* (123.26, 49.3 and 348 U/g solid, respectively) in SSF containing alkali pre-treated wheat bran, while maximum PGase activity was obtained by *T. reesei* (499.9 U/g solid) in SSF containing alkali pre-treated palm leaves. Among the different solid substrates screened, saccharification of alkali pre-treated wheat bran supported maximum yields of total carbohydrates and reducing sugars (45 and 38.92 mg/g solid, respectively) by *T. virens* compared with other *Trichoderma* sp. tested (Fig. 2). There was no relation between total carbohydrate/sugar and enzyme contents, except of wild grasses and wheat bran. This may be attributed to some fungi used the total carbohydrate and sugar as nutrients (produced by

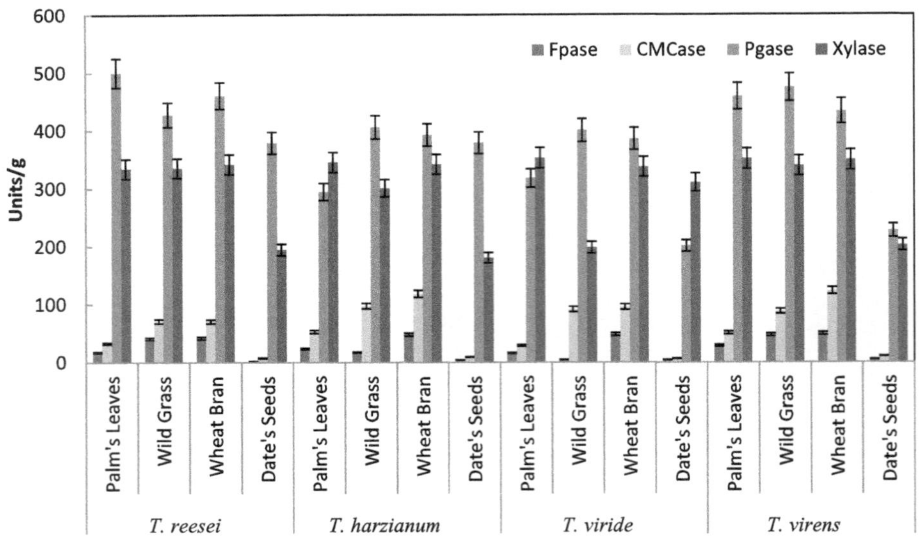

Fig. 1 Effect of different organic materials on the production of CMCase, FPase, PGase and Xylase by *Trichoderma* sp. in SSF. Process conditions: incubation times 72 h, initial moisture content 50 % (by volume per mass) and temperature 30 °C. The data presented were averages of three experiments

enzymes act on certain agricultural residues as palm leaves and date seeds) for growth of fungi, resulted in the depletion of the total carbohydrate and sugar. From these results, the optimization of the production of total carbohydrate, reducing sugar, CMCase, FPase, PGase and Xylase of *T. virens* using alkali pre-treated wheat bran in SSF was performed in the following studies.

Figure 3 shows the time course experiments of CMCase, FPase, Xylase, PGase and protein production by *T. virens* grown on alkali pretreated wheat bran in SSF. All enzymes and protein exhibited their maximum activities 58.2, 65.6, 372.4, 474 U/g and 8.5 mg/g at 72 hours. However, time course of enzyme cellulase production by *Trichoderma* spp. was studied using steamed alkali-treated sugarcane bagasse at 30 °C for six days. CMCase biosynthesis was not detected up to 24 hours of incubation and then the enzyme activity increased sharply up to 72 hours and highest CMCase activity was found 1.31 U/ml at 144 hours. FPase production started after a large lag period (about 72 hours) and thereafter the enzyme synthesis increase sharply. The final FPase activity was 0.110 U/ml at 144 hours. Initially, Xylase

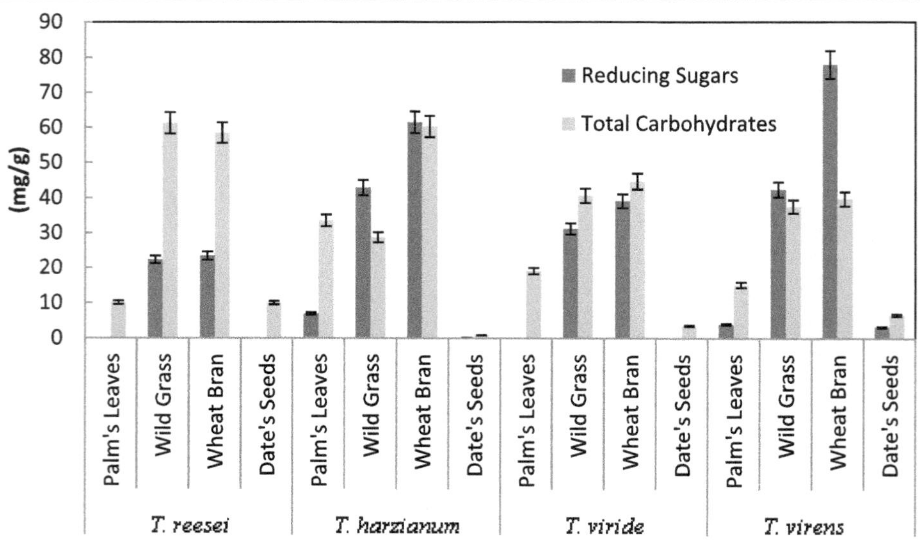

Fig. 2 Effect of different organic materials on Saccharification content by *Trichoderma* sp. in SSF. Process conditions: incubation times 72 h, initial moisture content 50 % (by volume per mass) and temperature 30 °C. The data presented were averages of three experiments

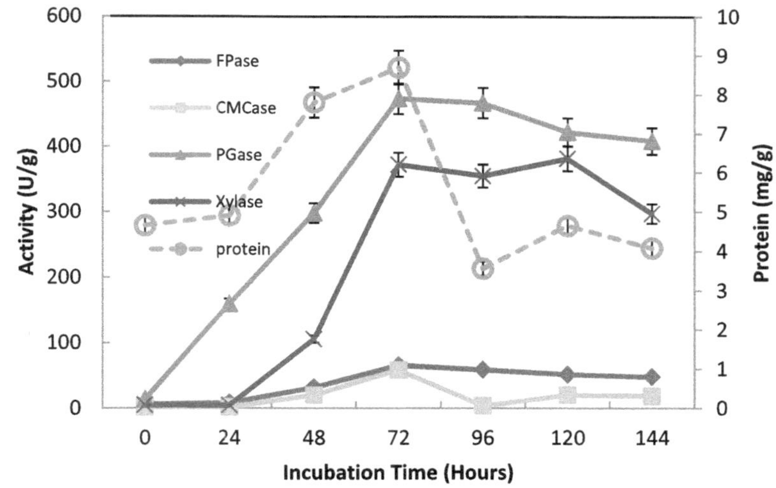

Fig. 3 Effect of incubation time on the production of CMCase, FPase, PGase, Xylase and protein by *T. virens* in SSF using alkali pre-treated wheat bran as a substrate. Process conditions: incubation times 72 h, initial moisture content 50 % (by volume per mass) and temperature 30 °C. The data presented were averages of three experiments

production was low and at 96 hours a sharp increase in enzyme production was observed and after that the enzyme activity remained constant. The highest Xylase activity was found after 144 hours of incubation which was 13.250 U/ml [29]. Of all four strains of fungi, *T. reesei* and *T. viride* were the best cellulase-producing fungi after 72 h offer mutation. A co-culture of *T. reesei* and *T. viride* gave better results with wheat straw; the CMCase activities were 560.9 and 212.7 IU, respectively [30]. Another study reported that PGase and Xylase of *T. virens*, exhibited their maximum activity at day 4 (132 and 75 units/g solid, respectively) [31]. Similarly, *Penicillium decumbens* was grown on a mixture of corn straw (90 %) and wheat bran (10 %) the maximum activity of Xylase was measured after 4 days of fermentation [32]. Couri et al. [33] also studied the production of Xylase by *A. niger* using different agroindustrial residues – mango peel and wheat bran – as the solid substrate. The maximum of xylase activity was reached after 72 h and 24 h of fermentation using wheat brain and mango peel, respectively.

The increase in glucose production depends on availability of cellulose in the medium and also due to the specific binding of the enzymes with substrates [34]. Saccharification of alkali pre-treated wheat bran in SSF was shown in Fig. 4 with a maximum yield 49.1 mg/g reducing sugars and 71.9 mg/g total carbohydrates obtained at 72 h by *T. virens*. Begum and Alimon [35] obtained the maximum amount of reducing sugar (4.15 mg/g) measured in sugarcane bagasse after alkali pre-treatment at 24 h by *Aspergillus oryzae* ITCC-4857.01. They also obtained the highest saccharification for alkali-reated sugarcane bagasse at 96 hrs when water hyacinth induced enzyme was used. They concluded that the saccharification rates of alkali pre-treated substrates were higher than those of enzymatic treated substrates. Similarly, Ja'afaru and Fagade [36] recorded the highest reducing sugar 5.55 mg/ml for alkali pre-treated corn cob at 48 hrs. Time course of enzymatic saccharification of alkali pre-treated bagasse showed rapid initial increase of reducing sugar concentration (up to 8 h) and the rate of this increase was substantially reduced at later stages by *T. viride* [37].

The moisture content was an important factor that influences the growth and product yield in SSF [38]. Moisture was reported to cause swelling of the substrates, thereby facilitating better utilization of the substrate by microorganisms [39,40]. The data presented in the Figs. 5 and 6, clearly indicated a maximum production of the enzymes CMCase, FPase, Xylase and PGase ranged from 50 to 70 % moisture. The highest level of protein production was detected at 50 % moisture. For CMCase production moisture level was optimized and found that 40 % moisture level was best for production by *T. viridei* [41]. Any further increase in the ratio resulted in the decrease of enzyme yields may be due to clumping of solid particles which results in the decrease of interparticle space leading to decreased diffusion of nutrients [40,42]. In contrast, the low moisture content leads to the decreased solubility of nutrients present in the wheat bran thereby decreases enzyme yields [43]. The optimum moisture contents for Xylase production by *T. longibrachiatum* and *A. tereus* were 55 and 75 %, respectively [44]. A high production of Xylase of *Aspergillus* species was detected at 40–50 % moisture using dry koji as substrate [45]. An initial moisture content of 40 % provided

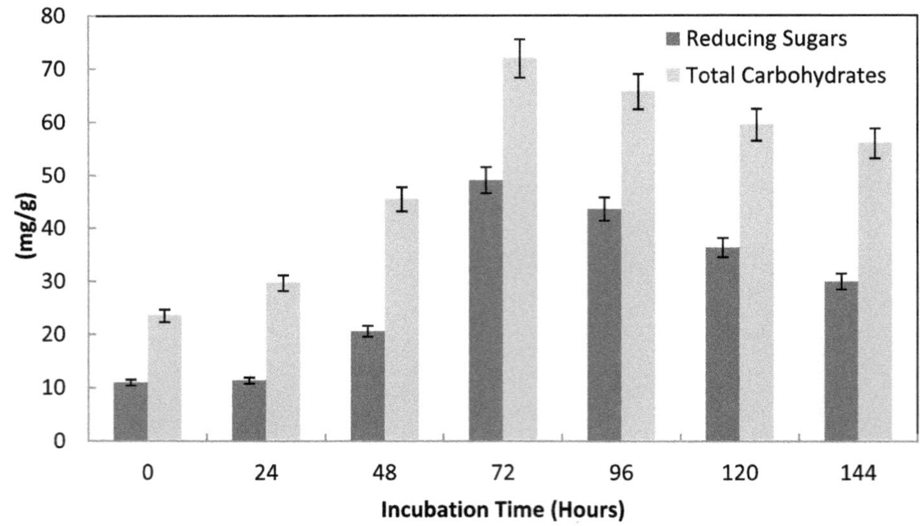

Fig. 4 Effect of incubation time on the saccharification content by *T. virenss* in SSF using alkali pre-treated wheat bran. Process conditions: initial moisture content 50 % (by volume per mass) and temperature 30 °C. The data presented were averages of three experiments

better conditions for production of PGase from *A. niger* than those of 25, 55, and 70 % [46].

The incubation temperature is very important in enzyme production [47]. The usual temperature maintained in SSF systems was in the range of 25-32 °C, depending on the growth kinetics of microorganism employed for fermentation purposes [48]. In the present study, the optimum temperature for maximum production of both CMCase and PGase was 30 °C; and 25 °C for FPase and xylase at 35 °C (Fig. 7). The protein content was ranged from 7 to 8 mg/g for all temperatures tested with high level at 35 °C. This was confirmed by maximum reducing sugar (49.2 mg/g) at 25 °C and total carbohydrates (55.5 mg/g) at 30 °C (Fig. 8). In contrast, maximum hydrolysis of substrates occurred at 50 °C [49]. Other studies had reported that 40 °C was found best for CMCase secretion by *T. viride* [41]. Optimal PGase and Xylase production (130 and 74 units/g solid, respectively) was obtained at 35 °C and 28 °C for *T. harzianum*, respectively [31]. On the contrary, the maximum activity of PGase and Xylase (120 and 55 units/g solid, respectively) of *T. virens* was detected at 28 °C and 35 °C, respectively. Similar optimal temperatures of production of PGase and xylanase from *Penicillium decumbens* [32], *A. niger* [33], *A. oryzae* [47], and *A. awamori* [50] were ranged from 28 °C to 32 °C. In the steady state operation for production of Xylase by

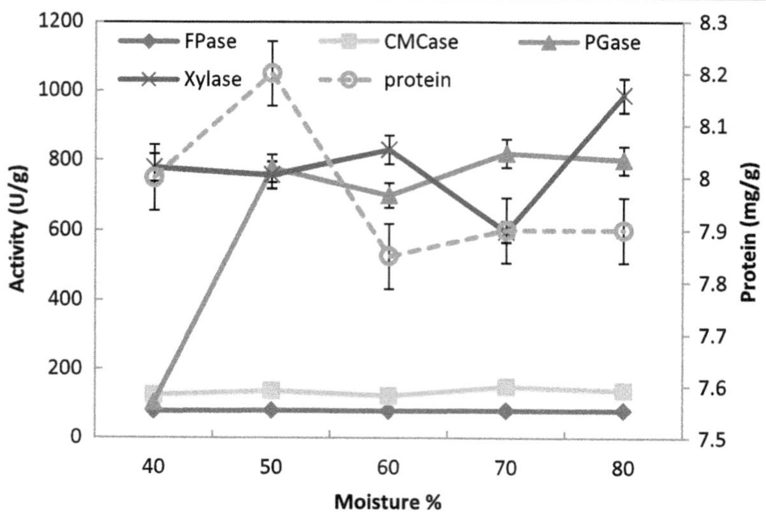

Fig. 5 Effect of moisture % on the production of CMCase, FPase, PGase, Xylase and protein by *T. virens* in SSF using alkali pre-treated wheat bran as a substrate. Process conditions: incubation times 72 h, and temperature 30 °C. The data presented were averages of three experiments

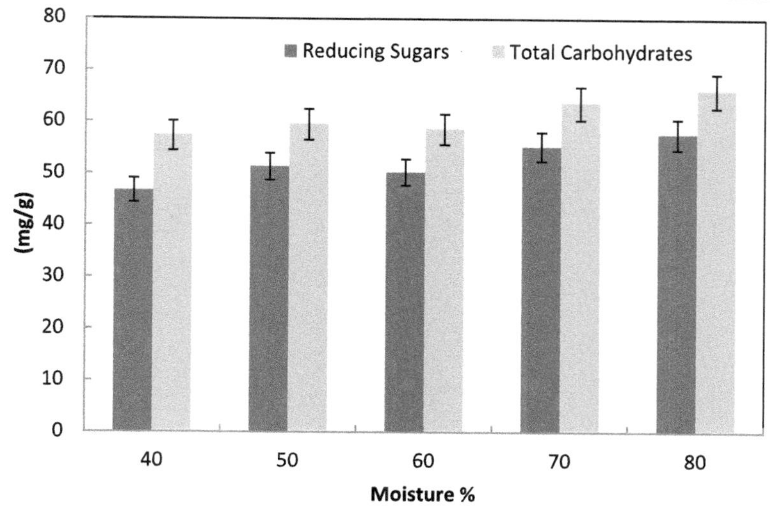

Fig. 6 Effect of moisture % on the saccharification content by *T. virens* in SSF using alkali pre-treated wheat bran as substrate. Process conditions: incubation times 72 h and temperature 30 °C. The data presented were averages of three experiments

A. niger the optimum temperature was 28 °C. Results showed that higher incubation temperature favors biomass growth and lower temperature favors the biosynthesis of xylanase [51]. The maximal PGase production by mixed culture of *A. niger* and *Saccharomyces cerevisiae* was detected at 37 °C [52].

Figure 9 showed a maximum activity of CMCase (110 U/g), FPase (95 U/g), Xylase (910 U/g), PGase (820 U/g) and protein (9 mg/g) for *T. virens* at pH 5.0. This was reinforced by a maximum saccharification at the same pH 5 (Fig. 10). Another study reported a maximum degree of saccharification at pH 5.0 by *Trichoderma* sp. [29]. Maximum CMCase activity (16.2 U/ml) was also obtained at pH 5.5 by *T. viride* [41]. Optimal PGase and Xylase production of *T. harzianum* was obtained at pH 7 and 6 with 120 and 70 units/g solid, respectively. On the contrary, the maximum activity of PGase and xylanase of *T. virens* was detected at pH 6.0 and 7.0 with 140 and 60 units/g solid, respectively [31]. At the medium pH (6.0), the maximal xylanase production by *A. terreus* under SSF using palm as substrate was reported [53]. Patil and Dayanand [54] reported that pH 5.0 was

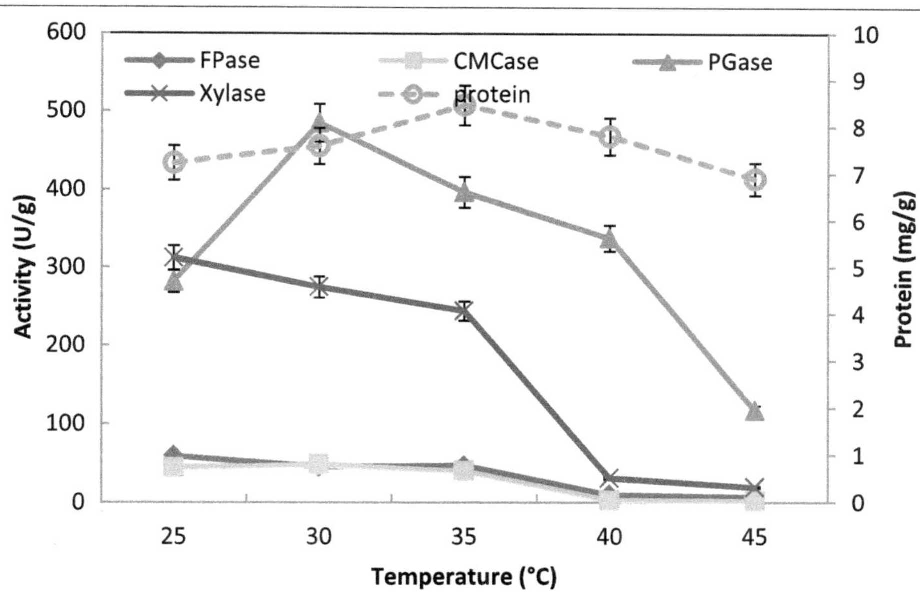

Fig. 7 Effect of incubation temperature on the production of CMCase, FPase, PGase, Xylase and protein by *T. virens* in SSF using alkali pre-treated wheat bran as a substrate. Process conditions: incubation times 72 h and initial moisture content 50 % (by volume per mass). The data presented were averages of three experiments

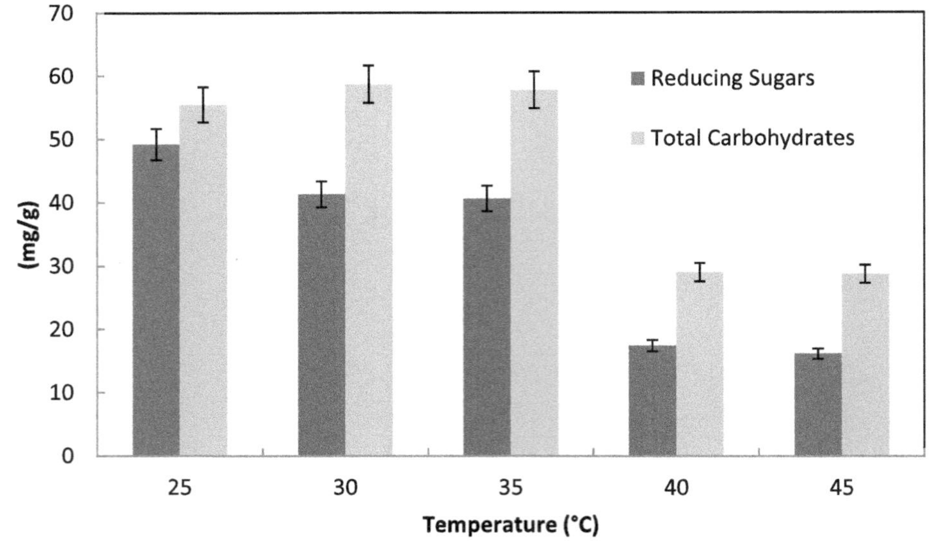

Fig. 8 Effect of incubation temperature on the saccharification content by *T. virens* in SSF using alkali pre-treated wheat bran as substrate. Process conditions: incubation times 72 h and initial moisture content 50 % (by volume per mass).The data presented were averages of three experiments

optimum for the maximum production of pectinases of *A. niger* using deseeded sunflower head in both submerged and solid state fermentation.

The influence of supplementary carbon sources such as starch, sucrose, maltose, lactose or glucose at 1 % (by mass) on production of CMCase, FPase, Xylase, PGase and protein was studied. The protein content was slightly increased in presences of the carbon sources. While lactose and sucrose enhanced the activity of PGase from 79.2 to 582.9 and 632.6 U/g solid, respectively, starch inhibited all other enzymes (Fig. 11). This was confirmed by maximum saccharification content of 54.9 and 63.4 mg/g reducing sugars respectively by lactose and sucrose, and minimum reducing sugars of 28.1 mg/g by starch (Fig. 12). However, starch and sucrose enhanced the Xylase activities from 40 to 55–60 units/g solid, while all carbon sources exhibited slightly effect on PGase activities for *Trichoderma* spp. using cantaloupe and watermelon rinds [31]. Botella et al. [50] reported that when 6 % glucose was added as extra

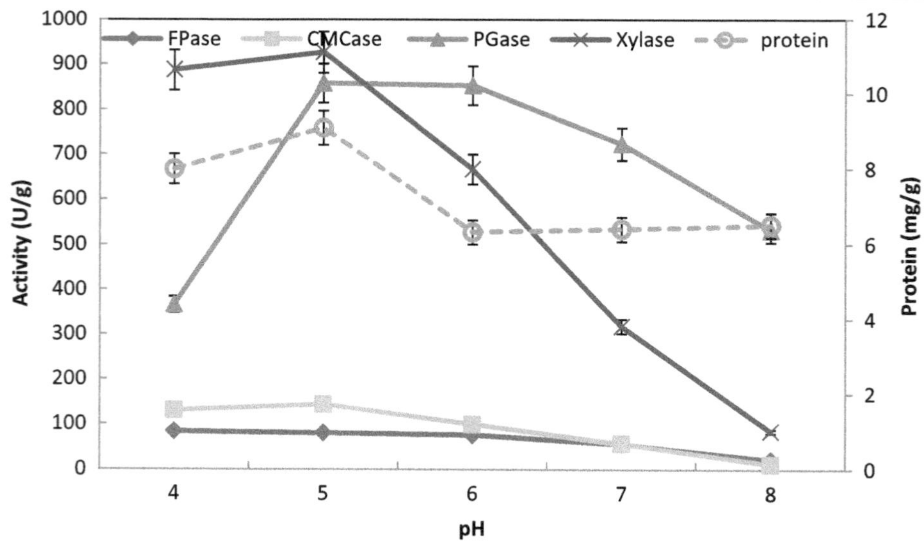

Fig. 9 Effect of pH on the production of CMCase, FPase, PGase, Xylase and proein by *T. virens* in SSF using alkali pre-treated wheat bran as a substrate. Process conditions: incubation times 72 h, initial moisture content 50 % (by volume per mass) and temperature 30 °C. The data presented were averages of three experiments

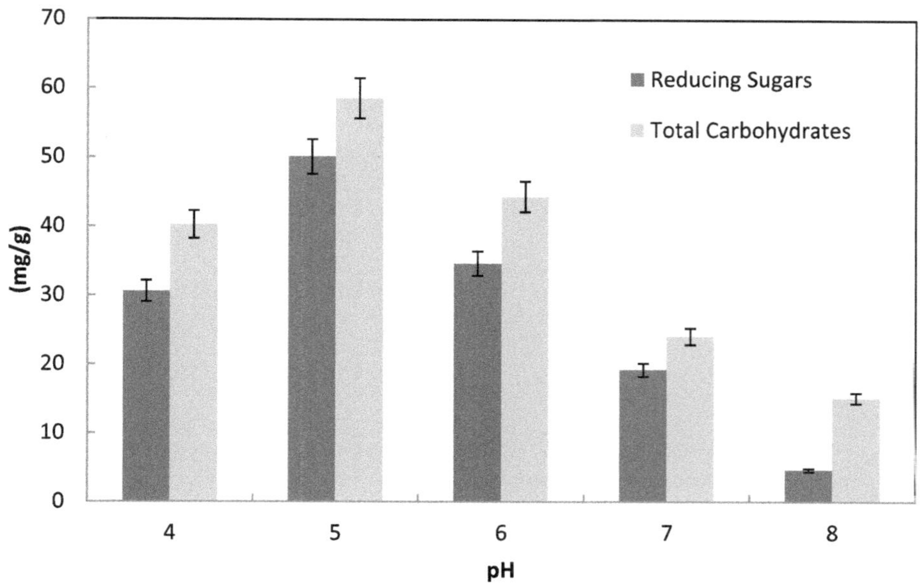

Fig. 10 Effect of pH on the saccharification content by *T. virens* in SSF using alkali-pre-treated wheat bran. Process conditions: incubation times 72 h, initial moisture content 50 % (by volume per mass) and temperature 30 °C. The data presented were averages of three experiments

carbon source, the production of Xylase and PGase by *A. awamori* using grape pomace as substrate increased significantly. However, at 8 % both enzyme activities declined. In contrast, when wheat bran was used as the solid substrate, Xylase by *A. tamari* was resistant to catabolic repression even at 10 % glucose [55]. Alternatively, D-glucose, D-mannose, maltose, sucrose and cellobiose significantly repressed CMCase formation in case of using rice straw and sugar cane bagasse as carbon sources [56].

Studies on supplementation of nitrogen sources such as ammonium sulphate, ammonium nitrate, ammonium chloride, yeast extract or urea at 1 % concentration to the solid substrates showed various effects on CMCase, FPase, Xylase, PGase and protein production by *T. virens* (Fig. 13). All nitrogen sources enhanced the saccharification enzymes with significant enhancement for CMCase (from 86.9 to 204 U/g), PGase (from 491.1 to 939.4 U/g) and Xylase (from 455 to 1064 U/g) by yeast extract and urea. This was confirmed by maximum saccharification

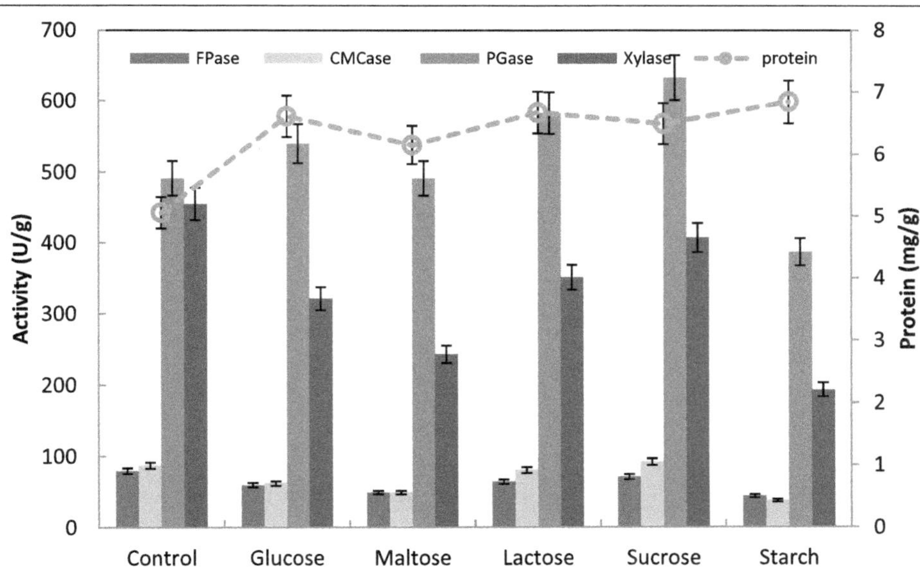

Fig. 11 Effect of carbon source (1 %) supplementation on the production of CMCase, FPase, PGase, Xylase and protein by *T. virens* in SSF using alkali pre-treated wheat bran as a substrate. Process conditions: incubation times 72 hr, initial moisture content 50 % (by volume per mass) and temperature 30 °C. The data presented were averages of three experiments

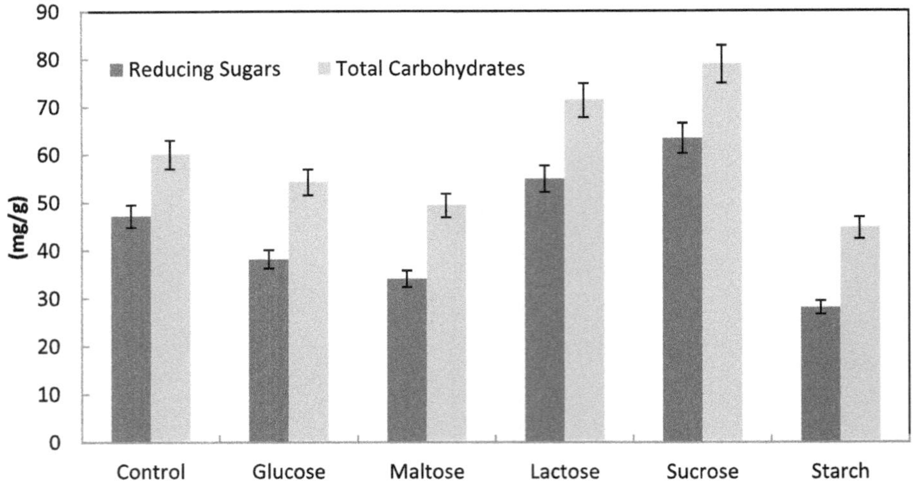

Fig. 12 Effect of carbon source (1 %) supplementation on the saccharification content by *T. virens* in SSF using alkali pre-treated wheat bran. Process conditions: incubation times 72 h, initial moisture content 50 % (by volume per mass) and temperature 30 °C. The data presented were averages of three experiments

from 47.2 to 157.1 and 162.1 by yeast extract and urea respectively (Fig. 14). However, ammonium sulphate, ammonium nitrate, yeast extract and urea increased PGase activities of *T. harzianum* from 90 to 110–113 units/g solid and decreased PGase activities of *T. virens.* The high level of protein was detected in presence of KNO_3. However, urea increased Xylase activities of *T. harzianum* and *T. virens* [30]. Among nitrogen sources yeast extract was the best to enhance the enzyme activity FPase (0.281 ± 0.13 IU/g) and CMCase (3.66 ± 0.02 IU/g) by *A. fumigatus* grown on alkali-pretreated sawdust [57]. Another study reported the highest level of enzyme formation expressed in terms of specific activity with ammonium chloride with both rice straw and sugar cane bagasse by *A. terreus* DSM 826. Other nitrogen sources namely ammonium sulphate, potassium nitrate and ammonium oxalate gave also considerable amounts of CMCase with both wastes as compared with sodium nitrate [56].

Comparison of CMCase, FPase, PGase and xylase production pattern by *T. virens* used in this study and other microorganisms confirmed the potential of *T. virens* for economic hydrolytic enzymes production (Table 1). In fact, the present microorganism was better producer of

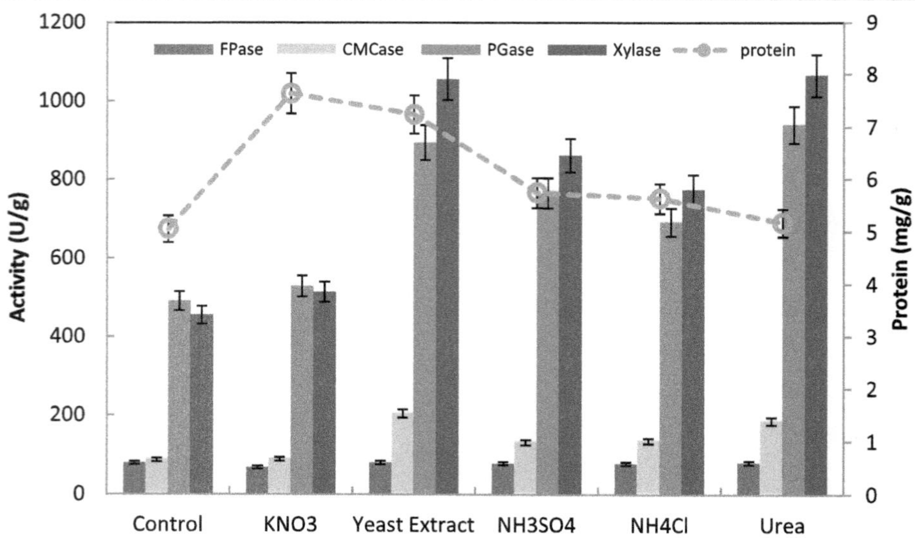

Fig. 13 Effect of nitrogen source (1 %) supplementation on the production of CMCase, FPase, PGase, Xylase and protein by *T. virens* in SSF using alkali pre-treated wheat bran as a substrate. Process conditions: incubation times 72 h, initial moisture content 50 % (by volume per mass) and temperature 30 °C. The data presented were averages of three experiments

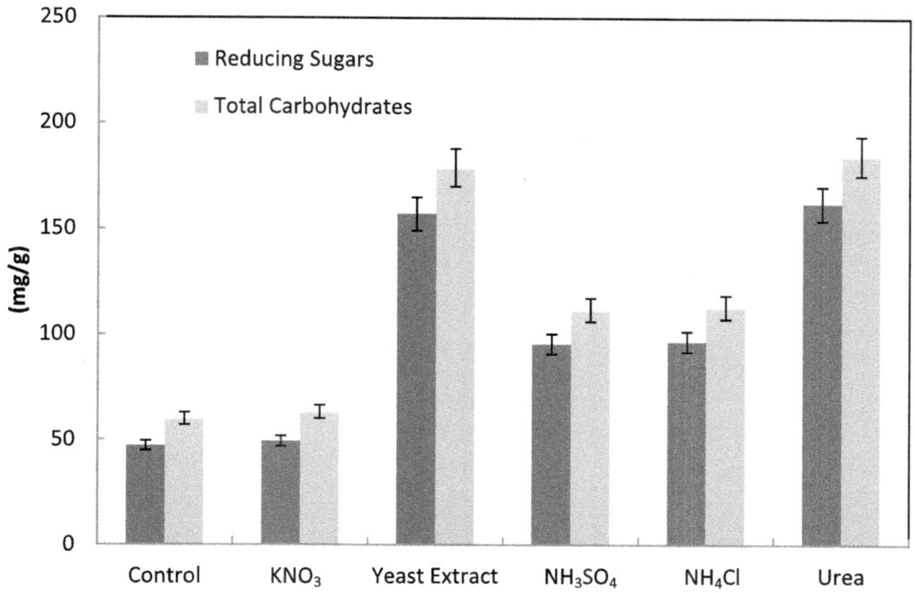

Fig. 14 Effect of nitrogen source (1 %) supplementation on the saccharification content by *T. virens* in SSF using alkali-pre-treated wheat bran. Process conditions: incubation times 72 h, initial moisture content 50 % (by volume per mass) and temperature 30 °C. The data presented were averages of three experiments

Table 1 Sacchrification enzyme activities of different fungal isolates grown on lignocellulosic substrates under solid state fermentation

Enzyme source	Carbon source	Enzyme activities (U/g dry substrate)				References
		FPase	CMCase	Xylase	PGase	
Aspergillus ustus	Rice straw	5.82	12.58	740	-	Shamala and Sreekantiah [58]
	Wheat bran	3.78	11.84	615.26	-	
Aspergillus soja	Crushed maize	-	-	-	30	Ustck et al. [59]
Aspergillus awamori	Grape pomace	-	-	40	25	Botella et al. [60]
Aspergillus terreus M11	Corn stover	243	581	-	-	Gao et al. [61]
Aspergillus niger KK2	Rice straw	19.5	129	5070	-	Kang et al. [62]
Aspergillus niger	Ordos Plateau	-	-	-	36	Debing et al. [63]
Aspergillus niger	Citrus peel	-	-	65	18	Rodriguez-fernandez et al. [64]
Aspergillus niger	Deseeded sunflower	-	-	-	34	Patil and Dayanand [54]
Myceliophthora sp. IMI 387099	Rice straw	2.44	32.9	900.2	-	Badhan et al. [65]
	Wheat straw	1.37	30.8	656.6	-	
	Bagasse	0.7	6.62	620.1	-	
	Corn cob	0.31	11.38	411.6	-	
	Wheat bran	0.74	26.6	128.9	-	
Thermoascus aurantiacus	Wheat straw	4.3	956	2973	-	Kalogeris et al. [66]
Trichoderma harzianum and *Trichoderma virens*	cantaloupe and watermelon	-	-	80	140	Mohamed et al. [31]
Trichoderma harzianum SNRS3	Rice straw	6.25	111.31	433.75	-	Rahnama et al. [67]
Trichoderma reesei MCG77	Rice bran	2.314	-	-	-	Latifan et al. [68]
Trichoderma virens	Wheat bran	49.3	123.26	348	499.9	Present study

FPase, CMCase, Xylase and PGase (approximately 49.3, 123.26, 348 and 499.9 units/g solid, respectively) compared to other microorganisms.

Conclusion

The present study revealed the saccharification potential of *T. virens* on alkali pre-treated wheat bran as an agricultural waste in SSF. The optimal conditions for production of CMCase, FPase, PGase and xylase and sccharification content utilizing alkali pre-treated wheat bran as the solid substrate in SSF included incubation for 72 h, temperature at 25-35 °C, substrate moisture content of 50-70 % and pH at 5.0.

Competing interests

The authors declare that they have no competing interests.

Authors' contributions

El-R, MS, AA, GM, II and AL-H performed all experiments and read and approved the final manuscript.

Acknowledgement

This project was funded by the National Plan for Science, Technology and Innovation (MAARIFAH) – King Abdulaziz City for Science and Technology - the Kingdom of Saudi Arabia – award number (11-ENE1527-03). The authors also, acknowledge with thanks Science and Technology Unit, King Abdulaziz University for technical support.

Author details

[1]Chemistry Department, Faculty of Science, King Abdulaziz University, P.O. Box 80203, Jeddah 21589, Saudi Arabia. [2]Biochemistry Department, Faculty of Science, King Abdulaziz University, Jeddah, Kingdom of Saudi Arabia. [3]The Center of Excellence for Advanced Materials Research, King Abdulaziz University, Jeddah 21589, Saudi Arabia. [4]Biology Department, Faculty of Science, King Abdulaziz University, Jeddah, Kingdom of Saudi Arabia.

References

1. Kumakura M, Kasai N, Tameda M, Kaetsu I. Method of pretreatment in saccharification and fermentation of waste cellulose resource. US Pat. 1988;4:769–082.
2. Howard RL, Abotsi E, Jansen Van REL, Howard S. Lignocellulose biotechnology: issues of bioconversion and enzyme production. Afric J Biotechnol. 2003;2:602–19.
3. Heck JX, Hertz PF, Ayub MAZ. Cellulase and xylanase production by isolated Amazon *Bacillus* strains using soybean industrial residue based solid state cultivation. Braz J Microbiol. 2002;33:213–8.
4. Baig MMV, Baig MLB, Baig MIA, Yasmeen M. Saccharification of banana agro-waste by cellulolytic enzymes. Afric J Biotechnol. 2004;3:447–50.
5. Katzen R, Fowler DE. Ethanol of lignocellulosic waste with utilization of recombinant bacteria. Appl Biochem Biotechnol. 1994;45:697–707.
6. van Wyk JPH, Leogale PB. Saccharification of wastepaper mixtures with cellulase from *Penicillium funiculosum*. Biotechnol Lett. 2001;23:1849–52.
7. Raimbault M. General and microbiological aspects of solid substrate fermentation. Elec J Biotech. 1998;1:1–15.
8. Martins ES, Silva D, Da Silva R, Gomes E. Solid state production of thermostable pectinases from thermophilic *Thermoascusurantiacus*. Process Biochem. 2002;37:949–54.
9. Couto SR, Sanroman MA. Application of solid-state fermentation to food industry – a review. J Food Eng. 2005;22:211–9.
10. Pandey A. Aspects of fermenter design for solid-state fermentations. Process Biochem. 1991;26:355–61.
11. Sukumaran RK, Singhania RR, Mathew GM, Pandey A. Cellulase production using biomass feed stock and its applicationin lignocellulose saccharification for bio-ethanol production. Renew Energ. 2009;34:421–4.
12. Fan LT, Lee YH, Gharpuray MM. The nature of lignocellulosics and their pretreatments for enzymatic hydrolysis. Advan Biochem Eng. 1982;23:157–87.
13. Weimer P, Weston W. Relationship between the fine structure of native cellulose and cellulose degradability by the cellulase complexes of *Trichoderma reesei* and *Clostridium thermocellum*. Biotechnol Bioeng. 1985;27:1540–7.
14. Broda P, Birch PRJ, Brooks PR, Sims PFG. Lignocellulose degradation by *Phanerochaete chrysosporium*: gene families and gene expression for a complex process. Mol Microbiol. 1996;19:923–32.
15. Mach RL, Zeilinger S. Regulation of gene expression in industrial fungi: Trichoderma. Appl Microbiol Biotechnol. 2003;60:515–22.
16. Tolan JS, Finn RK. Fermentation of D-xylose to ethanol by genetically *modified Klebsiella planticola*. Appl Environ Microbiol. 1987;53:2039–44.
17. Beck MJ. Fermentation of pentoses from wood hydrolysates. Wallingford, UK: CAB International; 1993.
18. Hahn-Hgerdal B, Hallborn J, Jeppsson H, Olsson L, Skoog K, Walfridsson M. Pentose fermentation to alcohol. In: Saddler JN, editor. Bioconversion of forest and agricultural plant residues. Wallingford: CAB International; 1993. p. 231–90.
19. El-Shishtawy RM, Mohamed SA, Asiri AM, Gomaa AM, Ibrahim IH, Al-Talhi HA. Solid fermentation of wheat bran for hydrolytic enzymes production and saccharification content by a local isolate *Bacillus megatherium*. BMC Biotechnol. 2014;14:29.
20. Brilluet JM, Mercier C. Fractionation of wheat bran carbohydrates. J Sci Food Agric. 1981;32:243–51.
21. Hui YH. Fruit and fruit processing. Iowa: Black Publishing. Anes; 2006. p. 391–441.
22. Okia Y, Saito T, Isogai A. TEMPO-mediated oxidation of soft wood thermomechanical pulp. Holzforschung. 2009;63:529–35.
23. Detroy RW, Cunningham RL, Bothast RJ, Bagby MO, Herman A. Bioconversion of wheat straw cellulose/hemicellulose to ethanol by Saccharomyces uvarum and *Pachysolen tannophilus*. Biotechnol Bioeng. 1982;24:1105–13.
24. Lynd LR, Wolkin RH, Grethlein HE. Continuous fermentation of pretreated hardwood and avicel by Clostridium thermocellum. Biotechnol Bioeng Symposium. 1987;17:265–74.
25. Duenas R, Tengerdy RP, Correa MG. Cellulase production by mixed fungi in solid-substrate fermentation of bagasse. World J Microbiol Biotechnol. 1995;11:333–7.
26. Miller GL. Use of dinitrosalicylic acid reagent for determination of reducing sugar. Anal Chem. 1959;31:426–9.
27. Bradford MM. A rapid sensitive method of quantitation micro gram quantities of proteins utilizing the principles of protein-dye binding. Anal Biochem. 1976;72:248–54.
28. Dubois M, Gilles K, Hamilton J, Rebers P, Smith F. Colorimetric methodfor determination of sugars and related substances. Anal Chem. 1956;28:350–6.
29. Mahamudand MR, Gomes DJ. Enzymatic saccharification of sugar cane bagasse by the crude enzyme from indigenous fungi. J Sci Res. 2012;4:227–38.
30. Zia-ullah Khokhar QS, Nadeem M, Irfan M, Wu J, Samra ZQ, Gul I, et al. Enhanced Production of Cellulase by *Trichodermareesei* Using Wheat Straw as a Carbon Source. World Appl Sci J. 2014;30:1095.
31. Mohamed SA, Al-Malki AL, Khan JA, Kabli SA, Al-Garni SM. Solid state production of polygalacturonase and xylanase by Trichoderma species using cantaloupe and watermelon rinds. J Microbiol. 2013;51:605–11.
32. Yang X, Chen H, Gao H, Li Z. Bioconversion of corn straw by coupling ensiling and solid-state fermentation. Biores Technol. 2001;78:277–80.
33. Couri S, Terzi S, Pinto GS, Freitas S, Da Costa ACA. Hydrolytic enzyme production in solid state fermentation by *Aspergillus niger* 3T5B8. Process Biochem. 2000;36:255–61.
34. Omojasola PF, Jilani OP, Ibiyemi SA. Cellulase production by some fungi cultured on pineapple waste. Nature. 2008;6:64–79.
35. Begum MF, Alimon AR: Bioconversion and saccharification of some lignocellulosic wastes by *Aspergillus oryzae* ITCC-4857.01 for fermentable sugar production. *Elec J Biotechnol* 2011;14(5). www.ejbiotechnology.info/index.php/ejbiotechnology/article/view/v14n53/1355.
36. Ja'afaru MI, Fagade OE. Cellulase production and enzymatic hydrolysis of some selected local lignocellulosic substrates by a strain of *Aspergillus niger*. Res J Biol Sci. 2007;2:13–6.
37. Ahmed FM, Rahman SR, Gomes DJ. Saccharification of sugarcane bagasse by enzymatic treatment for bioethanol production. Malays J Microbiol. 2012;8:97–103.

38. Ramesh MV, Lonsane BK. Criticalimportance of moisture content of the medium in α-amylase by *Bacillus licheniformis* M27 in a solid-state fermentation system. Appl Microbiol Biotechnol. 1990;33:501–5.
39. Kim JH, Hosobuchi M, Kishimoto M, Seki T, Ryu DDY. Cellulase production by a solidstateculture system. Biotechnol Bioeng. 1985;27:1445–50.
40. Nagendra PG, Chandrasekharan M. Lglutaminase production by marine *Vibrio costicola* under solid-state fermentation using different substrates. J Marine Biotechnol. 1996;4:176–9.
41. Irfan M, Syed Q, Yousaf M, Nadeem M, Baig S, Jafri SA. Studies on the Pretreatment of wheat straw for improve production of Carboxymethyl Cellulase by *Trichoderma viride* FBL1 in Solid State fermentation. Acad Arena. 2010;2:18–30.
42. Babu KR, Satyanarayana T. Production of bacterial enzymes by solid-state fermentation. J Sci Ind Res. 1996;55:464–7.
43. Feniksova RV, Tikhomirova AS, Rakhleeva BE. Conditions for forming amylase andproteinase in surface culture of *Bacillus subtilis*. Mikrobiol. 1960;29:745–8.
44. Gervais P, Molin P. The role of the water in solid state fermentation. Biochem Eng J. 2003;13:85–101.
45. Lu W, Li D, Wu Y. Influence of water activity and temperature on xylanase biosynthesis in pilot-scale solid state fermentation by *Aspergillus sulphurous*. Enzyme Microb Technol. 2003;32:305–11.
46. Castilho LR, Medronho RA, Alves TLM. Production and extraction of pectinases obtained by solid state fermentation of agroindustrial residues with *Aspergillus niger*. Biores Technol. 2000;71:45–50.
47. Smits JP, Rinzema A, Tramper J, van Sonsbeek HM, Knol W. Solid-state fermentation of wheat bran by *Trichoderma reesei*QM9414: Substrate composition changes, C balance, enzyme production, growth and kinetics. Appl Microbiol Biotechnol. 1996;46:489–96.
48. Lonsane BK, Ghildyal NP, Budiatman S, Ramakrishna SV. Engineering aspects of solid state fermentation. Enzyme Microb Technol. 1985;7:258–65.
49. Yamane Y, Fujita J, Shimizu R, Hiyoshi A, Fukuda H, Kizaki Y, et al. Production of cellulose- and xylan-degrading enzymes by a koji mold, *Aspergillus oryzae*, and their contribution to the maceration of rice endosperm cell wall. J Biosci Bioeng. 2002;93:9–14.
50. Botella C, Diaz A, de Ory I, Webb C, Blandino A. Xylanase and pectinase production by *Aspergillusawamori*ongrape pomace in solid state fermentation. Process Biochem. 2007;42:98–101.
51. Yuan QP, Wang JD, Zhang H, Qian ZM. Effect of temperature shift on production of xylanase by *Aspergillus niger*. Process Biochem. 2005;40:3255–7.
52. Zhou JM, Ge XY, Zhang WG. Improvement of polygalacturonase production at high temperature by mixed culture of *Aspergillus niger* and *Saccharomyces cerevisiae*. Biores Technol. 2011;102:10085–8.
53. Lakshmi GS, Rao CS, Rao RS, Hobbs PJ, Prakasham RS. Enhanced production of xylanase by a newly isolated *Asperigullus terreus* under solid state fermentation using palm industrial optimization. Biochem Eng J. 2009;48:51–7.
54. Patil SR, Dayanand A. Optimization of process for the production of fungal pectinases from deseeded sunflower head in submerged and solid-state conditions. Bioresour Technol. 2006;97:2340–4.
55. Farani De Souza D, Marques De Souza CG, Peralta RM. Effect of easily metabolizable sugars in the production of xylanase by *Aspergillus tamarii* in solid state fermentation. Process Biochem. 2001;36:835–8.
56. Abdel-Fatah OM, Hassan MM, Elshafei AM, Haroun BM, Atta HM, Othman AM: Physiological studies on carboxymethyl cellulase formation by *Aspergillus terreus* DSM 826. *Braz J Microbiol* 2012;43:01–11.
57. Gilna W, Khaleel KM. Cellulase enzyme activity of *aspergillus fumigatus* from mangrove soil on lignocellulosic substrate. Recent Res Sci Technol. 2011;3:132–4.
58. Shamala TR, Sreekantiah KR. Production of cellulases and D-xylanase by some selected fungal isolates. Enzyme Microb Technol. 1986;8:178–82.
59. Ustok FI, Canan TC, Gogus N. Solid-state production of polygalacturonase by *Aspergillus sojae* ATCC 20235. J Biotechnol. 2007;127:322–34.
60. Botella C, de Ory I, Webb C, Cantero D, Blandino A. Hydrolytic enzyme production by *Aspergillus awamori* on grape pomace. Biochem Eng J. 2005;26:100–6.
61. Gao J, Weng H, Zhu D, Yuan M, Guan F, Xi Y. Production and characterization of cellulolytic enzymes from thermoacidophilic fungal *Aspergillus Terreus* M11 under solid state cultivation of corn stover. Biores Technol. 2008;99:7623–9.
62. Kang SW, Park YS, Lee JS, Hong SI, Kim SW. Production of cellulose and hemicellulases by *Aspergillus niger* KK2 from lignocellulosic biomass. Biores Technol. 2004;91:151–6.
63. Debing J, Peijun L, Stagnitti F, Xianzhe X, Li L. Pectinase production by solid fermentation from *Aspergillus niger* by a new prescription experiment. Ecotoxicol Environ Safety. 2006;64:244–50.
64. Rodriguez-Fernndez DE, Rodriguez-Len JA, de Carvalho JC, Sturm W, Soccol CR. The behavior of kinetic parameters in production of pectinaseand xylanase by solidstate fermentation. Biores Technol. 2011;102:10657–62.
65. Badhan AK, Chadha BS, Kaur Saini HS, Bhat MK. Production of multiple xylanolytic and cellulolytic enzymes by thermophilic fungus Myceliophthora sp. IMI 387099. Biores Technol. 2007;98:504–10.
66. Kalogeris E, Fountoukides G, Kekos D, Macris BJ. Design of a solid state bioreactor for thermophilic microorganisms. Biores Technol. 1999;67:313–5.
67. Nooshin R, Suhaila M, Kalsom MSU, Hooi LF, Nor Aini AR, Arbakariya AB. Effect of Alkali Pretreatment of Rice Straw on Cellulase and Xylanase Production By Local *Trichoderma harzianum* SNRS3 Under Solid State Fermentation. BioRes. 2013;8:2881–96.
68. Latifian M, Hamidi-Esfahani Z, Barzegar M. Evaluation of culture conditions for cellulose production by two *Trichoderma reesei* mutants under solid state fermentation conditions. Biores Technol. 2007;9:3634–7.

Permissions

All chapters in this book were first published in BMCB, by BioMed Central; hereby published with permission under the Creative Commons Attribution License or equivalent. Every chapter published in this book has been scrutinized by our experts. Their significance has been extensively debated. The topics covered herein carry significant findings which will fuel the growth of the discipline. They may even be implemented as practical applications or may be referred to as a beginning point for another development.

The contributors of this book come from diverse backgrounds, making this book a truly international effort. This book will bring forth new frontiers with its revolutionizing research information and detailed analysis of the nascent developments around the world.

We would like to thank all the contributing authors for lending their expertise to make the book truly unique. They have played a crucial role in the development of this book. Without their invaluable contributions this book wouldn't have been possible. They have made vital efforts to compile up to date information on the varied aspects of this subject to make this book a valuable addition to the collection of many professionals and students.

This book was conceptualized with the vision of imparting up-to-date information and advanced data in this field. To ensure the same, a matchless editorial board was set up. Every individual on the board went through rigorous rounds of assessment to prove their worth. After which they invested a large part of their time researching and compiling the most relevant data for our readers.

The editorial board has been involved in producing this book since its inception. They have spent rigorous hours researching and exploring the diverse topics which have resulted in the successful publishing of this book. They have passed on their knowledge of decades through this book. To expedite this challenging task, the publisher supported the team at every step. A small team of assistant editors was also appointed to further simplify the editing procedure and attain best results for the readers.

Apart from the editorial board, the designing team has also invested a significant amount of their time in understanding the subject and creating the most relevant covers. They scrutinized every image to scout for the most suitable representation of the subject and create an appropriate cover for the book.

The publishing team has been an ardent support to the editorial, designing and production team. Their endless efforts to recruit the best for this project, has resulted in the accomplishment of this book. They are a veteran in the field of academics and their pool of knowledge is as vast as their experience in printing. Their expertise and guidance has proved useful at every step. Their uncompromising quality standards have made this book an exceptional effort. Their encouragement from time to time has been an inspiration for everyone.

The publisher and the editorial board hope that this book will prove to be a valuable piece of knowledge for researchers, students, practitioners and scholars across the globe.

List of Contributors

John Gräsvik
Department of Chemistry, Umeå University, Umeå SE-901 87, Sweden

Sandra Winestrand
Department of Chemistry, Umeå University, Umeå SE-901 87, Sweden

Monica Normark
Department of Chemistry, Umeå University, Umeå SE-901 87, Sweden

Leif J Jönsson
Department of Chemistry, Umeå University, Umeå SE-901 87, Sweden

Jyri-Pekka Mikkola
Department of Chemistry, Umeå University, Umeå SE-901 87, Sweden
Industrial Chemistry & Reaction Engineering, Åbo Akademi University, Åbo-Turku FI-20500, Finland

Jianjun Wang
State Key Laboratory of Microbial Resources, Institute of Microbiology, Chinese Academy of Sciences, Beijing 100101, PR China

Junge Zhu
State Key Laboratory of Microbial Resources, Institute of Microbiology, Chinese Academy of Sciences, Beijing 100101, PR China

Cong Min
State Key Laboratory of Microbial Resources, Institute of Microbiology, Chinese Academy of Sciences, Beijing 100101, PR China

Sheng Wu
State Key Laboratory of Microbial Resources, Institute of Microbiology, Chinese Academy of Sciences, Beijing 100101, PR China

Xiaoyun Li
Guangdong Provincial Key Lab of Biotechnology for Plant Development, College of Life Science, South China Normal University, Guangzhou 510631, P R China

Jiabao Lu
Guangdong Provincial Key Lab of Biotechnology for Plant Development, College of Life Science, South China Normal University, Guangzhou 510631, P R China

Shuai Liu
Guangdong Provincial Key Lab of Biotechnology for Plant Development, College of Life Science, South China Normal University, Guangzhou 510631, P R China

Xu Liu
Key Laboratory of Plant Resources Conservation and Sustainable Utilization, South China Botanical Garden, Chinese Academy of Science, Guangzhou 510650, P R China

Yingying Lin
Guangdong Provincial Key Lab of Biotechnology for Plant Development, College of Life Science, South China Normal University, Guangzhou 510631, P R China

Ling Li
Guangdong Provincial Key Lab of Biotechnology for Plant Development, College of Life Science, South China Normal University, Guangzhou 510631, P R China

Meritxell Zurita-Turk
Department of General Biology, Institute of Biological Sciences, Federal University of Minas Gerais, Belo Horizonte, Brazil

Silvina del Carmen
Reference centre for Lactobacilli (CERELA-CONICET), San Miguel de Tucumán, Argentina

Ana CG Santos
Department of Biochemistry and Immunology, Institute of Biological Sciences, Federal University of Minas Gerais, Belo Horizonte, Brazil

Vanessa B Pereira
Department of General Biology, Institute of Biological Sciences, Federal University of Minas Gerais, Belo Horizonte, Brazil

Denise C Cara
Department of Morphology, Institute of Biological Sciences, Federal University of Minas Gerais, Belo Horizonte, Brazil

Sophie Y Leclercq
Ezequiel Dias Foundation (FUNED), Belo Horizonte, Brazil

Alejandra dM de LeBlanc
Reference centre for Lactobacilli (CERELA-CONICET), San Miguel de Tucumán, Argentina

Vasco Azevedo
Department of General Biology, Institute of Biological Sciences, Federal University of Minas Gerais, Belo Horizonte, Brazil

Jean-Marc Chatel
INRA, UMR1319 Micalis, Domaine de Vilvert, F-78350 Jouy-en-Josas, France

Jean G LeBlanc
Reference centre for Lactobacilli (CERELA-CONICET), San Miguel de Tucumán, Argentina

Anderson Miyoshi
Department of General Biology, Institute of Biological Sciences, Federal University of Minas Gerais, Belo Horizonte, Brazil

Yadan Li
College of Bioscience and Biotechnology, Hunan Agricultural University, Changsha 410128, China
Hunan Agricultural Bioengineering Research Institute, Changsha 410128, China

Ning Liu
Hunan Agricultural Bioengineering Research Institute, Changsha 410128, China
College of Life Science, Hunan Normal University, Changsha 410181, China

Hui Yang
College of Bioscience and Biotechnology, Hunan Agricultural University, Changsha 410128, China
Hunan Agricultural Bioengineering Research Institute, Changsha 410128, China

Fei Zhao
College of Bioscience and Biotechnology, Hunan Agricultural University, Changsha 410128, China
Hunan Agricultural Bioengineering Research Institute, Changsha 410128, China

Ye Yu
College of Bioscience and Biotechnology, Hunan Agricultural University, Changsha 410128, China
Hunan Agricultural Bioengineering Research Institute, Changsha 410128, China
Departments of Physiology, Biochemistry, and Molecular Cell Biology, Institute of Medical Science, Shanghai Jiao Tong University School of Medicine, Shanghai 200025, China

Yun Tian
College of Bioscience and Biotechnology, Hunan Agricultural University, Changsha 410128, China
Hunan Agricultural Bioengineering Research Institute, Changsha 410128, China

Xiangyang Lu
College of Bioscience and Biotechnology, Hunan Agricultural University, Changsha 410128, China
Hunan Agricultural Bioengineering Research Institute, Changsha 410128, China

Pornanong Aramwit
Bioactive Resources for Innovative Clinical Applications Research Unit and Department of Pharmacy Practice, Faculty of Pharmaceutical Sciences, Chulalongkorn University, PhayaThai Road, Phatumwan, Bangkok 10330, Thailand

Nipaporn Bang
Bioactive Resources for Innovative Clinical Applications Research Unit and Department of Pharmacy Practice, Faculty of Pharmaceutical Sciences, Chulalongkorn University, PhayaThai Road, Phatumwan, Bangkok 10330, Thailand

Matthias Buntru
Fraunhofer Institute for Molecular Biology and Applied Ecology IME, Forckenbeckstr. 6, 52074 Aachen, Germany

Simon Vogel
Fraunhofer Institute for Molecular Biology and Applied Ecology IME, Forckenbeckstr. 6, 52074 Aachen, Germany

Holger Spiegel
Fraunhofer Institute for Molecular Biology and Applied Ecology IME, Forckenbeckstr. 6, 52074 Aachen, Germany

Stefan Schillberg
Fraunhofer Institute for Molecular Biology and Applied Ecology IME, Forckenbeckstr. 6, 52074 Aachen, Germany

Sidra Pervez
The Karachi Institute of Biotechnology & Genetic Engineering (KIBGE), University of Karachi, Karachi -75270, Pakistan

Afsheen Aman
The Karachi Institute of Biotechnology & Genetic Engineering (KIBGE), University of Karachi, Karachi -75270, Pakistan

Samina Iqbal
The Karachi Institute of Biotechnology & Genetic Engineering (KIBGE), University of Karachi, Karachi -75270, Pakistan

Nadir Naveed Siddiqui
The Karachi Institute of Biotechnology & Genetic Engineering (KIBGE), University of Karachi, Karachi -75270, Pakistan

Shah Ali Ul Qader
The Karachi Institute of Biotechnology & Genetic Engineering (KIBGE), University of Karachi, Karachi -75270, Pakistan

Richard B McQualter
Australian Institute for Bioengineering and Nanotechnology, the University of Queensland, Brisbane QLD 4072, Australia

Maria N Somleva
Metabolix, Inc, 21 Erie St, Cambridge, MA 02139, USA

Leigh K Gebbie
Australian Institute for Bioengineering and Nanotechnology, the University of Queensland, Brisbane QLD 4072, Australia

Xuemei Li
Metabolix, Inc, 21 Erie St, Cambridge, MA 02139, USA
Agrivida, 200 Boston Ave, Medford, MA 02155, USA

Lars A Petrasovits
Australian Institute for Bioengineering and Nanotechnology, the University of Queensland, Brisbane QLD 4072, Australia

Kristi D Snell
Metabolix, Inc, 21 Erie St, Cambridge, MA 02139, USA

Lars K Nielsen
Australian Institute for Bioengineering and Nanotechnology, the University of Queensland, Brisbane QLD 4072, Australia

Stevens M Brumbley
Australian Institute for Bioengineering and Nanotechnology, the University of Queensland, Brisbane QLD 4072, Australia
Department of Biological Sciences, The University of North Texas, 1155 Union Circle #305220, Denton, TX 76203-5017, USA

Aditya Bhalla
Department of Chemical and Biological Engineering, South Dakota School of Mines and Technology, Rapid City, SD 57701, USA
Great Lakes Bioenergy Research Center, Michigan State University, East Lansing, MI 48824, USA

Kenneth M Bischoff
Renewable Product Technology Research Unit, Agricultural Research Service, National Center for Agricultural Utilization Research, U.S. Department of Agriculture, Peoria, IL 61604, USA

Rajesh K Sani
Department of Chemical and Biological Engineering, South Dakota School of Mines and Technology, Rapid City, SD 57701, USA

Thomas Paschos
Biotechnology Laboratory, School of Chemical Engineering, National Technical University of Athens, 9 Iroon Polytechniou Str, Zografou Campus, Athens 5780, Greece

Charilaos Xiros
Industrial Biotechnology, Department of Chemical and Biological Engineering, Chalmers University of Technology, Kemivägen 10, Gothenburg 41296, Sweden

Paul Christakopoulos
Biochemical Process Engineering, Division of Chemical Engineering, Department of Civil, Environmental and Natural Resources Engineering, Luleå University of Technology, Luleå SE-971 87, Sweden

Longmei Zhao
Laboratory of Feed Biotechnology, State Key Laboratory of Animal Nutrition, College of Animal Science and Technology, China Agricultural University, Beijing 100193, China

Jiang Geng
Laboratory of Feed Biotechnology, State Key Laboratory of Animal Nutrition, College of Animal Science and Technology, China Agricultural University, Beijing 100193, China

Yaoqi Guo
Laboratory of Feed Biotechnology, State Key Laboratory of Animal Nutrition, College of Animal Science and Technology, China Agricultural University, Beijing 100193, China

Xiudong Liao
Laboratory of Feed Biotechnology, State Key Laboratory of Animal Nutrition, College of Animal Science and Technology, China Agricultural University, Beijing 100193, China

Xuhui Liu
Laboratory of Feed Biotechnology, State Key Laboratory of Animal Nutrition, College of Animal Science and Technology, China Agricultural University, Beijing 100193, China

Rujuan Wu
Laboratory of Feed Biotechnology, State Key Laboratory of Animal Nutrition, College of Animal Science and Technology, China Agricultural University, Beijing 100193, China

Zhaojun Zheng
Laboratory of Feed Biotechnology, State Key Laboratory of Animal Nutrition, College of Animal Science and Technology, China Agricultural University, Beijing 100193, China

Rijun Zhang
Laboratory of Feed Biotechnology, State Key Laboratory of Animal Nutrition, College of Animal Science and Technology, China Agricultural University, Beijing 100193, China

David A Brummell
The New Zealand Institute for Plant & Food Research Limited, Food Industry Science Centre, Private Bag 11600, Palmerston North 4442, New Zealand

Lyn M Watson
The New Zealand Institute for Plant & Food Research Limited, Food Industry Science Centre, Private Bag 11600, Palmerston North 4442, New Zealand

Jun Zhou
The New Zealand Institute for Plant & Food Research Limited, Food Industry Science Centre, Private Bag 11600, Palmerston North 4442, New Zealand

Marian J McKenzie
The New Zealand Institute for Plant & Food Research Limited, Food Industry Science Centre, Private Bag 11600, Palmerston North 4442, New Zealand

Ian C Hallett
The New Zealand Institute for Plant & Food Research Limited, Mount Albert Research Centre, Private Bag 92169, Auckland 1142, New Zealand

Lyall Simmons
The New Zealand Institute for Plant & Food Research Limited, Canterbury Agriculture & Science Centre, Private Bag 4704, Christchurch 8140, New Zealand

Margaret Carpenter
The New Zealand Institute for Plant & Food Research Limited, Canterbury Agriculture & Science Centre, Private Bag 4704, Christchurch 8140, New Zealand

Gail M Timmerman-Vaughan
The New Zealand Institute for Plant & Food Research Limited, Canterbury Agriculture & Science Centre, Private Bag 4704, Christchurch 8140, New Zealand

Eugenia S. Mardanova
Centre 'Bioengineering', Russian Academy of Sciences, Prosp. 60-letya Oktyabrya, bld 7-1, 117312 Moscow, Russia

Roman Y. Kotlyarov
Centre 'Bioengineering', Russian Academy of Sciences, Prosp. 60-letya Oktyabrya, bld 7-1, 117312 Moscow, Russia

Victor V. Kuprianov
Centre 'Bioengineering', Russian Academy of Sciences, Prosp. 60-letya Oktyabrya, bld 7-1, 117312 Moscow, Russia

Liudmila A. Stepanova
Research Institute of Influenza, Russian Federation Ministry of Health, 15/17 Prof. Popova str., 197376 St. Petersburg, Russia

Liudmila M. Tsybalova
Research Institute of Influenza, Russian Federation Ministry of Health, 15/17 Prof. Popova str., 197376 St. Petersburg, Russia

George P. Lomonosoff
Department of Biological Chemistry, John Innes Centre, Norwich Research Park, NR4 7UH Norwich, UK

Nikolai V. Ravin
Centre 'Bioengineering', Russian Academy of Sciences, Prosp. 60-letya Oktyabrya, bld 7-1, 117312 Moscow, Russia

Satiander Rana
Plant Biotechnology Division, CSIR-Indian Institute of Integrative Medicine, Canal Road, Jammu Tawi-180001, India

Wajid Waheed Bhat
Plant Biotechnology Division, CSIR-Indian Institute of Integrative Medicine, Canal Road, Jammu Tawi-180001, India

Niha Dhar
Plant Biotechnology Division, CSIR-Indian Institute of Integrative Medicine, Canal Road, Jammu Tawi-180001, India

Shahzad A Pandith
Plant Biotechnology Division, CSIR-Indian Institute of Integrative Medicine, Canal Road, Jammu Tawi-180001, India

Sumeer Razdan
Plant Biotechnology Division, CSIR-Indian Institute of Integrative Medicine, Canal Road, Jammu Tawi-180001, India

Ram Vishwakarma
Medicinal Chemistry Division, CSIR-Indian Institute of Integrative Medicine, Canal Road, Jammu Tawi-180001, India

Surrinder K Lattoo
Plant Biotechnology Division, CSIR-Indian Institute of Integrative Medicine, Canal Road, Jammu Tawi-180001, India

Narendra Tuteja
Plant Molecular Biology Group, International Centre for Genetic Engineering and Biotechnology (ICGEB), Aruna Asaf Ali Marg, New Delhi 110067, India

Panchanand Mishra
Stress Biology Laboratory, Gene Function and Regulation, Institute of Life Sciences, Bhubaneswar 751023, Odisha, India

Sandep Yadav
Plant Molecular Biology Group, International Centre for Genetic Engineering and Biotechnology (ICGEB), Aruna Asaf Ali Marg, New Delhi 110067, India

Marjan Tajrishi
Plant Molecular Biology Group, International Centre for Genetic Engineering and Biotechnology (ICGEB), Aruna Asaf Ali Marg, New Delhi 110067, India

Sudhir Baral
Stress Biology Laboratory, Gene Function and Regulation, Institute of Life Sciences, Bhubaneswar 751023, Odisha, India

Surendra Chandra Saba
Stress Biology Laboratory, Gene Function and Regulation, Institute of Life Sciences, Bhubaneswar 751023, Odisha, India

Yongbo Liu
State Key Laboratory of Environmental Criteria and Risk Assessment, Chinese Research Academy of Environmental Sciences, Beijing 100012, China

Feng Ge
State Key Laboratory of Integrated Management of Pest and Rodents, Institute of Zoology, the Chinese Academy of Sciences, 8 Dayangfang, Beijing 100101, China

Yuyong Liang
Institute of Plant Protection, Jiangxi Academy of Agricultural Sciences, Nanchang 330200, China

Gang Wu
Hubei Insect Resources Utilization and Sustainable Pest Management Key Laboratory, College of Plant Science and Technology, Huazhong Agricultural University, Wuhan 430070, China

Junsheng Li
State Key Laboratory of Environmental Criteria and Risk Assessment, Chinese Research Academy of Environmental Sciences, Beijing 100012, China

Xiang Li
GMO Detection Laboratory, Shanghai Entry-Exit Inspection and Quarantine Bureau, 1208 Minsheng Road, Shanghai 200135, PR China

Xiuxiu Wang
Institute of Biotechnology, East China University of Science and Technology, 130 Meilong Road, Shanghai 200237, PR China

Jielin Yang
GMO Detection Laboratory, Shanghai Entry-Exit Inspection and Quarantine Bureau, 1208 Minsheng Road, Shanghai 200135, PR China

Yueming Liu
GMO Detection Laboratory, Shanghai Entry-Exit Inspection and Quarantine Bureau, 1208 Minsheng Road, Shanghai 200135, PR China

Yuping He
GMO Detection Laboratory, Shanghai Entry-Exit Inspection and Quarantine Bureau, 1208 Minsheng Road, Shanghai 200135, PR China

Liangwen Pan
GMO Detection Laboratory, Shanghai Entry-Exit Inspection and Quarantine Bureau, 1208 Minsheng Road, Shanghai 200135, PR China

Basti Bergdahl
Division of Applied Microbiology, Department of Chemistry, Lund University,P.O. Box 124, Lund SE-22100, Sweden

Marie F Gorwa-Grauslund
Division of Applied Microbiology, Department of Chemistry, Lund University,P.O. Box 124, Lund SE-22100, Sweden

Ed WJ van Niel
Division of Applied Microbiology, Department of Chemistry, Lund University,P.O. Box 124, Lund SE-22100, Sweden

Eridan Orlando Pereira
Agriculture and Agri-Food Canada, 1391 Sandford Street, London, ON N5V 4T3, Canada Department of Biology, Western University, London, ON N6A 5B7, Canada
Universidade Estadual do Ceará, Av. Dr. Silas Munguba, 1700 - Campus do Itaperi, Fortaleza 60714-903, Brazil

Igor Kolotilin
Agriculture and Agri-Food Canada, 1391 Sandford Street, London, ON N5V 4T3, Canada

Andrew Jonathan Conley
VTT Technical Research Centre of Finland, Tietotie 2, Espoo 02044, Finland

Rima Menassa
Agriculture and Agri-Food Canada, 1391 Sandford Street, London, ON N5V 4T3, Canada Department of Biology, Western University, London, ON N6A 5B7, Canada

Gerben P Voshol
Molecular Microbiology and Biotechnology, Institute of Biology Leiden, Leiden University, Leiden, The Netherlands
International Research Centre "Biotechnologies of the Third Millennium", ITMO University, Saint-Petersburg, Russia

Vera Meyer
Molecular Microbiology and Biotechnology, Institute of Biology Leiden, Leiden University, Leiden, The Netherlands
Applied and Molecular Microbiology, Institute of Biotechnology, Berlin University of Technology, Berlin, Germany

Cees A M J J van den Hondel
Molecular Microbiology and Biotechnology, Institute of Biology Leiden, Leiden University, Leiden, The Netherlands

Blake L Joyce
Department of Plant Sciences, The University of Tennessee, Knoxville, Tennessee 37996, USA

Shigetoshi Eda
Center for Wildlife Health, Department of Forestry, Wildlife and Fisheries, the University of Tennessee, Knoxville, Tennessee 37996, USA

John Dunlap
Advanced Microscopy and Imaging Center, The University of Tennessee, Knoxville, TN 37996, USA

C Neal Stewart Jr
Department of Plant Sciences, The University of Tennessee, Knoxville, Tennessee 37996, USA

Reda M. El-Shishtawy
Chemistry Department, Faculty of Science, King Abdulaziz University, P.O. Box 80203, Jeddah 21589, Saudi Arabia

Saleh A. Mohamed
Biochemistry Department, Faculty of Science, King Abdulaziz University, Jeddah, Kingdom of Saudi Arabia

Abdullah M. Asiri
Chemistry Department, Faculty of Science, King Abdulaziz University, P.O. Box 80203, Jeddah 21589, Saudi Arabia
The Center of Excellence for Advanced Materials Research, King Abdulaziz University, Jeddah 21589, Saudi Arabia

Abu-bakr M. Gomaa
Biology Department, Faculty of Science, King Abdulaziz University, Jeddah, Kingdom of Saudi Arabia

Ibrahim H. Ibrahim
Biochemistry Department, Faculty of Science, King Abdulaziz University, Jeddah, Kingdom of Saudi Arabia

Hasan A. Al-Talhi
Biochemistry Department, Faculty of Science, King Abdulaziz University, Jeddah, Kingdom of Saudi Arabia